P_m	air-gap power	W
P_{mech}	mechanical power	W
q	charge	C
q	circumferential current density	A/m
Q	reactive power	VAR
r	radius	m
R	resistance	Ω
$\mathscr{R}$	reluctance	A/Wb
$\mathscr{R}_d$	direct-axis reluctance	A/Wb
$\mathscr{R}_q$	quadrature-axis reluctance	A/Wb
S	complex or apparent power	VA
t	time	s
T	period	s
T	torque	N·m
v, V	potential difference	V
V	volume	m³
w	energy density	J/m³
W	energy	J
x	linear displacement	m
X	reactance	Ω
Z	impedance	Ω
α	operator $1\lfloor 2\pi/3$	-
β	angular displacement	deg, rad
η	efficiency	-
θ	angular position, phase angle	deg, rad
λ	flux linkage	Wb
μ_0	magnetic constant	T·m/A
μ_r	relative permeability	-
ν	velocity	m/s
ρ	electrical resistivity	Ω·m
τ	time constant	s
ϕ, Φ	magnetic flux	Wb
ϕ	flux per pole	Wb
Φ_r	residual flux per pole	Wb
ω	angular frequency	rad/s
ω_m	angular velocity	rad/s
Ω_0	rated speed	rad/s

Electric Machines

Electric Machines

G. R. SLEMON
Professor of Electrical Engineering
UNIVERSITY OF TORONTO

A. STRAUGHEN
Emeritus Professor of Electrical Engineering
UNIVERSITY OF TORONTO

ADDISON-WESLEY PUBLISHING COMPANY

READING, MASSACHUSETTS

MENLO PARK, CALIFORNIA

LONDON · AMSTERDAM

DON MILLS, ONTARIO

SYDNEY

Library of Congress Cataloging in Publication Data

Slemon, Gordon R.
  Electric machines.

  Includes index.
  1. Electric machinery. I. Straughen, A.,
joint author. II. Title.
TK2181.S54   621.31′04′2   79-16369
ISBN 0-201-07730-2

Second printing, December 1980

ISBN 0-201-07730-2
  BCDEFGHIJ-MA-89876543210

Since Pekuah was taken from me, said the princess, I have no pleasure to reject or to retain. She that has no one to love or trust has little to hope. She wants the radical principle of happiness. We may, perhaps allow that what satisfaction this world can afford, must arise from the conjunction of wealth, knowledge and goodness: wealth is nothing but as it is bestowed, and knowledge nothing but as it is communicated: they must therefore be imparted to others, and to whom could I now delight to impart them? Goodness affords the only comfort which can be enjoyed without a partner, and goodness may be practised in retirement.

Samuel Johnson
The History of Rasselas
Prince of Abissinia

Preface

This book is intended for the somewhat reduced courses in electromechanical energy conversion that are given in today's departments of electrical engineering, where so many new and vital topics crowd the curriculum. There are many books already available on this subject, and new ones appear at intervals. It is therefore incumbent upon the authors of such a text to justify their wish to have their own treatment of a seemingly well-covered topic published.

The books that have appeared during the past few years have been original, comprehensive, often well written, rich in new insights into the subject matter—and difficult. Any lecturer faced with the problem of finding a text for a core course in an electrical engineering curriculum or, in descending scale, for a class of electrical engineering students whose members intend to specialize in electronics, or a class of students of a branch of engineering other than electrical, will realize the frustration that arises from trying to employ what is at present available. He or she will end up virtually writing a textbook in the form of lecture notes. We hope that the needs of both student and lecturer can be met by an appropriate selection from what appears in the following pages. Our guiding principle has been that of selectivity. Our care has been to ensure that complete and self-contained courses at a number of different levels of difficulty and covering a number of different areas of interest can be selected by the simple process of discussing some sections of the text and omitting all reference to the remainder.

Chapter 1 is devoted to a discussion of the nature and behavior of magnetic systems as they appear in electric machines. For those readers requiring a review of the basic principles of electromagnetism, these are discussed in Appendix A. In view of the increasing importance of permanent-magnet machines, we have included a discussion of permanent-magnet systems in this chapter.

Chapter 2 deals with transformers and transformer-like devices and describes a general method for developing an equivalent electric circuit for any electromagnetic system.

Chapter 3 is the core of the book and must be fully understood, since it deals with the basic electromechanical energy conversion process as it occurs in all types of electric machines.

Chapters 4 through 6 deal with the dc and ac machines that are most frequently encountered in industry. Mastery of these chapters should enable the reader to specify an appropriate machine for a given application and predict its performance under anticipated working conditions.

We are convinced that the essence of electrical engineering today is system design, and since "system" is a very elastic word, we must make clear that in this context we consider such devices as transformers, motors, and generators to be system elements. An understanding of the physical behavior of these elements is essential for the engineer wishing to apply them.

A large part of the book is devoted to developing models for devices. All models are by nature approximate in that they retain only that information considered necessary for adequate performance prediction. We have chosen to emphasize the use of equivalent circuit models because these are also encountered in parallel studies of other electric and electronic systems. A further advantage of this emphasis on equivalent circuits is that individual circuit parameters generally can be related to the dimensions and material properties of the device.

At appropriate points in the text, we have introduced brief discussions of the basic principles of motor control by solid-state or power-semiconductor converters. We do not claim that these discussions will enable the reader to design a control system embodying a machine and a converter; rather, their limited objective is to show how controlled variable speed may be achieved using combinations of solid-state converters and machines. For this purpose the converters have been idealized. Also, many of the important differences in the operation of machines on pulsating direct current, rather than smooth direct current, or on nonsinusoidal alternating current, rather than sinusoidal current, have been neglected; readers who wish to design motor control systems will need a sound knowledge of the nature and functioning of solid-state power converters, as well as the fundamentals of electric machines as presented here.

Much has been written in recent years about generalized models for electric machines. In many of the proposed models, linearity has been assumed. But most devices have operating regions in which magnetic nonlinearity is significant. Moreover, although a certain model may be appropriate for the steady-state performance of the device, something substantially different and usually more complicated is required to simulate its dynamic behavior. To cover all these aspects would require a much longer book than this, one that would inevitably overwhelm the student with less advanced interests. We have, therefore, for the most part, contented ourselves with stating clearly the limitations of our models, indicating in broad terms the consequences of exceeding their applicable operating bounds.

The student using this book should have taken an introductory course in electricity and magnetism, as well as a basic course in electric circuit analysis. We have assumed that parallel courses in differential equations and systems analysis will provide an adequate range of analytical techniques. The methods of control-system analysis have not been included, since it is assumed that most curricula provide a general course in control systems.

Sections of the text and problems dealing with material that may be considered in any way "advanced" or specialized have been marked with an asterisk. Sections not marked with asterisks constitute the material for a one-semester introductory course. Where time is particularly short, the treatment of synchronous machines given in Chapter 3 may be considered sufficient and Sections 6.1.1 through 6.1.5 omitted. Such a shortened course is given to students of mechanical engineering at the University of Toronto. A wide range of choices may be made from the remaining material in accordance with the interests of lecturers able to devote more than a single-semester course to this subject. This material may also provide a convenient and familiar reference for engineers in later practice.

Two consecutive one-semester courses are given to all electrical-engineering students in the third or junior year at the University of Toronto. The following material is covered in the first course:

Chapter 1—Sections 1.1 through 1.19.

Chapter 2—Sections 2.1 and 2.2; 2.4 through 2.6.5; 2.8 through 2.10.2; 2.12 through 2.12.2.

Chapter 3—Complete.

In the second course the following material is covered:

Chapter 4—Sections 4.1 through 4.6.2; 4.7 and 4.9 through 4.11.

Chapter 5—Sections 5.1, 5.2, 5.4 and 5.4.3.

Chapter 6—Sections 6.1 through 6.1.7.

Some of the remaining sections, particularly those concerning solid-state motor control, are covered and enlarged upon in a senior elective course.

Two points relating to symbols and notation are of sufficient importance to mention here. In general, most publications show vector quantities in boldface type. We consider this practice unfortunate, for it presents instructors with a symbol they cannot conveniently represent on the chalkboard and students with a symbol they cannot write in notebooks. Rather than disconcert readers accustomed to boldface vectors by using normal italic letters, we have added an arrow above the boldface symbol. A normal letter with an arrow can easily be written on the chalkboard or in the notebook and will be unambiguous. The second point concerns the writing of very large and very small numbers. In expressing data or solutions, we have employed International System (SI) engineering notation, in which powers of ten are always multiples of ± 3. This notation is now being built into electronic hand calculators and expresses directly the mega-, kilo-, milli-, micro-, etc., series of subunits normally used in engineering. In the expression of physical constants, such as the magnetic constant ($4\pi \times 10^{-7}$) or the electron charge (1.603×10^{-19}), we have adhered to the familiar forms.

The number of symbols is unavoidably large. Many of them are employed with several different subscripts and superscripts. For convenience in referring to them, lists of many of these symbols have been printed on the endpapers. Where subscripts have a limited application, they are defined where they are used and are not included in a list. As far as possible, lower-case letters have been used to signify instantaneous values of variables, upper-case letters to signify rms or constant direct values. Exceptions to this practice have been made to avoid confusion with other frequently employed symbols (T for torque; t for time) and where common usage makes an exception (B for flux density).

ACKNOWLEDGMENTS

We are indebted to our colleagues—Professors P. P. Biringer, P. E. Burke, S. B. Dewan, J. D. Lavers and S. D. T. Robertson—who have used various drafts of this textbook in a variety of courses and have helped us with their comments. The problem solutions have come from several hands, but many have

We are grateful to a number of firms who have been kind enough to provide photographs of the various types of machines we have discussed. In particular, we wish to acknowledge the generosity of John Wiley & Sons, Inc. for permitting us to incorporate in this text, without the ungrateful labor of rewriting to "disguise" their origin, a limited number of sections and a substantial number of problems that appear in their book *Magnetoelectric Devices,* Copyright 1966, by G.R. Slemon.

Finally, we express our sincere thanks to Mrs. Shirley DesLauriers, Mrs. Sandy Langill, and Miss Amelia Chung, who, in typing the manuscript, have wrestled long and patiently with two quite distinct kinds of bad handwriting.

Toronto   G.R.S.
January 1980   A.S.

Contents

4 DIRECT-CURRENT MACHINES   265

APPENDIXES

ENDPAPERS

1 / Magnetic Systems

Apart from essentially mechanical components, such as shafts and bearings, the parts of an electromagnetic machine may be divided into two groups: the electrical system and the magnetic system. The purpose of the machine is the conversion of energy from electrical to mechanical form, or the reverse. The engineer wishing to apply such machines is therefore concerned with the relationships between the electrical and mechanical terminal variables: potential difference, current, power factor, frequency, speed, and torque or force. The character of these relationships, however, may be affected to a significant degree by the behavior of the magnetic system. If the applications engineer does not possess a basic understanding of this behavior, the features of the relationships that he or she wishes to exploit will appear to be mysterious and inexplicable. The purpose of this chapter, therefore, is to give this basic understanding. Its breadth, that is to say, the proportion of this chapter which should be studied, will depend upon the range of types of machine and of applications that are of concern.

As the preceding paragraph implies, the process of electromechanical energy conversion depends upon intermediate energy conversion processes, from electrical to magnetic and from magnetic to mechanical form. Electromagnetic energy conversion depends upon the production of a magnetic field, either by means of electric currents in coils of wire or by means of permanent magnets. An essential preliminary to the treatment of energy conversion is therefore a discussion of the methods and materials employed to produce a magnetic field in the form, location, and intensity in which it is required. A useful concept in this discussion is that of the equivalent magnetic circuit, which is analogous to the electric circuit. An essential tool of the applications engineer is an equivalent electric circuit that may be employed to model the properties of the combined magnetic and electric systems. These two concepts are therefore first discussed.

The physical principles underlying the discussion that follows are reviewed in Appendix A.

1

1.1 EQUIVALENT MAGNETIC AND ELECTRIC CIRCUITS

Figure 1.1 represents a coil of N turns of wire uniformly wound on a wooden or plastic torus of circular cross section. The coil is carrying current i. A magnetic field is produced within the torus with the direction shown by the vectors $\vec{\mathbf{H}}$, and the magnitude of the magnetic field intensity, may be determined by employing *Ampere's Circuital Law*. This law may be expressed by the equation

$$\oint \vec{\mathbf{H}} \cdot \vec{\mathbf{dl}} = \int_A \vec{\mathbf{J}} \cdot \vec{\mathbf{dA}} \quad \text{A} \tag{1.1}$$

(see Appendix A.5). The expression on the left-hand side of Eq. (1.1) is a line integral of magnetic-field intensity H around a closed path in a magnetic field. The expression on the right-hand side of Eq. (1.1) is a surface integral of current density J over any surface bounded by the closed path.

Let the closed path in the magnetic field produced by the toroidal winding in Fig. 1.1 be the center line of the torus cross section, which is a circle of radius a, and let the surface bounded by this path be flat. Each turn of the conductor will penetrate the surface perpendicularly, so that from Eq. (1.1),

$$H(2\pi a) = Ni \quad \text{A} \tag{1.2}$$

from which

$$H = \frac{Ni}{2\pi a} = \frac{Ni}{l} \quad \text{A/m} \tag{1.3}$$

where l is the length of the flux path. Application of the circuital law also shows that H outside the torus is everywhere zero in the plane of the torus, since for a

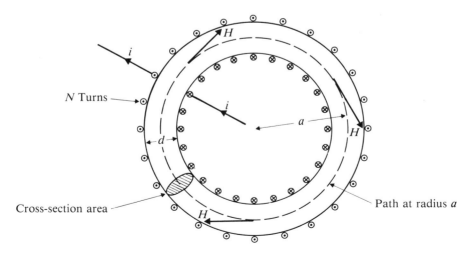

Fig. 1.1 Coil wound on a wooden torus.

circular path of radius smaller than that of the inner radius of the coil, no current is enclosed, while for a circular path of radius greater than that of the outer radius of the coil, the net current enclosed is also zero.

Applying Eq. (1.2) to circular paths within the torus of greater or lesser radius than a shows that H (and consequently B) within the torus will not be uniform. However, if the diameter of the torus cross-section d is much smaller than a, H may be considered to be sensibly constant over the cross section, and the value of H at the mean radius a may be used as a good approximation to the average.

The average value of the magnetic flux density is then given by

$$\vec{\mathbf{B}}_{av} = \mu_0 \vec{\mathbf{H}}_{av} \quad \text{T} \tag{1.4}$$

where μ_0 is the magnetic constant ($4\pi \times 10^{-7}$), so that to a close approximation the flux produced in the torus is

$$\phi = B_{av} \frac{\pi d^2}{4} \quad \text{Wb} \tag{1.5}$$

The flux ϕ may be considered to be the result of the total current (Ni) encircling the flux path. This current is called the *magnetomotive force* (mmf)

$$\mathscr{F} = Ni \quad \text{A} \tag{1.6}$$

The unit of $\mathscr{F}$ is often called an "ampere-turn," but since N has no dimensions, $\mathscr{F}$ is properly expressed in amperes

For the magnetic system presented by the coil of Fig. 1.1, the circuital law of Eq. (1.1) may be employed to yield

$$\mathscr{F} = Ni = \int_A \vec{\mathbf{J}} \cdot \vec{\mathbf{dA}} = \oint \vec{\mathbf{H}} \cdot \vec{\mathbf{dl}} \quad \text{A} \tag{1.7}$$

That is, the mmf around a closed path is equal to the current enclosed by that path. From Eqs. (1.2), (1.4), and (1.5),

$$\mathscr{F} = H(2\pi a) = \frac{B}{\mu_0}(2\pi a) = \frac{\phi}{(\pi d^2/4)} \cdot \frac{2\pi a}{\mu_0} \quad \text{A} \tag{1.8}$$

Thus, ϕ is directly proportional to $\mathscr{F}$ and may be considered to be the result of $\mathscr{F}$. The constant of proportionality between these two quantities is called the *reluctance* $\mathscr{R}$ of the magnetic path and is defined by the relation

$$\mathscr{R} = \frac{\mathscr{F}}{\phi} \quad \text{A/Wb} \tag{1.9}$$

Thus, $\mathscr{R}$ is expressed in amperes per weber, and for the toroidal system,

$$\mathscr{R} = \frac{2\pi a}{\mu_0(\pi d^2/4)} \quad \text{A/Wb} \tag{1.10}$$

Fig. 1.2 Equivalent magnetic circuit.

Reluctance is seen to be directly proportional to the length of the magnetic path and inversely proportional to its cross-section area. As a property of a magnetic system, reluctance is analogous to the resistance in an electric system in which the resistance of the current path in ohms, or volts per ampere, is

$$R = \frac{\rho \cdot l_{\text{wire}}}{A_{\text{wire}}} \quad \Omega \tag{1.11}$$

where ρ is the volume resistivity of the conductor material.

In many situations, it is useful to have a model in the form of an *equivalent magnetic circuit* analogous to an electric circuit. Such a model of the system in Fig. 1.1 is shown in Fig. 1.2.

In the system of Fig. 1.1, if current i is varied, then flux ϕ will also be varied, and by Faraday's law an *electromotive force* (emf) will be induced in each turn of the coil. This emf per turn will be

$$e_{\text{turn}} = \frac{d\phi}{dt} \quad \text{V} \tag{1.12}$$

The total emf induced in the coil is thus

$$e = \frac{Nd\phi}{dt} = \frac{d\lambda}{dt} \quad \text{V} \tag{1.13}$$

where λ is the *flux linkage* of the coil defined by

$$\lambda = N\phi \quad \text{Wb} \tag{1.14}$$

For the system shown in Fig. 1.1, substitution for ϕ in Eq. (1.14) using Eqs. (1.3), (1.4), and (1.5) yields

$$\lambda = \frac{N^2\mu_0}{2\pi a} \cdot \frac{\pi d^2}{4} \cdot i \quad \text{Wb} \tag{1.15}$$

so that flux linkage is directly proportional to the coil current. The *inductance L* of a coil is defined as the flux linkage per ampere of current in the coil and is expressed in *henries*. Thus

$$L = \frac{\lambda}{i} \quad \text{H} \tag{1.16}$$

and for the system of Fig. 1.1 the magnitude of L in terms of the system dimensions may be obtained from Eq. (1.15). From Eqs. (1.15), (1.16), and (1.10) it is seen that L is related to $\mathcal{R}$ by

$$L = \frac{N^2}{\mathcal{R}} \quad H \tag{1.17}$$

Since coils such as that illustrated in Fig. 1.1 may be used in electric circuits, it is convenient to model them using ideal circuit elements. Circuit analysis is then carried out in terms of the potential difference and current at the terminals of such elements and the relation between these two variables. For an ideal coil with no resistance, the terminal potential difference is equal to the emf induced in the coil, so that at the terminals

$$v = e = \frac{d\lambda}{dt} \quad V \tag{1.18}$$

and substitution for λ from Eq. (1.16) yields

$$v = L\frac{di}{dt} \quad V \tag{1.19}$$

The resistance-free coil may thus be represented by the purely inductive circuit element shown in Fig. 1.3(a). In practice, a coil will have resistance, and in many circumstances the effect of this resistance cannot be neglected without introducing serious inaccuracy. A model of a practical coil including this resistive effect is given by the equivalent circuit of Fig. 1.3(b). In that model,

$$v = Ri + L\frac{di}{dt} \quad V \tag{1.20}$$

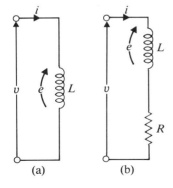

(a)   (b)

Fig. 1.3 Equivalent electric circuits.
(a) Resistance-free coil.
(b) "Practical" coil.

Example 1.1 Figure 1.4 shows a toroidal coil wound on a plastic ring of rectangular cross section. The coil has 200 turns of round copper wire which is 3 mm in diameter.

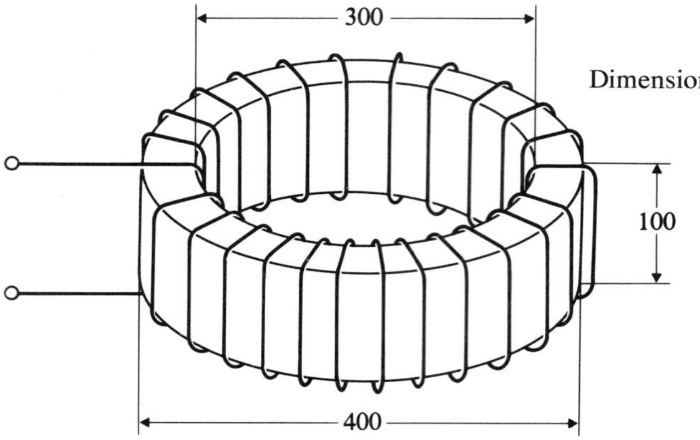

Fig. 1.4 Diagram for Example 1.1.

Dimensions in mm

a) For a coil current of 50 A, find the magnetic flux density at the mean diameter of the coil.

b) Find the inductance of the coil, assuming that the flux density within it is uniform and equal to that at the mean diameter.

c) Determine the percentage error incurred in assuming uniform flux density in the coil.

d) Given that the volume resistivity of copper is 17.2×10^{-9} ohm-meter, determine the parameters of the approximate electric circuit of Fig. 1.3(b).

Solution

a) At the mean diameter:

$$H = \frac{Ni}{2\pi r} = \frac{200 \times 50}{0.35\pi} = 9095 \quad \text{A/m}$$

$$B = \mu_0 H = 4\pi \times 10^{-7} \times 9095 = 11.43 \times 10^{-3} \quad \text{Wb/m}^2$$

b) Assume $B_{\text{av}} = 11.43 \times 10^{-3}$

$$\phi = BA = 11.43 \times 10^{-3} \times 0.1 \times 0.05 = 57.15 \times 10^{-6} \quad \text{Wb}$$

$$\lambda = N\phi = 200 \times 57.15 \times 10^{-6} = 11.43 \times 10^{-3} \quad \text{Wb}$$

$$L = \frac{\lambda}{i} = \frac{11.43 \times 10^{-3}}{50} = 0.2286 \times 10^{-3} \quad \text{H}$$

Alternatively, $\mathscr{R} = \dfrac{1}{\mu_0 A} = \dfrac{0.35\pi}{4\pi \times 10^{-7} \times 0.1 \times 0.05} = 175.0 \times 10^6$ A/Wb

$$L = \frac{N^2}{\mathscr{R}} = \frac{200^2}{175.0 \times 10^6} = 0.2286 \times 10^{-3} \quad \text{H}$$

c) At radius r, where $0.15 < r < 0.20$ m,

$$B = \frac{\mu_0 Ni}{2\pi r} = \frac{4\pi \times 10^{-7} \times 200i}{2\pi r} \quad \text{T}$$

$$\phi = \int_{0.15}^{0.20} B \times 0.1 \; dr \quad \text{Wb}$$

$$\lambda = N\phi = \frac{0.1\mu_0 N^2 i}{2\pi} \int_{0.15}^{0.20} \frac{dr}{r} \quad \text{Wb}$$

$$L = \frac{\lambda}{i} = \frac{0.1 \times 4\pi \times 10^{-7} \times 200^2}{2\pi} \, ln \left(\frac{0.2}{0.15}\right) = 0.2301 \times 10^{-3} \quad \text{H}$$

$$\text{Error} = \frac{0.2301 - 0.2286}{0.2301} \times 100\% = 0.651\%$$

d)
$$R = \frac{\rho l_{\text{wire}}}{A_{\text{wire}}} = \frac{17.2 \times 10^{-9} \times 200 \times 0.3}{\pi \times 3^2 \times 10^{-6}/4} = 0.1460 \quad \Omega$$

The parameters of an approximate equivalent circuit are therefore
$$R = 0.1460 \; \Omega; \quad L = 0.2286 \; \text{mH}$$

1.2 FERROMAGNETISM

In the toroidal system of Fig. 1.1, the coil current produces a magnetic field intensity H inside the torus. This in turn produces a collinear magnetic flux density B. With a vacuum inside the coil, the relation between B and H at any point is expressed by

$$B = \mu_0 H \quad \text{T} \tag{1.21}$$

If the toroidal space inside the coil is filled with air, the flux density B is increased by about 0.4 part per million. Indeed, for most nonferromagnetic substances, the departure of the B–H relationship from that given in Eq. (1.21) is so small as to be completely negligible. Moreover, if dimension a in Fig. 1.1 is much larger than dimension d, it may be assumed that H and B are uniform over the cross section of the torus. In discussions of such a system, therefore, Eq. (1.21) may be employed in place of Eq. (1.4). If the wooden torus of Fig. 1.1 is replaced by a torus of iron of identical dimensions, it is found that the total flux produced by the same coil current is enormously increased. This increase is due to the phenomenon of *ferromagnetism*, so named since it was first observed in iron. It is a very important factor in the process of energy conversion by electromagnetic machines.

In order to understand the underlying principles of the phenomenon of ferromagnetism, it is sufficient to employ the relatively simple model of an atom, which consists of a nucleus surrounded by a cloud of electrons. Each electron has a charge of 1.6×10^{-19} coulomb, and this charge may be considered to be concentrated in a small sphere. The electrons are in orbit around the nucleus, and each orbiting electron spins upon its own axis while it moves around its orbit.

Figure 1.5(a) illustrates an electron in orbit around the atomic nucleus, where the lower sector of the orbit may be considered as passing between the axis of the orbit and the reader, while the upper sector lies beyond the axis of the orbit. For the direction of orbital motion indicated in the diagram, the negative charge on the

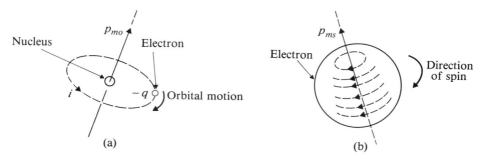

Fig. 1.5 (a) Orbital motion of an electron. (b) Spin of an electron.

electron is equivalent to a positive current i in a loop formed by the orbit and flowing in the direction shown on the diagram. Applying the right-hand screw rule will show that the direction of the resulting magnetic flux, and consequently the direction of the resulting *magnetic moment* along the axis of the orbit is that indicated by the vector $\vec{p}_{mo}$.

Each electron also has a magnetic moment that is independent of its orbital motion. This moment may be visualized as arising from the rotation of the charge of the electron on its own axis—that is, from its spin. Figure 1.5(b) illustrates the effect of spin. If the electronic charge is visualized as spherically distributed about its axis, then the argument applied to an orbiting electron in the preceding paragraph may also be applied to this situation. The electronic charge rotating clockwise about the axis of the electron is equivalent to a loop of current directed counterclockwise. Just as an electron has a specific amount, or quantum, of electric charge, it also has a specific magnetic moment. The *spin magnetic moment* p_{ms} has a magnitude of 9.27×10^{-24} ampere·metre², and the *orbital magnetic moment* is either zero or an integral multiple of the same value.

There are thus two factors that may combine to produce the magnetic moment p_m of an atom. However, in the atoms of many elements, the electrons are arranged symmetrically, so that the magnetic moments due to the spin and the orbital motion cancel out, leaving the atom with no net magnetic moment. Nevertheless, the atoms of more than one-third of the known elements lack this symmetry, so that they do possess a net magnetic moment.

While the atoms of many elements have net magnetic moments, the arrangement of the atoms in most materials is such that the magnetic moment of one atom is cancelled out by that of an oppositely directed (antiparallel) near neighbor. It is only in five of the elements that the atoms are arranged with their magnetic moments in parallel so that they supplement, rather than cancel, one another. The five ferromagnetic elements are iron, nickel, cobalt, dysprosium, and gadolinium; the last two are metals of the rare earths and have limited industrial application. A number of alloys of these five elements, which include nonferromagnetic elements in their composition, also possess the property of ferromagnetism.

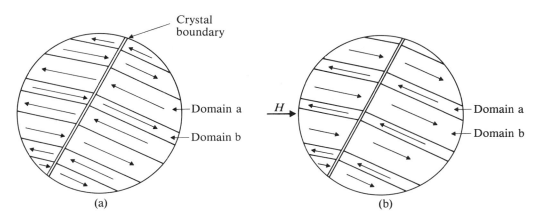

Fig. 1.6 Ferromagnetic domains. (a) No applied magnetic field. (b) Applied magnetic field intensity H.

The direction of alignment of the magnetic moments in a ferromagnetic material is normally along one of the crystal axes. It has been shown experimentally that a specimen of ferromagnetic material is divided into so-called *magnetic domains,* usually of microscopic size, in each of which the atomic moments are aligned. The alignment direction differs, however, from one domain to another. This is illustrated in Fig. 1.6(a) where the arrows are intended to indicate the magnetic-moment direction in each domain. However, it must be appreciated that the domain alignments may be randomly distributed in three dimensions. The size of the domains is such that a single crystal may contain many domains, each aligned with an axis of the crystal.

When a specimen of ferromagnetic material is placed in a magnetic field, the magnetic moments of its atoms tend to rotate into alignment with the direction of the applied field. Domains in the specimen in which the magnetic moments are more or less aligned with the applied magnetic field increase in size at the expense of neighboring domains that are more or less oppositely aligned to the applied field. This phenomenon is known as *domain wall motion,* and is illustrated in Fig. 1.6. The consequence of wall motion is that the specimen of material as a whole acquires a magnetic moment that may be considered as the resultant of all of its atomic moments. The magnetic moment of the specimen provides a measure of the degree of alignment of the atomic moments.

1.3 MAGNETIZATION

If the coil of Fig. 1.1 is wound on a torus of cast iron, then a measured magnetization curve for cast iron may be obtained by determining a series of sets of values for B and H over a suitable range of coil current (see Appendix B). Typically, the measured relationship has the form of the lowest curve in Fig. 1.7. If the coil is

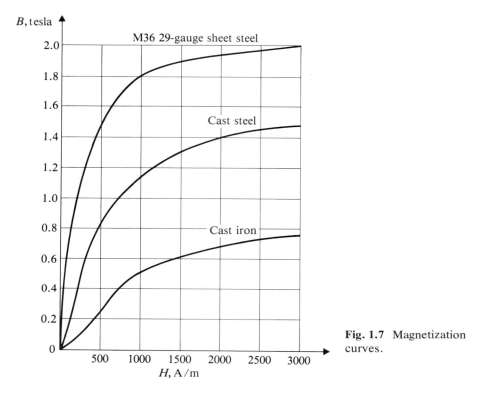

Fig. 1.7 Magnetization curves.

wound on a torus of cast steel of the sort used in electric machinery, then the measured magnetization curve is typically that of the middle curve of Fig. 1.7.

It is convenient to consider that the flux density produced in the ferromagnetic torus is made up of two components:

$$B = B_0 + B_\mu \text{ T} \tag{1.22}$$

where B_0 is the flux density which would occur in a coil containing a vacuum, and B_μ is the additional flux density due to the presence of the ferromagnetic torus. At $H = 1000$ amperes per meter, the value of B_0 would be $4\pi \times 10^{-4} \simeq 0.00125$ T. From Fig. 1.7, the value of B for a cast-iron torus at this value of H is seen to be 0.513 T; thus, from Eq. (1.22), $B_\mu \simeq 0.512$ T. The presence of the cast iron results in a flux density about 500 times as great as would be produced in air.

The flux density B_μ is a result of the partial alignment of the atomic magnetic moments of the iron with the direction of the applied magnetic field H. If the atomic moments are visualized as due to minute current loops arising from electron spin, a cross section of the iron torus may be represented by the diagram of Fig. 1.8(a), although in fact many millions of current loops should be shown in any practical cross-section area. Inside the iron, each minute loop current is adjacent to a similar loop current flowing in the opposite direction. The effects of the adjacent currents

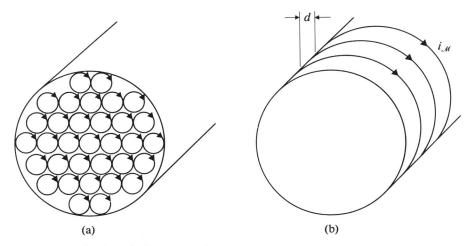

Fig. 1.8 Magnetization of a ferromagnetic torus.

may therefore be considered to cancel. Only on the outer surface is there no cancellation, and there the effect is identical with that of a fictitious surface current.

It is known that each atom of iron has a magnetic moment 2.2 times the basic quantum of magnetic moment $(9.27 \times 10^{-24}$ A $\cdot$ m^2). The spacing between atoms of metallic iron is $d = 2.27 \times 10^{-10}$ meter. The area occupied by a single atom in a cross section of an iron specimen will thus be d^2, and the atomic magnetic moment may be visualized as due to a current i_M flowing in a loop of area d^2. Thus, for one atom,

$$p_m = i_M d^2 = 2.2 \times 9.27 \times 10^{-24} \, \text{A} \cdot \text{m}^2 \tag{1.23}$$

and

$$i_M = \frac{2.2 \times 9.27 \times 10^{-24}}{2.27^2 \times 10^{-20}} = 394 \times 10^{-6} \quad \text{A} \tag{1.24}$$

The effect of the *totally* aligned magnetic moments of a single layer of atoms in the cross section of the torus shown in Fig. 1.8(a) is now seen to be equivalent to a current i_M completely encircling the cross section of the torus. Other layers of atoms, spaced distance d from one another, will be equivalent to further encircling currents, and the effect of all the atoms in the torus will be that of a toroidal coil carrying current i_M, and having turns spaced $d = 2.27 \times 10^{-10}$ meter from one another. The equivalent magnetic field intensity due to this fictitious coil will be

$$H_{\text{equiv}} = \frac{i_M}{d} = \frac{i_M d^2}{d^3} = \frac{pm}{d^3} \quad \text{A/m} \tag{1.25}$$

Since d^3 is the volume of an atom, H_{equiv} expresses the maximum possible magnetic moment per unit volume of the iron.

Magnetic moment per unit volume of a specimen of ferromagnetic material is called the *magnetization* and is seen from Eq. (1.25) to have the same dimensions as magnetic field intensity. Thus, for iron the maximum possible magnetization is

$$\mathcal{M}_{max} = H_{equiv} = \frac{394 \times 10^{-6}}{2.27 \times 10^{-10}} = 1.73 \times 10^6 \quad A/m \tag{1.26}$$

The flux density produced in vacuum by such a magnetic field intensity is

$$B_{\mathcal{M} max} = \mu_0 \mathcal{M}_{max} = 4\pi \times 10^{-7} \times 1.73 \times 10^6 = 2.18 \quad T \tag{1.27}$$

The magnetization curves of Fig. 1.7 show that the flux density B rises rapidly as the magnetic field intensity H is increased from zero. This indicates that only a small applied field is required to cause the domain boundaries to move and allow more of the atomic moments to come into partial alignment with H. As H is further increased, the slope of the $B-H$ curve is reduced, indicating that the domain walls are moving less readily. Further magnetization depends mainly upon the application of a value of H large enough to rotate the atomic moments away from the crystal axes to a direction more nearly in alignment with H. This flattening of the $B-H$ curve is said to be due to *saturation* of the iron. Complete saturation, corresponding to magnetization $\mathcal{M}_{max}$, would occur if all atomic moments were brought into complete alignment with the direction of the applied field. For any lower value of magnetization, from Eqs. (1.22) and (1.27).

$$B = B_0 + B_{\mathcal{M}} = \mu_0(H + \mathcal{M}) \quad T \tag{1.28}$$

As may be seen in Fig. 1.7, even at $H = 3000$ A/m, the flux density for cast iron is only 0.76 tesla, a value much less than the ideal 2.18 tesla that would occur in perfectly aligned iron. This material is therefore very difficult to saturate completely. If H were greatly increased, B would continue to increase, and the curve would eventually approach a slope of μ_0 as maximum magnetization of the material was reached. The M-36 sheet steel of Fig. 1.7, a material designed specially for magnetic systems, reaches a flux density of 1.8 tesla at $H = 1000$ A/m. At this point, only 0.0012 tesla is due to the applied field, the remainder being due to the magnetization of the steel. At 3000 A/m, 1.99 tesla is achieved, which is some 92% of $\mathcal{M}_{max}$ for iron, the chief constituent of the alloy. Thus, some ferromagnetic materials can closely approach the theoretical limit of $\mathcal{M}$ with the application of relatively small values of H.

The atomic moments of materials are said to be held in parallel or antiparallel by *exchange forces,* thought to be due to the sharing or exchange of electrons between neighboring atoms in the crystal structure of the material. When the temperature of a material is increased each atom oscillates about its mean position in the crystal lattice, and this oscillation disturbs the alignment of the spin magnetic moments. Consequently, as the temperature of a ferromagnetic material is increased, its magnetization decreases in the manner illustrated in Fig. 1.9. At a temperature known as the *Curie temperature, T_c*, the parallel alignment of atomic moments completely disappears, and the moments become randomly aligned. The

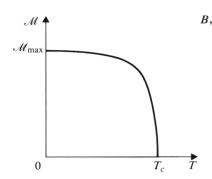

Fig. 1.9 Decrease of ferromagnetic induction with temperature.

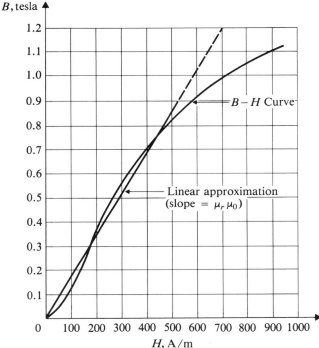

Fig. 1.10 Approximation to the magnetization curve for cast steel in Fig. 1.7.

Curie temperature for iron is 770°C. Since the temperatures in most electric machines are usually below 150°C, the effect of temperature on the ferromagnetic property of iron is small. Temperature effects are more important in nickel, which has a Curie temperature of 348°C. In the rare-earth elements, the Curie temperatures are below normal room temperature.

The $B-H$ curves in Fig. 1.7 are highly nonlinear. However, for reasons that will become apparent in the following sections, magnetic materials are normally operated at points on their $B-H$ curves that are not far into the saturated region. It is therefore often convenient to approximate by a straight line that part of the $B-H$ curve that *is* used. Such an approximation for the cast-steel curve of Fig. 1.7 is shown in Fig. 1.10. In this case, the approximation is acceptable up to a value of $B \cong 0.9$ T. Above this value, the inaccuracies the approximation would introduce into any calculations would become serious.

Within the acceptable "linear" range, the $B-H$ curve may be described by the relation

$$B = \mu_r \mu_0 H \quad \text{T} \tag{1.29}$$

where μ_r is known as the *relative permeability* of the material—that is, the factor by which the flux density is multiplied due to the presence of the ferromagnetic material. In the linear approximation of Fig. 1.10, $\mu_r \cong 1350$.

Example 1.2   The coil in Fig. 1.1 has 1000 turns. The torus is of cast steel and has a mean radius of 250 mm and a cross-section diameter of 25 mm. Employ the $B-H$ curve in Fig. 1.7 to determine $\mathcal{M}$ and μ_r, when the coil current is 1.2 amperes.

Solution

$$H = \frac{Ni}{l} = \frac{1000 \times 1.2}{0.5\pi} = 764 \quad \text{A/m}$$

From Fig. 1.7,

$$B = 1.03 \text{ T}$$
$$B_0 = \mu_0 H = 4\pi \times 10^{-7} \times 764 = 0.960 \times 10^{-3} \quad \text{T}$$
$$B_{\mathcal{M}} = B - B_0 \approx 1.03 \quad \text{T}$$

From Eq. (1.28),

$$\mathcal{M} = \frac{B_{\mathcal{M}}}{\mu_0} = \frac{1.03}{4\pi \times 10^{-7}} = 0.820 \times 10^6 \text{ A/m}$$

$$\mu_r = \frac{1.03}{4\pi \times 10^{-7} \times 764} = 1070$$

Example 1.3   For the magnetic system of Example 1.2, determine:

a) The coil current to produce a flux density of 1.2 tesla in the torus.

b) The relative permeability for a flux density of 0.9 tesla.

c) The inductance of the coil using a straight line through the point on the curve for 0.9 tesla as an approximation to the curve.

Solution

a) From Fig. 1.7, at $B = 1.2$, $H = 1140$ A/m
    From Eq. (1.3),

$$i = \frac{0.5\pi \times 1140}{1000} = 1.79 \quad \text{A}$$

b) From Fig. 1.10, at $B = 0.9$, $H = 580$ A/m
    From Eq. (1.29):

$$\mu_r = \frac{0.9}{4\pi \times 10^{-7} \times 580} = 1235$$

c) Flux in the ring for $B = 0.9$ is

$$\phi = \frac{\pi}{4} \times 25^2 \times 10^{-6} \times 0.9 = 0.491 \times 10^{-3} \quad \text{Wb}$$

$$L = \frac{\lambda}{i} = \frac{N\phi}{i} = \frac{1000 \times 0.491 \times 10^{-3}}{1.79} = 0.274 \quad \text{H}$$

1.4 HYSTERESIS

If the coil in Fig. 1.1 is wound on an iron torus and excited with an alternating current of *very low frequency*, then H varies between peak values $\hat{H}$ and $-\hat{H}$ as in Fig. 1.11(a). If the iron is initially unmagnetized, then the variation of B is as illustrated in Fig. 1.11(b). After a few cycles of H have taken place, the variation of flux density B in the iron is found to be a double-valued function of H forming the $B-H$ *loop abcdef* shown in Fig. 1.11(c). The arrowheads on this loop indicate the direction of movement of the point representing the magnetic state of the iron as H alternates slowly.

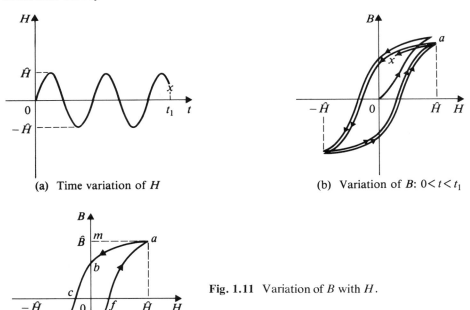

(a) Time variation of H

(b) Variation of B: $0 < t < t_1$

(c) Steady-state loop

Fig. 1.11 Variation of B with H.

In varying H to obtain the closed $B-H$ loop shown in Fig. 1.11(c), the change of magnetic field intensity from $\hat{H}$ to $-\hat{H}$ and from $-\hat{H}$ back to $\hat{H}$ must be unidirectional. At no time, for instance, may the continuous increase in magnetic field intensity which yields the lower limb of the loop be halted and reversed to retrace part of the curve *defa*. If this were done, a minor hysteresis loop like the one illustrated in Fig. 1.12 would be traced out, and several cycles of variation of H would be required to reestablish the stable closed loop *abcdef*.

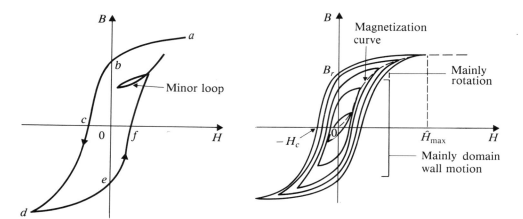

Fig. 1.12 Minor hysteresis loop.

Fig. 1.13 Family of steady-state hysteresis loops.

At point b in Fig. 1.11(c), the iron is magnetized, even though the current in the coil is zero. To remove this magnetization the current must be increased in a negative direction. Thus, at points b and e on the loop, the iron possesses magnetization that does not disappear merely upon removal of the magnetic field that produced it. In fact, throughout the whole cycle of variation, the value of B lags behind the value of H which would be expected to produce it if the relationship between B and H were single-valued. This lagging is called *hysteresis*.

Figure 1.13 shows a family of hysteresis loops for several values of $\hat{H}$. *In each loop,* when H has been increased to zero from a negative value, B is negative. This indicates that more domains in the iron have atomic magnetic moments partly aligned with the negative direction of $\vec{H}$ than with the positive direction. As H is increased, domain walls move, causing domains that have a more or less positive alignment to grow at the expense of those that have a more or less negative alignment. When high values of H are applied, the magnetic moments of a domain may also rotate together toward alignment with $\vec{H}$, without or with only small wall movement. This rotation is an elastic phenomenon—that is, removal of H permits the moments to rotate back into alignment with the crystal axes. The effect of domain rotation as opposed to domain wall motion can be seen in the tendency of the $B-H$ loop to approach a single-valued relation at high values of H.

After H has been increased to magnitude $\hat{H}$ and then reduced to zero, some domain walls move spontaneously toward the positions they took when H was initially zero. This reduces the extent of the positively aligned domains, but the movement is not complete, and the consequence is the flux density B_r shown in Fig. 1.13 on the loop for $\hat{H} = \hat{H}_{max}$. This is known as the *remanence* or *residual flux density* for this particular value of peak magnetic field intensity. The negative magnetic field intensity $-H_c$ required to remove the residual flux density is known as the *coercivity* or *coercive force* of the iron.

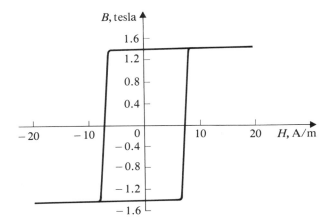

Fig. 1.14 B–H loop for deltamax (50% Ni : 50% Fe).

If the points at the first-quadrant tips of a family of loops produced by various values of $\hat{H}$ are joined, the result is a curve like the dashed line shown in Fig. 1.13. This is the magnetization curve for the iron, already illustrated in Fig. 1.7. (The reason for the method of obtaining this curve, described in Appendix B, should now be appreciated.)

Ferromagnetic alloys have been developed for special purposes, which have B–H loops differing markedly in shape from the loop in Fig. 1.11(c). An example of such a loop for a material consisting of 50% iron and 50% nickel is shown in Fig. 1.14. In this material the domains are naturally aligned in the direction of the desired magnetic field. Almost all the change in flux density B is achieved by domain wall motion. If the amplitude of variation of H for this material is increased above the peak value shown in Fig. 1.14, the loop does not change appreciably, since the material is already saturated at the peak value shown in the diagram.

1.5 ENERGY IN THE MAGNETIC FIELD

If a varying potential difference is applied to the terminals of the N-turn coil in Fig. 1.1, then the current in the coil varies also. If the coil is wound on a nonmagnetic torus, then a varying value of flux density B_0 is produced within it. The flux linkage of the coil at any instant is

$$\lambda = NB_0A \quad \text{Wb} \tag{1.30}$$

where A is the cross-section area of the coil. Due to the variation of this flux linkage, an emf is induced in the coil, of which the instantaneous magnitude is

$$e = \frac{d\lambda}{dt} = NA\frac{dB_0}{dt} \quad \text{V} \tag{1.31}$$

If the instantaneous potential difference at the coil terminals is v, then

$$v = Ri + NA \frac{dB_0}{dt} \quad \text{V} \tag{1.32}$$

where R is the resistance of the coil.

The instantaneous power developed at the coil terminals is

$$p = vi = Ri^2 + NAi \frac{dB_0}{dt} \quad \text{W} \tag{1.33}$$

Of this power, the part represented by the term Ri^2 is dissipated as heat in the coil. The remaining part is

$$p_B = NAi \frac{dB_0}{dt} \quad \text{W} \tag{1.34}$$

Substitution from Eq. (1.3) in Eq. (1.34) yields the alternative form

$$p_B = AlH \frac{dB_0}{dt} \quad \text{W} \tag{1.35}$$

The power p_B represents a flow of energy into the magnetic field within the coil. If this power is negative, it represents a release of energy from the magnetic field.

Let W_B be the energy in the magnetic field. When flux density B_0 is zero, the field energy also will be zero. As the flux density is increased, the field energy may, from Eq. (1.35), be expressed as

$$W_B = \int p_B \, dt = \int_0^{B_0} AlH dB_0 = \int_0^{B_0} \frac{AlB_0}{\mu_0} \, dB_0 = \frac{AlB_0^2}{2\mu_0} \quad \text{J} \tag{1.36}$$

If the flux density is decreased from B_0 to zero, the energy returned to the coil from the magnetic field also has the value given by Eq. (1.36). W_B therefore represents energy stored in the magnetic field—that is, energy which is recoverable. Since the product Al in Eq. (1.36) is the volume of the space enclosed by the coil, the density of the stored energy of the magnetic field within the coil is

$$w_B = \int_0^{B_0} \frac{B_0}{\mu_0} \, dB_0 = \frac{1}{2} \frac{B_0^2}{\mu_0} \quad \text{J/m}^3 \tag{1.37}$$

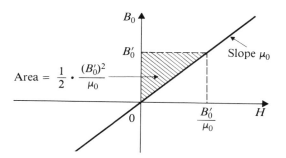

Fig. 1.15 $B\text{–}H$ "loop" for a nonmagnetic material.

The magnetization curve of the nonmagnetic ring may be represented by the straight line through the origin of the B_0–H axes shown in Fig. 1.15. From that diagram and from Eq. (1.37), it may be seen that the energy density corresponding to a specific flux density B_0' is given by the area enclosed between the line representing the B_0–H relationship and the B_0 axis.

1.6 HYSTERESIS LOSS

In general, it may be said that for any magnetic material for which part of the B–H relationship for B increasing may be represented by the curve

$$H = f_i(B) \quad \text{A/m} \tag{1.38}$$

shown in Fig. 1.16(a), an increase of the flux density B from B_1 to B_2 results in an input of magnetic field energy per unit volume expressed by

$$\Delta w_{B\text{incr}} = \int_{B_1}^{B_2} H dB \quad \text{J/m}^3 \tag{1.39}$$

This input is represented by the area enclosed between the B–H curve and the B axis as shown in Fig. 1.16(a). On the other hand, if the curve $H = f_d(B)$ of Fig. 1.16(b) represents the B–H relationship for B decreasing, then a decrease of the flux density from B_2 to B_1 results in a decrease of energy density expressed by

$$\Delta w_{B\text{decr}} = \int_{B_2}^{B_1} H dB \quad \text{J/m}^3 \tag{1.40}$$

If the energy returned during the decrease in flux density is less than the energy supplied, then some of the field energy is lost.

For a magnetic material represented by the B–H loop in Fig. 1.11(c), point a corresponds to a flux density $\hat{B}$ produced by magnetic field intensity $\hat{H}$. The energy per unit volume of the material released by the magnetic field as H is reduced from $\hat{H}$ to zero is given by area amb. This energy is returned to the source supplying the

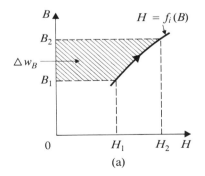
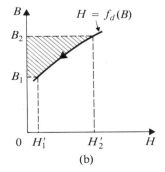

Fig. 1.16 B–H curves for a ferromagnetic material.

toroidal coil, with some loss due to coil resistance. If H is now increased in a negative direction to value $-\hat{H}$, the energy per unit volume supplied by the source to the magnetic field is given by the area $bcdn$.

If H is again reduced to zero, the energy per unit volume released by the field is given by the area dne. If, finally, H is again increased to $\hat{H}$, the energy per unit volume supplied to the field is given by the area $efam$. Since the cycle of H has now returned to its starting point, the stored energy at the end of the cycle must be the same as it was at the beginning. Thus there is a loss of energy per unit volume during this cycle of magnetization, which is given by

$$bcdn + efam - amb - dne = abcdef = \text{Area of loop} \qquad (1.41)$$

This "lost" energy is dissipated as heat in the magnetic material and represents the work done per unit volume in reorienting the magnetic moments of the material as it is carried through a cycle of magnetization. The power dissipated in this manner is called the *hysteresis loss*. It is a function of the peak flux density $\hat{B}$.

The hysteresis loop may be rescaled to represent the relationship between the magnetic flux ϕ and the magnetomotive force $\mathscr{F}$ of the torus by multiplying the ordinates by the cross-section area A and the abscissas by the length of the flux path l. The net effect of this is to multiply any area on the B–H diagram by the volume of the torus Al. The area of the ϕ–$\mathscr{F}$ loop represents hysteresis loss in joules per cycle in the entire torus.

For magnetic materials employed in the construction of electric machinery an approximate relation is

$$\text{Area of loop} = k(\hat{B})^n \qquad 1.5 < n < 2.5 \qquad (1.42)$$

where k and n are empirically determined constants. If the material is carried through f cycles of magnetization per second, then the energy dissipated per second in hysteresis loss is directly proportional to f.

Thus, for the entire torus, the power dissipated in hysteresis loss is given by

$$P_h = K_h f(\hat{B})^n \quad \text{W} \qquad (1.43)$$

where K_h is a constant determined by the nature of the ferromagnetic material and the dimensions of the torus.

Example 1.4   A B–H loop for a type of electric steel sheet is shown in Fig. 1.17. Determine approximately the hysteresis loss per cycle in a torus of 300 mm mean diameter and a square cross section of 50×50 mm.

Solution   The area of each square in Fig. 1.17 represents

$$(0.1 \text{ tesla}) \times (25 \text{ amperes/meter}) = 2.5 \, \frac{\text{weber}}{\text{meter}^2} \times \frac{\text{ampere}}{\text{meter}}$$

$$= 2.5 \, \frac{\text{volt} \times \text{second} \times \text{ampere}}{\text{meter}^3} = 2.5 \, \frac{\text{joule}}{\text{meter}^3}$$

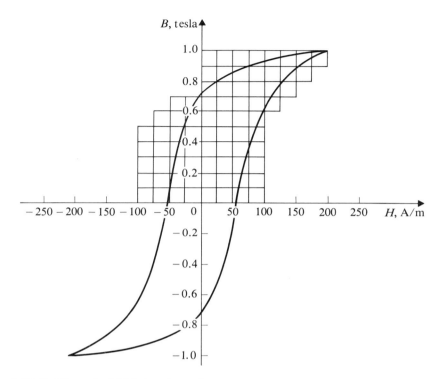

Fig. 1.17 *B–H* loop for M-36 electric steel.

If a square that is more than half within the loop is regarded as totally enclosed, and one that is more than half outside the loop is disregarded, then the area of the loop is

$$2 \times 43 \times 2.5 = 215 \quad \text{J/m}^3$$

The volume of the torus is

$$0.05^2 \times 0.3\pi = 2.36 \times 10^{-3} \quad \text{m}^3$$

Energy loss in the torus per cycle is thus

$$2.36 \times 10^{-3} \times 215 = 0.507 \quad \text{J}$$

1.7 ALTERNATING FLUX

A coil with or without a ferromagnetic core is frequently called an *inductor,* since it is introduced into an electric circuit for the express purpose of providing inductance. An inductor designed for use in an alternating-current circuit may also be

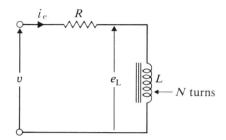

Fig. 1.18 Inductor with ac excitation.

called an *inductive reactor*. Figure 1.18 shows an electric circuit representing an N-turn coil with a ferromagnetic core to whose terminals an alternating potential difference is applied. The coil resistance and inductance are represented as lumped parameters R and L. Let the alternating *exciting current* i_e that flows in the coil be such as to produce a value of flux in the core that is a sinusoidal function of time:

$$\phi = \hat{\phi} \sin \omega t \quad \text{Wb} \tag{1.44}$$

The emf induced in the coil by this flux variation will be

$$e_L = N \frac{d\phi}{dt} = \omega N \hat{\phi} \cos \omega t \quad \text{V} \tag{1.45}$$

The rms value of e_L is related to the rms value of ϕ by the equation

$$E_L = \omega N \Phi = 2\pi f N \Phi \quad \text{V} \tag{1.46}$$

where E_L and Φ are rms values, and f is the frequency of the applied potential difference in hertz. If the potential difference across the resistive element in Fig. 1.18 is small, then

$$v \simeq e_L \quad \text{V} \tag{1.47}$$

and

$$\Phi \simeq \frac{V}{2\pi f N} \quad \text{Wb} \tag{1.48}$$

where V is the rms applied potential difference. Equation (1.48) expresses an approximate relationship, but a very useful one, since it indicates that when a sinusoidal potential difference is applied to an inductor a sinusoidal flux is established inducing an emf that is essentially equal to the applied potential difference. The ferromagnetic properties of the core material and the configuration of the core determine the exciting current required to produce the flux.

1.8 THE EDDY CURRENT LOSS

Once more consider the magnetic system of Fig. 1.1, in which the N-turn coil is wound on a solid torus of ferromagnetic material. The flux linkage of the coil is given by

$$\lambda = NAB \quad \text{Wb} \tag{1.49}$$

where A is the cross-section area of the torus, and B is the flux density, which is assumed to be uniform over the entire cross section. The current in the coil, which may now be called the exciting current i_e, may be expressed as

$$i_e = \frac{Hl}{N} \quad A \tag{1.50}$$

where H is the magnetic field intensity, assumed uniform over the cross section, and l is the mean length of the flux path in the toroidal core. Since N, A, and l are constants, it is possible to draw a $\lambda-i_e$ loop for a particular *core*, which will be similar in shape to the $B-H$ loop for the core material and will merely be rescaled in units of λ and i_e. Such a loop is shown in full line in Fig. 1.19.

It is found that if the frequency of alternation of exciting current i_e and hence of magnetic field intensity H is increased, while the peak value of flux density $\hat{B}$, and hence λ, is held constant, then the $\lambda-i_e$ loop becomes broader. This effect is due to currents induced in the core material which, in the case of ferromagnetic cores, is a conductor of electricity, even if a poor one.

Figure 1.20 shows a cross section of a toroidal core where the direction of i_e produces flux density B entering the cross section normally. Indicated in the cross section is a circular path of elementary width that is concentric with the boundary

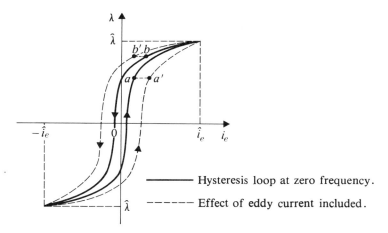

Fig. 1.19 $\lambda-i_e$ loops for the system of Fig. 1.1.

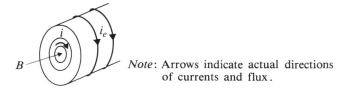

Fig. 1.20 Eddy currents in a torus—i_e decreasing.

of the cross section. This path may be considered to extend around the entire torus. Initially, it will be discussed as if it were isolated from the remainder of the toroidal core.

If i_e is reduced, then the flux density in the core is reduced, and, in particular, the flux linking the elementary path is reduced. The path acts as a single closed loop of conductor, and by Faraday's law an emf is induced in it. By Lenz's law this emf tends to cause a current opposing the change in flux linking the path. The direction of induced emf and resulting current is therefore that shown in Fig. 1.20. This argument may be applied to any other circular path of greater or lesser radius that is concentric with that shown in Fig. 1.20.

The magnetic flux density at any point in the core depends on the exciting current i_e in the coil and the induced currents i in the core. At the surface of the core, B is due to i_e only, since this is the only current that encircles the path of B. The effect of the induced currents is greatest at the center of the core where, if each concentric elementary path is considered as a "turn," the number of such turns encircling the path of B is greatest. The result is that H and, therefore, B are inhibited from changing in magnitude toward the center of the cross section shown in Fig. 1.20, while at the periphery, H and B can change readily in response to change in i_e. This has a feedback effect, in that the reduction in the change of B toward the center of the cross section causes a reduction of the induced emf's and the resultant currents in that region. The net effect is that both the flux and the induced circulating currents tend to be concentrated in the outer part of the core.

If i_e is increased, the direction of the induced emf and the resulting current is opposite to that shown in Fig. 1.20. If i_e is alternating, the induced currents are also alternating, and at very high frequency the interior of the core remains virtually unused due to the large circulating currents induced and their inhibiting effect. This phenomenon is called the *magnetic skin effect*. The circulating currents in the core are called *eddy currents*.

Figure 1.19 illustrates the effect of eddy currents upon the $\lambda - i_e$ loop of a solid iron core. The hysteresis loop in full line is obtained when i_e, and hence λ, is being varied very slowly. Since this means that the flux density in the core is changing very slowly, the induced emf's and resulting eddy currents will be negligibly small. If i_e is varied more rapidly, as when an alternating current is employed to produce flux in a core, the eddy currents may no longer be negligible.

When i_e in Fig. 1.19 is increasing from $-\hat{i}_e$ to $\hat{i}_e$, the eddy currents will produce an mmf opposing the increase in core flux. In order to produce a given value of λ, the exciting current i_e must be increased by the amount necessary to overcome the eddy-current mmf. A point a on the original loop will therefore be replaced by a point a' on the loop that results from a rapidly alternating excitation. This causes the rising limb of the loop to be moved to the right. When i_e is decreasing from $\hat{i}_e$ to $-\hat{i}_e$, the eddy current will produce an mmf opposing the reduction in core flux. In order to produce a given value of λ, therefore, i_e will be reduced by an amount corresponding to the eddy-current mmf. This causes the falling, or reducing, limb of the loop to be moved to the left. Thus, if i_e is alternating, the $\lambda - i_e$ loop will be broadened due to the eddy currents. This broadening will increase as the frequency of the alternating flux linkage is increased.

Eddy currents have two major effects in cores subjected to alternating magnetic fields. As already explained, the alternating flux tends to be forced toward the outer surface of the core. Also, the core material has resistance, and energy is dissipated as heat in the core due to the Ri^2 loss caused by the eddy currents. In devices designed to operate efficiently with alternating flux, the loss effect is more significant. This loss may be reduced in two ways:

1. Core material having a high resistivity may be used. The resistivity of iron is increased substantially by the addition of a few percent of silicon.

2. The cross section of the core may be divided into a number of smaller areas, each insulated from its neighbors. In these smaller areas, smaller eddy currents flowing in shorter paths are induced, and the result is an overall reduction in Ri^2 loss.

There are a number of ways of partitioning the cross section of the core. These include forming the core of a bundle of insulated iron wires (a very old method now being reintroduced for certain special devices); molding the core of nonconducting material in which ferromagnetic particles are suspended; or, most commonly, laminating the core in the plane of the flux and insulating the laminations from one another. In transformers and electric machines, the iron parts carrying varying flux are normally laminated. The use of laminations in toroidal cores usually dictates a rectangular cross section.

Figure 1.21(a) shows a toroidal core of rectangular cross section that is made up of n laminations of ferromagnetic sheet insulated from one another. An enlarged cross section of the core is shown in Fig. 1.21(b). If the lamination of the core is

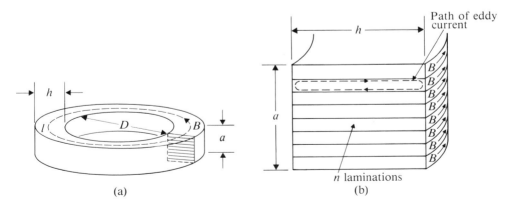

Fig. 1.21 Laminated toroidal core.

sufficiently effective in controlling eddy currents to permit consideration of a flux density that is uniform over the entire cross section of the core, then the flux in each lamination is

$$\phi = \frac{Bha}{n} \quad \text{Wb} \tag{1.51}$$

(The eddy currents may still produce significant loss even if they have little effect on the flux density distribution.) The emf induced around the periphery of each lamination is

$$e = \frac{d\phi}{dt} = \frac{ha}{n}\frac{dB}{dt} \quad \text{V} \tag{1.52}$$

The resulting eddy current flows around a path of resistivity ρ, length approximately $2h$, and cross-section area half that of the entire circular lamination—that is, $al/2n$—where l is the mean circumference of the torus. The resistance of this path is

$$R = k\rho \frac{2h}{al/2n} \quad \Omega \tag{1.53}$$

where k is a constant introduced to account for the fact that paths in the interior of the lamination will have smaller induced emf's than those near the surface. The power loss in a lamination as B varies in time is then given by

$$p = \frac{e^2}{R} = \frac{la}{4k\rho hn}\left(\frac{h^2a^2}{n^2}\right)\left(\frac{dB}{dt}\right)^2 \quad \text{W} \tag{1.54}$$

For the whole core of n laminations

$$P_{\text{core}} = \frac{a^2}{4k\rho n^2}\left(\frac{dB}{dt}\right)^2 (lah) \quad \text{W} \tag{1.55}$$

where lah is the volume of the core. Thus, the core loss per unit volume is

$$\frac{P_{\text{core}}}{lah} = \frac{a^2}{4k\rho n^2}\left(\frac{dB}{dt}\right)^2 \quad \text{W/m}^3 \tag{1.56}$$

Equation (1.56) shows that the core loss per unit volume is proportional to the square of the lamination thickness $a/n = d$. The thinner and more numerous the laminations, the less the loss.

For an alternating flux density

$$B = \hat{B} \sin \omega t \quad \text{T} \tag{1.57}$$

$$\frac{dB}{dt} = \omega\hat{B} \cos \omega t = 2\pi f \hat{B} \cos (2\pi ft) \quad \text{T/s} \tag{1.58}$$

Equation 1.56 shows that the core loss per unit volume is also proportional to the square of the excitation frequency f. Thus the average loss due to eddy currents in a core may be expressed by

$$P_e = K_e(\hat{B}fd)^2 \quad \text{W/m}^3 \text{ or W/kg} \tag{1.59}$$

where K_e is a constant determined by the nature of the ferromagnetic material and the dimensions of the core. As in the case of Eq. (1.43), which expresses the hysteresis loss in a core, this is not an exact relationship but provides an indication of how eddy current loss is affected by the various factors involved.

Electrical steel sheets manufactured for the purpose of building laminated cores normally are covered with a thin surface layer of oxide, which insulates one lamination from its neighbors in a stack of laminations forming a core. This layer may be supplemented by a coat of varnish or inorganic material of high resistivity.

Eddy current loss can occur even if actual flux reversal does not take place in the core. If the flux density is changed from one value to another, eddy currents will be induced, losses will occur, and the change in flux density may take longer than without eddy currents.

Lamination of iron cores is an adequate means of limiting eddy currents up to frequencies of about 50 kHz when laminations as thin as 25 μm may be used. For higher frequencies, the core may be made of powdered iron suspended in a nonconducting material. Alternatively, ferrite cores which have inherently high resistivity are frequently used (see Section 1.17).

1.9 CORE LOSSES

The hysteresis loss described in Section 1.6 and the eddy current loss described in the preceding section are lumped together as the *core losses* of the inductor. These losses are normally measured for given values of lamination thickness, frequency, and peak value of sinusoidally alternating flux density. It is not possible in any single measurement to determine how much of the loss is due to hysteresis and how much to eddy currents, but fortunately it is not necessary to do so. The expressions in Eqs. (1.43) and (1.59) describe approximately the manner in which each component of core loss varies with the core parameters, but give no indication of their relative magnitudes. Manufacturers of electric sheet steel publish experimentally determined curves showing the power loss per kilogram for various values of d, f, and $\hat{B}$.

Figure 1.22 shows curves of peak flux density $\hat{B}$ versus core loss p_c in watts/kilogram for two grades of 0.635-mm thick steel sheet. The alternating excitation is at 60 hertz. M-19 is used for core structures in high-quality rotating machines, small transformers for the electronics industry, and larger transformers designed to operate at power frequencies. M-36 is a more general-purpose steel used for good-quality small motors and for larger motors and generators of medium efficiency. In smaller lamination thicknesses, M-36 is also used for small intermittent-duty transformers, as well as reactors and relays. To obtain the desired magnetic properties in the finished product, the steel-sheet parts must be annealed after being cut or punched from the sheets supplied by the manufacturer. A number of other grades of steel sheet are available.

Figure 1.23 shows core–loss curves for three thicknesses of M-36 electrical sheet. As would be expected from Eq. (1.59) due to the reduction in eddy-current

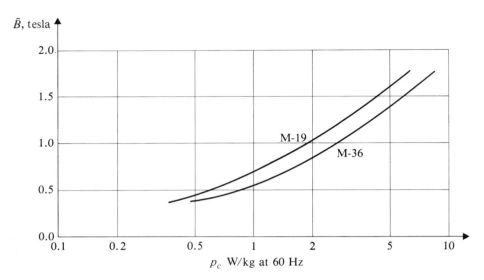

Fig. 1.22 Core loss for two types of electrical steel sheet—0.635-mm thickness.

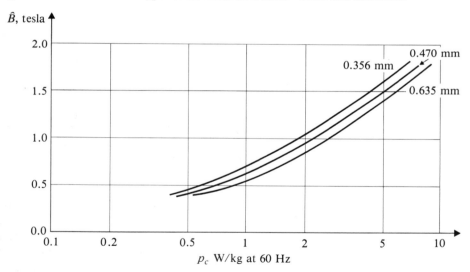

Fig. 1.23 Core loss for three thicknesses of M-36 steel sheet.

loss, the total core loss is less for the thinner sheet. Finally, Fig. 1.24 shows curves of core losses as a function of frequency of excitation for M-36, 0.356-mm sheet. For frequencies higher than 60 hertz, core losses can be kept within acceptable limits either by reducing the sheet thickness or by reducing the flux density.

What grade of sheet and what thickness should be employed in a particular application is an economic question. Sheets with low core loss per kilogram tend to

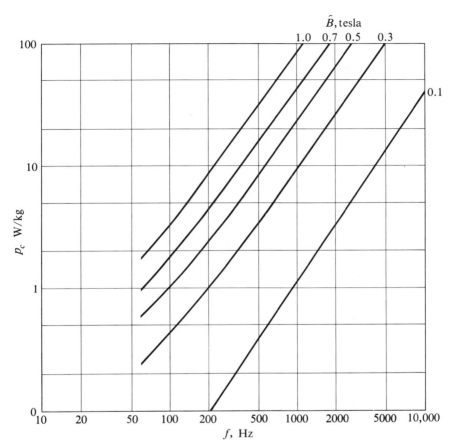

Fig. 1.24 Core losses for M-36, 0.356-mm steel sheet.

cost more. There is little advantage in using thinner laminations when the core loss is due predominantly to hysteresis. Also, a core formed of thin laminations requires more time and labor to assemble than a core of thick laminations, and the sheets are more subject to damage during assembly. On the other hand, the increased losses associated with thicker laminations reduce the efficiency of the device and represent heat that must be removed in order to limit the temperature of the device.

1.10 INDUCTOR EXCITING CURRENT

Due to the nonlinear relationship between B and H for ferromagnetic materials, the current from an inductor excited from a sinusoidal potential source is not a sinusoidal function of time. This phenomenon will be initially discussed for the case of an inductor in which winding resistance is negligible. An equivalent electric circuit of such an inductor is shown in Fig. 1.25.

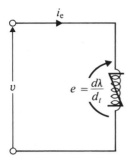

Fig. 1.25 Equivalent circuit of a nonlinear inductor with negligible winding resistance.

The waveshape of the exciting current may be determined graphically from a $\lambda - i_e$ loop for the core and winding. At very low frequencies, the $\lambda - i_e$ loop is essentially the same shape as the $B-H$ loop for the core material. At power frequencies, the $\lambda - i_e$ loop is somewhat broadened due to the effect of eddy currents, as explained in Section 1.8. The method by which the current waveform may be determined is illustrated in Fig. 1.26, from which it is seen that the steady-state current wave has alternating symmetry and is therefore made up of a fundamental fre-

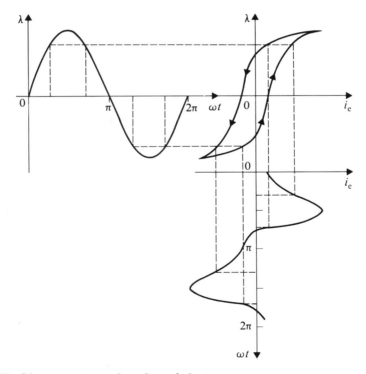

Fig. 1.26 Exciting-current waveform for an inductor.

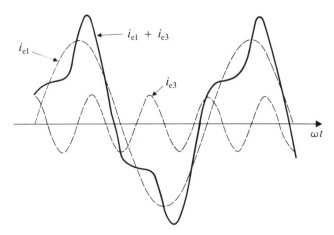

Fig. 1.27 Fundamental and third harmonic components of exciting current.

quency component and odd harmonics. The exciting current may be expressed by the series

$$i_e = i_{e1} + i_{e3} + i_{e5} + \cdots \quad A \tag{1.60}$$

Figure 1.27 shows a diagram of i_{e1}, i_{e3}, and $(i_{e1} + i_{e3})$. This last curve closely resembles the curve of i_e in Fig. 1.26. It may also be seen from Fig. 1.26 that the zero values of the i_e wave occur earlier on the ωt scale than do the zero values of the λ wave. This is due to the double-valued relationship between λ and i_e. In addition, the fundamental component of i_e leads the waveform of λ and lags that of the terminal potential difference v by less than 90°. A phasor diagram representing the relationships between the fundamental components of v, λ, and i_{e1} is shown in Fig. 1.28. In that diagram, V, Λ, and I_{e1} are the rms magnitudes of the three variables. The phasor $\bar{I}_{e1}$ may be resolved into two components, $\bar{I}_c$ in phase with $\bar{V}$, and $\bar{I}_{m1}$ in phase with $\bar{\Lambda}$.

I_c represents a sinusoidally varying current i_c in phase with v, and since v is sinusoidal, accounts for all power dissipated in core losses. I_{m1} represents a sinusoidally varying current i_{m1} in phase with λ, which is necessarily sinusoidal, since

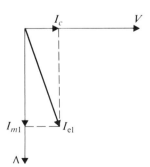

Fig. 1.28 Phasor diagram of exciting-current components.

$$v = \frac{d\lambda}{dt} \quad \text{V} \tag{1.61}$$

The rms value of the exciting current is given by

$$I_e = \sqrt{I_{e1}^2 + I_{e3}^2 + I_{e5}^2 + \cdots} \quad \text{A} \tag{1.62}$$

where I_{e1}, I_{e3}, I_{e5}, etc., are the rms values of i_{e1}, i_{e3}, i_{e5}, etc. In iron-cored devices the difference between I_e and I_{e1} is often small enough to allow the more easily measured value I_e to be used for I_{e1} in a phasor diagram such as Fig. 1.28.

An inductor which is excited from a sinusoidal potential source can be approximately represented by an equivalent circuit of the form shown in Fig. 1.29. In this circuit, the winding resistance and the harmonics of the exciting current have been ignored. The resistance R_c provides for the core loss while the inductive reactance X_m carries the fundamental magnetizing current I_{m1}.

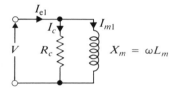

Fig. 1.29 Fundamental-frequency equivalent circuit for an inductor.

1.11 MAGNETIC SYSTEMS

From the $B–H$ loops shown in Fig. 1.13, it is apparent that the flux density in a ferromagnetic material depends upon its magnetic history as well as upon the applied magnetic field intensity. The position of the operating point on the $B–H$ diagram describing the magnetic state of a specimen of the material depends upon whether that point has been reached by means of an increase or a decrease in H and how great that change has been. Thus the single-valued magnetization curve gives only an approximate indication of the value of B resulting from a given H. This approximation is, however, adequate for many engineering calculations. It is quite accurate when H is large and the material is approaching saturation. Its greatest inaccuracy occurs when H is small enough to be neglected in many systems.

The coil on the torus in Fig. 1.1 extends around the entire circumference, and the magnetic flux density outside the coil is essentially zero. Figure 1.30 shows a ferromagnetic torus with a coil that extends only part of the way around its circumference. For this arrangement, it is found experimentally that as long as the material is not near magnetic saturation, the flux density outside the torus is so small that for many practical purposes it may be assumed to be zero. Also, if $a \gg d$, then the flux density in the torus may be considered uniform over the entire cross-section area, and hence uniform throughout the whole volume of the torus.

If B is uniform throughout the whole volume of the torus, then H is uniform throughout the space occupied by the torus. This illustrates an important property of ferromagnetic materials, since if the torus were removed and only the coil

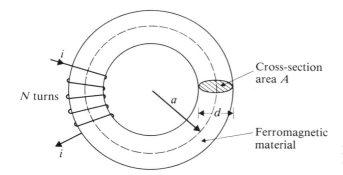

Fig. 1.30 Partially wound iron torus.

remained in its original location, the value of H throughout the space formerly occupied by the torus would not be uniform.

If the circuital law, Eq. (1.1), is applied to the path at the mean radius of the torus in Fig. 1.30, then once again, since H is constant

$$H = \frac{NI}{2\pi a} = \frac{Ni}{l} \quad \text{A/m} \tag{1.63}$$

If the values of N, i and l are known for a particular torus and coil, and a magnetization curve for the material of the torus has been obtained experimentally, then H may be calculated and B obtained from the magnetization curve. The total magnetic flux in the torus is then

$$\phi = BA \quad \text{Wb} \tag{1.64}$$

where A is the cross-section area of the torus.

The preceding analysis may be applied to nontoroidal magnetic systems such as that illustrated in Fig. 1.31. In this system a rectangular core takes the place of the torus. In practice, the flux density outside the core is not zero, since some flux

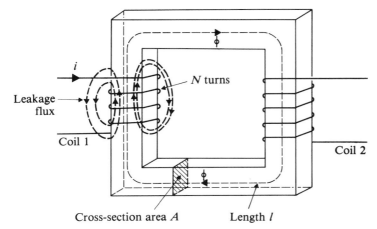

Fig. 1.31 Nontoroidal core.

leaks out of it, as illustrated in Fig. 1.31. However, if the core material is not saturated, this *leakage flux* is negligible in comparison with the flux in the core, so that virtually the whole of the flux produced by the current in coil 1 of Fig. 1.31 links the turns of coil 2. Leakage flux will therefore be disregarded in the discussion that follows.

Suppose that the state of the material of the core in Fig. 1.31 at a particular coil current can be described by the linear relationship of Eq. (1.29),

$$B = \mu_r \mu_0 H \quad \text{T} \tag{1.65}$$

The magnetic flux in the core is

$$\phi = \mu_r \mu_0 H A = \frac{\mu_r \mu_0 A N i}{l} \quad \text{Wb} \tag{1.66}$$

The cause of the magnetic flux ϕ in Eq. (1.66) is the magnetomotive force defined by Eq. (1.6) as $\mathscr{F} = Ni$ amperes. The reluctance of the path followed by the flux is then, from Eq. (1.9),

$$\mathscr{R} = \frac{\mathscr{F}}{\phi} = \frac{l}{\mu_r \mu_0 A} \quad \text{A/Wb} \tag{1.67}$$

Equation (1.67) may be illustrated by the equivalent magnetic circuit of Fig. 1.2. It must be remembered, however, that this may be a nonlinear circuit or a linear circuit applicable over a limited range of flux. Just as the appropriate value of μ_r in Eq. (1.65) is a function of flux density B, so also is the value of $\mathscr{R}$ in Eq. (1.67) a function of the flux ϕ.

A resistive electric circuit may be made up of more than one resistive element. Similarly, a magnetic system may be made up of more than one section of ferromagnetic material. The system may consist of several different kinds of material in parts with different dimensions. A simple example is shown in Fig. 1.32. If a magnetic flux

$$\phi = B_i A_i \quad \text{Wb} \tag{1.68}$$

is produced in the cast-iron section of the magnetic system, what then are the magnetic flux and flux density in the cast-steel section? This question is answered by the continuity law for magnetic flux explained in Appendix A. This law states that for a *closed* surface in a magnetic field,

$$\int_A \vec{B} \cdot \overrightarrow{dA} = 0 \tag{1.69}$$

Consider now that a closed cubic surface envelops the upper junction of the two materials in such a way that two faces of the cube coincide with the shaded cross sections of the two materials indicated by symbols A_i and A_s, which denote the cross-section areas. Let B_s be the flux density in the cast-steel section directed out of the closed surface perpendicular to the area A_s, while B_i is directed into the

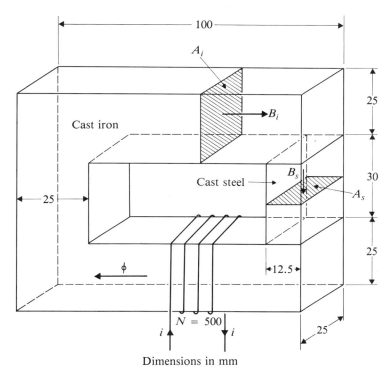

Fig. 1.32 Magnetic system of two different materials for Example 1.5.

closed surface perpendicular to the area A_i. Then over the closed cubic surface, flux density will exist over the two areas A_i and A_s only. Thus, from Eq. (1.69),

$$\int_A \vec{\mathbf{B}} \cdot \overrightarrow{\mathbf{dA}} = -B_i A_i + B_s A_s = 0 \tag{1.70}$$

from which

$$B_s = \frac{A_i}{A_s} B_i \quad \text{T} \tag{1.71}$$

This conclusion might well have been reached intuitively from a consideration of the analogous situation at a junction between two conductors in an electric circuit.

The magnetic system of Fig. 1.32 may be represented by the equivalent magnetic circuit of Fig. 1.33, where $\mathscr{R}_i$ is the reluctance of the cast-iron part, and $\mathscr{R}_s$ that of the cast-steel part. The magnetomotive force $\mathscr{F}$ is equal to the total current encircling the path of the magnetic flux ϕ. From the circuital law,

$$\oint \vec{\mathbf{H}} \cdot \overrightarrow{\mathbf{dl}} = H_i l_i + H_s l_s = Ni = \mathscr{F} \quad \text{A} \tag{1.72}$$

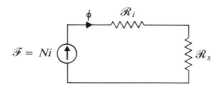

Fig. 1.33 Equivalent magnetic circuit for the system of Fig. 1.32.

The term $H_i l_i$ is the magnetic potential difference across the cast-iron part of the system. The reluctance of that part is

$$\mathcal{R}_i = \frac{H_i l_i}{\phi} = \frac{H_i l_i}{B_i A_i} = \frac{l_i}{\mu_{ri} \mu_0 A_i} \quad \text{A/Wb} \tag{1.73}$$

The reluctance of the cast-steel part is

$$\mathcal{R}_s = \frac{l_s}{\mu_{rs} \mu_0 A_s} \quad \text{A/Wb} \tag{1.74}$$

The total reluctance of the system as seen by the source of magnetomotive force $\mathcal{F}$ is

$$\mathcal{R} = \frac{\mathcal{F}}{\phi} = \mathcal{R}_i + \mathcal{R}_s \quad \text{A/Wb} \tag{1.75}$$

Example 1.5 In the magnetic system shown in Fig. 1.32, employ the magnetization curves of Fig. 1.7 to determine:

a) The coil current required to produce total flux $\phi = 0.25 \times 10^{-3}$ Wb.

b) The reluctance of the entire flux path.

c) Relative permeability μ_r for each material under these conditions.

d) The reluctance of each part, cast iron and cast steel, of the magnetic system.

Leakage flux may be neglected.

Solution The magnetic system may be represented by the equivalent magnetic circuit of Fig. 1.33, where $\mathcal{R}_i$ is the reluctance of the cast-iron part and $\mathcal{R}_s$ that of the cast-steel part.

It will be assumed that the flux density is uniform in each part of the system, so that H will be uniform in each part. Then by the circuital law,

$$\oint \vec{H} \cdot \vec{dl} = H_i l_i + H_s l_s = Ni \quad \text{A}$$

The value of H needed to produce 0.25×10^{-3} Wb in each part must be determined.

$$A_i = 25 \times 25 \times 10^{-6} \quad \text{m}^2$$
$$A_s = 12.5 \times 25 \times 10^{-6} \quad \text{m}^2$$

a)
$$B_i = \frac{\phi}{A_i} = \frac{0.25 \times 10^{-3}}{625 \times 10^{-6}} = 0.4 \quad \text{T}$$

From Fig. 1.7,

$$H_i = 710 \text{ A/m}$$
$$B_s = \frac{\phi}{A_s} = \frac{0.25 \times 10^{-3}}{312.5 \times 10^{-6}} = 0.8 \quad \text{T}$$

From Fig. 1.7,

$$H_s = 480 \text{ A/m}$$

$$l_i \simeq (80 - 25) + 2\left(100 - \frac{25 + 12.5}{2}\right) + \frac{2 \times 25}{2} = 242.5 \text{ mm} = 0.2425 \quad \text{m}$$

$$l_s = 30 \times 10^{-3} \quad \text{m}$$

$$i = \frac{H_i l_i + H_s l_s}{N} = \frac{710 \times 0.2425 + 480 \times 3 \times 10^{-2}}{500} = 0.373 \quad \text{A}$$

b) $$\mathscr{R} = \frac{\mathscr{F}}{\phi} = \frac{Ni}{\phi} = \frac{500 \times 0.373}{0.25 \times 10^{-3}} = 746 \times 10^3 \quad \text{A/Wb}$$

c) $$\mu_r \mu_0 = \frac{B}{H}$$

$$\mu_{ri} = \frac{0.4}{4\pi \times 10^{-7} \times 710} = 448$$

Thus, cast iron is 448 times as effective as air in producing a magnetic field of this flux density.

$$\mu_{rs} = \frac{0.8}{4\pi \times 10^{-7} \times 480} = 1330$$

Cast steel is more effective than cast iron.

d) $$\mathscr{R}_i = \frac{l_i}{\mu_{ri}\mu_0 A_i} = \frac{0.2425}{448 \times 4\pi \times 10^{-7} \times 25^2 \times 10^{-6}} = 690 \times 10^3 \quad \text{A/Wb}$$

$$\mathscr{R}_s = \frac{l_s}{\mu_{rs}\mu_0 A_s} = \frac{3 \times 10^{-2}}{1330 \times 4\pi \times 10^{-7} \times 12.5 \times 25 \times 10^{-6}}$$
$$= 57.7 \times 10^3 \quad \text{A/Wb}$$

Check: $$\mathscr{R}_i + \mathscr{R}_s = (690 + 58)10^3 = 748 \times 10^3 = \mathscr{R}$$

1.12 MORE COMPLEX MAGNETIC SYSTEMS

Equivalent magnetic circuits may be analyzed by methods similar to those employed for electric circuits, provided that the hysteresis loops of the materials are narrow and may be satisfactorily represented by the magnetization curves. In some systems, the magnetization curve of a material may be approximately represented by a straight line through the origin, as illustrated in Fig. 1.10, so that

$$\frac{B}{H} = \mu_r \mu_0 \tag{1.76}$$

where μ_r is constant. In such cases, the magnetic system may be represented by a linear equivalent magnetic circuit, and the techniques of linear electric circuit analysis may be used.

Example 1.6   In the magnetic system shown in Fig. 1.34,

$$l_1 = l_3 = 300 \text{ mm} \qquad l_2 = 100 \text{ mm}$$
$$A_1 = A_3 = 200 \text{ mm}^2 \qquad A_2 = 400 \text{ mm}^2$$
$$\mu_{r1} = \mu_{r3} = 2250 \qquad \mu_{r2} = 1350$$
$$N = 25$$

Determine the flux densities B_1, B_2, and B_3 in the three branches of the circuit when the coil current is 0.5 A.

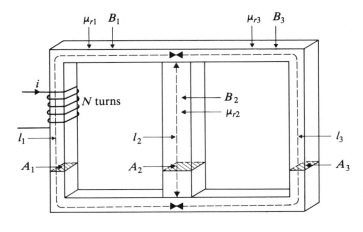

Fig. 1.34 Magnetic circuit for Example 1.6.

Solution

$$\mathcal{R}_1 = \mathcal{R}_3 = \frac{300 \times 10^{-3}}{2250 \times 4\pi \times 10^{-7} \times 200 \times 10^{-6}} = 0.531 \times 10^6 \quad \text{A/Wb}$$

$$\mathcal{R}_2 = \frac{100 \times 10^{-3}}{1350 \times 4\pi \times 10^{-7} \times 400 \times 10^{-6}} = 0.148 \times 10^6 \quad \text{A/Wb}$$

The equivalent magnetic circuit for this system is shown in Fig. 1.35, and the problem may be solved by writing mmf equations for the two loops employing branch fluxes. Thus

$$\mathcal{F} = \mathcal{R}_1 \phi_1 + \mathcal{R}_2 \phi_2$$
$$0 = \mathcal{R}_3 \phi_3 - \mathcal{R}_2 \phi_2$$

These are analogous to equations of potential difference for a dc circuit. Also,

$$\phi_1 = \phi_2 + \phi_3$$

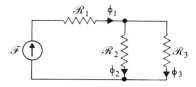

Fig. 1.35 Equivalent magnetic circuit for the system of Fig. 1.34.

This is analogous to a current equation at a node. Substitution of the values of mmf and reluctances in these equations gives

$$12.5 \times 10^{-6} = 0.531 \, \phi_1 + 0.148 \, \phi_2$$
$$0 = -0.148\phi_2 + 0.531 \, \phi_3$$
$$0 = -\phi_1 + \phi_2 + \phi_3$$

Solution of these equations yields

$$\phi_1 = 19.3 \times 10^{-6} \quad \text{Wb}$$
$$\phi_2 = 15.1 \times 10^{-6} \quad \text{Wb}$$
$$\phi_3 = 4.21 \times 10^{-6} \quad \text{Wb}$$

from which the following values are obtained:

$$B_1 = \frac{\phi_1}{A_1} = \frac{19.3 \times 10^{-6}}{200 \times 10^{-6}} = 0.0965 \quad \text{T}$$

$$B_2 = \frac{\phi_2}{A_2} = \frac{15.1 \times 10^{-6}}{400 \times 10^{-6}} = 0.0377 \quad \text{T}$$

$$B_3 = \frac{\phi_3}{A_3} = \frac{4.21 \times 10^{-6}}{200 \times 10^{-6}} = 0.0210 \quad \text{T}$$

1.13 MAGNETIC FLUX IN AIR

The flux density produced in air at the center of a closely packed circular N-turn coil carrying current i is given approximately by the expression

$$B = \frac{\mu_0 N i}{d} \quad \text{T} \tag{1.77}$$

where d is the diameter of the coil. If $N = 1000$, $i = 1.5$ A, $d = 175$ mm, and a current density of 5 A/mm² were employed, then the approximate dimensions of the coil would be those shown in Fig. 1.36(a), where allowance has been made for using round insulated wire to form the coil.

The flux density at the center of this coil would then be, from Eq. (1.77),

$$B = \frac{4\pi \times 10^{-7} \times 10^3 \times 1.5}{175 \times 10^{-3}} = 10.77 \times 10^{-3} \quad \text{T}$$

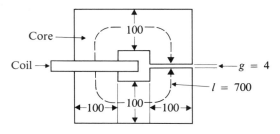

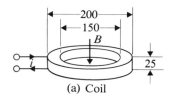

Fig. 1.36 Magnetic system for Example 1.7. (Dimensions in mm. Not to scale.)

(a) Coil

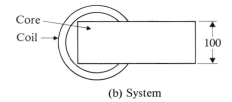

(b) System

A much higher flux density may be produced in air by the same coil and current employed in conjunction with ferromagnetic material, as illustrated in the following example.

Example 1.7   The 1000-turn coil discussed in the preceding paragraph is located on a ferromagnetic core that has an air gap, as illustrated in Fig. 1.36(b). If the coil current is again 1.5 amperes and the relative permeability of the core is 1450,

a) Determine the proportion of the total mmf required to overcome the reluctance of the air gap.

b) Determine the flux density produced in the air gap.

c) Determine the ratio of the flux density in the air gap to the flux density produced at the center of the coil in the absence of the core.

d) Determine the magnetic field intensity in the core and in the air gap.

Leakage flux and fringing at the air gap may be neglected.

Solution   The equivalent magnetic circuit for the system is shown in Fig. 1.37, where $\mathcal{R}_i$ is the reluctance of the core, and $\mathcal{R}_a$ is that of the air gap.

a) $$\mathcal{R}_i = \frac{700 \times 10^{-3}}{1450 \times 4\pi \times 10^{-7} \times 100^2 \times 10^{-6}} = 38.42 \times 10^3 \quad \text{A/Wb}$$

$$\mathcal{R}_a = \frac{4 \times 10^{-3}}{4\pi \times 10^{-7} \times 100^2 \times 10^{-6}} = 318.3 \times 10^3 \quad \text{A/Wb}$$

The total reluctance of the flux path is

$$\mathcal{R} = \mathcal{R}_i + \mathcal{R}_a = (38.42 + 318.3)10^3 = 356.7 \times 10^3 \quad \text{A/Wb}$$

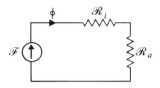

Fig. 1.37 Equivalent magnetic circuit of the system of Fig. 1.36(b).

The proportion of the total mmf required to overcome the air-gap reluctance is

$$\frac{\mathscr{R}_a}{\mathscr{R}} = \frac{318.3}{356.7} = 0.892$$

b)
$$\phi = \frac{\mathscr{F}}{\mathscr{R}} = \frac{10^3 \times 1.5}{356.7 \times 10^3} = 4.205 \times 10^{-3} \quad \text{Wb}$$

$$B = \frac{\phi}{A} = \frac{4.205 \times 10^{-3}}{100^2 \times 10^{-6}} = 0.4205 \quad \text{T}$$

c)
$$\text{Ratio} = \frac{0.4205}{10.77 \times 10^{-3}} = 39.0$$

d) In the air gap,

$$H_a = \frac{B}{\mu_0} = \frac{0.4205}{4\pi \times 10^{-7}} = 0.3346 \times 10^6 \quad \text{A/m}$$

In the core,

$$H_i = \frac{B}{\mu_r \mu_0} = \frac{0.4205}{1450 \times 4\pi \times 10^{-7}} = 230.8 \quad \text{A/m}$$

*1.14 LINEARIZING EFFECT OF AN AIR GAP

Suppose that in Example 1.7, which concerned the system of Fig. 1.36, the actual $B-H$ loop for the ferromagnetic material was employed in determining the flux produced in the system. The total mmf of the coil is, from the circuital law,

$$\mathscr{F} = H_i l_i + H_a l_a \quad \text{A} \tag{1.78}$$

and this may be written

$$\mathscr{F} = \mathscr{F}_i + \mathscr{F}_a \quad \text{A} \tag{1.79}$$

where $\mathscr{F}_i$ is the mmf required to produce a flux density of 0.4205 tesla in the core alone, and $\mathscr{F}_a$ is the mmf required to produce the same flux density in the air gap.

The $B-H$ loop for the ferromagnetic material with $\hat{B} = 0.4205$ T is shown in Fig. 1.38(a), and this may be rescaled as a $\phi - \mathscr{F}_i$ relationship by multiplying B by the cross-section area of the core, and multiplying H by l_i. This gives $(162, 4.205 \times 10^{-3})$ as coordinates of the peak of the $\phi - \mathscr{F}_i$ loop in the first quadrant, shown in Fig. 1.38(a). The $B-H$ relationship for the air gap is $B = \mu_0 H$, and

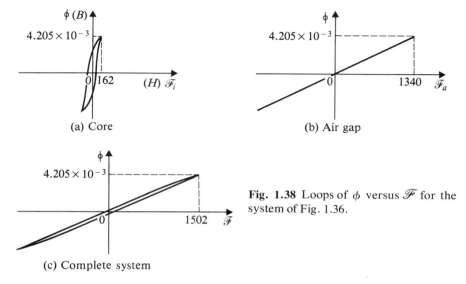

(a) Core

(b) Air gap

(c) Complete system

Fig. 1.38 Loops of ϕ versus $\mathscr{F}$ for the system of Fig. 1.36.

this also may be rescaled as a $\phi - \mathscr{F}_a$ diagram. This gives $(1340, 4.205 \times 10^{-3})$ as the coordinates of the limit of the $\phi - \mathscr{F}_a$ straight-line relationship in the first quadrant, as shown in Fig. 1.38(b).

The $\phi-\mathscr{F}$ relationship for the complete magnetic system may now be obtained by adding the values of $\mathscr{F}_i$ and $\mathscr{F}_a$ for all values of ϕ within the range

$$-4.205 \times 10^{-3} < \phi < 4.205 \times 10^{-3}$$

The result is the loop shown in Fig. 1.38(c). From this loop it may be seen that only a small error (about 12%) would occur if the reluctance of the core were ignored and only the reluctance of the air gap included in the equivalent magnetic circuit. Although the flux path through the core is much longer than that across the air gap, the relative permeability of the ferromagnetic material (often in the range $1000 < \mu_r < 100,000$) is so great that the iron core reluctance is negligible. This approximation of ideal or perfect iron will frequently be made in analyzing magnetic systems. However, it does not apply when the iron is nearing saturation.

In magnetic systems where the reluctance of the ferromagnetic core is a small but significant proportion of that of the whole system (as in Example 1.7), it is convenient to replace the $\phi - \mathscr{F}_i$ loop by a straight line of slope $1/\mathscr{R}_i$, which passes through the peaks of the loop in Fig. 1.38(a). The error involved in ignoring hysteresis and the nonlinearity of the $\phi - \mathscr{F}_i$ characteristic is small in terms of the overall $\phi-\mathscr{F}$ relationship, which becomes a straight line of slope $1/(\mathscr{R}_i + \mathscr{R}_a)$.

*1.15 INDUCTANCE AND STORED ENERGY IN AN AIR-GAPPED SYSTEM

Due to the linear relationship between ϕ and $\mathscr{F}$ which exists, or is assumed to exist, in magnetic systems which are "linearized" by the presence of an air gap, the

inductance of a coil exciting such a system is a constant quantity. Thus, by definition of inductance

$$L = \frac{\lambda}{i} = \frac{N\phi}{i} = \frac{N^2\phi}{Ni} = \frac{N^2}{\mathscr{R}} \quad \text{H} \tag{1.80}$$

The energy stored in a magnetic system excited by a coil of inductance L carrying
current i is

$$W_B = \tfrac{1}{2}Li^2 \quad \text{J} \tag{1.81}$$

Example 1.8   Figure 1.39 shows a cross section of the magnetic system of a direct-
current machine taken at right angles to the rotor axle. On each of the four stator
poles there is a 500-turn coil; and as the four coils are connected in series, all carry
the same current. The stator poles are made of many laminations of M-36 0.356-mm
sheet steel; they are 100 mm radially along their center lines, 90 mm circumferentially, and 110 mm axially—that is, into the plane of the diagram. The rotor also is
of sheet steel and is 200 mm in diameter. The effective axial length of the rotor is
the same as that of the stator poles. The stator yoke is of cast steel. It has a mean
diameter of 460 mm and a cross section of 150 × 60 mm. The air gaps are 1.5 mm
in length.

a) Draw an equivalent magnetic circuit for this system.

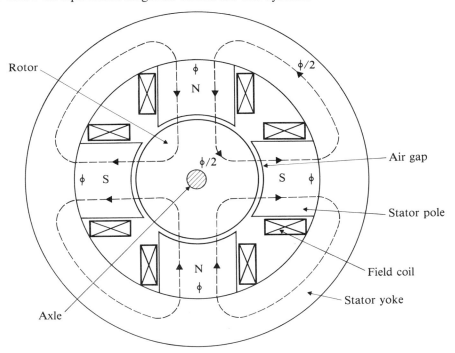

Fig. 1.39 Magnetic system of a dc machine for Example 1.8.

b) Employ the curves of Fig. 1.7 to determine the coil current required to produce a flux density of 1.0 tesla in the air gap.

c) Calculate the total flux linkage of the field coils.

On the assumption that the linearizing effect of the air gap in this system is such as to justify neglect of hysteresis and assumption of linear B–H relationships for the ferromagnetic materials, calculate:

d) The inductance of the entire field circuit.

e) The energy stored in the magnetic system.

f) The energy stored in the air gaps.

Leakage flux and fringing at the air gaps may be neglected.

Solution

a) A partial equivalent magnetic circuit is shown in Fig. 1.40. This includes only two of the field poles. The subscripts p, a, r, and y indicate "pole," "air-gap," "rotor," and "yoke," respectively.

b) The relative permeabilities of the ferromagnetic parts of the system are not known and will in any event be functions of flux density. However, the circuital law may be applied to one of the paths indicated in broken line in Fig. 1.39. Thus, for any one of the flux paths shown in Fig. 1.39:

$$\oint \vec{H} \cdot \vec{dl} = \int \vec{J} \cdot \vec{dA} \quad A \tag{1.82}$$

That is,

$$2H_p l_p + 2H_a l_a + H_r l_r + H_y l_y = 2\,Ni \quad A \tag{1.83}$$

On the assumption that no fringing takes place at the air gap, the cross-section area of the flux path in the pole will be the same as that in the air gap; consequently,

$$B_p = B_a = 1.0 \quad T$$

and the flux in each pole is

$$\phi = B_a \times A_a = 1.0 \times 0.11 \times 0.09 = 9.9 \times 10^{-3} \quad Wb \tag{1.84}$$

$$H_a = \frac{B_a}{\mu_0} = \frac{1.0}{4\pi \times 10^{-7}} = 0.7958 \times 10^6 \quad A/m \tag{1.85}$$

From the curve for sheet steel in Fig. 1.7

$$H_p = 210 \quad A/m$$

In the rotor, the flux ϕ from each pole divides equally between two paths, as indicated in Fig. 1.40. Thus,

$$B_r = \frac{\phi}{2A_r}$$

Assuming that A_r = rotor radius $\times$ axial length of the stator pole, then

$$A_r = 0.1 \times 0.11 = 11 \times 10^{-3} \quad m^2$$

$$B_r = \frac{9.9 \times 10^{-3}}{2 \times 11 \times 10^{-3}} = 0.450 \quad T$$

From Fig. 1.7,

$$H_r = 40 \quad A/m$$

In the yoke also ϕ is divided between two paths, and

$$B_y = \frac{\phi}{2A_y} = \frac{9.9 \times 10^{-3}}{2 \times 0.15 \times 0.06} = 0.550 \quad T$$

From Fig. 1.7,

$$H_y = 295 \quad A/m$$
$$2l_a = 2 \times 1.5 \times 10^{-3} = 3 \times 10^{-3} \quad m$$
$$2l_p = 2 \times 0.1 = 0.2 \quad m$$
$$l_r \simeq \frac{\pi D_r}{4} = \frac{\pi \times 0.2}{4} = 0.1571 \quad m$$
$$l_y \simeq \frac{\pi D_y}{4} = \frac{\pi \times 0.46}{4} = 0.3613 \quad m$$

Substitution in Eq. (1.65) yields

$$210 \times 0.2 + 0.7958 \times 10^6 \times 3 \times 10^{-3} + 40 \times 0.1571 + 295 \times 0.3613 = 2 \times 500\, i$$

or

$$42 + 2387 + 6 + 107 = 1000\, i$$

Note the dominating effect of the air gap.

$$i = 2.54 \quad A$$

This operation could be repeated for a series of values of B in the air gap. A magnetization curve of ϕ versus i for the machine could then be plotted.

c) The total flux linkage of the four coils connected in series is

$$\lambda = 4N\phi = 4 \times 500 \times 9.9 \times 10^{-3} = 19.8 \quad Wb$$

d) On the basis of the linearizing assumption specified,

$$L = \frac{\lambda}{i} = \frac{19.8}{2.54} = 7.80 \quad H$$

e)

$$W_B = \tfrac{1}{2} L i^2 = \frac{7.80 \times 2.54^2}{2} = 25.2 \quad J$$

f) The energy density in the air gap is, from Eq. 1.37,

$$w_a = \frac{1}{2}\frac{B_a^2}{\mu_0} = \frac{1}{2} \times \frac{1^2}{4\pi \times 10^{-7}} = 0.398 \times 10^6 \quad \text{J/m}^3$$

The energy stored in the four air gaps is thus

$$W_a = 4 \times 0.09 \times 0.11 \times 1.5 \times 10^{-3} \times 0.398 \times 10^6 = 23.6 \quad \text{J}$$

The results of (e) and (f) appear to imply that the energy stored in the ferromagnetic parts of the system is $25.2 - 23.6 = 1.60$ J. However, it must be borne in mind that the result in (e) was arrived at on the basis of the assumption that the $B-H$ curves for the materials were straight lines through the origin and the points on the curves of Fig. 1.7 for the actual flux density in the materials. The area enclosed between the $B-H$ curve and the B-axis has been shown in Fig. 1.16 to express the energy density in the material; since the true characteristic is a curved line, the straight-line approximation gives a high value for the energy stored in the materials. Since the materials in this system are far from saturated, however, the error involved is small. The exact figure could be obtained by determining areas from Fig. 1.7.

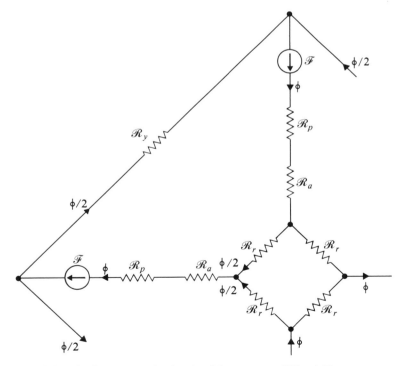

Fig. 1.40 Partial equivalent magnetic circuit of the system of Fig. 1.39.

*1.16 EQUIVALENT CIRCUIT OF AN INDUCTOR

If the behavior of an inductor is to be analyzed, then a mathematical model of it must be constructed, and this involves two stages. In the first stage, a mental picture of the physical system presented by the inductor must be formed. Even for so simple a device as that illustrated in Fig. 1.31 with coil 2 omitted, this is a very complicated system of electric and magnetic fields. In the second stage, the physical system must be described by suitable mathematical equations. In order that these equations shall not be utterly unmanageable, if not unwritable, certain assumptions and approximations must be made.

After the physical system has been visualized in all its complexity, it must then be simplified. First, it must be decided which of its features may be ignored. Then it must be determined what modifications may be assumed that will simplify the final mathematical description, while still giving a sufficiently accurate model. What is "sufficiently accurate" depends upon the purpose for which the analysis is required. In a particular case, when an inductor is available for testing, its behavior may be measured under certain operating conditions. A mathematical model of the inductor may be constructed, and its behavior under the same operating conditions may be predicted by calculation. A comparison of the measured and predicted sets of results then gives a measure of the accuracy of the model.

For many engineering purposes, particularly for preliminary calculations of systems and installations, "sufficiently accurate" means that calculated results should not differ by more than five percent from measured results. For such purposes a number of significant assumptions and approximations may drastically modify the mental picture formed of the physical system before the describing equations are written. In the limit, these assumptions and approximations lead to what may be called an *ideal inductor*.

For the ideal inductor it is assumed that

1. The electric fields produced by the winding are negligible.

2. The winding resistance is negligible.

3. All magnetic flux is confined to the ferromagnetic core.

4. The relative permeability of the core material is constant.

5. The core losses are negligible.

Assumptions 4 and 5 imply that the $B-H$ loop for the ferromagnetic core material may be represented by a straight line through the origin similar to that shown in Fig. 1.10.

If a potential difference v is applied to the terminals of the N-turn winding of the inductor, then current i_e will flow in that winding. The resulting mmf of magnitude Ni_e will produce flux ϕ in the core and establish flux linkage λ in the winding.

If v varies in time, then i_e, ϕ, and λ will vary in time, and an emf e will be induced in the winding. Since the winding resistance is negligible, this emf will be equal in magnitude to the applied potential difference v. Thus

$$v = e = \frac{d\lambda}{dt} \quad \text{V} \tag{1.86}$$

Since the permeability of the core is constant, the flux linkage is given by

$$\lambda = Li_e \quad \text{Wb} \tag{1.87}$$

where L is the inductance of the inductor, and

$$v = L\frac{di_e}{dt} \quad \text{V} \tag{1.88}$$

The inductor may, therefore, be represented by the ideal inductive element of Fig. 1.41(a). This differs from the model of Fig. 1.3(a) only in that it refers to a winding with a ferromagnetic core, rather than to a winding with an *air core*.

If v is a sinusoidal function of time of angular frequency ω, then the following phasor equation may be written:

$$\overline{V} = \overline{E} = j\omega L\overline{I}_e = jX_L\overline{I}_e \quad \text{V} \tag{1.89}$$

where X_L is the inductive reactance of the inductor.

The ideal inductor is not a sufficiently accurate model for all practical purposes. When a more accurate prediction of behavior is required, then a model that corresponds more closely to the physical system must be employed. This means that less radical assumptions and approximations must be made. What these are will become apparent as the required modifications to the model of Fig. 1.41(a) are introduced.

A practical inductor dissipates some energy in the form of heat; that is, it has power losses. Part of these losses occur in the winding. They may be accounted for by the introduction of an element R_w into the equivalent circuit, which is the lumped winding resistance. This is shown in Fig. 1.41(b), which may be compared with Fig. 1.3(b).

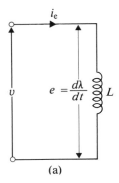 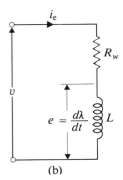

(a)   (b)

Fig. 1.41 Equivalent circuit of inductor. (a) Ideal. (b) With winding resistance.

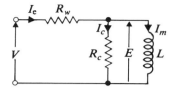

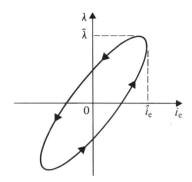

Fig. 1.42 Equivalent circuit of inductor with winding resistance and core losses.

Fig. 1.43 Approximation to the λ-i_e loop.

Further elaboration of the model of Fig. 1.41(b) is necessitated by two factors that did not exist in the system represented by Fig. 1.3. The first factor is the presence of the ferromagnetic core; the second is the nontoroidal nature of the system, which could be represented by Fig. 1.31 with coil 2 removed.

The ferromagnetic core introduces hysteresis and eddy-current loss into the system. It has been shown in Section 1.10 that if the flux linkage of the inductor coil is a sinusoidal function of time, then the rms exciting current I_e may be divided into a magnetizing component I_m and a core loss component I_c. This loss may be accounted for in the equivalent circuit by a resistive element R_c in parallel with the inductive element L, as shown in Fig. 1.42. This corresponds to the replacement of the λ–i_e loop of Fig. 1.26 by the ellipse of Fig. 1.43. The area of this ellipse is proportional to $(\hat{\lambda})^2$; that is, it is proportional to $(\hat{e})^2$. It is therefore appropriate that the power dissipated in R_c should be proportional to E^2.

In practice, inductors are usually constructed with an air gap in the core, as illustrated in Fig. 1.44. This air gap has the linearizing effect discussed in Section 1.14. The result is that the inductor can often be satisfactorily represented in a system by the equivalent circuit model of Fig. 1.41. However, the inductor of Fig. 1.44 departs from ideal behavior in yet another respect, which is that the increased reluctance of the main flux path results in the appearance of leakage flux ϕ_l as indicated in the diagram. This flux takes a path of reluctance $\mathscr{R}_l$ through the air

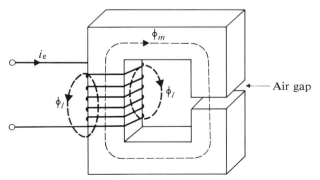

Air gap

Fig. 1.44 Practical inductor.

around the coil rather than through the core and the air gap. The total flux ϕ linking the coil may therefore be divided into two components:

$$\phi = \phi_m + \phi_l \quad \text{Wb} \tag{1.90}$$

Since, as has been shown in Example 1.7, the core reluctance will in practice be very small in comparison with the air-gap reluctance $\mathscr{R}_g$, then to a close approximation

$$\phi_m = \frac{\mathscr{F}}{\mathscr{R}_g} \quad \text{Wb} \tag{1.91}$$

and

$$\phi_l = \frac{\mathscr{F}}{\mathscr{R}_l} \quad \text{Wb} \tag{1.92}$$

where

$$\mathscr{F} = Ni_e \quad \text{A} \tag{1.93}$$

A magnetizing inductance L_m and a leakage inductance L_l may be defined as follows

$$N\phi_m = \frac{N(Ni_e)}{\mathscr{R}_g} = L_m i_e \quad \text{Wb} \tag{1.94}$$

$$N\phi_l = \frac{N(Ni_e)}{\mathscr{R}_l} = L_l i_e \quad \text{Wb} \tag{1.95}$$

so that substitution in Eq. (1.90) yields

$$\lambda = N\phi = L_m i_e + L_l i_e \quad \text{Wb} \tag{1.96}$$

and

$$e = \frac{d\lambda}{dt} = L_m \frac{di_e}{dt} + L_l \frac{di_e}{dt} \quad \text{V} \tag{1.97}$$

This permits the modification of the equivalent circuit of Fig. 1.42 to the form shown in Fig. 1.45.

The whole equivalent circuit model of Fig. 1.45 is rarely required in order to give a sufficiently accurate representation of the effect of an inductor in a system. Which circuit elements can be dispensed with is a matter of judgment for the system designer. However, the model of Fig. 1.45 will be found to be of great service in discussing the behavior of transformers.

The assumptions and approximations, which have been made step-by-step as the equivalent circuit of Fig. 1.45 was evolved, may now be summarized. They are as follows:

1. The electric fields produced by the winding are negligible.
2. The winding resistance may be represented by a lumped parameter at one terminal.
3. The flux produced by the winding may be divided into two distinct parts:
    i) Main flux, confined to the ferromagnetic core and air gap.

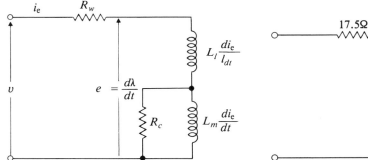

Fig. 1.45 Inductor equivalent circuit including leakage.

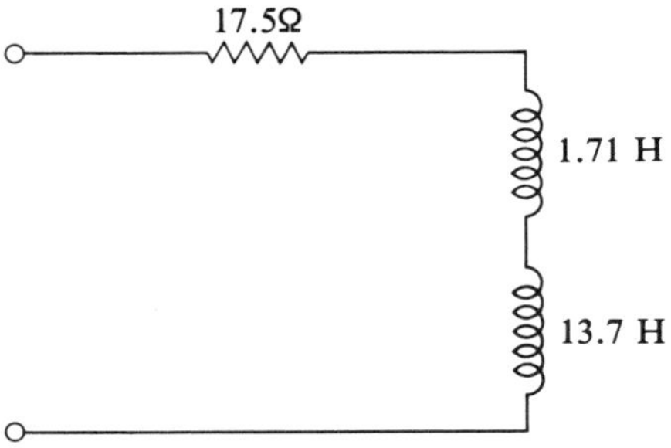

Fig. 1.46 Inductor equivalent circuit for Example 1.9.

ii) Leakage flux, which is produced in a path through the air around the coil.

4. The λ–i_e loop for the core and air gap may be represented by an ellipse whose area equals that of the loop.

Example 1.9   An inductor with an air gap in the core, as illustrated in Fig. 1.44, has a 2000-turn coil of measured resistance 17.5 Ω. At a coil current of 0.7 A, a search coil placed on the core immediately adjacent to the air gap indicates a flux of 4.8×10^{-3} Wb, while a search coil wound on the outside of the inductor coil indicates a flux of 5.4×10^{-3} Wb.

Assuming that the core losses in this inductor are negligible, draw an equivalent circuit and determine the circuit parameters.

Solution

$$\phi = 5.4 \times 10^{-3}\,\text{Wb}; \qquad \phi_m = 4.8 \times 10^{-3}\quad\text{Wb}$$
$$\phi_l = \phi - \phi_m = 0.6 \times 10^{-3}\quad\text{Wb}$$

From Eq. (1.94),

$$L_m = \frac{N\phi_m}{i_e} = \frac{2000 \times 4.8 \times 10^{-3}}{0.7} = 13.71\quad\text{H}$$

From Eq. (1.95),

$$L_l = \frac{N\phi_l}{i_e} = \frac{2000 \times 0.6 \times 10^{-3}}{0.7} = 1.741\quad\text{H}$$

By measurement,

$$R_w = 17.5\ \Omega$$

The equivalent circuit is shown in Fig. 1.46.

*1.17 FERRITES AND FERRIMAGNETISM

The naturally occurring magnets known to the ancients were pieces of "lodestone," or magnetite, an oxide of iron. In recent years, a new class of magnetic materials has been developed, which consists of mixed oxides of iron and other metals. These are known as *ferrites*. They are closely related to the ferromagnetic metals, but because the magnetic mechanism is somewhat different, they are said to be *ferrimagnetic*.

The fundamental particle considered in discussing ferr*o*magnetism was the atom of the ferromagnetic element. The fundamental particle that must be considered in discussing ferr*i*magnetism is the molecule of the ferrimagnetic compound which, like the ferromagnetic atom, possesses a magnetic moment. In addition, the ferrimagnetic materials are mixtures of two different compounds with differing molecular magnetic moments. Coupling forces, similar to the exchange forces of the atom, exist between the molecules of each of the compounds. These forces hold the magnetic moments of like molecules in parallel and those of unlike molecules in antiparallel alignment. Since the moments of the two kinds of molecules are unequal, there is a net magnetic moment for the mixture.

When ferrite cores are manufactured, the two oxides are milled together in the correct proportions and "sintered"—that is, they are heated to a temperature in the neighborhood of 1300°C and kept at that temperature for several hours. The resulting material is chemically homogeneous and mechanically extremely hard. Ferrites are reasonably good electrical insulators. Consequently, eddy currents are almost completely absent, so that these materials can be employed at very high frequencies. With higher temperatures, ferrites lose their ferrimagnetic property, but since the two different compounds of the mixture may have different Curie temperatures, the variation of magnetization with temperature does not necessarily follow a curve such as that shown in Fig. 1.9 for a single ferromagnetic material. The temperature at which the magnetization of a ferrite material falls to zero is called the *Néel* temperature, the value of which is usually in the range 250°–500°C for practical ferrite materials.

Due to the antiparallel alignment of the two groups of magnetic moments and the presence of nonferromagnetic oxygen in the mixture, a ferrite cannot have as high a magnetization as a ferromagnetic material. Typical values of $B_{M\,\text{max}}$ are about 0.3 to 0.5 tesla, as compared with 2.18 tesla for pure iron; however, the raw materials are cheap and the finished material is lighter, volume for volume, than an alloy steel.

Ferrite cores are available in a large range of standard sizes, shapes, and magnetic characteristics. They include an internationally standardized series of so-called *pot cores,* commonly made of manganese–zinc or nickel–zinc ferrite. Figure 1.47 illustrates a pot core in two similar sections that are clamped together by means of specially designed hardware, usually with adhesive between the matching faces. Diameter *a* may be as much as 70 mm. The coil is wound on a former and fills the space between the inner and outer cylinder formed by the two halves of the core.

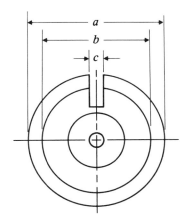

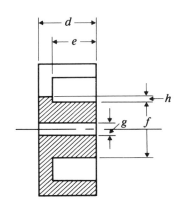

Fig. 1.47 Ungapped pot-core half.

The leads to the coil emerge through the slot shown at the top of the core sections. Pot cores may be manufactured with an air gap between the central bosses of the two core halves. Figure 1.48 shows $B-H$ curves for two materials suitable for cores like those illustrated in Fig. 1.47. Figure 1.49 shows curves of core loss as a function of maximum flux density at various frequencies for the material whose $B-H$ loop is shown in Fig. 1.48(b).

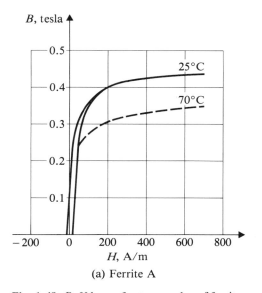

(a) Ferrite A

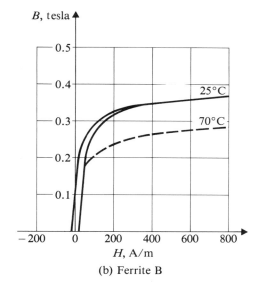

(b) Ferrite B

Fig. 1.48 $B-H$ loops for two grades of ferrite.

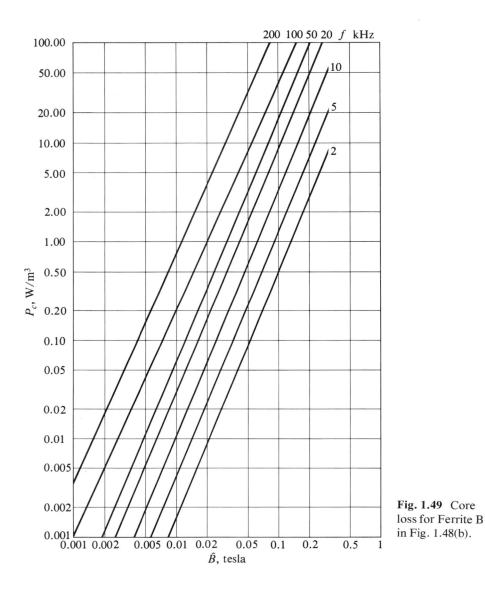

Fig. 1.49 Core loss for Ferrite B in Fig. 1.48(b).

Example 1.10 A pot core is made up of two halves, each of which is as illustrated in Fig. 1.47. The dimensions in millimeters are

a: 67.5	*c:* 7.5	*e:* 21.0	*g:* 6.5
b: 55.5	*d:* 30.0	*f:* 29.0	*h:* 1.0

The core is made of ferrite B, whose $B-H$ loop is given in Fig. 1.48(b). Assume that a satisfactory linear approximation to this loop is given by a straight line passing through the origin and the point (50, 0.25), and that this straight line may be used up to a peak flux density of 0.3 T. The central bosses of the two core halves are ground down to give a total air gap of 6 mm when the halves are clamped together.

Assuming a conductor current density of 2.5 A/mm^2 is acceptable, and the coil is required to carry a current of 0.5 A, determine the maximum possible inductance that can be provided by this core. It may be assumed that 50% of the available winding space is occupied by conductor material, the remainder being taken up by insulation and necessary mechanical clearances.

Solution The total cross-section area available for coil conductor is

$$A_w = 0.5 \times 2e(b - f) = 21(55.5 - 29) = 556.5 \quad \text{mm}^2$$

The cross-section area of the conductor must be $0.5/2.5 = 0.2$ mm^2. The number of turns is thus

$$N = \frac{556.5}{0.2} = 2780$$

For the straight-line approximation of the B–H loop,

$$\mu_r \mu_0 = \frac{\hat{B}}{\hat{H}} = \frac{0.25}{50} = 5 \times 10^{-3}$$

so that

$$\mu_r = \frac{5 \times 10^{-3}}{4\pi \times 10^{-7}} = 3980$$

In determining the reluctance of the core, the effect of the slot (dimensions c and h) will be ignored. The length of the flux path through the skirts of the core halves is

$$2\left(e + \frac{d - e}{2}\right) = e + d = 51 \quad \text{mm}$$

The length of the flux path through the central bosses of the core halves is this amount less the air-gap length—that is, $51 - 6 = 45$ mm. The skirt cross-section area is

$$\frac{\pi}{4}(a^2 - b^2) = \frac{\pi}{4}(67.5^2 - 55.5^2) = 1159 \quad \text{mm}^2$$

The boss cross-section area is

$$\frac{\pi}{4}(f^2 - g^2) = \frac{\pi}{4}(29^2 - 6.5^2) = 627.3 \quad \text{mm}^2$$

The length of the radial flux paths through the webs between the boss and skirt is

$$2\left(\frac{a + b}{4} - \frac{f + g}{4}\right) = \frac{1}{2}(a + b - f - g)$$

$$= \frac{1}{2}(67.5 + 55.5 - 29 - 6.5) = 43.75 \quad \text{mm}$$

The mean cross-section area of the flux path through the web will be taken to occur at diameter $(b + f)/2$. The web cross-section area is therefore

$$\pi \left(\frac{b + f}{2} \right) (d - e) = \pi \left(\frac{55.5 + 29}{2} \right) (30 - 21) = 1195 \quad \text{mm}^2$$

The reluctance of the core is thus

$$\mathcal{R}_c = \frac{1}{5 \times 10^{-3}} \left(\frac{51}{1159} + \frac{5}{627.3} + \frac{43.75}{1195} \right) \times 10^3$$

$$= 30.47 \times 10^3 \quad \text{A/Wb}$$

Ignoring fringing of flux around the air gap, the reluctance of the air gap is

$$\mathcal{R}_a = \frac{6 \times 10^3}{4\pi \times 10^{-7} \times 627.3} = 7.611 \times 10^6 \quad \text{A/Wb}$$

Total reluctance of the flux path is thus

$$\mathcal{R} = \mathcal{R}_c + \mathcal{R}_a = 30.47 \times 10^3 + 7.611 \times 10^6 = 7.641 \times 10^6 \quad \text{A/Wb}$$

The effect of the core reluctance is thus seen to be negligible. The inductance is

$$L = \frac{N^2}{\mathcal{R}} = \frac{2780^2}{7.641 \times 10^6} = 1.011 \quad \text{H}$$

Finally, the flux density in the core should be checked to ensure that the limits to the straight-line approximation to the $B-H$ loop have not been exceeded. Ignoring H in the ferrite, the air-gap mmf is

$$H_{\text{gap}} \times 0.006 = 2780 \times 0.5 \quad \text{A/m}$$

from which

$$H_{\text{gap}} = 232 \times 10^3 \quad \text{A/m}$$

For the central bosses,

$$B_{\text{boss}} = B_{\text{gap}} = 4\pi \times 10^{-7} \times 232 \times 10^3 = 0.291 \quad \text{T}$$

which is satisfactory.

*1.18 GRAIN-ORIENTED MATERIALS

The permeability of electrical steel sheet is normally found to be greater in the direction in which the sheet is rolled than perpendicular to that direction. This is due to the fact that the crystals or grains that form the basic structure of the material are oriented by the rolling process into the direction of rolling. Core laminations of such material may be cut so that the magnetic flux is along the grain for the greater

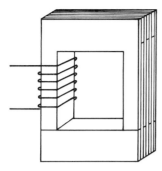

Fig. 1.50 Core formed of grain-oriented stampings.

part of its path through the core. When flux must be produced perpendicular to the direction of grain, the stampings may be cut so that the cross-section area of the flux path is increased, as illustrated in Fig. 1.50.

Cores may also be formed from lengths of grain-oriented steel strip, wound to shape and impregnated with shellac or other insulating adhesive to hold them in shape. With the core still in one piece, the coil may be wound on to it by special machinery. Alternatively, the coil may be formed first and the core wound on the coil. Either of these two methods of manufacture results in a magnetic system with no air gap and, thus, very low reluctance. Another way to assemble the coil and core is to cut the core into two parts and grind and etch the cut faces to give a very small air gap when the two parts are banded together. Such cut cores are called *C-cores,* and one such is illustrated in Fig. 1.51. For power-frequency operation the thickness of the strip is commonly about 0.3 mm. The magnetization curve for a typical grain-oriented strip is shown in Fig. 1.52 and a curve of core loss at 60 hertz is shown in Fig. 1.53.

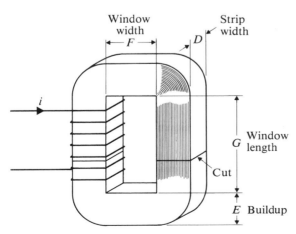

Fig. 1.51 C-core of grain-oriented strip.

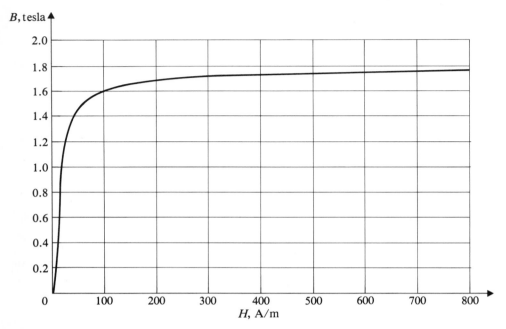

Fig. 1.52 Magnetization curve for 0.305-mm, grain-oriented, steel strip.

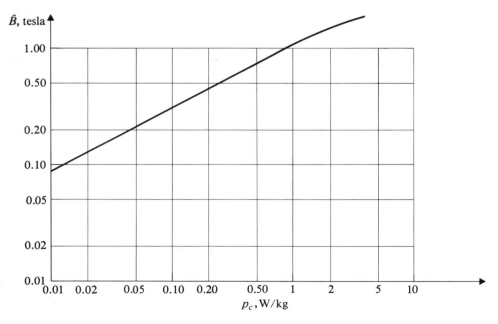

Fig. 1.53 Core losses at 60 Hz for 0.305-mm, grain-oriented, steel strip.

*1.19 INDUCTOR CONSTRUCTION AND DESIGN

The preceding sections have shown that an inductor is essentially a magnetic system consisting of a ferromagnetic or ferrimagnetic core with an air gap and an exciting coil. Its purpose is that of introducing inductance into an electric system. On the other hand, a similar magnetic system may be devised in order to produce a specified magnetic field in the air gap, in which case it is properly called an electromagnet. In the highly specialized electromagnets employed in materials handling, the magnetic flux path is completed through the material (sheet steel, scrap metal, etc.) which is to be lifted, moved, and released by interruption of the exciting current. It is usual for inductors to be linear within the operating range, and this necessitates the air gap, unless the core is a ferrite with constant but low permeability. In electromagnets the cores are usually operated well into saturation.

When an inductor is required as a component in an electric system, the inductance and the coil current will be specified, and it will be necessary to determine the following quantities:

$$A_w = \text{Conductor cross-section area of the coil;}$$
$$A_i = \text{Steel cross-section area of the core;}$$
$$N = \text{Number of turns on the coil;}$$
$$g = \text{Length of air gap.}$$

The winding factor is defined by

$$k_w = \frac{\text{Total conductor cross-section area}}{\text{Core window area}} \tag{1.98}$$

For small coils this may be as low as 0.4 to 0.5. In large inductors this can be improved upon, particularly if a square or rectangular conductor is used.

As a consequence of the lamination of the core, the cross-section area of the magnetic circuit cannot be completely made up of steel, no matter how closely the laminations may be clamped or wound together. This gives rise to a stacking factor defined by

$$k_i = \frac{\text{Steel cross-section area}}{\text{Gross core cross-section area}} \tag{1.99}$$

This factor becomes lower as thinner laminations are employed. With the thicknesses of sheet or strip used in power devices it is normally 0.9 or higher.

If a laminated core built of stampings is employed, it will normally have the form shown in Fig. 1.54. This is referred to as *shell-type* construction, since the core to some extent forms a "shell" enclosing the coil. The cross-section of the center leg that carries the coil is normally made square so that the maximum core cross-section area is enclosed within a minimum length of conductor.

A C-core may also be employed to make an inductor, with nonferromagnetic spacers between the faces of the core sections to provide the "air" gaps. The coil is wound on one of the long limbs of the core, as indicated in Fig. 1.51, or two coils in series may be wound on the two limbs.

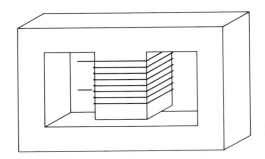

Fig. 1.54 Laminated inductor core.

Other factors that the designer of the inductor will know before beginning the design are the peak flux density $\hat{B}$, determined by the nature of the core material, and the allowable current density in the coil. On the basis of experience with conventional insulation, it is known that, to avoid excessive heating, the rms current density J in the copper conductors of continuously operated power devices must not exceed a value in the range of 3–8 A/mm².

The following relationships exist between the four unknown quantities that are to be determined:

$$Ni = A_w J \quad \text{A} \tag{1.100}$$

$$L = \frac{N\phi}{i} = \frac{N\hat{B}A_i}{\hat{i}} \quad \text{H} \tag{1.101}$$

Since the reluctance of the air gap will form by far the greatest part of the entire reluctance of the flux path, it may be assumed initially that

$$L = \frac{N^2}{\mathscr{R}_a} = \frac{N^2\mu_0 A_i}{gk_i} \quad \text{H} \tag{1.102}$$

where

$$\frac{gk_i}{\mu_0 A_i} = \mathscr{R}_a \quad \text{A/Wb} \tag{1.103}$$

The stacking factor k_i appears in this last relationship because the cross-section area of the air gap is equal to the *gross* cross-section area of the core if fringing at the air gap is neglected.

Equations (1.100)–(1.103) provide three relationships among four unknowns. It is therefore possible to choose one unknown arbitrarily and solve for the remaining three unknowns analytically. However, a fourth relationship will usually exist. This may take the form of a requirement that the inductor shall have minimum bulk, weight, or cost; or it may take the form of specified core proportions. It is common practice for manufacturers of C-cores to include in their catalogs a quantity expressed in terms of the dimensions shown in Fig. 1.51. This is

$$D \times E \times F \times G = \frac{A_i A_w}{k_i k_w} \quad \text{m}^4 \tag{1.104}$$

A core with this product may therefore be chosen, and the core proportions are thus specified, giving the required fourth relationship between the four unknowns. The proportions for a typical core might be

$$G = 2F = 2D = 4E \quad \text{m} \tag{1.105}$$

In designing magnetic systems with air gaps, it is also desirable to make some allowance for fringing of magnetic flux around the edges of the air gap, which has the effect of increasing the air gap cross-section area. A common correction is to add the air-gap length to the length of each of the sides (or to the diameter, in the case of a round cross-section core).

Example 1.11 A 0.5-H inductor is required to carry a continuous current of 2.5 A at 60 Hz. The core is to be formed of the 0.305-mm, grain-oriented steel strip of Figs. 1.52 and 1.53 and may be assumed to have the proportions given in Eq. (1.105) for Fig. 1.51. Assume a stacking factor of 0.95 and a winding factor of 0.55. The rms current density is to be 5 A/mm². The density of the steel strip is 7640 kg/m³, and the resistivity of copper wire is 1.72×10^{-8} $\Omega \cdot$m.

Prepare an approximate design for this inductor, and determine the power dissipated in it when rated current is flowing.

Solution From Eqs. (1.100) and (1.101),

$$A_i A_w = \frac{L(\hat{\imath})^2}{\hat{B}\hat{J}} \tag{1}$$

From Eqs. (1.104) and (1.105),

$$DEFG = 16E^4 = \frac{A_i A_w}{k_i k_w} \tag{2}$$

Then from (1) and (2),

$$16E^4 = \frac{1}{k_i k_w} \frac{L(\hat{\imath})^2}{\hat{B}\hat{J}} \tag{3}$$

From Fig. 1.52, let $B = 1.2$ T. Then substitution of the specified values in (3) gives

$$E^4 = \frac{1}{16 \times 0.95 \times 0.55} \times \frac{0.5(2.5\sqrt{2})^2}{1.2 \times 5\sqrt{2} \times 10^6}$$

from which $E = 17.2$ mm. Thus, for the given proportions,

$$A_i = k_i \times 2E^2 = 0.95 \times 2 \times 17.2^2 \times 10^{-6} = 0.562 \times 10^{-3} \quad \text{m}^2$$
$$A_w = k_w \times 8E^2 = 0.55 \times 8 \times 17.2^2 \times 10^{-6} = 1.31 \times 10^{-3} \quad \text{m}^2$$

From Eq. (1.100),

$$N = \frac{A_w J}{i} = \frac{1.31 \times 10^{-3} \times 5 \times 10^6}{2.5} = 2612$$

The window is able to accommodate 2612 turns.

From Eqs. (1.102) and (1.103),

$$g = \frac{\mu_0 A_i N^2}{k_i L} = \frac{4\pi \times 10^{-7} \times 0.562 \times 10^{-3} \times 2612^2}{0.95 \times 0.5} = 10.14 \times 10^{-3} \quad \text{m}$$

Two air gaps each of 5.07 mm are required.

At this stage in the design process, an actual C-core would be chosen from a manufacturer's catalog, and the design would be repeated, possibly with more than one iteration, allowance made for the mmf absorbed in the core and the fringing at the air gaps. Provision would also be made for nonferromagnetic shims in the air gaps to make final adjustment of the inductance after a prototype has been assembled.

The volume of steel in the core is given by the mean length of the flux path, l_i, multiplied by A_i.

$$l_i = 2(F + G) + \pi E = (12 + \pi)17.2 \quad \text{mm}$$

so that the volume of the core is

$$V_i = \frac{(12 + \pi)17.2}{10^3} \times 0.562 \times 10^{-3} = 0.147 \times 10^{-3} \quad \text{m}^3$$

The weight of the core is

$$0.147 \times 10^{-3} \times 7640 = 1.12 \quad \text{kg}$$

From Fig. 1.53, core loss/kg at $\hat{B} = 1.2$ T is 1.25 watts. Total core loss is therefore

$$P_c = 1.25 \times 1.12 = 1.40 \quad \text{W}$$

Conductor cross-section area of the coil is

$$A_w = 1.31 \times 10^{-3} \quad \text{m}^2$$

Length of mean turn in the coil is

$$l_w = 2D + 2E + \pi E = (6 + \pi)E = 157 \quad \text{mm}$$

Winding loss is therefore

$$P_w = \rho J^2 A_w l_w$$
$$= 1.72 \times 10^{-8} \times (5 \times 10^6)^2 \times 1.31 \times 10^{-3} \times 0.157 = 88.4 \quad \text{W}$$

The power dissipated in the inductor is therefore

$$P = P_c + P_w = 1.4 + 88.4 = 89.8 \quad \text{W}$$

*1.20 PERMANENT MAGNETISM

A magnetic system that can produce flux in an air gap but requires no exciting coil, and therefore produces no dissipation of electric power, has obvious attractions, particularly in the field of energy conversion. *Permanent magnets,* which can provide such magnetic fields, have aroused human interest since pre-Socratic times, but only in recent years has it become possible to manufacture them in the qualities, sizes, and shapes that now permit their use in a wide range of industrial applications. "Permanent" magnetism of a precarious nature is produced in any ferromagnetic material when it is carried through a cycle of magnetization to a point on the hysteresis loop indicated by $\pm B_r$ in Fig. 1.13. This remanence, however, is far from truly permanent and may be reduced by mechanical vibration, thermal agitation of the atoms, or by application of a small reversed magnetic field intensity.

Nevertheless, some materials have been known for a good many years that have a permanent magnetism much more durable than the remanence of iron. They are normally alloys of iron, nickel, and cobalt, three of the ferromagnetic elements. These materials are usually subjected to heat treatment, which results in mechanical hardness of the finished magnet. From this feature arose the terminology that refers to permanent-magnet materials as *hard* and other magnetic materials as *soft*. Since this terminology was adopted, some materials that are magnetically hard and mechanically soft have been developed.

In soft magnetic materials, as was explained in Section 1.4, the early stage of easy magnetization of the material at low values of H is due to domain wall movement. Only at high values of H does domain rotation make an elastic contribution to magnetization, which disappears as soon as the magnetic field intensity is lowered. Remanence of the soft material is due to domains whose walls have not returned to the position that they occupied before H was applied. The composition and heat treatment of many hard magnetic materials produce materials that are formed of very small elongated crystals, a structure that inhibits domain wall movement. In hard magnetic materials subjected to a magnetic field, most of the magnetization occurs as a result of abrupt inelastic switching of the orientation of entire domains through 180°. The persistence of the domains in their switched orientations after H is removed gives rise to the high remanence of permanent magnet materials and to a coercivity that may be several thousand times as great as that of a soft material.

Figure 1.55 illustrates the stable hysteresis loop for a hard magnetic material obtained by applying a very large amplitude of variation of H. At point a in Fig. 1.55, most of the inelastic switching of domains has already occurred, and further increase of H results in an increase in B, which is due to (1) increase in $B_0 = \mu_0 H$, and (2) *elastic* rotation of domains similar to that encountered in soft materials. Unless the domains are very well oriented in the direction of H, this elastic rotation goes on as long as H is increased, but its contribution to B grows less and less. Consequently, the slope of the $B-H$ curve approaches the value μ_0 at very high values of H. It should be noted that the residual flux density B_r and the coercivity H_c in Fig. 1.55, the values of which are quoted in manufacturers' catalogs, are the values obtained *after the material has been driven past point* a *well into saturation.* From Eq. (1.28),

$$B = \mu_0(H + \mathcal{M}) = B_0 + B_{\mathcal{M}} \qquad \text{T} \qquad (1.106)$$

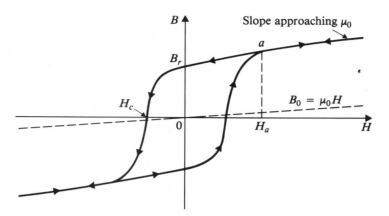

Fig. 1.55 B–H loop for a hard magnetic material.

Also it has been seen that the component B_M of the flux density B is due to the alignment, partial or complete, of the atomic or molecular magnetic moments of the material with the applied magnetic-field intensity H. An ideal permanent magnet material would be one in which, at $H = H_c$, all the domains switched simultaneously into complete alignment with H, and remained so aligned until the application of $-H_c$ caused complete reversal of magnetization. For such a material, $\mu_0 M$ would be equal to $B_{M\,max}$. A B_M–H loop of such an ideal material is illustrated in Fig. 1.56(a). The B–H loop of the ideal material would be that obtained by adding the ordinates of the straight line $B_0 = \mu_0 H$ to those of the ideal B_M–H loop. Two possible results, depending upon the magnitude of H_c, are shown in Fig. 1.56(b) and 1.56(c). The location of the point on an actual B–H loop corresponding to point b on the ideal loops of Fig. 1.56(b) and (c) is an important factor in determining the suitability of a material for practical applications.

In Fig. 1.57 the B–H loop of Fig. 1.55 is repeated, and superposed upon it is the corresponding loop of B_M–H, obtained by subtracting $B_0 = \mu_0 H$ from the ordinates of the B–H loop. As saturation increases, the slope of the B_M–H curve approaches zero. At $H = 0$, $B_M = B_r$, the remanence.

In the use of permanent magnets, it is the second-quadrant segment of the loop, the *demagnetization curve,* which is of main interest. If a toroidal core of hard magnetic material is enclosed in a coil, and a pulse of current is passed through the coil such that $\hat{H} > H_a$ in Fig. 1.55, then the first quadrant of Fig. 1.58 illustrates the B–H locus that might be traced out as the current rises from and falls again to zero.

Suppose a reverse magnetic field intensity of magnitude H_n is now applied to the toroidal core. The flux density will fall to magnitude B_1 shown in Fig. 1.58; and when H_n is removed, the flux density will return along a minor hysteresis loop to B_{r1}. Thus, application of a reverse field has reduced the remanence, or permanent magnetism, of the core. Reapplication of H_n will again reduce the flux density,

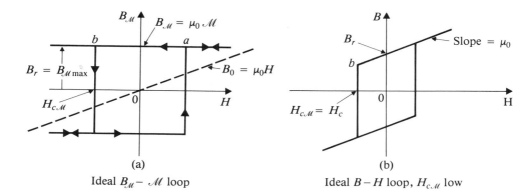

(a)
Ideal $B_\mathcal{M} - \mathcal{M}$ loop

(b)
Ideal $B - H$ loop, $H_{c.\mathcal{M}}$ low

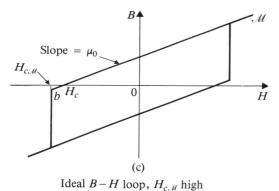

(c)
Ideal $B - H$ loop, $H_{c.\mathcal{M}}$ high

Fig. 1.56 $B_\mathcal{M}-H$ and $B-H$ loops for an ideal permanent-magnet material.

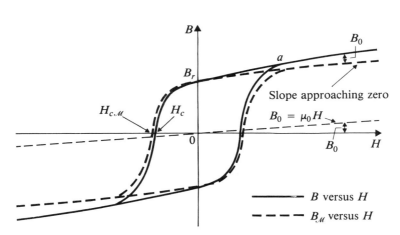

Fig. 1.57 $B-H$ and $B_\mathcal{M}-H$ loops for the material of Fig. 1.55.

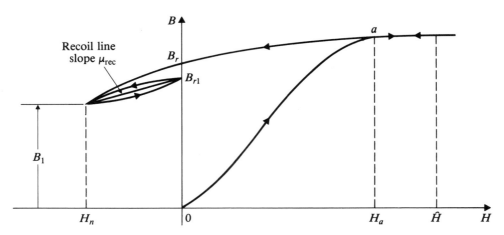

Fig. 1.58 *B–H* locus of permanent-magnet material during initial magnetization and subsequent application of a demagnetizing field.

completing the minor hysteresis loop by returning the core to approximately the same value of flux density B_1 as before. The minor hysteresis loop may usually be replaced with little error by a straight line called the *recoil line*. This line has a slope called the *recoil permeability*, μ_{rec}, which is approximately the same as that of the original $B-H$ curve at $H = 0$—that is, at $B = B_r$.

As long as the negative value of applied magnetic field intensity does not exceed H_n, the magnet may be regarded as reasonably permanent. If, however, a negative field intensity greater than H_n is applied, the flux density will be reduced to a value lower than B_1; and on removal of H, a new and lower recoil line will be established.

*1.21 PERMANENT-MAGNET SYSTEMS WITH AIR GAPS

Figure 1.59(a) shows a permanent magnet with an air gap. Suppose that the magnet with a closely fitting *keeper* of soft magnetic material in the air gap has been subjected to a magnetizing pulse as illustrated in Fig. 1.58. The state of the magnet and keeper at the end of the pulse is shown as point a in Fig. 1.59(b). If an increasing negative magnetic field intensity were then applied to the system, its varying state would be represented by the demagnetization curve in the second quadrant of Fig. 1.59(b). Consider that at the end of the magnetizing pulse, when the system is in state a, the keeper is removed from the air gap. The flux density in the magnet is reduced, since the magnetization $\mathcal{M}$ of the material must now produce flux in a magnetic circuit of increased reluctance. What will the new flux density be?

It will be assumed that, apart from the air gap, the magnetic flux is confined to the magnet and is uniformly distributed across its cross section. In the air gap, the existence of some fringing, illustrated in Fig. 1.59(c), will be allowed for by consid-

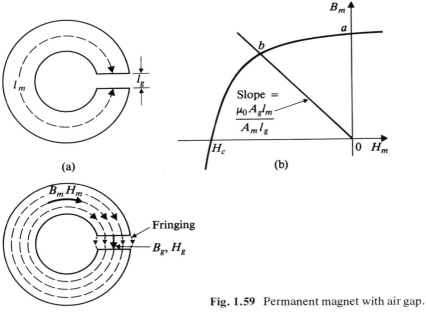

Fig. 1.59 Permanent magnet with air gap.

ering the air-gap cross-section area A_g to be somewhat greater than the magnet cross-section area A_m. Application of the circuital law around the path shown in Fig. 1.59(c) yields

$$H_m l_m + H_g l_g = 0 \tag{1.107}$$

from which

$$H_m = -H_g \frac{l_g}{l_m} \quad \text{A/m} \tag{1.108}$$

Thus the existence of an air gap is equivalent to the application of a negative field to the permanent-magnet material. Since the flux must be continuous around the path,

$$B_m A_m = B_g A_g \quad \text{Wb} \tag{1.109}$$

Thus, since $B_g = \mu_0 H_g$, from Eqs. (1.108) and (1.109),

$$B_m = -\mu_0 \frac{A_g}{A_m} \cdot \frac{l_m}{l_g} H_m \quad \text{T} \tag{1.110}$$

The straight line described by Eq. (1.110) is shown in Fig. 1.59(b), where its intersection with the demagnetization curve at point b represents the magnetic state of the core with the keeper removed.

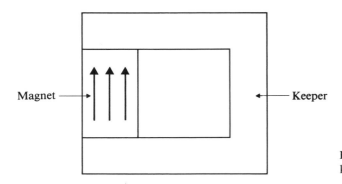

Fig. 1.60 Bar magnet with keeper.

This analysis shows that the operating state of a magnet with an air gap is determined by the demagnetization curve and the dimensions of the magnet. If a bar magnet and a keeper had been magnetized in the form shown in Fig. 1.60, the increased reluctance of the air path would simply have resulted in a state lower down on the demagnetization curve when the keeper was removed.

From Eqs. (1.107) and (1.109),

$$A_m l_m = \left(\frac{B_g A_g}{B_m}\right)\left(\frac{-H_g l_g}{H_m}\right) \tag{1.111}$$

so that the volume of the magnet may be expressed by

$$V_m = |A_m l_m| = \left|\frac{B_g^2 V_g}{\mu_0 B_m H_m}\right| \quad \text{m}^3 \tag{1.112}$$

where $V_g = A_g l_g$ is the volume of the air gap, and $B_g = \mu_0 H_g$. Thus, to produce a flux density B_g in an air gap of volume V_g, a minimum volume of magnet material is required if the material is operated in the state represented by the maximum value of the product $B_m H_m$. In the discussion of hysteresis loss in Section 1.6, it was shown that an area in a B–H diagram, that is, a product of B and H, has the dimensions of energy density. The quantity $B_m H_m$ is therefore known as the *energy product* of the material. It does not, however, represent the energy stored in the magnet. A typical curve of energy product $B_m H_m$ versus H_m is shown, with the demagnetization curve from which it was derived, in Fig. 1.61. It would appear from this diagram that the best magnet design would be obtained by operating the material at point b. This would be true for a magnet with a fixed air gap that would not be subjected to additional reversed fields. However, the system of Fig. 1.59 is a very simple one, rarely met in practice; and in order to obtain the optimum magnet design in a practical system, other factors must be taken into account.

Consider now a closer approximation to a practical system, such as that illustrated in Fig. 1.62. The behavior of this system will be affected by factors that were ignored in the discussion of the toroidal magnet of Fig. 1.59. When the keeper is

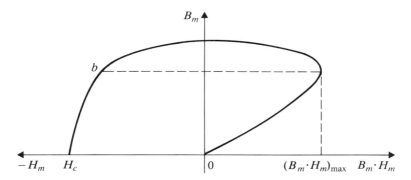

Fig. 1.61 Demagnetization curve and derived energy-product curve.

removed from the air gap, some appreciable leakage flux will bypass part of the main flux path, passing between the pole pieces without crossing the air gap, as indicated by the path for ϕ_l in Fig. 1.62. To allow for this, a leakage factor q may be introduced into Eq. (1.109), giving

$$B_m A_m = q B_g A_g \quad \text{Wb} \tag{1.113}$$

where

$$q = \frac{\text{Flux in magnet}}{\text{Flux in air gap}} \tag{1.114}$$

The value of a leakage factor q for a particular type of system may be determined either by field computation or by estimation on the basis of design experience with similar systems. In the case of some small electric machines, it may be as high as 4; that is to say, only a quarter of the magnet flux is being usefully employed. The relationship between B_m and H_m now becomes

$$B_m = \frac{-\mu_0 q A_g l_m}{A_m l_g} \cdot H_m \quad \text{T} \tag{1.115}$$

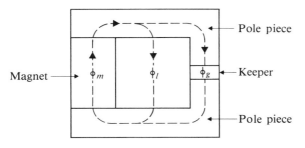

Fig. 1.62 Practical permanent-magnet system.

and the volume of the magnet is expressed by

$$V_m = |A_m l_m| = \left| \frac{qB_g^2 V_g}{\mu_0 B_m H_m} \right| \quad \text{m}^3 \qquad (1.116)$$

If the pole pieces of the system in Fig. 1.62 were operated in a highly saturated condition, then their reluctance would further affect the relationships of Eqs. (1.115) and (1.116) since Eq. (1.107) would no longer be valid. The effect of such reluctances is normally negligible, however.

When a system is going to be subjected to demagnetizing mmf's during the normal course of its operation—as in the case of an electric machine—the permanent magnet will not operate on the demagnetization curve, but on a recoil line. The recoil line is established in practice by *stabilizing* the magnet—that is, by subjecting the system after its initial magnetization to a stabilizing mmf H_s at least as great as any that will be applied in service. The operating point of a practical system will therefore lie at some such point as b in Fig. 1.63. It may be remarked at this point that the degree of stabilization required may be determined, not by the demagnetizing mmf applied in service, but by the necessity to dismantle the system after magnetization, with a result similar to that described for the bar magnet of Fig. 1.60. The intersection of the *load line* described by Eq. (1.115) with the recoil line will now give the operating values of B_m and H_m. Their product will be less than $(B_m H_m)_{\max}$, and their substitution in Eq. (1.116) will give the required magnet volume.

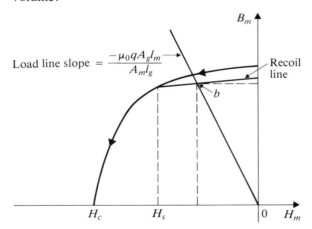

Load line slope $= \dfrac{-\mu_0 q A_g l_m}{A_m l_g}$

Fig. 1.63 Demagnetization curve with load line and recoil line.

*1.22 ALLOY PERMANENT-MAGNET MATERIALS

The permanent-magnet steel most commonly used in the early part of this century was simply an alloy of iron with 1% carbon. This had an energy product of about 1600 joules per cubic meter and a coercivity of some 400 amperes per meter. A much more expensive steel, with 35% cobalt and smaller quantities of chromium and tungsten, had an energy product of about 7500 J/m³ and a coercivity of some

20 kA/m. These steels could be shaped by normal methods and then quenched from a high temperature to give them their hard magnetic (and mechanical) property. Machining operations were possible after quenching.

In 1931 a new alloy of iron, nickel, and aluminum was discovered by I. Mishima. Its coercivity was about 40 kA/m, but the energy product was only some 10 kJ/m³. However this proved to be the first member of the very important Alnico (Al-Ni-Co) family of alloys. It was later discovered that adding 12% or more cobalt to the alloy and cooling the castings in a magnetic field yielded a much higher energy product when the material was subsequently magnetized in the direction in which the field was applied during cooling. At right angles to this direction, permanent-magnet properties were much poorer. Since, in general, permanent magnets are required to act in one direction only, the majority of magnets are now field treated. Figure 1.64 shows the demagnetization curve for Alnico 5 as well as loci for three values of energy product. The part of this demagnetization curve near the vertical

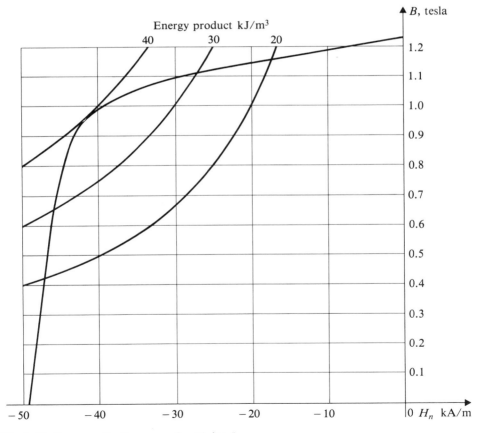

Fig. 1.64 Demagnetization curve for Alnico 5.

axis has a very low slope of about $3\mu_0$. The maximum energy product occurs in approximately the same region as that in which the demagnetization curve begins to depart markedly from an approximately straight line. Other Alnico alloys in which, in addition to field treatment, uniform crystal orientation is achieved, have still higher energy products than Alnico 5. The process of manufacture is more expensive, however, and most magnets are cast with random crystal orientation. Other Alnico alloys have lower residual flux density and higher coercive force than Alnico 5.

The mechanical properties of Alnicos are inconvenient. The magnetic hardness is accompanied by extreme mechanical hardness and brittleness. Moreover, the castings shrink substantially on cooling. To obtain close mechanical tolerances it is therefore necessary to grind the magnets to shape, or cut them with abrasive wheels. Alloy magnets are stabilized after magnetization as described in Section 1.21.

*1.23 FERRITE PERMANENT-MAGNET MATERIALS

Barium and strontium ferrites form good permanent magnets that can be manufactured in large sizes. While possessing lower remanence and energy product than the Alnicos and some other alloy materials, they have much greater coercivities and can withstand greater demagnetizing fields without appreciable reduction in permanent magnetism. For the purpose of energy conversion, the hard magnetic ferrites are of great practical use.

When ferrite permanent magnets are manufactured, the raw materials are milled together and sintered, after which the resulting powder is pressed into shape in a magnetic field. Sometimes the pressing is done with the materials in the form of a wet slurry. The applied magnetic field orients the materials for magnetization in the required direction(s); after this magnetic treatment, the molded shapes are again fired in an accurately controlled furnace.

Demagnetization curves for two permanent-magnet ferrite materials are shown in Fig. 1.65. Ferrite C is a barium ferrite designed for loudspeaker magnets. In such an application neither demagnetizing mmf nor dismantling after magnetization are to be expected. High coercivity is therefore not required, and the aim is to achieve as high a remanence and energy product as possible. Ferrite D is a strontium ferrite designed for use in electric motors where demagnetizing mmf will be applied during service. It is also probable that it will be necessary to remove the motor armature at some stage for servicing. High coercivity is here of the greatest importance. Two curves for constant energy product are also shown in Fig. 1.65.

The second-quadrant part of the demagnetization curve for Ferrite D is essentially a straight line with an incremental slope of about $1.05\mu_0$. In Section 1.20 it was shown that the flux density in a magnetic material arises from the magnetization $\mathcal{M}$ and the applied magnetic field intensity H according to the expression

$$B = \mu_0(H + \mathcal{M}) = B_0 + B_{\mathcal{M}} \qquad \text{T} \tag{1.117}$$

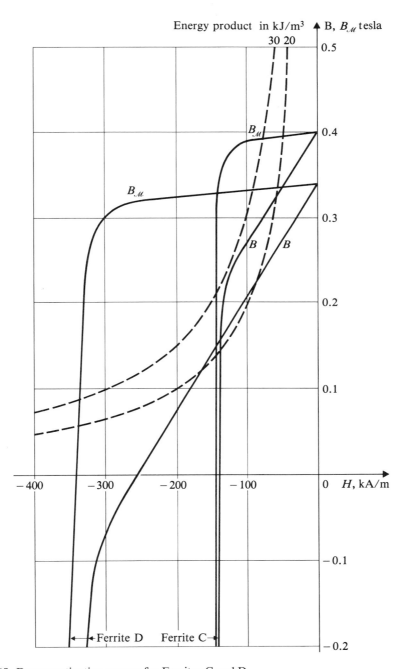

Fig. 1.65 Demagnetization curves for Ferrites C and D.

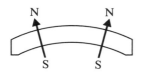
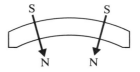

(a) Radially oriented segments for dc machines

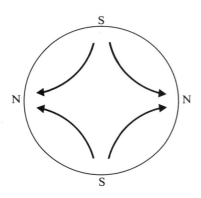
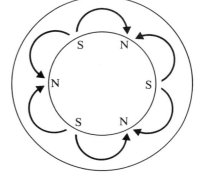

(b) Permanent-magnet rotor   (c) Permanent-magnet stator

Fig. 1.66 Magnetically oriented ferrite shapes.

A material in which the magnetization $\mathcal{M}$ remained absolutely constant would therefore have a $B_M - H$ relationship, which would be a horizontal straight line, and a $B - H$ relationship with an incremental slope of μ_0 (Fig. 1.56). Material D therefore comes very close to ideal constancy of magnetization and retains this for values of negative H in excess of H_c. Stabilization of such a ferrite is not necessary, since the recoil line virtually coincides with the demagnetization curve. Dismantling a magnet system embodying this ferrite would cause no loss of magnetization, since the demagnetization curve is a straight line down to $B = 0$.

Because of the very high value of coercivity of ferrite materials, the length required in the magnet is usually relatively small, typically of the same order of magnitude as the air-gap length. Some available shapes and directions of orientation designed for electric machines are shown in Fig. 1.66. Magnets are usually fixed in place in the assembled machine by clips or adhesives. They are not mechanically strong and cannot be used as structural members of a magnetic system.

*1.24 RARE-EARTH PERMANENT-MAGNET MATERIALS

Since 1960 a new class of permanent-magnet materials has been developed, combining the relatively high remanence of the Alnico-type materials with coercivities greater than those achieved in the ferrites. These materials are compounds of iron, nickel, and cobalt with one or more of the rare-earth elements. One of the most useful is samarium–cobalt, for which the demagnetization curve is shown in Fig. 1.67. The energy product for this material has a maximum value of about 160 kJ/m³, some four times that achievable with an Alnico.

The demagnetization curve begins to deviate from a straight line only with reversed flux density and a negative magnetic-field intensity exceeding 10^6 A/m in magnitude. The straight portion of the curve has a slope of some $1.06\mu_0$, which is also the recoil permeability, provided that reverse magnetization does not exceed 10^6 A/m. Within this limit, the recoil line is virtually identical with the demagnetization curve. Two similar magnets of this material may be brought face-to-face in opposition to one another without causing permanent demagnetization.

The cost of rare-earth magnets is relatively high, but is decreasing. Application to date has been restricted to small magnets where small size and weight are important requirements.

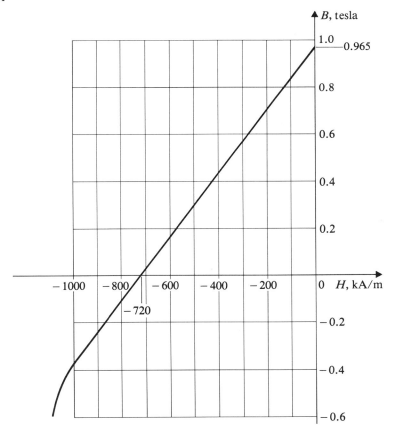

Fig. 1.67 Demagnetization curve for samarium–cobalt.

*1.25 DESIGN OF PERMANENT-MAGNET SYSTEMS

The choice of whether to use alloy or ferrite magnets in a given application depends upon such factors as the permissible volume and weight of the system, the acceptable cost, and the magnitude of demagnetizing mmf that will be encountered in service. The energy product obtainable with ferrites is substantially less than that obtainable with Alnico; as a consequence, magnet volume is greater when ferrites are employed. On the other hand, the relative density (specific gravity) of ferrites is only 0.6 of that of metallic magnets. This means that a system designed with ferrite magnets may have a smaller weight than an equivalent system designed with metallic magnets. The lower cost of ferrite magnets per unit of energy product also tends to give a more economical, even if more bulky, design when ferrites are used.

Because of its lower flux density, the cross-section area of a ferrite magnet will be greater than that of an equivalent metallic magnet. On the other hand, the high coercivity of ferrites reduces the necessary length of the magnet. Thus, ferrite magnets are both shorter and broader than metallic equivalents. This may or may not be an advantage, depending upon the application. Servicing requirements and the possible need to dismantle a magnet system must also be borne in mind when the choice of material is being made.

It has been seen that the operating locus of a permanent magnet is formed by a recoil line, as illustrated in Fig. 1.58. In the case of an alloy magnet, this recoil line is established by the process of stabilization. Afterward, the negative mmf applied to the magnet must not exceed in magnitude that employed in stabilization; otherwise a new recoil line is produced. In the case of a ferrite or rare-earth magnet, where stabilization is unnecessary, the recoil line coincides with the straight-line portion of the demagnetization curve.

Once the magnet material has been decided upon, design of the magnet system can proceed along the lines indicated in Eq. (1.112) or (1.116) and Fig. 1.62 of Section 1.21. On the other hand, the linearity of the $B–H$ relationship formed by the recoil line suggests that the magnet might be represented by a linear model. Such a model would simplify design calculations, particularly in complicated systems.

Figure 1.68(a) shows a permanent magnet of cross-section area A_m and length l_m. The magnetic path is completed by a section of soft magnetic material, and it will be assumed that the value of H required to produce magnetic flux in the soft material is negligible. An N-turn coil is wound around the magnetic path. The magnet may be assumed to have been magnetized and stabilized, yielding the required recoil line. Figure 1.68(b) shows the relationship between the magnet flux ϕ_m and the magnetomotive force $\mathscr{F}_n$. This relationship is obtained by rescaling the recoil line of the material, using

$$\phi_m = B_m A_m \quad \text{Wb} \tag{1.118}$$

$$\mathscr{F}_n = H_n l_m \quad \text{A} \tag{1.119}$$

where H_n is the demagnetizing magnetic field intensity.

If the rescaled recoil line extended to the left in Fig. 1.68(b) cuts the horizontal axis at $\mathscr{F}_n = -\mathscr{F}_0$, then the slope of the line is

$$\frac{\phi_{r1}}{\mathscr{F}_0} = \frac{1}{\mathscr{R}_0} \quad \text{Wb/A} \tag{1.120}$$

so that

$$\phi_m = \phi_{r1} + \frac{\mathscr{F}_n}{\mathscr{R}_0}, \qquad \phi_m > \phi_a \quad \text{Wb} \tag{1.121}$$

From Eqs. (1.120) and (1.121),

$$\mathscr{R}_0 \phi_m = \mathscr{F}_0 + \mathscr{F}_n, \qquad \mathscr{F}_n > -\mathscr{F}_a \quad \text{A} \tag{1.122}$$

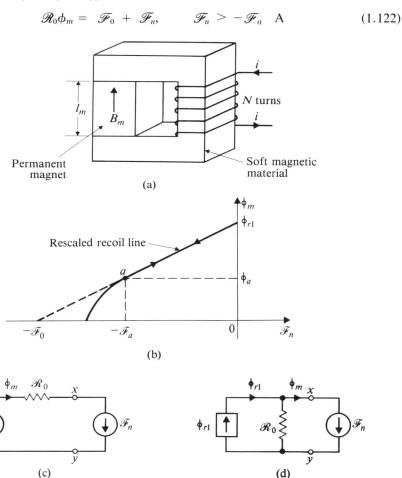

Fig. 1.68 Linear model for a permanent-magnet system. (a) Magnet system. (b) Graphical analysis. (c) Equivalent magnetic circuit with mmf source. (d) Equivalent magnetic circuit with flux source.

Equation (1.22) describes the equivalent magnetic circuit of Fig. 1.68(c). The magnet is represented as a source of mmf, $\mathscr{F}_0$ in series with reluctance $\mathscr{R}_0$. The remaining part of the equivalent magnetic circuit is simply the coil mmf

$$\mathscr{F}_n = Ni \quad \text{A} \tag{1.123}$$

The mmf source in series with a reluctance of Fig. 1.68(c) may be transformed into a flux source in parallel with a reluctance, just as in dc electric circuits a source of electromotive force with series resistance may be transformed into a current source with the same resistance in parallel. The magnitude of the equivalent flux source is

$$\frac{\mathscr{F}_0}{\mathscr{R}_0} = \phi_{r1} \quad \text{Wb} \tag{1.124}$$

and the equivalent magnetic circuit is described by Eq. (1.121) and also shown in Fig. 1.68(d).

These equivalent magnetic circuits are convenient for use in describing the behavior of magnetic systems, such as that shown in Fig. 1.69(a), which is similar to Fig. 1.68(a) with an air gap introduced. If the current in the coil of Fig. 1.69(a) is zero, the flux-source form of the equivalent magnetic circuit of the system is that in

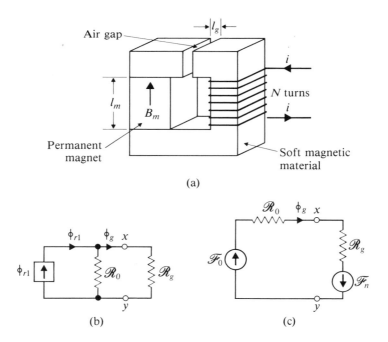

(a)

(b)   (c)

Fig. 1.69 Linear model for a permanent-magnet system with air gap. (a) Magnet system. (b) Equivalent magnetic circuit, $i = 0$. (c) Equivalent magnetic circuit, $i \neq 0$.

Fig. 1.69(b). If the recoil permeability (the slope of the recoil line) for the magnetic material is defined as $\mu_r\mu_0$, then the magnet reluctance is

$$\mathcal{R}_0 = \frac{l_m}{\mu_r\mu_0 A_m} \quad \text{A/Wb} \tag{1.125}$$

The reluctance of the air gap is

$$\mathcal{R}_g = \frac{l_g}{\mu_0 A_g} \quad \text{A/Wb} \tag{1.126}$$

and from the equivalent magnetic circuit

$$\phi_g = \frac{\mathcal{R}_0}{\mathcal{R}_0 + \mathcal{R}_g} \phi_{r1} \quad \text{Wb} \tag{1.127}$$

If, in addition, a current i flows in the N-turn coil, then

$$\mathcal{F}_n = Ni \quad \text{A} \tag{1.128}$$

Solution for air-gap flux in this case may be expedited by replacing the flux-source magnet model with the mmf source model, resulting in the equivalent magnetic circuit of Fig. 1.69(c).

Example 1.12   In the permanent magnet system of Fig. 1.69(a), the magnet is made of Alnico 5, whose demagnetization curve is shown in Fig. 1.64. The dimensions of the magnet are $l_m = 5$ cm, $A_m = 9$ cm². The air gap dimensions are $l_g = 1$ mm, $A_g = 8$ cm². The coil has 500 turns. The magnet is stabilized by a pulse of current giving $H_n = -40 \times 10^3$ A/m in the magnet. The permeability of the soft magnetic material is so high that the mmf required to produce flux in it may be neglected. Also, fringing around the air gap and leakage flux may be ignored.

a) Determine the flux densities in the air gap and the magnet.

b) Determine the maximum negative coil current that may be permitted without loss of permanent magnetization.

Solution

a) Assuming that the slope of the recoil line is equal to that of the demagnetization curve at $H_n = 0$, the recoil permeability is

$$\mu_r\mu_0 = \frac{1.230 - 1.125}{24 \times 10^3} = 4.375 \times 10^{-6}$$

The recoil line terminates at point $(-40 \times 10^3, 0.99)$ so that the minimum flux density permissible without danger of further demagnetization is 0.99 T. From the coordinates of the end point and the value of $\mu_r\mu_0$, the equation of the recoil line is

$$B = 1.165 + 4.375 \times 10^{-6} H_n \quad \text{T}, \qquad -40 \times 10^3 < H_n < 0 \quad \text{A}$$

From Eq. (1.125),

$$\mathscr{R}_0 = \frac{5 \times 10^{-2}}{4.375 \times 10^{-6} \times 9 \times 10^{-4}} = 12.70 \times 10^6 \quad \text{A/Wb}$$

From Eq. (1.126),

$$\mathscr{R}_g = \frac{1.0 \times 10^{-3}}{4\pi \times 10^{-7} \times 8 \times 10^{-4}} = 0.995 \times 10^6 \quad \text{A/Wb}$$

From the equation to the recoil line, the remanence for this degree of stabilization is $B_{r1} = 1.165$ T, so that from Eq. (1.118),

$$\phi_{r1} = 9 \times 10^{-4} \times 1.165 = 1.049 \times 10^{-3} \quad \text{Wb}$$

From the equivalent magnetic circuit of Fig. 1.69(b), or from Eq. (1.27),

$$\phi_g = \frac{12.70 \times 10^6}{(12.70 + 0.995)10^6} \times 1.049 \times 10^{-3} = 0.9728 \times 10^{-3} \quad \text{Wb}$$

$$B_g = \frac{\phi_g}{A_g} = \frac{0.9728 \times 10^{-3}}{8 \times 10^{-4}} = 1.216 \quad \text{T}$$

$$B_m = \frac{\phi_g}{A_m} = \frac{0.9728 \times 10^{-3}}{9 \times 10^{-4}} = 1.081 \quad \text{T}$$

b) To prevent loss of permanent magnetization, the magnet flux must not be reduced below the value

$$\phi_g = \phi_a = 0.99 \times 9 \times 10^{-4} = 0.891 \times 10^{-3} \quad \text{Wb}$$

The source mmf in Fig. 1.69(c) is

$$\mathscr{F}_0 = \phi_{r1}\mathscr{R}_0 = 1.049 \times 10^{-3} \times 12.7 \times 10^6 = 13.32 \times 10^3 \text{ A}$$

Then, from Fig. 1.69(c),

$$\mathscr{F}_n = \mathscr{F}_0 - (\mathscr{R}_0 + \mathscr{R}_g)\phi_a = 13.32 \times 10^3 - (12.7 + .995)10^6 \times 0.891 \times 10^{-3}$$
$$= 1119 \text{ A}$$

The coil current is

$$i = \mathscr{F}_n/N = \frac{1119}{500} = 2.24 \text{ A}$$

Example 1.13   Figure 1.70 shows a cross section of the magnetic system of a two-pole direct-current motor. The rotor diameter is 50 mm, the air gaps are 2.5 mm in length. The magnet sectors are to be radially magnetized and made of Ferrite D, whose demagnetization curve is given in Fig. 1.65. The leakage factor q, introduced in Eq. (1.115), may be taken as 1.10. It may be assumed that the recoil line for the magnet material is identical with the demagnetization curve. The stator yoke and

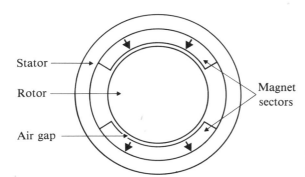

Fig. 1.70 Direct-current motor magnet system for Example 1.13.

the rotor core may be assumed to have infinite permeability. Fringing at the air gaps may be neglected.

Determine the thickness of the magnet sectors required to produce a flux density of 0.25 T at the surface of the rotor.

Solution   From Fig. 1.65, the remanence of the magnetic material is 0.340 T, and the recoil permeability is

$$\mu_r \mu_0 = \frac{0.34}{215 \times 10^3} = 1.58 \times 10^{-6}$$

The equation to the recoil line is therefore

$$B = 0.340 + 1.58 \times 10^{-6} H_n \quad \text{T}, \qquad 0 > H_n > -80 \times 10^3 \quad \text{A/m}$$

Since the magnet sectors will be relatively thin, it may be assumed that the cross-section area of one magnet is the same as that of one air gap. Thus $A_m = A_g$ m³. Substitution in Eq. (1.115) yields

$$B_m = \frac{-4\pi \times 10^{-7} \times 1.10 l_m}{2 \times 2.5 \times 10^{-3}} \times H_m \quad \text{T}$$

where l_m is the total thickness of two magnet sectors. The equation to the load line is therefore

$$B_m = -0.276 \times 10^{-3} l_m H_m \quad \text{T}$$

This line must intersect the recoil line at $B_m = 0.25$ T—that is, for H_m given by

$$0.25 = 0.340 + 1.58 \times 10^{-6} \quad H_m$$

From which

$$H_m = -56.9 \times 10^3 \quad \text{A/m}.$$

Substituting for B_m and H_m in the equation to the load line gives

$$0.25 = (-0.276 \times 10^{-3})(-56.9 \times 10^3)l_m$$

from which

$$l_m = 15.9 \times 10^{-3} \quad m$$

Each magnet sector should therefore be 7.95 mm in thickness.

PROBLEMS

1.1 On the average, the magnetic moment of an atom of metallic nickel is about 0.6 of the basic quantum of spin magnetic moment (9.27×10^{-24} A · m² = 1 Bohr magneton). Nickel has an atomic weight of 58.7 and its density is 8850 kg/m³.

    a) If all magnetic moments are aligned, what will be the flux density in the nickel?

    b) A useful magnetic material consists of 50% nickel and 50% iron, the latter having 2.2 Bohr magnetons per atom. Assume that all magnetic moments are aligned, and determine the flux density in the material and compare the result with the residual flux density in Fig. 1.14. *(Section 1.3)*

1.2 An ideal ferromagnetic material has a maximum magnetic moment per unit volume of 1.2×10^6 A/m. All domains are ideally oriented in the direction of magnetization and all domain walls remain fixed in position until a magnetic field intensity of 80 A/m is applied.

    a) Sketch hysteresis loops for this material

        i) for the case when it is driven well into saturation, and

        ii) for the case when the residual flux density is about half of its maximum value.

    b) Determine the magnetic field intensity required to produce a flux density 1.05 times the maximum residual value. *(Section 1.3)*

1.3 Refer to the plastic ring shown in Fig. 1.14.

    a) Determine the number of turns required to provide an inductance of 15 mH.

    b) Suppose round copper wire is used with an insulation thickness of 0.2 mm and that the coil is arranged in five layers. Determine the maximum wire diameter that can be used and the resistance of the coil.

    c) Determine the time constant of the coil.

    d) Draw an equivalent electric circuit for the coil.

    e) Draw an equivalent magnetic circuit for the coil showing the value of the reluctance and the flux for a current of 20 A. *(Section 1.3)*

1.4 A wooden torus has a mean diameter of 250 mm and a cross-section area of 1000 mm². It is uniformly wound with 1200 turns of wire in which a current of 2 A flows. Find

    a) The magnetic field intensity inside the coil.

    b) The total flux produced in the torus.

    c) The flux density in the torus.

    d) The inductance of the coil.

A similar torus of cast steel, whose magnetization curve is shown in Fig. 1.7, is wound with an identical coil in which a current of 2 A flows. Determine

e) The total flux produced in the steel torus.
f) The relative permeability of the cast steel under these conditions.
g) The inductance of the coil. *(Section 1.3)*

1.5 The toroidal core of Fig. 1.4 is made from a continuous strip of 29-gauge M-36 steel for which the magnetization curve is shown in Fig. 1.7. The coil has 300 turns.

a) Determine the coil current required to produce a flux density of 1.4 T in the core.
b) If the magnetization curve is approximated by a straight line through the origin and the point for which $B = 1.4$ T, what is the relative permeability of the core material and the effective inductance of the coil? *(Section 1.3)*

1.6 The plastic toroidal ring of Fig. 1.4 is wound with a coil of 1615 turns carrying a current of 20 A.

a) Determine the magnetic field energy per unit volume at the mean diameter of the ring.
b) Determine the total field energy stored in the ring.
c) Compare the value in (b) with that determined from the inductance of Problem 1.3 and the coil current. *(Section 1.5)*

1.7 For the nickel–iron magnetic material of which the limiting $B–H$ loop is given in Fig. 1.14:

a) Determine the hysteresis loss per cycle if the core shown in Fig. 1.4 were to be formed of this material.
b) Assuming a frequency of operation of 20 Hz, determine the hysteresis loss.
c) If the potential difference applied to the coil is a 20 Hz sinusoidal wave of 110 V rms and the coil resistance is negligible, what are the number of turns required to drive the core round the $B–H$ loop of Fig. 1.14? *(Section 1.6)*

1.8 A 350-turn coil is wound on a toroidal core having a cross-section area of 10^{-4} m² and a mean flux-path length of 0.15 m. The core material may be considered to have a relative permeability of 5000 up to a flux density of 1.4 T.

a) What rms value of sinusoidal emf can be induced in the coil at a frequency of 400 Hz if the core flux density is to be limited to a maximum value of 1.4 T?
b) Determine the current in part (a).
c) If the coil has a resistance of 1.5 ohms, what is the applied potential difference across the coil terminals for the condition of part (a)? *(Section 1.7)*

1.9 A magnetic core has a cross-section area of 500 mm² and can be assumed to have infinite permeability up to a flux density of 1.5 T. Determine the maximum amplitude of a rectangular waveform of potential difference having a period of 0.15 s that can be applied to the terminals of a 400-turn coil on this core without exceeding the flux density of 1.5 T.

(Section 1.8)

1.10 A magnetic core has a cross-section area of 0.015 m² and a magnetic path length of 0.9 m. The core material is M-36 sheet steel 0.356 mm in thickness with a density of 7800 kg/m³. A coil of 20 turns and negligible resistance is wound on the core.

a) Determine the core loss when a sinusoidal potential difference of 110 V (rms) at a frequency of 60 Hz is applied to the coil.

b) Repeat part (a) for the same potential difference at a frequency of 100 Hz.
(Section 1.9)

1.11 A uniformly wound ferromagnetic torus has a cross-section area of 100 mm² and a mean diameter of 50 mm. A family of cyclic hysteresis loops for the core material is shown in Fig. 1.71. The coil on the core has 200 turns and negligible resistance.

a) Suppose a rectangular wave of potential difference having a peak-to-peak value of 12 V is applied to the coil. What value of the period of the wave will produce a peak flux density of 1.6 T?

b) Sketch the current waveform for the condition of (a), and determine the peak value of the current in the coil.

c) What rms value of a 50 Hz sine wave of potential difference would produce a peak flux density of 1.2 T?

d) Sketch the current waveform for the condition of part (c) and determine the value of the current at the instant of peak potential difference. *(Section 1.10)*

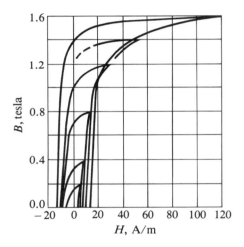

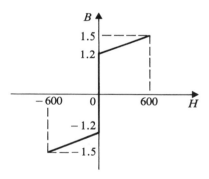

Fig. 1.71 Diagram for Problem 1.11.   **Fig. 1.72** Diagram for Problem 1.12.

1.12 A core with a cross-section area of 500 mm² and a magnetic path length of 200 mm has a B–H loop, which may be represented by the piecewise-linear approximation of Fig. 1.72.

a) If the core is wound with a 500-turn coil of negligible resistance, what is the rms value of the 300-Hz sinusoidal potential difference at the coil terminals that will produce a peak flux density of 1.5 T?

b) Sketch the waveform of the magnetizing current that will flow under the steady-state conditions of part (a).

c) Evaluate the first three terms in the Fourier series describing the current of (b).
(Section 1.10)

1.13 A core of a material whose B–H loop may be represented by the idealization in Fig. 1.73 is wound with a 200-turn coil. A 180-Hz sinusoidal potential difference that produces a peak flux density of 1.5 T is applied to the coil. The core has a magnetic path length of 150 mm. Determine the fundamental, third, and fifth harmonic components of the coil current. *(Section 1.10)*

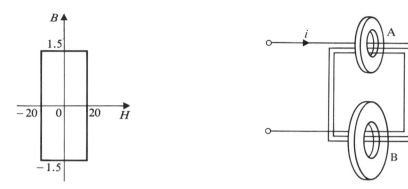

Fig. 1.73 Diagram for Problem 1.13.   **Fig. 1.74** Diagram for Problem 1.14.

1.14 A coil of 800 turns is wound through two toroidal cores as shown in Fig. 1.74. Core A has a mean diameter of 100 mm and a cross-section area of 150 mm². Core B has a mean diameter of 120 mm and a cross-section area of 250 mm². The relative permeability of the material of both cores is 2800.

   a) Determine the reluctance of each of the cores.
   b) Draw an equivalent magnetic circuit of the system.
   c) Determine the coil current required to produce a coil flux linkage of 0.2 Wb.
   (Section 1.11)

1.15 A flux of 0.4 mWb is to be produced in the magnetic system shown in Fig. 1.75 using a coil of 240 turns. The cast steel is to be operated at a flux density of 0.9 T and the sheet steel at 1.4 T. The cast steel is to be square in cross section. Any air gap at the junction of the materials may be ignored.

   a) Determine the required dimensions of the two sections of material.

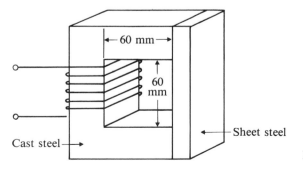

Figure 1.75

b) Determine the required coil current.

c) Determine the reluctance of each section of the magnetic system.

d) Draw an equivalent magnetic circuit of the system.   *(Section 1.12)*

1.16 The material of the magnetic system in Fig. 1.76 has a relative permeability of 1500, $N_1 = 1000$, $N_2 = 2000$. (Leakage flux is negligible.)

a) Draw a magnetic equivalent circuit for this system, showing on it the values of the reluctances.

b) Assuming that the current in coil N_1 is 50 mA, and that coil N_2 is open-circuited, determine the flux linkage of each of the coils.   *(Section 1.12)*

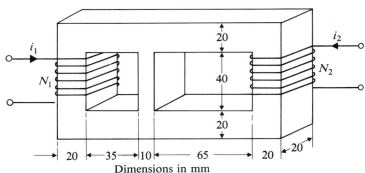

Dimensions in mm

Fig. 1.76 Diagram for Problem 1.16.

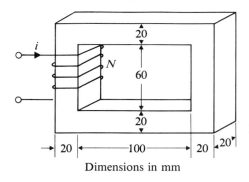

Dimensions in mm

Fig. 1.77 Diagram for Problem 1.17.

1.17 The material of the magnetic system shown in Fig. 1.77 has a relative permeability of 2500.

a) Find the reluctance of the magnetic circuit.

b) If $N = 1000$, what is the inductance of the coil?

c) Find the flux density in the core for a coil current of 0.1 A.   *(Section 1.12)*

*1.18 An air gap of 1.25 mm is cut in the cast-steel torus of Problem 1.4. Fringing at the air gap may be neglected.

a) Determine the current required to produce a flux density of 1.2 T in the torus and air gap.

b) By a method of trial and error, find the flux in the air gap when the winding current is 1.2 A. *(Section 1.13)*

*1.19 A toroidal magnetic core of rectangular cross section has an inside diameter of 120 mm, an outside diameter of 160 mm, and a thickness of 30 mm. The core material may be considered to have a constant relative permeability of 15,000 up to a flux density of 1.5 T. An air gap 0.2 mm long is cut across the core, and a coil of 250 turns is wound on it.

  a) Draw a magnetic equivalent circuit for this system and determine its reluctances.
  b) Determine the coil current required to produce a core flux of 0.5 mWb.
  c) For the condition of (b), determine the energy stored in the core and in the air gap.
  d) Determine the self-inductance of the coil.
  e) Determine the maximum value of current for which the linear model of (a) is valid. *(Section 1.13)*

*1.20 The coil in the magnetic system of Fig. 1.78 has 400 turns and a resistance of 10 ohms. The core material may be considered to have infinite permeability, and fringing around the air gaps may be neglected.

  a) Draw a magnetic equivalent circuit of the system, showing the appropriate reluctance values.
  b) Determine the flux in each of the air gaps as a function of the coil current.
  c) Neglecting core loss, draw an electric equivalent circuit of the form shown in Fig. 1.45 and determine its parameters. *(Section 1.16)*

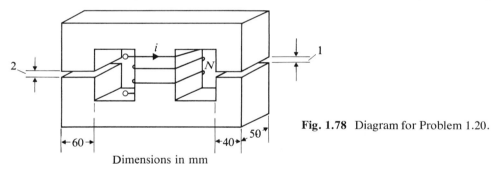

Fig. 1.78 Diagram for Problem 1.20.

Dimensions in mm

*1.21 A toroidal ferrite core has a cross-section area of 100 mm² and a mean diameter of 60 mm. The ferrite material has a constant relative permeability of 400 up to a flux density of 0.25 T.

  a) Determine the number of turns required in a coil on this torus to provide an inductance of 10 mH.
  b) What is the peak current in the coil if the flux density is not to exceed 0.25 T? *(Section 1.17)*

*1.22 In the pot core of Example 1.10, what is the length of the air gap required to provide an inductance of 0.4 H? Assume all other dimensions are the same. *(Section 1.17)*

*1.23 An iron-cored inductor is often used to smooth out the ripple in the current from a rectifier. To be effective, the inductor must have appreciable inductance when carrying the direct current of the rectifier. Saturation of the core is prevented by the inclusion of an air gap.

Design an inductor to have an inductance of 10 H when it carries a direct current of 0.5 A. The magnetic material may be assumed to require no appreciable field intensity up to a flux density of 1.4 T. A current density of 2 A/mm² may be used in the coil and a space factor of 0.5 assumed. The stacking factor for the core is 0.95. *(Section 1.19)*

1.24 An inductor is required to absorb a reactive power of 100 kVA when connected to an 11.2-kV, 60-Hz supply.

Design an appropriate inductor of the shape shown in Fig. 1.54 using the following constraints: The core material is to be the grain-oriented steel strip of Figs. 1.52 and 1.53 with a maximum flux density of 1.6 T, a mass density of 7640 kg/m³ and a stacking factor of 0.95. Winding space factor = 0.45; current density = 4A (rms)/mm²; conductor resistivity = 2.1 × 10⁻⁸ Ω·m.

Estimate the total loss in the inductor and determine its power factor. *(Section 1.19)*

1.25 A permanent magnet made of Alnico 5 has the shape shown in Fig. 1.59(a) with a mean diameter of 150 mm and a cross-section area of 500 mm². The demagnetization curve for the material is shown in Fig. 1.64.

Ignoring fringing, determine the length of air gap across which this magnet can produce a flux density of 1.05 T. *(Section 1.22)*

1.26 A permanent magnet is required to produce a flux density of 0.6 T in an air gap of length 4 mm and cross-section area 2000 mm². A magnet of the shape shown in Fig. 1.62 is to be used.

   a) Ignoring fringing and leakage flux, determine the minimum volume of Alnico 5 material that will be required and specify the dimensions of the magnet.
   b) Repeat part (a) assuming a leakage factor of 2.5. *(Section 1.22)*

1.27 Suppose that the air gap in the magnet system of Problem 1.26 is closed by a keeper of ideal ferromagnetic material requiring no magnetic field intensity. If the recoil permeability (relative) of the magnetic material is 3.5, what is the flux density in the magnet?

(Section 1.22)

1.28 Repeat Problem 1.26(a) using the Ferrite C material of Fig. 1.65. Compare the shape and volume of this magnet with that made of Alnico 5. *(Section 1.22)*

1.29 Figure 1.79 shows a small magnet assembly intended for use as a door holder. The keeper is attached to the door while the remainder is attached to the door frame. The magnet material is Ferrite D of Fig. 1.65. It may be assumed that the recoil characteristic is identical

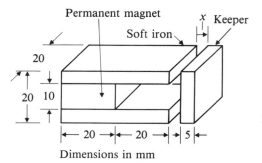

Figure 1.79

Dimensions in mm

with the straight portion of the demagnetization characteristic. The soft-iron pole pieces may be assumed to have infinite permeability.

a) Neglecting fringing and leakage, develop a linearized magnetic circuit for the magnet system.

b) If a flux density of 0.6 T is needed in the air gap to provide the necessary force, what is the maximum permissible length of the air gap?

c) Estimate the reluctance of the leakage path in the assembly by assuming a uniform magnetic field in the air space enclosed by the soft-iron pole pieces between the magnet and the keeper.

d) Draw a magnetic equivalent circuit including the leakage of part (c).

e) Repeat (b) using the circuit of (d).   *(Section 1.23)*

1.30 A rectangular block of permanent-magnet material is to be magnetized using an arrangement similar to that shown in Fig. 1.68(a). The block has a length $l_m = 50$ mm and a cross-section area of 1000 mm². The soft magnetic material may be assumed to have infinite permeability. The coil has 100 turns. Figure 1.80 shows the $B-H$ characteristic published by the manufacturer of the material.

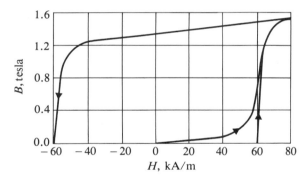

Figure 1.80

a) Suppose that the material has an initial flux density of zero. Determine the peak value of coil current i required to magnetize the magnet so that it will have a residual flux density of 1.35 T.

b) Determine approximately the energy required to magnetize the magnet under the conditions of (a). (Coil resistance may be neglected.)

c) One simple method of magnetizing a magnet consists of connecting a charged capacitor to the coil terminals. If the capacitance is 100 μF, to what terminal potential difference should the capacitor be charged to provide the necessary magnetization current? Coil resistance may again be neglected.

d) The method of magnetization discussed in (c) produces an oscillatory current that might tend to demagnetize the magnet. Show that this demagnetization can be prevented by placing a rectifier in series with the capacitor.   *(Section 1.25)*

1.31 Figure 1.81 shows a cross-section view of the permanent-magnet assembly of a loudspeaker. A magnetic flux density of 1.2 T is required in the cylindrical air gap. Permanent-magnet material having the $B-H$ characteristic of Fig. 1.80 may be used.

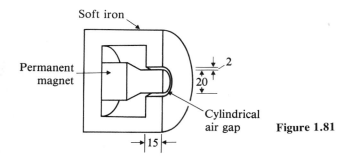

Figure 1.81

a) Neglecting leakage and fringing fluxes, determine appropriate dimensions for the various parts of the magnet assembly. The flux density in the soft iron should not exceed 1.2 T.

b) What is the mass of the permanent magnet if its material has mass density 7330 kg/m³?

(Section 1.25)

2 / Transformers

Before discussing electromagnetic energy conversion by means of electromechanical machines involving relative rotational or translational motion of their parts, it is first desirable to examine the behavior of a static electromagnetic machine, the transformer. This is so for two reasons: first, because the transformer is an important machine in its own right; second, because transformer action also takes place in electromechanical machines, and an understanding of the transformer is therefore a prerequisite to an understanding of these machines. In this chapter, therefore, an analysis of the behavior of a transformer viewed as a magnetic system will be carried out.

2.1 THE IDEAL TRANSFORMER

A transformer consists of a magnetic system in which a time-varying flux links two or more coils. Its purpose is to transfer electrical energy from one circuit to another, usually without any electrical connection between the two circuits. Usually the potential difference at which the energy leaves one circuit is different from the potential difference at which it enters the other. Thus, all that is strictly necessary for transformer action to take place is that two coils are so positioned that some of the flux produced by a current in one coil links some of the turns of the other coil. The arrangement in Fig. 2.1 therefore constitutes a transformer, and some transformers employed in communications equipment are no more elaborate than this.

In transformers employed in power circuits, it is necessary to ensure that energy shall not be wasted. To this end, the coils are arranged on a ferromagnetic core in

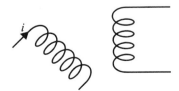

Fig. 2.1 Basic transformer action.

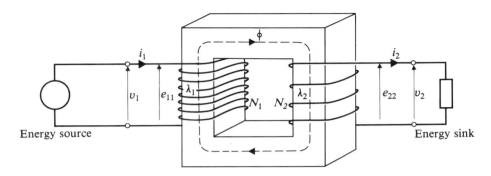

Fig. 2.2 Primitive two-winding power transformer.

such a way that a large flux is produced in the core by a current in any one coil, and as much of that flux as possible links as many of the turns as possible of the other coils on the core. Figure 2.2 represents a transformer consisting of two coils on a ferromagnetic core. In practice, a designer would not normally build a transformer as primitive as this, for reasons that will become clearer in the discussion to follow.

If the behavior of the transformer is to be analyzed, then a mathematical model of it must be constructed. The first model that will be derived for the transformer is the *ideal transformer,* for which it is assumed that:

1. The electric fields produced by the windings are negligible.
2. The winding resistances are negligible.
3. All magnetic flux is confined to the ferromagnetic core.
4. The relative permeability of the core material is so high that negligible mmf is required to establish the flux in the core.
5. The core losses are negligible.

If a potential difference v_1 is applied to the terminals of winding N_1 in Fig. 2.2, then current i_1 will flow in that winding. The resulting mmf $N_1 i_1$ will produce flux ϕ in the core and a flux linkage λ_1 of winding N_1. If v_1 varies in time, then i_1, ϕ, and λ_1 will vary in time, and an emf e_{11} will be induced in winding N_1, where

$$e_{11} = \frac{d\lambda_1}{dt} \quad \text{V} \tag{2.1}$$

If the current and therefore the flux are increasing, the emf induced in the coil will oppose the change in current (Lenz's law) and will act in the direction shown in Fig. 2.2. Since it is assumed that all flux is confined to the core, and hence is linking all turns of winding N_1, then

$$\lambda_1 = N_1 \phi \quad \text{Wb} \tag{2.2}$$

and

$$e_{11} = N_1 \frac{d\phi}{dt} \quad \text{V} \tag{2.3}$$

Moreover, since the resistance of winding N_1 is assumed to be negligible, then

$$v_1 = e_{11} \quad \text{V} \tag{2.4}$$

Flux ϕ will also link winding N_2, producing flux linkage λ_2. If the flux is increasing, an emf, e_{22}, will be induced in winding N_2 acting in the direction indicated in Fig. 2.2 and of magnitude

$$e_{22} = \frac{d\lambda_2}{dt} = N_2 \frac{d\phi}{dt} \quad \text{V} \tag{2.5}$$

If a passive external load circuit is connected to the terminals of winding N_2, then e_{22} will cause current i_2 to flow as shown in Fig. 2.2.

Since the resistance of winding N_2 is assumed to be negligible, it follows that

$$v_2 = e_{22} \quad \text{V} \tag{2.6}$$

Thus, by virtue of the first three assumptions,

$$\frac{v_1}{v_2} = \frac{e_{11}}{e_{22}} = \frac{N_1}{N_2} \tag{2.7}$$

—that is, the potential ratio is equal to the turns ratio, and

$$v_1 = \frac{N_1}{N_2} v_2 \quad \text{V} \tag{2.8}$$

The net mmf acting on the core at any instant is

$$\mathscr{F} = N_1 i_1 - N_2 i_2 \quad \text{A} \tag{2.9}$$

By assumption 4,

$$N_1 i_1 - N_2 i_2 = 0 \tag{2.10}$$

This assumption is equivalent to representing the λ–i loop of the core by a straight line coincident with the λ axis. From Eq. (2.10),

$$i_1 = \frac{N_2}{N_1} i_2 \tag{2.11}$$

Thus no current can exist in winding N_1 unless there is a corresponding current in winding N_2. Current i_2, so to speak, "calls current i_1 into existence." The current ratio is therefore the inverse of the turns ratio.

From Eqs. (2.8) and (2.11),

$$v_1 i_1 = v_2 i_2 \quad \text{W} \tag{2.12}$$

that is

$$\frac{\text{Instantaneous power}}{\text{input}} = \frac{\text{Instantaneous power}}{\text{output}} \qquad (2.13)$$

This is to be anticipated, since assumptions 1, 2, and 5 abolish all losses and energy storage in the transformer.

If potential difference v_1 is a sinusoidal function of time, then the system made up of the source, the transformer, and the connected passive circuit may be represented by a diagram showing the rms or effective magnitudes of the variables. Such a diagram is given in Fig. 2.3, where the ideal transformer is represented simply as two coils, N_1 and N_2. Omission of the ferromagnetic core from the diagram eliminates the information given in Fig. 2.2 concerning the directions in which coils N_1 and N_2 are wound on the core. This information is supplied in Fig. 2.3 by the dots at the ends of the two windings. The ends of the windings at which the dots are placed go positive in potential simultaneously with respect to the other ends of the two windings.

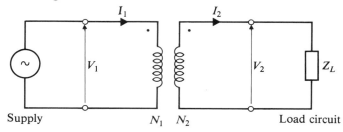

Fig. 2.3 Ideal transformer.

Phasor equations may be written directly from Eqs. (2.8) and (2.11). They are

$$\overline{V}_1 = \frac{N_1}{N_2} \overline{V}_2 \quad \text{V} \qquad (2.14)$$

$$\overline{I}_1 = \frac{N_2}{N_1} \overline{I}_2 \quad \text{A} \qquad (2.15)$$

from which

$$\frac{\overline{V}_1}{\overline{I}_1} = \left[\frac{N_1}{N_2}\right]^2 \frac{\overline{V}_2}{\overline{I}_2} = \left[\frac{N_1}{N_2}\right]^2 \overline{Z}_L \quad \Omega \qquad (2.16)$$

where

$$\overline{Z}_L = \frac{\overline{V}_2}{\overline{I}_2} \quad \Omega \qquad (2.17)$$

Z_L is the impedance of the passive circuit connected to the terminals of winding N_2.

Equation (2.16) shows that, as far as could be detected by instruments connected at the source, the system might equally well be that shown in Fig. 2.4, in

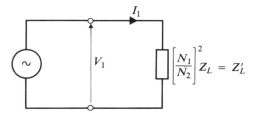

Fig. 2.4 Load-circuit impedance referred to supply side of ideal transformer.

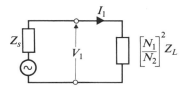

Fig. 2.5 Maximum power transfer.

which the ideal transformer and connected passive load circuit are replaced by an impedance

$$\overline{Z}'_L = \left[\frac{N_1}{N_2}\right]^2 \overline{Z}_L \quad \Omega \tag{2.18}$$

The impedance Z'_L is said to be "Z_L referred to the N_1-turn side of the transformer."

Thus the turns ratio of the transformer may be employed to change the effective impedance of the load imposed upon the source. This property of the transformer may therefore be exploited to obtain maximum power transfer from sources that themselves have appreciable internal impedance Z_s. This situation is illustrated in Fig. 2.5 where, in order to obtain maximum power transfer from a source of internal impedance Z_s to a load of resistive impedance Z_L, it is necessary to choose the turns ratio such that

$$Z'_L = \left[\frac{N_1}{N_2}\right]^2 Z_L = Z_s \quad \Omega \tag{2.19}$$

For maximum power transfer when Z_L may be complex,

$$\overline{Z}'_L = \overline{Z}^*_s \quad \Omega \tag{2.20}$$

where $\overline{Z}^*_s$ is the conjugate of $\overline{Z}_s$.

In a practical situation, where the direction of energy flow through a transformer is known, it is customary to designate one winding the *primary* and the other the *secondary*. The primary winding is the one to which the source of energy is connected. Energy flows into the primary winding. The secondary winding is the one to which the load network is connected. Energy flows out of the secondary winding. Normally, one winding is designed for a higher potential difference than the other, although *isolating* transformers, which are employed simply to separate two circuits electrically, may have two similar windings. In situations in which the energy flow may take place in either direction through the transformer, it is advisable to identify the windings as *high potential* and *low potential,* or in some other unambiguous fashion.

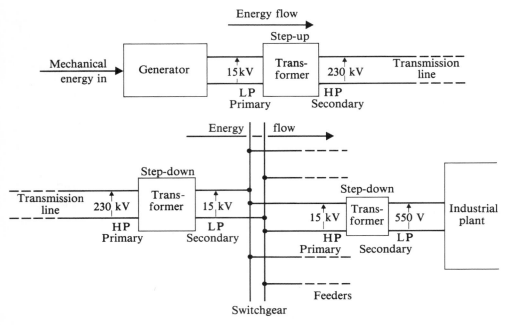

Fig. 2.6 Block diagram of a typical power distribution system.

When the low-potential winding is the primary, it is common to speak of a *step-up* transformer; when the high-potential winding is the primary, of a *step-down* transformer. Any transformer may be employed either to step up or to step down the potential difference, so that its location in a system determines its designation and those of its windings. These designations are illustrated in Fig. 2.6, where single-phase circuits without switchgear are indicated, whereas in practice three-phase circuits with elaborate switchgear would be employed.

The conditions of frequency, potential difference, current, and power under which a transformer is designed to operate are collectively termed the *rating* of the transformer. The rated potential difference is limited by the need to operate at less than saturation flux density in the core. The rated current is limited by the need to limit the losses to a value that will not cause excessive heating of the transformer. It will therefore be seen that the figure which defines the power-handling capacity of a transformer must refer to the apparent power S, measured in volt-amperes, not to the active power W, measured in watts. The stated rating of a transformer describes what it *can* do when operated at its maximum capacity; it does not describe what it *is* doing in a particular situation.

Example 2.1 A 60-Hz, 50-kVA, 2400/240-V transformer is used to step down the potential from that of a transmission line to that of a domestic distribution system.

a) What load impedance connected to the terminals of the low-potential winding will cause the transformer to be fully loaded?

b) What is the value of this impedance referred to the high-potential side of the transformer?

c) What is the current in the high-potential winding?

Consider the transformer ideal.

Solution

a) Since there are no losses in the transformer, the apparent power is the same in both windings. Thus,

$$S = V_1 I_1 = V_2 I_2 \quad \text{VA}$$

from which

$$I_2 = \frac{S}{V_2} = \frac{50 \times 10^3}{240} = 208 \quad \text{A}$$

$$Z_L = \frac{V_2}{I_2} = \frac{240}{208} = 1.15 \quad \Omega$$

This impedance may be resistive or reactive or a combination of the two.

b)
$$\frac{N_1}{N_2} = \frac{V_1}{V_2} = \frac{2400}{240} = 10$$

$$Z_L' = \left[\frac{N_1}{N_2}\right]^2 Z_L = 10^2 \times 1.15 = 115 \quad \Omega$$

c)
$$I_1 = \frac{S}{V_1} = \frac{50 \times 10^3}{2400} = 20.8 \quad \text{A}$$

2.2 LINEAR EQUIVALENT CIRCUIT OF A TWO-WINDING TRANSFORMER

The ideal transformer is not a sufficiently accurate model for all purposes; and when a more accurate prediction of transformer behavior is required, a model which corresponds more closely to the physical system must be employed. This means that the assumptions and approximations made must be less radical than those made in deriving the ideal transformer model. A model which is capable of giving very accurate prediction of transformer behavior under most operating conditions is obtained when the following assumptions are made:

1. The electric fields produced by the windings are negligible.
2. Winding resistances may be represented by lumped parameters at the terminals of the windings.
3. The flux produced by the mmf of one winding may be divided into two distinct parts:
   a) Leakage flux linking all of the turns of the winding producing the mmf, but none of the turns of the other winding.

    b) Mutual flux linking all turns of both windings.

4. The permeability of the core is constant.

5. Core losses are negligible.

    Assumptions 1 and 5 are identical with assumptions made in deriving the ideal transformer model. The consequences of assumptions 2 and 3 are illustrated in Fig. 2.7, where the magnetic system at the center of the diagram represents the transformer deprived of its winding resistance. The magnetic system is therefore no longer a pictorial representation of a real transformer. The leakage fluxes produced by the two windings are given the symbols ϕ_{l1} and ϕ_{l2}. Their directions are consistent with the directions of i_1 and i_2. The mutual flux, which links both windings and is produced by their resultant mmf, is given the symbol ϕ_m. If reluctance of the core is $\mathcal{R}_m$, then

$$\phi_m = \frac{N_1 i_1 - N_2 i_2}{\mathcal{R}_m} \quad \text{Wb} \tag{2.21}$$

The fluxes ϕ_m, ϕ_{l1}, and ϕ_{l2} in Fig. 2.7 are all positive when $N_1 i_1 > N_2 i_2$.

    The consequences of assumptions 4 and 5 are illustrated in Fig. 2.8, which represents a model of the $B-H$ "loop" of the core material or, alternatively, the $\phi-i$ loop for the core and one of the windings. Since core losses are assumed zero, the loop has zero area. Since permeability is assumed constant, there is a linear relationship between ϕ and i, and it is from this assumed linear relationship that the name *linear equivalent circuit* is taken.

    When both windings are carrying current, the flux linking winding N_1 is

$$\phi_1 = \phi_{l1} + \phi_m \quad \text{Wb} \tag{2.22}$$

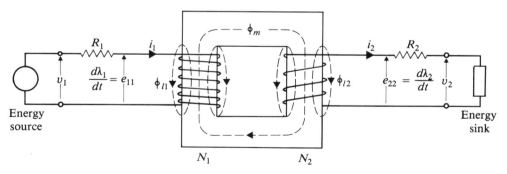

Fig. 2.7 Consequences of assumptions 2 and 3.

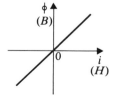

Fig. 2.8 Consequences of assumptions 4 and 5.

Similarly, the flux linking winding N_2 is

$$\phi_2 = -\phi_{l2} + \phi_m \quad \text{Wb} \tag{2.23}$$

Since the total flux linkages of windings N_1 and N_2 are

$$\lambda_1 = N_1\phi_1, \qquad \lambda_2 = N_2\phi_2, \quad \text{Wb} \tag{2.24}$$

then from Fig. 2.7,

$$v_1 = R_1 i_1 + e_{11} = R_1 i_1 + \frac{d\lambda_1}{dt} \quad \text{V} \tag{2.25}$$

and

$$v_2 = -R_2 i_2 + e_{22} = -R_2 i_2 + \frac{d\lambda_2}{dt} \quad \text{V} \tag{2.26}$$

The emf's induced in the two windings may, from Eqs. (2.22)–(2.24), also be expressed by

$$e_{11} = \frac{d\lambda_1}{dt} = N_1 \frac{d\phi_{l1}}{dt} + N_1 \frac{d\phi_m}{dt} \quad \text{V} \tag{2.27}$$

$$e_{22} = \frac{d\lambda_2}{dt} = -N_2 \frac{d\phi_{l2}}{dt} + N_2 \frac{d\phi_m}{dt} \quad \text{V} \tag{2.28}$$

Leakage inductances of the two windings may be defined as follows:

$$L_{l1} = \frac{N_1\phi_{l1}}{i_1}, \qquad L_{l2} = \frac{N_2\phi_{l2}}{i_2} \quad \text{H} \tag{2.29}$$

and the emf's induced in the windings by the mutual flux ϕ_m may be designated

$$N_1 \frac{d\phi_m}{dt} = e_1, \qquad N_2 \frac{d\phi_m}{dt} = e_2 \quad \text{V} \tag{2.30}$$

Substitution from Eqs. (2.27)–(2.30) in Eqs. (2.25) and (2.26) then yields

$$v_1 = R_1 i_1 + L_{l1} \frac{di_1}{dt} + e_1 \quad \text{V} \tag{2.31}$$

$$v_2 = -R_2 i_2 - L_{l2} \frac{di_2}{dt} + e_2 \quad \text{V} \tag{2.32}$$

These last two relationships are illustrated in Fig. 2.9, where the magnetic system at the center of the diagram represents the transformer deprived of its properties of winding resistance and of winding leakage inductance.

From Eq. (2.30),

$$\frac{e_1}{e_2} = \frac{N_1}{N_2} \tag{2.33}$$

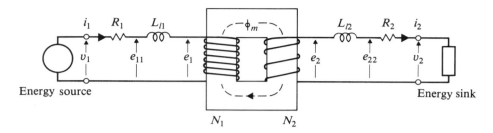

Fig. 2.9 Introduction of leakage inductances.

These emf's are in the same ratio as the terminal potential differences of an ideal transformer with turns N_1 and N_2. They may be considered to be the terminal potential differences of the "transformer" in Fig. 2.9. This "transformer" is not ideal, however, since the net mmf acting on the core is not zero, due to the fact that the permeability of the core is not infinite.

Let i'_m be the current that would be required in the N_1 winding alone to produce the mutual flux ϕ_m. This may be called the *magnetizing current* referred to winding N_1. Then

$$N_1 i'_m = N_1 i_1 - N_2 i_2 \quad A \tag{2.34}$$

from which

$$i_1 = i'_m + \frac{N_2}{N_1} i_2 \quad A \tag{2.35}$$

Since i'_m is a current that produces flux, it must flow in an inductive circuit. Thus Eqs. (2.33) and (2.35) may be considered to describe a circuit consisting of an ideal transformer with an inductance in which i'_m flows connnected across the N_1 winding. This circuit is shown in Fig. 2.10. The requirement that the resultant mmf on the core of the ideal transformer shall be zero is met, since

$$N_1 \left[\frac{N_2}{N_1} i_2 \right] - N_2 i_2 = 0 \tag{2.36}$$

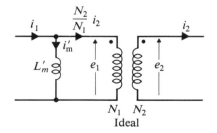

Fig. 2.10 Circuit described by Eqs. 2.33 and 2.35.

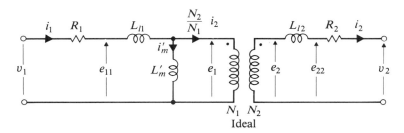

Fig. 2.11 Transformer linear equivalent circuit.

The inductance in Fig. 2.10 may be defined as

$$L'_m = \frac{N_1 \phi_m}{i'_m} = \frac{N_1^2}{\mathscr{R}_m} \quad \text{H} \tag{2.37}$$

and this is the *magnetizing inductance* of the transformer referred to the N_1 winding.

The complete equivalent circuit may now be produced by substituting the circuit shown in Fig. 2.10 for the "nearly ideal" transformer shown in Fig. 2.9, thus obtaining the circuit shown in Fig. 2.11.

Had it been assumed initially that the magnetizing current was i''_m flowing in winding N_2 alone, then an inductance L''_m would have been obtained, defined by the equation:

$$L''_m = \frac{N_2 \phi_m}{i''_m} = \frac{N_2^2}{\mathscr{R}_m} \quad \text{H} \tag{2.38}$$

and this inductance would have been connected across the N_2-turn winding. Since, on the basis of the assumptions made,

$$N_1 i'_m = N_2 i''_m \quad \text{A} \tag{2.39}$$

then from Eqs. (2.37)–(2.39),

$$\frac{L'_m}{L''_m} = \left[\frac{N_1}{N_2}\right]^2 \tag{2.40}$$

Thus, L''_m is the magnetizing inductance L'_m referred to winding N_2.

*2.3 COUPLED-CIRCUIT REPRESENTATION OF A TRANSFORMER

The two-winding transformer of Fig. 2.7 may also be represented by a linear circuit consisting of two branches with mutual inductance between them. This alternative circuit may be developed by considering that the mutual flux ϕ_m that links all turns of both windings consists of two components each produced by the mmf of one of the windings. Thus, from Eq. (2.21),

$$\phi_m = \frac{N_1 i_1}{\mathcal{R}_m} - \frac{N_2 i_2}{\mathcal{R}_m} = \phi_{m1} - \phi_{m2} \quad \text{Wb} \tag{2.41}$$

and from Eqs. (2.22)–(2.24):

$$\lambda_1 = N_1(\phi_{l1} + \phi_{m1}) - N_1\phi_{m2} \quad \text{Wb} \tag{2.42}$$

$$\lambda_2 = -N_2(\phi_{l2} + \phi_{m2}) + N_2\phi_{m1} \quad \text{Wb} \tag{2.43}$$

In each of these two equations, the first term on the right-hand side represents the flux linkage produced in a winding by the current in that same winding, while the second term represents the flux linkage produced in one winding by the current in the other. Self-inductances L_{11} and L_{22} and mutual inductances L_{12} and L_{21} of the two windings may therefore be defined as follows:

$$L_{11} = \frac{N_1(\phi_{l1} + \phi_{m1})}{i_1}, \qquad L_{22} = \frac{N_2(\phi_{l2} + \phi_{m2})}{i_2} \quad \text{H} \tag{2.44}$$

$$L_{12} = \frac{N_1\phi_{m2}}{i_2} \qquad L_{21} = \frac{N_2\phi_{m1}}{i_1} \quad \text{H} \tag{2.45}$$

But from Eq. (2.41),

$$\phi_{m1} = \frac{N_1 i_1}{\mathcal{R}_m} \qquad \phi_{m2} = \frac{N_2 i_2}{\mathcal{R}_m} \quad \text{Wb} \tag{2.46}$$

so that substitution for ϕ_{m1} and ϕ_{m2} in Eq. (2.45) yields

$$L_{12} = \frac{N_1 N_2}{\mathcal{R}_m} \qquad L_{21} = \frac{N_2 N_1}{\mathcal{R}_m} \quad \text{H} \tag{2.47}$$

Thus

$$L_{12} = L_{21} \quad \text{H} \tag{2.48}$$

The inductance parameters defined in the foregoing equations may now be combined with the lumped winding resistances to form the circuit shown in Fig. 2.12.

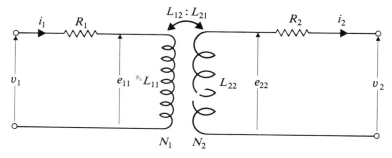

Fig. 2.12 Coupled-circuit representation of a transformer.

The variables in this circuit correspond to those in the equivalent circuit of Fig. 2.11. It should of course be understood that the circuit in Fig. 2.12 is every bit as "equivalent" to the transformer as the circuit in Fig. 2.11; however, it has become common usage to apply the name "equivalent circuit" to the latter only.

Substitution from Eqs. (2.44) and (2.45) in Eqs. (2.42) and (2.43) yields:

$$\lambda_1 = L_{11}i_1 - L_{12}i_2 \quad \text{Wb} \tag{2.49}$$

$$\lambda_2 = -L_{22}i_2 + L_{12}i_1 \quad \text{Wb} \tag{2.50}$$

in which L_{12} is adopted as the symbol for the mutual inductance of the two windings. Thus

$$e_{11} = \frac{d\lambda_1}{dt} = L_{11}\frac{di_1}{dt} - L_{12}\frac{di_2}{dt} \quad \text{V} \tag{2.51}$$

$$e_{22} = \frac{d\lambda_2}{dt} = -L_{22}\frac{di_2}{dt} + L_{12}\frac{di_1}{dt} \quad \text{V} \tag{2.52}$$

and from Fig. 2.12

$$v_1 = R_1 i_1 + L_{11}\frac{di_1}{dt} - L_{12}\frac{di_2}{dt} \quad \text{V} \tag{2.53}$$

$$v_2 = -R_2 i_2 - L_{22}\frac{di_2}{dt} + L_{12}\frac{di_1}{dt} \quad \text{V} \tag{2.54}$$

Since the circuits in Figs. 2.11 and 2.12 are alternative models of the same device, there must be a relationship between the two sets of parameters. Thus, from Eq. (2.44),

$$L_{11} = \frac{N_1\phi_{l1}}{i_1} + \frac{N_1\phi_{m1}}{i_1} \quad \text{H} \tag{2.55}$$

$$L_{22} = \frac{N_2\phi_{l2}}{i_2} + \frac{N_2\phi_{m2}}{i_2} \quad \text{H} \tag{2.56}$$

The two terms on the right-hand side of Eq. (2.55) are components of inductance L_{11}. The first has already been defined by Eq. (2.29), and is the leakage inductance of winding N_1. The second term, which expresses the mutual flux linkage in winding N_1 per ampere of current in N_1, has also been defined, albeit in different symbols, in Eq. (2.37); it is the magnetizing inductance of the transformer referred to winding N_1. Thus from Eq. (2.55),

$$L_{11} = L_{l1} + L'_m \quad \text{H} \tag{2.57}$$

and similarly

$$L_{22} = L_{l2} + L''_m \quad \text{H} \tag{2.58}$$

From Eqs. (2.37) and (2.47),

$$L'_m = \frac{N_1^2}{\mathcal{R}_m} = \frac{N_1}{N_2}L_{12} \quad \text{H} \tag{2.59}$$

From Eqs. (2.38) and (2.47),

$$L''_m = \frac{N_2^2}{\mathscr{R}_m} = \frac{N_2}{N_1} L_{12} \quad \text{H} \tag{2.60}$$

The coupled circuit model of Fig. 2.12 is often used to represent a two-winding air-core transformer. It is rarely used with iron-core transformers because the relative permeability of the core is very high, leading to a small value of the core reluctance $\mathscr{R}_m$ and large values of L_{11}, L_{22}, and L_{12}. If the core is assumed to be perfect—that is, of infinite permeability—all inductances become infinite and Eqs. (2.53) and (2.54) become indeterminate.

2.4 TRANSFORMER WITH SINUSOIDAL EXCITATION

When the potential difference v_1 applied by an energy source to the terminals of winding N_1 of a transformer is a sinusoidal function of time, and a linear network is connected to the terminals of winding N_2, then the entire system may be represented by the circuit of Fig. 2.13(a), in which all the variables are sinusoidal functions of time. The rms values of the variables are shown on the circuit, and the inductances have been replaced by the reactances for the frequency at which the transformer is operating. For example,

$$X_{l1} = 2\pi f L_{l1} = \omega L_{l1} \quad \Omega \tag{2.61}$$

For the direction of energy flow shown, winding N_1 may be referred to as the "primary" winding, and N_2 as the "secondary."

The analysis of the circuit may be simplified by referring the variables and parameters on one side of the ideal transformer to the other. For example, when all the secondary quantities are referred to the primary side, the circuit shown in Fig. 2.13(b) is obtained. The magnitudes of the referred quantities may be obtained from Eqs. (2.8), (2.11), and (2.18).

$$\overline{E}'_2 = \frac{N_1}{N_2} \overline{E}_2 \equiv \overline{E}_1 \quad \text{V} \tag{2.62}$$

$$\overline{V}'_2 = \frac{N_1}{N_2} \overline{V}_2 \quad \text{V} \tag{2.63}$$

$$\overline{I}'_2 = \frac{N_2}{N_1} \overline{I}_2 \quad \text{A} \tag{2.64}$$

$$\overline{Z}'_L = \left[\frac{N_1}{N_2}\right]^2 \overline{Z}_L \quad \Omega \tag{2.65}$$

$$X'_{l2} = \left[\frac{N_1}{N_2}\right]^2 X_{l2} \quad \Omega \tag{2.66}$$

$$R'_2 = \left[\frac{N_1}{N_2}\right]^2 R_2 \quad \Omega \tag{2.67}$$

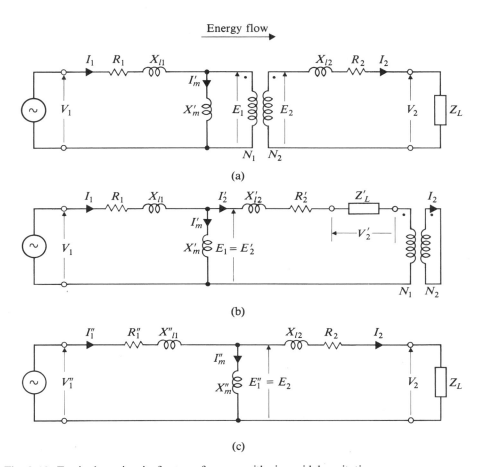

Fig. 2.13 Equivalent circuits for transformers with sinusoidal excitation.

After these transformations have been made, the secondary terminals of the ideal transformer are short-circuited, making the potential difference between its primary terminals zero. Thus these terminals also may be considered to be short-circuited, and the ideal transformer is eliminated from the equivalent circuit.

In a converse manner, everything on the primary side of the transformer, including the source, may be referred to the secondary side, giving the circuit in Fig. 2.13(c), where the ideal transformer again disappears from the equivalent circuit. The reader should write equations corresponding to Eqs. (2.62)–(2.67) to determine the values of the variables and parameters in this last circuit. The choice of whether to refer all quantities to the primary, Fig. 2.13(b), or to the secondary, Fig. 2.13(c), depends on the problem that is being solved. A phasor diagram for the equivalent

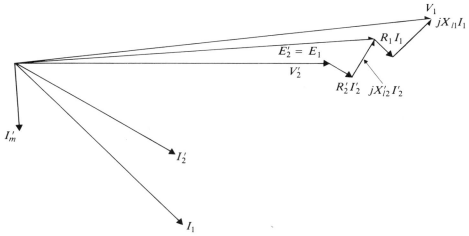

Fig. 2.14 Phasor diagram for system of Fig. 2.13(b).

circuit referred to the primary side of the transformer is shown in Fig. 2.14. In drawing the phasor diagram it has been assumed that the load circuit possesses both resistance and inductance, and therefore operates at a lagging power factor. Since, in carrying out calculations concerning a transformer, the point of chief interest is the effect of the transformer on the load circuit, the phasor of V_2' has been employed as the reference phasor in the diagram. That is,

$$\overline{V}_2' = V_2' \underline{|0} \quad V \tag{2.68}$$

The effects of winding resistances and leakage reactances are greatly exaggerated in Fig. 2.14, as also is the magnitude of the magnetizing current. In practice these quantities are so small that the phasors representing their effects are of the same order of magnitude as would be the drafting error in attempting to draw Fig. 2.14 to scale. The phasor diagram is therefore merely a useful guide to the calculation of the magnitudes of the variables in the circuit.

Example 2.2   A 20-kVA, 2200:220-V, 60-Hz, single-phase transformer has the following equivalent-circuit parameters referred to the high-potential side of the transformer.

$$R_1 = 2.51 \quad \Omega \qquad\qquad R_2' = 3.11 \quad \Omega$$
$$X_{l1} = 10.9 \quad \Omega \qquad\qquad X_{l2}' = 10.9 \quad \Omega$$
$$X_m' = 25,100 \quad \Omega$$

The transformer is supplying 15 kVA at 220 volts and a lagging power factor of 0.85.
    Determine the required potential difference at the high-potential terminals of the transformer.

Solution   The equivalent circuit of Fig. 2.13(b) and phasor diagram of Fig. 2.14 are applicable.

Note that the transformer is not supplying its rated output power. Also note that the rating gives the nominal ratio of terminal potential differences—that is, it gives the turns ratio of the ideal transformer. To a close approximation, this is also the ratio of potential differences on no load—that is, with the low-potential terminals open circuited.

$$\text{Let } \overline{V}_2 = 220\underline{|0} \text{ V}$$

$$\overline{V}_2' = \frac{2200}{220}\overline{V}_2 = 2200\underline{|0} \text{ V}$$

$$I_2 = \frac{15 \times 10^3}{220} = 68.2 \text{ A}$$

$$\cos^{-1} 0.85 = 31.7°$$

$$I_2 = 68.2\underline{|-31.7°} \text{ A}$$

$$\overline{I}_2' = \frac{220}{2200} \times 68.2\underline{|-31.7°} = 6.82\underline{|-31.7°} \text{ A}$$

From Fig. 2.13(b),

$$\overline{E}_1 = \overline{V}_2' + (R_2' + jX_{l2}')\overline{I}_2' = 2200\underline{|0} + (3.11 + j\,10.9)\,6.82\underline{|-31.7°} = 2260\underline{|1.3°} \text{ V}$$

$$\overline{I}_m' = \frac{\overline{E}_1}{jX_m'} = \frac{2260\underline{|1.3°}}{25,100\underline{|90°}} = 0.090\underline{|-88.7°} \text{ A}$$

Note that I_m' is very small compared with I_2'.

$$\overline{I}_1 = \overline{I}_2' + \overline{I}_m' = 6.82\underline{|-31.7°} + 0.090\underline{|-88.7°}$$

$$= 6.87\underline{|-32.3°} \text{ A}$$

$$\overline{V}_1 = \overline{E}_1 + (R_1 + jX_{l1})\overline{I}_1$$

$$= 2260\underline{|1.3°} + (2.51 + j\,10.9)6.87\underline{|-32.3°}$$

$$= 2311\underline{|2.6°} \text{ V}$$

Thus $V_1 = 2311$ V, as compared with the rated or nameplate value of 2200 V. The additional potential of 111 V is needed to overcome the impedance of the transformer.

2.5 TRANSFORMER CORE LOSSES

Core losses occur in a transformer for exactly the same reasons as they occur in an inductive reactor, and this process has been described in Sections 1.6, 1.8, and 1.9. Furthermore, a transformer on no load simply behaves as a high-impedance inductive reactor with no air gap. Under these conditions the transformer exciting current is similar to that of an inductor, being made up of two components: I_m', the magnetizing component; and I_c', the core-loss component. Assuming the components to be sinusoidal, a phasor diagram showing their relationship with the emf induced in the

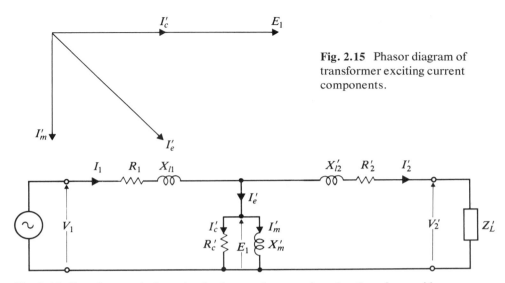

Fig. 2.15 Phasor diagram of transformer exciting current components.

Fig. 2.16 Complete equivalent circuit of a transformer referred to the primary side.

primary winding is therefore that shown in Fig. 2.15, and this may be compared with Fig. 1.28.

A circuit element R'_c may be added to the transformer equivalent circuit to absorb a power corresponding to that dissipated in core losses. This again corresponds to the core-loss element introduced into the equivalent circuit shown in Fig. 1.29. The complete equivalent circuit of a transformer in which the zero-core-loss assumption is no longer being made is therefore that shown in Fig. 2.16. The corresponding phasor diagram for a transformer supplying a load circuit with lagging power factor is shown in Fig. 2.17 and should be compared with Fig. 2.14. The magnitude of the current I'_c flowing in core-loss element R'_c is greatly exaggerated in the diagram.

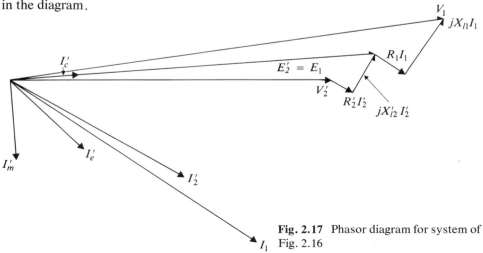

Fig. 2.17 Phasor diagram for system of Fig. 2.16

2.6 TRANSFORMER PERFORMANCE

The losses in a transformer are dissipated in the core and windings as heat, and this heat must be conveyed away from the transformer if it is not to overheat, with resulting accelerated deterioration of insulation. Since the possibilities of heat dissipation from any given transformer have a limit, then an equal limit must be placed on the power losses that are permitted to occur. As may be seen from the equivalent circuit in Fig. 2.16.

$$\text{Core loss} = P_C = \frac{E_1^2}{R_c'} \quad \text{W} \tag{2.69}$$

$$\text{Copper loss} = P_w = R_1 I_1^2 + R_2 I_2^2 \quad \text{W} \tag{2.70}$$

These equations show that the heat to be dissipated depends both upon the potential difference at which the transformer is operated and upon the currents in the windings, but is not dependent upon the power factor of the load circuit. For this reason, as has already been remarked, the rating of a transformer (as well as any other alternating-current device) is expressed in terms of the apparent power in volt-amperes at some specified terminal potential difference rather than in terms of the active power in kilowatts.

2.6.1 Determination of Equivalent-Circuit Parameters

The designer of a transformer is able to estimate from his or her design the parameters of the equivalent circuit as well as the ability of the transformer to dissipate heat by the chosen method of cooling, and is, therefore, able to predict the permissible output of any design and its detailed performance. Once the transformer is built, however, it is desirable to confirm the accuracy of the predictions, and this can be done by means of tests designed to determine the parameters of the equivalent circuit.

Tests may be carried out using either winding as the primary, depending upon the potential differences for which the windings are designed and upon available sources. In a test department it is preferable to work with potential differences that are neither very high nor very low. The first are dangerous; the second are inconvenient. For a test carried out at rated potential difference, it is often convenient to use the low-potential winding, since a suitable source is more likely to be available. For a test carried out at a reduced potential difference, it is often convenient to use the high-potential winding, since again a suitable source is likely to be available. In the following description, it will be assumed that one winding is used as the primary for both types of test. The measurements obtained will therefore give directly the equivalent-circuit parameters referred to the primary side of the transformer.

a) **Open-Circuit Test** Rated potential difference at rated frequency is applied to the primary terminals with the secondary terminals open-circuited. Since R_c' and X_m' are very much larger than R_1 and X_{l1}, the equivalent circuit for this test is effectively that shown in Fig. 2.18(a), where X_{l2}' and R_2' may also be omitted, since

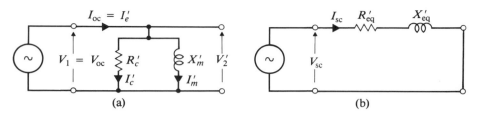

Fig. 2.18 Equivalent circuits for (a) open-circuit test, (b) short-circuit test.

I'_2 is zero. Under these conditions, the primary current is simply the exciting current I'_e of the transformer. V_1, V_2, I'_e, and P_{oc}, the power input to the transformer, are measured. From these measurements, the turns ratio V_1/V_2 may be checked, while R'_c and X'_m may be calculated. P_{oc} gives effectively the core losses of the transformer under the operating conditions, since E_1 varies only very slightly with variations in transformer current.

b) Short-Circuit Test   Potential V_{sc}, a fraction of the rated potential difference sufficient to produce rated primary current at rated frequency, is applied to the primary terminals with the secondary terminals short-circuited. Since R'_c and X'_m are very much larger than R'_2 and X'_{l2}, the equivalent circuit for this test is effectively that shown in Fig. 2.18(b), where R'_{eq} and X'_{eq} are called the *equivalent resistance* and *equivalent leakage reactance* and are defined by:

$$R'_{eq} = R_1 + R'_2 \quad \Omega \tag{2.71}$$

$$X'_{eq} = X_{l1} + X'_{l2} \quad \Omega \tag{2.72}$$

V_{sc}, I_{sc}, and P_{sc}, the power input to the transformer, are measured. From these measurements, R'_{eq} and X'_{eq} may be calculated. It is then usually assumed that

$$X_{l1} = X'_{l2} = \frac{X'_{eq}}{2} \quad \Omega \tag{2.73}$$

The implication of this is that the paths for leakage flux of both windings have the same reluctance, $\mathcal{R}_l$. Then

$$X_{l1} = \frac{\omega N_1^2}{\mathcal{R}_l} \qquad X_{l2} = \frac{\omega N_2^2}{\mathcal{R}_l} \quad \Omega \tag{2.74}$$

so that

$$X'_{l2} = \left[\frac{N_1}{N_2}\right]^2 X_{l2} = \frac{\omega N_1^2}{\mathcal{R}_l} = X_{l1} \quad \Omega \tag{2.75}$$

c) Winding-Resistance Measurements R_1 and R_2 may be measured directly by means of a Wheatstone or Kelvin bridge. Such measurements give the resistance of the windings to direct current, and it may be that these differ appreciably from the resistances to alternating current owing to nonuniform distribution of alternating currents in the conductors. This may be checked by determining R'_{eq} from the short-circuit test and comparing it with the equivalent dc winding resistance referred to the primary side of the transformer. If R_1 and R_2 are the measured dc values, then the equivalent dc resistance is

$$R_{dc} = R_1 + \left[\frac{N_1}{N_2}\right]^2 R_2 \quad \Omega \tag{2.76}$$

Should R_{dc} differ appreciably from R'_{eq}, then the ac winding resistances referred to the primary side may be determined by dividing R'_{eq} in the ratio of the two terms on the right-hand side of Eq. (2.76).

Example 2.3 The following measurements were obtained from tests carried out on a 10-kVA, 2300:230-V, 60-Hz distribution transformer:

Open-circuit test, with the low-potential winding excited:

$$\text{Applied potential difference, } V_{oc} = 230 \quad \text{V}$$
$$\text{Current, } I_{oc} = 0.45 \quad \text{A}$$
$$\text{Input power, } P_{oc} = 70 \quad \text{W}$$

Short-circuit test, with the high-potential winding excited:

$$\text{Applied potential difference, } V_{sc} = 120 \quad \text{V}$$
$$\text{Current, } I_{sc} = 4.5 \quad \text{A}$$
$$\text{Input power, } P_{sc} = 240 \quad \text{W}$$

Winding resistances, measured by dc bridge:

$$R_{hp} = 5.80 \quad \Omega, \qquad R_{lp} = 0.0605 \quad \Omega$$

a) Determine the equivalent circuit of the transformer referred to the low-potential side.

b) Express the exciting current of the transformer as a percentage of the rated full-load current.

Solution

a) For the open-circuit test, the copper loss in the low-potential winding is

$$P_w = R_{lp} I_{oc}^2 = 0.0605 \times 0.45^2 = 0.0163 \quad \text{W}$$

This is a very small proportion of the input power on open circuit and may be neglected. The power factor on open circuit is

$$\cos\theta = \frac{P_{oc}}{V_{oc} I_{oc}} = \frac{70}{230 \times 0.45} = 0.676$$
$$\theta = 47.4°$$

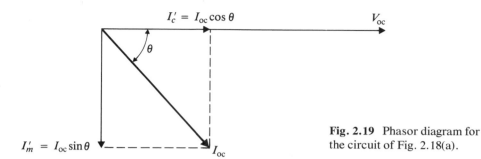

Fig. 2.19 Phasor diagram for the circuit of Fig. 2.18(a).

The phasor diagram for this test, corresponding to the circuit diagram of Fig. 2.18(a), is shown in Fig. 2.19. From this diagram,

$$R'_c = \frac{V_{oc}}{I_{oc}\cos\theta} = \frac{230}{0.45 \times 0.676} = 756\ \Omega$$

$$X'_m = \frac{V_{oc}}{I_{oc}\sin\theta} = \frac{230}{0.45 \times 0.737} = 694\ \Omega$$

and these parameters are measured on, and therefore referred to, the low-potential side of the transformer.

The circuit diagram for the short-circuit test is that of Fig. 2.18(b). From that diagram,

$$R'_{eq} = \frac{P_{sc}}{I_{sc}^2} = \frac{240}{4.5^2} = 11.85\ \Omega$$

The transformer impedance on short circuit is

$$Z_{sc} = \frac{V_{sc}}{I_{sc}} = \frac{120}{4.5} = 26.7\quad \Omega$$

so that

$$X'_{eq} = \left[Z_{sc}^2 - (R'_{eq})^2 \right]^{1/2} = [26.7^2 - 11.8^2]^{1/2} = 24.0\quad \Omega$$

and these parameters are measured on, and therefore referred to, the high-potential side of the transformer. Referred to the low-potential side,

$$X''_{eq} = \left(\frac{230}{2300}\right)^2 X'_{eq} = 0.240\quad \Omega$$

so that

$$X_{l2} = X''_{l1} = \frac{0.240}{2} = 0.120\quad \Omega$$

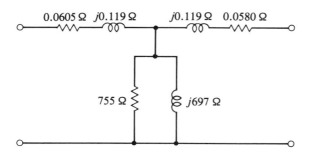

Fig. 2.20 Equivalent circuit of transformer of Example 2.3.

The resistance of the high-potential winding referred to the low-potential side is

$$R''_{hp} = \left(\frac{230}{2300}\right)^2 R_{hp} = 0.058 \ \Omega$$

The equivalent circuit of the transformer referred to the low-potential side is shown in Fig. 2.20.

b) The rated output S of the transformer is 10 kVA, thus the rated full-load current in the low-potential winding is

$$I_{1p} = \frac{S}{V_{1p}} = \frac{10 \times 10^3}{230} = 43.5 \ \text{A}$$

The exciting current is I_{oc}; thus, as a percentage of full-load current, the exciting current is

$$\frac{I_{oc}}{I_{1p}} \times 100\% = \frac{0.45}{43.5} \times 100\% = 1.04\%$$

2.6.2 Approximate Equivalent Circuits

It is rarely necessary to employ the complete equivalent circuit of Fig. 2.16 in order to predict with sufficient accuracy the performance of a transformer. The calculations involved in using the equivalent circuit are much reduced if further approximations are made. A series of approximate equivalent circuits appropriate to various required degrees of accuracy of performance prediction are shown in Fig. 2.21. Circuits (a) and (b) are based on the assumption that

$$V_1 \simeq E_1 = E'_2 \simeq V'_2 \quad \text{V} \tag{2.77}$$

The magnetizing branch consisting of R'_c and X'_m in parallel may therefore be connected across the circuit at the point that is most convenient for the problem being solved, and the magnitude of I'_e will not be greatly affected. Circuit (c) is based on the assumption that the exciting current I'_e is negligible in comparison with the winding current I_1. This corresponds to assumption of negligible core loss and very high

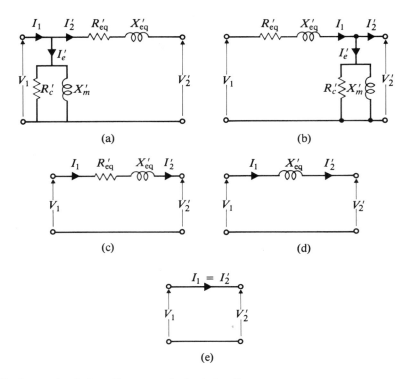

Fig. 2.21 Approximate transformer equivalent circuits.

core permeability. Circuit (c) is usually adequate for solving for the relationships between V_1 and V_2'. In larger transformers, the resistance R'_{eq} is generally much smaller than the reactance X'_{eq}. Circuit (d) is therefore an adequate model for the determination of the relationships between V_1 and V_2'. Finally, the potential difference across the leakage reactance may be small enough relative to the applied potential difference to allow the transformer to be represented by the ideal model of circuit (e).

2.6.3 Efficiency

In a well-designed transformer, both core losses and copper loss are extremely small, so that efficiency is very high—in the neighborhood of 99% for very large transformers. The percentage efficiency may be defined as follows:

$$\eta = \frac{\text{Output active power}}{\text{Input active power}} \times 100\% \qquad (2.78)$$

$$= \frac{\text{Output Power} \times 100\%}{\text{Output power} + \text{Core loss} + \text{Copper loss}} \qquad (2.79)$$

The ratio of output power to input power cannot be determined accurately by measuring these two quantities directly, since the difference between them is of the same order of magnitude as the errors made in measuring them. However, the copper loss may be determined for any condition of operation if the winding currents and their resistances are known. The core loss depends upon the peak flux density produced in the core, and this in turn depends upon the terminal potential difference. Since a power transformer operates at virtually constant terminal potential difference, core loss is nearly constant and may usually be taken as equal to the no-load input power to the transformer, as shown in Example 2.3. Thus if the equivalent-circuit parameters of a transformer are known, efficiency under any operating conditions may be determined.

The ratio of core loss to copper loss at rated output may be different in transformers designed for different types of service. An idea of why this should be so may be obtained by considering the approximate equivalent circuit shown in Fig. 2.21(a). On the basis of that circuit,

$$\text{Core loss} = P_c = \frac{V_1^2}{R_c'} = \text{constant} \quad \text{W} \tag{2.80}$$

$$\text{Copper loss} = P_w = R_{eq}' \, (I_2')^2 \quad \text{W} \tag{2.81}$$

If the power factor of the load circuit is $\cos \theta$, then the transformer efficiency is

$$\eta = \frac{V_2' \, I_2' \cos \theta \times 100\%}{V_2'I_2' \cos \theta + P_c + R_{eq}' \, (I_2')^2} \quad \text{W} \tag{2.82}$$

If $\cos \theta$ is fixed by the nature of the load, then for maximum efficiency,

$$\frac{d\eta}{dI_2'} = 0 \tag{2.83}$$

from which it may be shown that

$$P_c = R_{eq}' \, (I_2')^2 = P_w \quad \text{W} \tag{2.84}$$

Thus for maximum operating efficiency, core loss should be equal to the copper loss.

For a power-system transformer, which usually operates near its rated capacity and is switched out of circuit when not required, it is usual to design for maximum efficiency at or near rated output. For a distribution transformer, which is connected to the system for twenty-four hours a day and is operating well below rated output for most of those hours, it is desirable to design for maximum efficiency at or near the average output. This implies that the ratio of core loss to copper loss at full load shall be much lower in a distribution transformer than in a transformer designed for a constant load. An important figure of merit for a distribution transformer is thus the "all-day" or *energy efficiency* of the transformer, given by

$$\eta_{AD} = \frac{\text{Energy output over 24 hours}}{\text{Energy input over 24 hours}} \times 100\% \tag{2.85}$$

This can be determined if the load cycle of the transformer is known.

Example 2.4 The distribution transformer of Example 2.3 is supplying a load at 230 volts and a power factor of 0.85.

a) Determine the percentage of full-load rating at which the transformer efficiency is a maximum, and calculate the efficiency at that load.

b) The transformer operates at constant power factor on a load cycle that may be approximately represented by the following:

90% full load for 6 hours
50% full load for 10 hours
no load for 8 hours

Determine the all-day or energy efficiency of the transformer under these conditions.

Solution

a) At maximum efficiency the core loss is equal to the copper loss. The core loss is given by the input power for the open-circuit test. The copper loss may be expressed to a sufficiently close approximation for this purpose by employing the approximate equivalent circuit of Fig. 2.21(c) referred to the secondary or low-potential side of the transformer. Equivalent resistance $R''_{eq} = 0.118\ \Omega$. Equivalent reactance $X''_{eq} = 0.238\ \Omega$. Thus at maximum efficiency,

$$P_{oc} = R''_{eq}\ I_2^2 \quad \text{W}$$

from which

$$I_2 = \left(\frac{70}{0.118}\right)^{1/2} = 24.3 \quad \text{A}$$

The apparent power output at this efficiency is

$$S_2 = 230 \times 24.3 = 5680 \quad \text{VA}$$

and the percentage of full-load rating S is

$$\frac{S_2}{S} \times 100\% = \frac{5680}{10^4} \times 100\% = 56.8\%$$

This is appropriate for a distribution transformer for the reasons explained in Section 2.6.3. Output active power at $0.568S$ is

$$P = 0.568S \times \text{Power factor} = 5680 \times 0.85 = 4820 \quad \text{W}$$

$$\text{Core loss} = \text{Copper loss} = P_{oc} = 70 \quad \text{W}$$

Thus the maximum efficiency is

$$\eta_{max} = \frac{4820}{4820 + 70 + 70} \times 100\% = 97.2\%$$

b) Energy output during 24 hours, expressed in kilowatt-hours is,

$$6 \times 0.9 \times 10 \times 0.85 + 10 \times 0.5 \times 10 \times 0.85 = 88.3 \quad \text{kWh}$$
$$\text{Core loss during 24 hours is } 24 \times 70 = 1.68 \quad \text{kWh}$$

Full-load current referred to the low-potential side was seen to be 43.5 A. Thus copper loss during 24 hours is

$$6 \times (0.9 \times 43.5)^2 \times 0.118 + 10 \times (0.5 \times 43.5)^2 \times 0.118 = 1.65 \text{ kWh}$$
$$\eta_{\text{AD}} = \frac{88.3 \times 100\%}{88.3 + 1.68 + 1.65} = 96.3\%$$

2.6.4 Regulation

The primary potential difference of a transformer must be such that the transformer will deliver rated output at the anticipated power while maintaining rated potential difference at the secondary terminals. This means that, on no load, the secondary terminal potential difference may vary from the rated value due to the effect of the impedance of the transformer. The percentage regulation of a transformer is defined as

$$\text{Regulation} = \frac{V_{2(\text{no load})} - V_{2(\text{rated load})}}{V_{2(\text{rated load})}} \times 100\% \tag{2.86}$$

The regulation may be determined by calculating the magnitude of the required primary potential difference to produce the rated secondary value at rated load, and then calculating V_2 on no load for that same primary potential difference. For this purpose it is usually sufficiently accurate to employ the approximate equivalent circuit of Fig. 2.21(c), where

$$I_1 \equiv I_2' \quad \text{A} \tag{2.87}$$

and on no load

$$V_2 \equiv V_1 \quad \text{V} \tag{2.88}$$

Figure 2.22 shows phasor diagrams for the equivalent circuit of Fig. 2.21(c) with the transformer on load. As usual, the phasor of the secondary potential difference is employed as a reference, and the effects of the equivalent-circuit parameters are greatly exaggerated. In Fig. 2.22(a), the load has a lagging power factor, and $V_2' < V_1$. In Fig. 2.22(b), the load has a low leading power factor, and $V_2' > V_1$. Thus, the regulation may be either positive or negative depending upon the nature of the load.

Transformers are often designed with taps on one winding so that the turns ratio may be varied over a small range. In large transformers this tap-changing is often made automatic and maintains a sensibly constant secondary potential difference as the magnitude and power factor of the load vary. In distribution transformers, the tap changing is usually manual and must be adjusted to a setting that gives optimum performance over the anticipated load cycle. Tap changing can also compensate for variations in primary terminal difference due to the impedance of feeders.

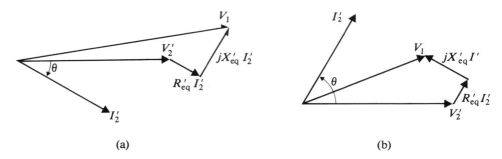

Fig. 2.22 Phasor diagrams for the approximate equivalent circuit of Fig. 2.21(c). (a) Lagging power factor. (b) Leading power factor.

Example 2.5 The transformer of Example 2.3 is supplying full load at 230 volts and a power factor of 0.8 lagging. Determine

a) The primary potential difference required.

b) The power factor at the primary terminals.

c) The transformer regulation.

d) The approximate change in turns ratio required if the primary potential difference is fixed at 2300 V.

Solution Since $\cos^{-1} 0.8 = 36.8°$, the equivalent circuit and phasor diagram for the specified condition of operation are as shown in Fig. 2.21(c) and 2.22(a), except the quantities will be referred to the secondary winding.

a) Since
$$\overline{V}_2 = 230\underline{|0°}, \text{ then } \overline{I}_2 = 43.5\underline{|-36.8°}$$
$$\overline{V}_1'' = \overline{V}_2 + (R_{eq}'' + jX_{eq}'') \overline{I}_2$$

From Example 2.3
$$R_{eq}'' + jX_{eq}'' = 0.118 + j0.238 = 0.266\underline{|63.5°}$$

Thus
$$\overline{V}_1'' = 230\underline{|0°} + 0.266\underline{|63.5°} \times 43.5\underline{|-36.8°} = 240.4\underline{|1.24°}$$

In Fig. 2.22(a) the angle between $\overline{V}_1''$ and $\overline{V}_2$ is 1.24°.

$$V_1 = \frac{2300}{230} \times V_1'' = 10 \times 240.4 = 2404 \quad V$$

b) The power factor at the primary terminals is cos (36.8° + 1.24°) = 0.788.

c) When load is removed from the transformer, V_2 will rise to the value of V_1'' on full load; that is, $V_2 = 240.4$ volts. Thus

$$\text{Percentage regulation} + \frac{240.4 - 230}{230} \times 100\% = 4.52\%$$

d) The result in (c) shows that the turns ratio of the transformer must be reduced on load by 4.52% if the secondary potential difference is to remain unaltered. That is,

$$\frac{N_1}{N_2} = \frac{100 - 4.52}{100} \times 10 = 9.548$$

2.6.5 Per-Unit Values

Such quantities as efficiency and regulation may be expressed in per-unit, as opposed to percentage, values. For this purpose the 100% may be omitted from Eqs. (2.78), (2.79), and (2.86). Variables and parameters may also be expressed in per-unit values; and this is very useful since, for example, the product of a per-unit current and a per-unit impedance yields a per-unit potential difference, whereas the same is not true of percentage quantities. The per-unit (pu) quantity is defined as follows:

$$\text{Quantity in pu} = \frac{\text{Actual quantity}}{\text{Base-value quantity}} \tag{2.89}$$

Clearly everything depends upon the base value of the quantity, and to some extent this may be chosen arbitrarily. Base values of apparent power and potential difference are usually chosen first, and these automatically establish the base values of all other parameters and variables in the transformer equivalent circuit. Usually

$$V_{\text{base}} = \text{Transformer rated terminal potential difference}$$

and

$$S_{\text{base}} = \text{Transformer volt-ampere rating}$$

Necessarily

$$P_{\text{base}} = Q_{\text{base}} = S_{\text{base}}$$

Note that there is a V_{base} for each winding of the transformer.

$$I_{\text{base}} = \frac{S_{\text{base}}}{V_{\text{base}}} \text{ (one for each winding)}$$

from which

$$R_{\text{base}} = X_{\text{base}} = Z_{\text{base}} = \frac{V_{\text{base}}}{I_{\text{base}}} \text{ (one for each winding)}$$

Example 2.6   For the transformer of Examples 2.3–2.5, express as per-unit quantities the maximum efficiency, the regulation, and the parameters of the equivalent circuit.

From Example 2.4, maximum efficiency is 0.972 pu
From Example 2.5, regulation is 0.0452 pu
Base value of apparent power is $S_{base} = 10 \times 10^3$ volt-amperes.

Since, in Example 2.3, the equivalent circuit is referred to the secondary side, the secondary rated potential difference is employed as the base quantity. Thus

$$V_{base} = 230 \quad V$$

$$I_{base} = \frac{S_{base}}{V_{base}} = \frac{10 \times 10^3}{230} = 43.5 \quad A$$

$$Z_{base} = \frac{V_{base}}{I_{base}} = \frac{230}{43.5} = 5.30 \quad \Omega$$

from which

$$R_1'' = \frac{0.058}{5.30} = 0.0109 \text{ per unit}$$

$$R_2 = \frac{0.0605}{5.30} = 0.0114 \text{ per unit}$$

$$X_{l1}'' = X_{l2} = \frac{0.119}{5.30} = 0.0224 \text{ per unit}$$

$$R_c'' = \frac{755}{5.30} = 142 \text{ per unit}$$

$$X_m'' = \frac{694}{5.30} = 131 \text{ per unit}$$

Note that the potential difference across the resistance or leakage reactance at rated current (one per unit) is equal to the per-unit value of the resistance or reactance. The per-unit values are therefore often more useful than the ohmic values in estimating the performance of a transformer. Per-unit values of some quantities such as leakage reactance vary very little with large changes in the size or rating of the transformer.

*2.7 TRANSIENT INRUSH CURRENT OF A TRANSFORMER

It has already been remarked that power transformers are so designed that the exciting current under normal steady-state operating conditions is usually a negligible fraction of the rated current. In the transient period, however, between the instant of applying the rated potential difference to the primary terminals and the eventual establishment of a steady-state condition, it is possible for the instanta-

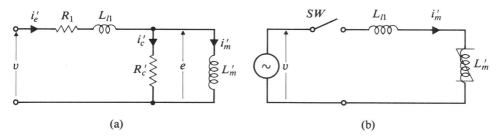

Fig. 2.23 No-load equivalent circuit of transformer. (a) Steady-state condition. (b) Switching-transient operation.

neous value of the exciting current to be very high, even greater than the amplitude of the short-circuit current of the transformer, (that is, the steady sinusoidal primary current that would flow if, in the distribution system, a fault occurred which short-circuited the secondary terminals of the transformer and was not cleared by the opening of a circuit breaker). A knowledge of the possible instantaneous magnitude of this transient current is of importance in determining the maximum mechanical stresses that could occur in the transformer windings. It is also important in designing the protection system for the transformer.

A transformer, operating on no load under steady-state conditions, can be represented by the equivalent circuit of Fig. 2.23(a). In steady-state operation, the exciting current i'_e is small, and its components i'_c and i'_m are of the same order of magnitude. An equivalent circuit, which gives an approximate model of the transformer when there is reason to think that i'_m will not be small, is shown in Fig. 2.23(b). In this circuit, L'_m is shown as a nonlinear inductance because of the anticipated saturation of the core. The core-loss resistance R'_c has been neglected on the assumption that $i'_m \gg i'_c$ during the period of saturation. Also, the coil resistance R_1 has been neglected on the assumption that its effect is much less than that of the leakage inductance L_{l1}. In this circuit, after switch SW is closed,

$$v = \frac{d\lambda_1}{dt} \quad V \tag{2.90}$$

and from Eqs. (2.22), (2.24), and (2.29),

$$\lambda_1 = N_1(\phi_{l1} + \phi_m) = L_{l1}i'_m + \lambda_m \quad Wb \tag{2.91}$$

where λ_m is the flux linkage of the primary winding due to the flux confined to the ferromagnetic core. Let

$$v = \hat{v}\sin \omega t \quad V \tag{2.92}$$

Then, if switch SW in Fig. 2.23(b) is closed at instant $t = 0$, the total flux linkage of the primary winding will be

$$\lambda_1 = \int \hat{v}\sin \omega t\, dt = -\frac{\hat{v}}{\omega}\cos \omega t + K \quad Wb \tag{2.93}$$

If there is no residual magnetism in the core, then at $t = 0$, $\lambda_1 = 0$, and

$$K = \frac{\hat{v}}{\omega} \quad \text{Wb} \tag{2.94}$$

Thus

$$\lambda_1 = \frac{\hat{v}}{\omega} - \frac{\hat{v}}{\omega} \cos \omega t = \hat{\lambda}_1 (1 - \cos \omega t) \tag{2.95}$$

where

$$\hat{\lambda}_1 = \frac{\hat{v}}{\omega} \quad \text{Wb} \tag{2.96}$$

In contrast, under steady-state conditions the flux linkage varies sinusoidally with zero average value. That is,

$$\lambda_1 = \int \hat{v} \sin \omega t \, dt = -\frac{\hat{v}}{\omega} \cos \omega t = -\hat{\lambda}_1 \cos \omega t \quad \text{Wb} \tag{2.97}$$

Figure 2.24(a) shows the transient flux linkage variation on the left and the steady-state flux linkage variation on the right. Comparing Eqs. (2.95) and (2.97), it is seen

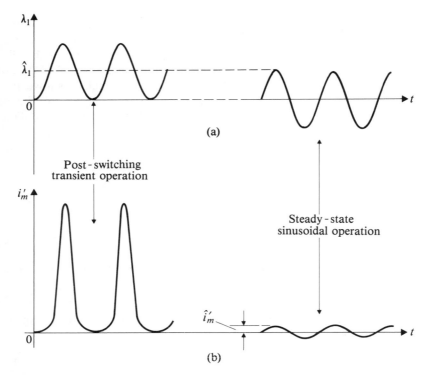

Fig. 2.24 Time variation of λ_1 and i'_m on no load.

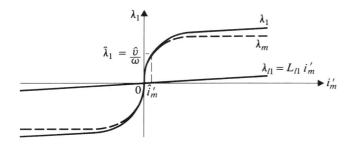

Fig. 2.25 $\lambda - i'_m$ curves for a transformer core.

that the maximum flux linkage during the initial transient, at $\omega t = \pi$, is twice the magnitude of the peak steady-state flux linkage.

The relation between the total flux linkage λ_1 and the current i'_m has the form shown in Fig. 2.25. It is made up of the magnetization characteristic for the core material scaled to give λ_m as a function of i'_m plus the leakage flux linkage $L_{l1}i'_m$. Using this relationship, the waveform of i'_m corresponding to the transient flux linkage of Eq. (2.95) can be found. The result is shown in the early part of Fig. 2.24(b). As indicated previously, the peak value of this current may be many times the rated current of the transformer.

In practice, the effect of the winding resistance R_1 on the initial peak of transient current is small. However, the potential difference across this resistance causes the peaks of the flux-linkage waveform to decrease in magnitude as the transient progresses and results in a gradual transition (lasting a second or two in a large transformer) to the steady-state waveforms of λ_1 and i'_m (shown in the later parts of Fig. 2.24(a) and (b)).

Two factors modify the performance of the transformer described in Figs. 2.23–2.25. One of these is the instant in the cycle of applied potential difference at which switch SW is closed. If the terminal impedance of the linear equivalent circuit shown in Fig. 2.23 is $\overline{Z} = R + jX$, and the switch is closed at the instant when $\omega t = \tan^{-1}(X/R)$, the current i'_e will begin to vary sinusoidally from zero, and the transformer operates immediately in the steady-state condition. Depending on the instant of switching, then, operation immediately after SW is closed may vary between the two extreme conditions shown in the early and late parts of Fig. 2.24(a) and (b).

The second modifying factor is the existence of some residual magnetism in the transformer core. This will give a residual flux linkage λ_r, which should have been included in the solution of Eq. (2.93). Usually λ_r will be small in comparison with $\hat{\lambda}$ except in transformers with no gaps in the cores. Its effect will be that of making the switching transient somewhat greater or less than it would have been if no residual magnetism had been present.

*2.8 POWER TRANSFORMER CONSTRUCTION

The foregoing discussion has shown that power transformers must consist of coils mounted in some way on a laminated core. However, a transformer such as that

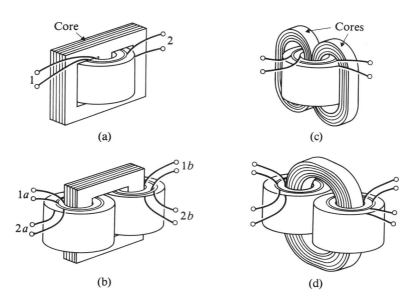

Fig. 2.26 Winding and core arrangements. (a) Shell-type with core of laminated sheets. (b) Core-type with core of laminated sheets. (c) Shell-type with wound core. (d) Core-type with wound core.

illustrated in Fig. 2.2, consisting of two separate coils, one on each of two opposite legs of a square core, would usually be a very unsatisfactory design. Its principal fault is very large leakage flux and hence large regulation. Leakage flux may be greatly reduced by employing the shell-type construction shown in Fig. 2.26(a) and (c) with the two coils wound concentrically. This ensures that almost all of the flux produced by one coil links the other. Another winding arrangement with low leakage consists of interleaved disks of primary and secondary winding placed on top of each other. Alternatively, each winding may be divided into two coils, the four coils then being mounted in the *core-type* construction shown in Fig. 2.26(b) and (d). This arrangement is particularly useful for laboratory purposes, since each pair of coils may be connected in series or in parallel, and four combinations of primary and secondary potential difference may be obtained.

Cores may be built of laminations cut from alloy sheet steel. They are assembled within the ready-wound coils and must be at least partly disassembled if a coil is to be removed for repair. The core for the shell-type transformer of Fig. 2.26(a) is normally made up of E- and I-shapes, as illustrated in Fig. 2.27(a). The positions of the two stampings are interchanged in alternate layers so that the air gaps formed at three places in each layer where the E butts onto the I are bridged by solid iron in the adjacent layers. The laminations for the core-type transformer of Fig. 2.26(b) may be made up of U- and I-shapes as illustrated in Fig. 2.27(b). If grain-oriented stampings are employed, they may have proportions such as are illustrated in Fig.

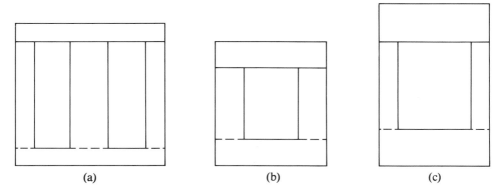

Fig. 2.27 Construction of transformer cores from stampings.

2.27(c). Figures 2.26(c) and (d) show that C-cores may be employed in transformers. Alternatively, the cores of Fig. 2.26(c) and (d) may be wound from a continuous steel strip fed through the coil using special winding machinery.

In small transformers, cores with limbs of square cross section may be used with either square or circular coils fitted on them. In large transformers either of these arrangements is wasteful of copper and increases copper loss. Stepped cores of cross sections like those illustrated in Fig. 2.28 are then employed to embrace the maximum cross-section area of steel with the minimum length of copper conductor.

The heat produced in a transformer must be removed to avoid overheating and consequent breakdown of insulation. Small transformers may be air-cooled, the heat being removed by radiation and convection either immediately from the core

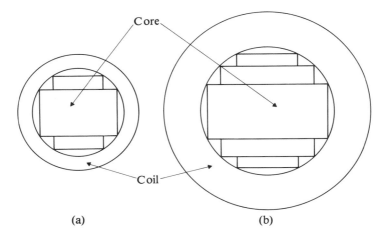

Fig. 2.28 Stepped transformer cores.

Fig. 2.29 25-kVA distribution transformer. *(Moloney Electric Company of Canada, Limited)*

Fig. 2.30 Transformer tanks. *(Moloney Electric Company of Canada, Limited)*

and coil surfaces or from the protective enclosure surrounding them. Larger transformers are immersed in oil tanks, where the oil removes the heat by convection and conveys it to the walls of the tank from which it is transferred to the atmosphere. In very large transformers, radiators and forced circulation of oil in the tank and air outside it are employed. Figure 2.29 shows a 25-kVA distribution transformer with a hand-operated tap-changing switch. It is designed to be enclosed in a tank such as that marked "25" in Fig. 2.30, where three different sizes of tanks are shown.

*2.9 ELEMENTS OF TRANSFORMER DESIGN

An essential part of the design of a transformer consists of the determination of the conductor cross-section area of the two windings A_w and the steel cross section of the core A_i. These areas are determined from estimates of suitable values for the peak flux density $\hat{B}$, the full-load rms current density in the windings J, the winding space factor k_w, and the stacking factor k_i. These last two factors are defined in Eqs. (1.98) and (1.99).

For the primary winding, from Eq. (1.70),

$$V_1 = 2\pi f N_1 \Phi \quad \text{V} \tag{2.98}$$

where V_1 and Φ are the rms values of the primary potential difference and the flux produced in the core. So,

$$V_1 = 2\pi f N_1 B A_i \quad \text{V} \tag{2.99}$$

where B is the rms value of the core flux density. The copper cross-section area in the window is

$$A_w = k_w A_0 = N_1 a_1 + N_2 a_2 \quad \text{m}^2 \tag{2.100}$$

where A_0 is the window area, while a_1 and a_2 are the conductor cross-section areas. Also

$$N_1 a_1 \simeq N_2 a_2 \quad \text{m}^2 \tag{2.101}$$

thus

$$A_w = 2N_1 a_1 \quad \text{m}^2 \tag{2.102}$$

Primary full-load current is

$$I_1 = a_1 J = \frac{A_w}{2N_1} J \quad \text{A} \tag{2.103}$$

The transformer volt-ampere rating is

$$S = V_1 I_1 = 2\pi f B A_i \times \frac{A_w J}{2} \quad \text{VA} \tag{2.104}$$

from which

$$S = \pi f B A_i A_w J \quad \text{VA} \tag{2.105}$$

The frequency f is known and the values of B, k_i, k_w, and J may be estimated from past experience. Thus, Eqs. (2.99) and (2.105) relate three unknown quantities, N_1, A_i, and A_w. A series of values of N_1 could now be taken, a design worked out for each, and the best of the resulting designs chosen from them. This, however, would be a tedious procedure, and it is usual to choose a figure for N_1 on the basis of previous experience of the volts per turn, V_t, normally employed in transformers of the type considered. One relationship embodying such experience is

$$V_t = \frac{\sqrt{S}}{C}, \qquad 20 < C < 70 \quad \text{V} \tag{2.106}$$

where C depends upon the type of construction and intended service of the transformer.

Example 2.7   Determine the core dimensions for a core-type 2-kVA, 220:55-V, 60-Hz transformer of the configuration shown in Fig. 2.26(b). Assume the following parameters:

Permissible peak core flux density, $\hat{B} = 1$ T
Stacking factor, $k_i = 0.9$
Winding space factor, $k_w = 0.25$
Permissible rms current density $J = 2 \times 10^6$ A/m². Volts/turn, $V_t = 1$ V.

The core should be square in cross section. The window should be approximately twice as high as it is wide. The transformer is to be air-cooled by convection.

Solution   The rms flux density $B = 1/\sqrt{2} = 0.707$ T. The low value of winding space factor is to allow ample space for cooling air to circulate round the winding. The rms current density is made low to reduce copper loss in view of the cooling method.

Assume that this is a step-down transformer, so that $V_1 = 220$ V. Then since $V_t = 1$ V,

$$N_1 = 220, \qquad N_2 = 55$$

From Eq. (2.99), the gross core cross-section area is

$$\frac{A_i}{k_i} = \frac{V_1}{2\pi f N_1 \, B k_i} = \frac{220}{2\pi \times 60 \times 220 \times 0.707 \times 0.9} = 0.00417 \text{ m}^2.$$

Length of core side $= (A_i/k_i)^{1/2} = 64.6$ mm. From Eq. (2.105), the window area is

$$\frac{A_w}{k_w} = \frac{S}{\pi f B k_i A_i k_w J} = \frac{2 \times 10^3}{\pi \times 60 \times 0.707 \times 0.9 \times 0.00417 \times 0.25 \times 2 \times 10^6}$$
$$= 0.00798 \text{ m}^2$$

Let $d =$ window width; then

$$\frac{A_w}{k_w} = 2d^2 = 0.00798$$

from which

$$d = 63.2 \text{ mm}$$
$$\text{Window height} = 2d = 126.4 \text{ mm}$$

*2.10 EQUIVALENT CIRCUITS FOR MORE COMPLEX MAGNETIC SYSTEMS

In Sections 1.7 and 1.8 it has been shown that complicated magnetic systems may be represented and analyzed by means of magnetic equivalent circuits. However, if the winding or windings on the magnetic system are connected to other electric elements, it is desirable to have an electric equivalent circuit for the electromagnetic system from which the relationships between the electric circuit variables can be obtained directly. Such electric equivalent circuits have been obtained for an inductor in Section 1.16 and for a primitive transformer in Section 2.2. In this section it is shown how electric equivalent circuits may be obtained for more realistic transformer configurations, such as are illustrated in Fig. 2.26, and indeed for any other more complicated electromagnetic system.

*2.10.1 Derivation of Magnetic Equivalent Circuits

A magnetic equivalent circuit will first be derived for a simple magnetic system. Figure 2.31(a) shows a torus of magnetic material with a winding of N turns. The magnetomotive force around the magnetic path is

$$\mathscr{F} = Ni \quad \text{A} \tag{2.107}$$

This magnetomotive force establishes a magnetic field intensity H, which in turn produces a magnetic flux density B in the material. Integration of this flux density over the cross-sectional area A of the core gives the magnetic flux ϕ. The cause-effect relationship between the magnetomotive force $\mathscr{F}$ and the magnetic flux ϕ may be expressed symbolically by the equivalent magnetic circuit of Fig. 2.31(b).

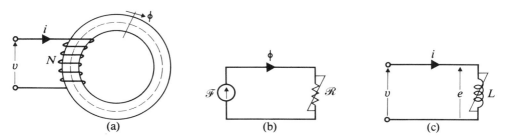

Fig. 2.31 (a) A simple magnetic element. (b) A magnetic equivalent circuit for the element. (c) An electric equivalent circuit for the element.

The magnetic properties of the material and the dimensions of the core determine its reluctance $\mathcal{R}$. Under the idealized condition where the relative permeability can be regarded as constant, that is,

$$B = \mu_r\mu_0 H \quad \text{T} \tag{2.108}$$

the reluctance can be expressed as

$$\mathcal{R} = \frac{\mathcal{F}}{\phi} = \frac{l}{\mu_r\mu_0 A} \quad \text{A/Wb} \tag{2.109}$$

In general, the reluctance of a ferromagnetic core is nonlinear. The reluctance symbol is then used in a magnetic equivalent circuit merely to denote a magnetic element for which an $\mathcal{F}$–ϕ relation exists. This relation may be obtained by rescaling the B–H characteristic for the material, using the expressions

$$\phi = BA \quad \text{Wb} \tag{2.110}$$

and

$$\mathcal{F} = Hl \quad \text{A} \tag{2.111}$$

For purposes of analysis, approximate models for the B–H characteristic may be used.

Consider now, as an example, the magnetic system of Fig. 2.32(a). This system consists of a three-legged magnetic core, two of the legs having windings and the third leg having an air gap. Basically, this is a complex three-dimensional magnetic field problem. If simplifying assumptions are used, the magnetic field can be reduced to a magnetic circuit of lumped reluctances. It will be assumed that, except in the air gap, all magnetic flux is confined to the magnetic material. The leakage flux in the air paths around the windings is considered negligible.

The magnetic system may now be divided into four sections, each of which has a uniform flux over its length. Three of these sections represent magnetic paths in the material, and the fourth represents the air-gap path. Each section may be rep-

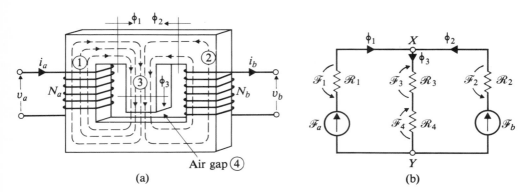

Fig. 2.32 (a) A magnetic system. (b) A magnetic circuit for the system.

resented by a reluctance that relates the flux to the magnetomotive force required to establish that flux along the length of the section.

Figure 2.32(b) shows the magnetic equivalent circuit that results from the foregoing assumptions. Reluctances $\mathcal{R}_1$, $\mathcal{R}_2$, and $\mathcal{R}_3$ represent the three paths in the magnetic material carrying magnetic fluxes ϕ_1, ϕ_2, and ϕ_3, respectively. The air gap is represented by the linear reluctance $\mathcal{R}_4$, and its magnetic flux is ϕ_3. The circuital law of Eq. (1.1) applies to any closed path in a magnetic-field system. In a magnetic equivalent circuit, this law is represented by the following relation: Around any closed path, the total magnetomotive force of the windings is equal to the sum of the products of reluctance and flux.

$$\sum \mathcal{F} = \sum \mathcal{R} \, \phi_{\text{around closed path}} \tag{2.112}$$

The continuity of magnetic flux in the magnetic field is represented by equating the sum of the fluxes entering any junction of magnetic paths in the equivalent circuit to zero:

$$\sum \phi_{\text{into junction}} = 0 \tag{2.113}$$

Normal methods of circuit analysis may now be employed to determine the fluxes in Fig. 2.32(b) for a given set of magnetomotive forces. If the relative permeability of the magnetic material can be considered constant, the reluctance of the sections may be determined by use of Eq. (2.109), in which l is the mean length of the flux path in each section and A is the cross-sectional area. If the permeability cannot be considered constant, each reluctance element may be represented by a graph of the relation between its flux and its magnetomotive force. Graphical or trial-and-error methods may then be used for analysis of the equivalent circuit.

It should be noted that all the assumptions are introduced in the process of deriving the magnetic equivalent circuit from the magnetic system. With different assumptions, a different equivalent circuit is obtained. For example, if the leakage fluxes in the air paths around the windings in Fig. 2.32(a) had not been neglected, the reluctances of these air leakage paths would have been connected across the respective magnetomotive forces in Fig. 2.32(b). There is therefore no unique equivalent magnetic circuit for a magnetic system. The chosen circuit should contain just the information required for the solution of the problem to the desired degree of accuracy.

*2.10.2 Derivation of Electric Equivalent Circuits

First, consider the simple magnetic circuit of Fig. 2.31(b), which relates two variables—the coil magnetomotive force $\mathcal{F}$ and the flux ϕ—by reluctance parameter $\mathcal{R}$.

$$\mathcal{F} = \mathcal{R}\phi \quad \text{A} \tag{2.114}$$

In the equivalent electric circuit, the variables are the potential difference v between the coil terminals and the current i in the coil. Constant core reluctance and negligible coil resistance are assumed. The electric circuit variables are related to the magnetic circuit variables by the two relations:

$$i = \frac{\mathscr{F}}{N} \quad \text{A} \tag{2.115}$$

and

$$e = N \frac{d\phi}{dt} \quad \text{V} \tag{2.116}$$

By substituting from Eqs. (2.114) and (2.115) into Eq. (2.116), the relation between the two electric circuit variables is

$$V = N \frac{d}{dt}\left[\frac{\mathscr{F}}{\mathscr{R}}\right] = \frac{N}{\mathscr{R}} \frac{d\mathscr{F}}{dt} = \frac{N^2}{\mathscr{R}} \frac{di}{dt} = L \frac{di}{dt} \quad \text{V} \tag{2.117}$$

Thus, the reluctance parameter $\mathscr{R}$ in the magnetic circuit is replaced by an inductance parameter L in the electric equivalent circuit of Fig. 2.31(c). The value of the inductance is inversely proportional to the value of the reluctance.

If the reluctance parameter is not constant, it can be represented by a curve relating magnetomotive force and flux. The corresponding nonlinear inductance can be repesented by a curve relating the core flux linkage $\lambda = N\phi$ to the current i. Approximations may be used for this relation where appropriate.

Now consider the more complex magnetic circuit of Fig. 2.32(b). Suppose for the present that each of the fluxes ϕ_1, ϕ_2, and ϕ_3 links an N-turn coil. The corresponding emf's e_1, e_2, and e_3 induced in these coils are given by

$$e_1 = N \frac{d\phi_1}{dt} \qquad e_2 = N \frac{d\phi_2}{dt} \qquad e_3 = N \frac{d\phi_3}{dt} \quad \text{V} \tag{2.118}$$

At node X in Fig. 2.32(b), the flux variables are related, from Eq. (2.113), by

$$\sum \phi_{\text{into node}} = \phi_1 + \phi_2 - \phi_3 = 0 \quad \text{Wb} \tag{2.119}$$

The flux variables at node Y are related by the same equation. From Eq. (2.118), the corresponding emf variables must be related by the expression

$$e_1 + e_2 - e_3 = 0 \quad \text{V} \tag{2.120}$$

For the left-hand mesh of the magnetic circuit of Fig. 2.32(b), from Eq. (2.112), the magnetomotive force variables are related by

$$\mathscr{F}_a = \mathscr{F}_1 + \mathscr{F}_3 + \quad_4 \quad \text{A} \tag{2.121}$$

Consider each of these magnetomotive force components to be produced by corresponding components of current in N-turn coils. These current components are then related by

$$i'_a = i_1 + i_3 + i_4 \quad \text{A} \tag{2.122}$$

Around the right-hand mesh of Fig. 2.32(b) the magnetomotive force relation is

$$\mathscr{F}_b = \mathscr{F}_2 + \mathscr{F}_3 + \mathscr{F}_4 \quad \text{A} \tag{2.123}$$

The relation between the corresponding current variables is

$$i'_b = i_2 + i_3 + i_4 \quad A \tag{2.124}$$

[The prime symbols are added to i'_a and i'_b to distinguish them from i_a and i_b in Fig. 2.32(a).]

Each reluctance in the magnetic circuit relates a magnetic flux variable ϕ and a magnetomotive force variable $\mathscr{F}$. From Eqs. (2.115)–(2.117), the corresponding emf variable e and current variable i can be related by an inductance parameter. For example, the relation $\mathscr{F}_1 = \mathscr{R}_1\phi_1$ in the magnetic circuit corresponds to the relation

$$e_1 = L_1 \frac{di_1}{dt} \quad V \tag{2.125}$$

Equations (2.120), (2.122), (2.124), and (2.125) describe the electric equivalent circuit shown in Fig. 2.33(a). For each of the two independent meshes in the mag-

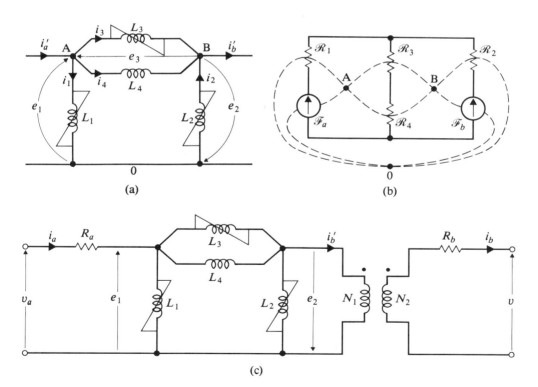

Fig. 2.33 Equivalent circuit for the magnetic circuit of Fig. 2.32. (a) Elementary form of circuit. (b) Topological technique of derivation. (c) Circuit including ideal transformer and winding resistances.

netic circuit, there is an independent node in the electric circuit. The branch currents entering these two nodes (designated as A and B) are related by Eqs. (2.122) and (2.124). For each independent node in the magnetic circuit there is a corresponding mesh in the electric circuit. The branch emf's around the one independent mesh in the electric circuit are related by Eq. (2.120). For each reluctance branch in the magnetic circuit there is a corresponding inductance branch in the electric circuit. For each magnetomotive force source in the magnetic circuit there is a corresponding coil current in the electric circuit.

The form of the electric circuit of Fig. 2.33(a) may be derived directly from the magnetic circuit of Fig. 2.32(b) by the topological principle of duality. This topological technique is demonstrated in Fig. 2.33(b). A node is marked within each mesh of the magnetic circuit, and a reference node is marked outside the circuit. These nodes are then joined by branches, one of which passes through each element of the magnetic circuit. It is observed that the form of the resulting network of branches is identical to the form of the electric circuit of Fig. 2.33(a). For each reluctance in a mesh of the magnetic circuit, there is an inductance connected to the corresponding node of the electric circuit. Where a reluctance is common to two meshes in the magnetic circuit, the corresponding inductance connects the corresponding nodes of the electric circuit. For each magnetomotive force there is a corresponding driving current; for each flux there is a corresponding emf between nodes.

The reluctances corresponding to ferromagnetic parts of the magnetic system may represent nonlinear relationships between their fluxes and magnetomotive forces. Each reluctance element in the magnetic circuit has a corresponding inductance element in the electric circuit. Thus each inductance element may represent a similar nonlinear relationship between the flux linkage λ in a coil of N turns encircling the particular branch of the magnetic system and the current i in an N-turn coil that produces the magnetomotive force for the branch. The rate of change of flux linkage produces the emf variable in the electric circuit. The nonlinearities in the magnetic circuit are therefore preserved in the electric equivalent circuit.

Various nonlinear models may be used to represent these nonlinear inductance elements. The choice depends on the problem under study. For use with alternating currents each nonlinear inductance element can be represented by a circuit model of the form shown in Fig. 1.29. Both R_c and L_m are generally nonlinear functions of the emf induced in inductance L_m.

The electric equivalent circuit of Fig. 2.33(a) was developed with the assumption that all windings had N turns. Since the numbers of turns are generally different for the various windings, it is necessary to add ideal transformers at the terminals of the equivalent electric circuit to obtain the actual induced emf's and the actual currents in the windings. The reference number of turns N is normally made equal to the number of turns in one of the windings; no ideal transformer is then required for this winding. In Fig. 2.33(c), N has been made equal to N_a.

The resistances R_a and R_b of the two windings have also been added to the equivalent circuit in Fig. 2.33(c). The terminal potential differences of the two windings are v_a and v_b; the induced emf's in the windings are e_1 and $-e_2(N_2/N_1)$.

When an appropriate equivalent circuit has been developed for a magnetic device, the performance of the device may be predicted by use of the techniques employed in circuit analysis. When convenient, elements may be transformed across the ideal transformers, as described in Section 2.1.

*2.10.3 Equivalent Circuit for Two-Winding Shell-Type Transformer

The principles developed and illustrated so far in Section 2.10 may now be applied to the two-winding shell-type transformer, a cross section of which is shown in Fig. 2.34(a). The first step in producing an equivalent electric circuit is to reduce the significant features of the magnetic system to a magnetic equivalent circuit. The pertinent assumptions are introduced at this stage.

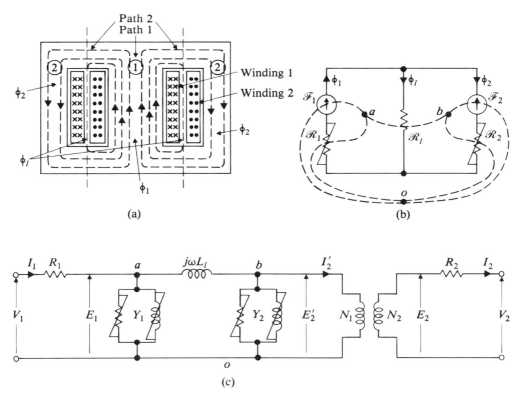

Fig. 2.34 (a) A two-winding, shell-type transformer. (b) Magnetic circuit. (c) Equivalent electric circuit.

The flux pattern in Fig. 2.34(a) occurs when the magnetomotive force $\mathscr{F}_1$ of the inner winding 1 is slightly greater than the magnetomotive force $\mathscr{F}_2$ of the outer winding 2. A magnetic flux ϕ_1 passes upward through the central leg and divides into two equal parts that proceed outward along the yokes. A part, ϕ_l, of this flux takes the path down through the air space between the windings to avoid the oppositely directed magnetomotive force of winding 2. The remainder, ϕ_2, passes down the outer legs and thereby links winding 2. This flux pattern suggests that the ferromagnetic core consists of two significant paths, each of which carries a distinctive magnetic flux. Path 1, carrying the flux ϕ_1, includes the central leg and about half of each of the yoke sections. Path 2, carrying the total flux ϕ_2, consists of the outer legs and the remainder of the yoke sections. This second path consists of two physically separate parts; but, because of the symmetry, they may be regarded as being, in parallel magnetically.

Fig. 2.34(b) shows, in solid line, a magnetic equivalent circuit in which the assumed paths of flux are represented by their reluctances. In this circuit, winding 1 (represented by its magnetomotive force $\mathscr{F}_1$) is linked by the flux ϕ_1, which exists in the reluctance $\mathscr{R}_1$ of path 1. This reluctance may be dependent on the value of the flux ϕ_1 and is therefore indicated as nonlinear in the magnetic equivalent circuit. The magnetic flux ϕ_l is established in the air space between the windings. As this path is assumed to be entirely in the air space, its reluctance is constant. The flux $\phi_2 = \phi_1 - \phi_l$ is established in the reluctance $\mathscr{R}_2$ of the combined parts of path 2. This flux, ϕ_2, links winding 2, which is represented by its magnetomotive force $\mathscr{F}_2$. Since path 2 is ferromagnetic, its reluctance is regarded as nonlinear.

In developing an electric circuit that is equivalent to the magnetic circuit of Fig. 2.34(b), the number N_1 of the turns of winding 1 may be used as a reference number. The form of the electric circuit may be derived by use of the simple topological technique described in Section 2.10.2. The result is the set of dashed lines joining the nodes a, b, and o in Fig. 2.34(b).

Suppose the transformer is to operate at an angular frequency of ω radians per second. The equivalent electric circuit of Fig. 2.34(c) includes an inductive impedance $j\omega L_l$ connected between nodes a and b. The value of the leakage inductance L_l is related to the leakage reluctance by

$$L_l = \frac{N_1^2}{\mathscr{R}_l} \quad \text{H} \qquad (2.126)$$

(The reference winding has N_1 turns.) The equivalent circuit includes two magnetizing branches representing the reluctances $\mathscr{R}_1$ and $\mathscr{R}_2$ of the magnetic circuit. If these reluctances could be considered constant, the corresponding magnetizing branches would consist of constant inductive impedances. To include the effects of nonlinearity, hysteresis, and eddy-current loss in paths 1 and 2, each branch of the electric circuit corresponding to a nonlinear reluctance is represented by parallel nonlinear resistive and inductive impedances. For convenience in notation, these nonlinear branches are identified by their admittances, Y_1 and Y_2.

The induced emf phasor, $\overline{E}_1$ in winding 1 is given by

$$\overline{E}_1 = j\omega N_1 \overline{\Phi}_1 \quad \text{V} \tag{2.127}$$

where $\overline{\Phi}_1$ is the phasor representing the magnetic flux ϕ_1. The terminal potential difference V_1 differs from the induced emf by the potential difference across the winding resistance R_1. The emf $\overline{E}'_2$ is given by

$$\overline{E}'_2 = j\omega N_1 \overline{\Phi}_2 \quad \text{V} \tag{2.128}$$

This is the induced emf that would apply for winding 2 if $N_2 = N_1$. The actual induced emf in winding 2 is $\overline{E}_2$, which appears on the other side of the ideal transformer of ratio $N_1 : N_2$. Thus

$$\overline{E}_2 = \frac{N_2}{N_1} \overline{E}'_2 = j\omega N_2 \overline{\Phi}_2 \quad \text{V} \tag{2.129}$$

Winding 2 has a resistance R_2 and a current phasor $\overline{I}_2$. If winding 2 were replaced by a winding of N_1 turns, its current would have the value noted on the other side of the ideal transformer,

$$\overline{I}'_2 = \frac{N_2}{N_1} \overline{I}_2 \quad \text{A} \tag{2.130}$$

The emf E_{ab} in Fig. 2.34(c) is equal to the induced emf in a winding of N_1 turns caused by the rate of change of the leakage flux ϕ_l.

Note that the equivalent circuit of Fig. 2.34(c) is not the only one that could have been developed for this transformer. In fact, the leakage paths are not as simple as indicated in Fig. 2.34(a). Some of the leakage flux links only a part of each winding and passes through parts of the core legs. To represent the increased complexity in flux paths, more complex equivalent magnetic and electric circuits are required. Fortunately, such complex models are seldom needed. The circuit may, in fact, be simplified by assuming that the difference between the fluxes ϕ_1 and ϕ_2 is so small that a single reluctance $\mathcal{R}_1 + \mathcal{R}_2$ may be used to represent the whole magnetic core. In the electric equivalent circuit, this has the effect of paralleling the admittances Y_1 and Y_2. This resultant admittance may be connected between nodes a and 0 if the whole core is assumed to carry the flux ϕ_1. This gives the equivalent circuit of Fig. 2.35(a) on page 138. In some cases, the core reluctances are so small that Y_1 and Y_2 may be omitted, leaving the approximate equivalent circuit of Fig. 2.35(b), which is also that of Fig. 2.21(c).

*2.11 VARIABLE-FREQUENCY OPERATION

In electronic and communication systems, transformers are often used over a wide range of frequencies. An example is the transformer used in an audio amplifier to couple the output stage to the speaker. Ideally, this audio transformer should operate well over the audible frequency range of about 20–20,000 hertz. A pulse transformer is another example. A train of pulses may be represented as a Fourier series

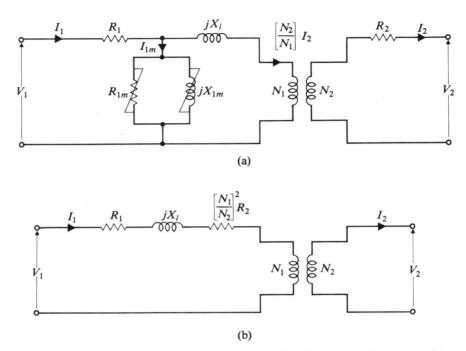

Fig. 2.35 (a) Equivalent circuit for a transformer. (b) Circuit with negligible magnetizing current.

of frequency components. A further example is a variable frequency source containing a transformer and supplying power to a variable speed motor.

The relationship between the potential ratio of a transformer and frequency can be found by using a general equivalent circuit such as that of Fig. 2.16. In certain frequency ranges several of the parameters are negligible, and the analysis may be simplified. At high frequencies, the effects of interwinding capacitances may have to be included to obtain an adequate analysis.

Figure 2.36(a) shows an equivalent circuit that is applicable over the middle of the frequency range for which the transformer is designed. In this range, the leakage reactance and the magnetizing current generally can be neglected. In small transformers, the winding resistances may be significant, while in large transformers they may be negligible. In Fig. 2.36, the source (for example, an electronic amplifier) is represented as a variable-frequency source E_s in series with an internal resistance R_s. The load (for example, a speaker) is represented by its resistance R_L. In the mid-frequency range, the ratio of the load potential difference to the source emf is given approximately by

$$\frac{V_L}{E_s} = \frac{nR_L}{R_{11} + R_{22}} \tag{2.131}$$

where $R_{11} = R_1 + R_s$, $R_{22} = n^2(R_2 + R_L)$, and n is the turns ratio N_1/N_2.

To obtain maximum power input to the load, the transformer parameters should be chosen so that the load resistance R_L is equal to the magnitude of the equivalent Thévenin impedance as seen from the load terminals. In the mid-frequency range, the load resistance for maximum load power is given by

$$R_L = \frac{R_s + R_1}{n^2} + R_2 \quad \Omega \tag{2.132}$$

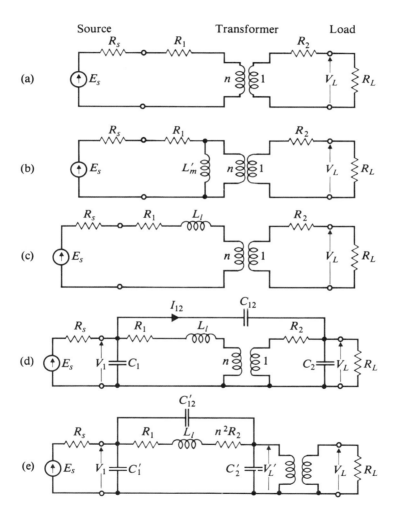

Fig. 2.36 Equivalent circuits for variable-frequency transformer. (a) Mid-frequency range. (b) Low-frequency range. (c) High-frequency range including interwinding capacitances. (e) Circuit equivalent of (d).

Under this condition,

$$\frac{V_L}{E_s} = \frac{1}{2n} \qquad (2.133)$$

As the frequency decreases, the magnetizing current of the transformer becomes significant, and the circuit of Fig. 2.36(b) applies. At these low frequencies, core loss normally may be neglected. If the magnetizing inductance L'_m is regarded as constant, the ratio of load potential difference to source emf can be expressed approximately as

$$\frac{\overline{V_L}}{\overline{E_s}} = \frac{nR_L}{R_{11} + R_{22}}\left(\frac{1}{1 - jR_q/\omega L'_m}\right) \qquad (2.134)$$

where $R_q = R_{11}R_{22}/(R_{11} + R_{22})$. At the frequency $\omega_l = R_q/L'_m$ radians/second, the magnitude of the V_L/E_s ratio is $1/\sqrt{2}$ of its value in the mid-frequency range, and the load power for a given source potential is reduced to one-half.

In using this low-frequency equivalent circuit, note that excessive magnetizing current drives the core of the transformer into saturation and thus reduces the effective value of the magnetizing inductance L'_m. Equation (2.127) applies only if the source emf is low enough to allow the magnetizing inductance to be regarded as constant.

In the high-frequency range the leakage inductance of the transformer becomes significant and the equivalent circuit of Fig. 2.36(c) is applicable. The load-to-source ratio for this range may be expressed as

$$\frac{\overline{V_L}}{\overline{E_s}} = \frac{nR_L}{R_{11} + R_{22}}\left[\frac{1}{1 + j\ L_l/(R_{11} + R_{22})}\right] \qquad (2.135)$$

At the frequency $\omega_h = (R_{11} + R_{22})L_l$ radians/second, the magnitude of the ratio V_L/E_s is again $1/\sqrt{2}$ of its value in the mid-frequency range, and the power is reduced to one-half.

Figure 2.37 shows the ratio V_L/E_s relative to its value in the mid-frequency range (Eq. (2.133)) as a function of frequency plotted on a logarithmic scale. The

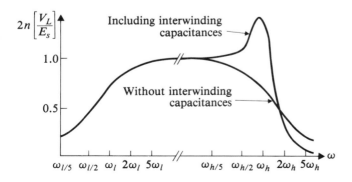

Fig. 2.37 The normalized ratio of a transformer as a function of frequency ω.

width of the transformer is defined as the frequency range from ω_l to ω_h—that is, between the frequencies at which the load potential difference is reduced to 0.707 and the load power is reduced to 0.5 of the values obtained in the midfrequency range.

In the high-frequency range, the effects of distributed capacitances across and between the windings may become significant. Figure 2.36(d) shows an equivalent circuit in which these distributed capacitances are represented approximately by capacitances C_1 and C_2 across the transformer terminals, and a capacitance C_{12} between the primary and secondary terminals. The values of these capacitances depend on the arrangement of the windings. For example, if the primary and secondary windings are interleaved to reduce the leakage inductance, the capacitances are increased. The capacitances also depend on the connections made between the windings; their effects can be minimized by connecting a primary and a secondary terminal of like polarity.

Analysis of the equivalent circuit of Fig. 2.36(d) is facilitated if all parameters are referred to one side of the ideal transformer, as shown in Fig. 2.36(e). As seen from the primary winding, the current I_{12} in the interwinding capacitance can be expressed as

$$\bar{I}_{12} = j\omega C_{12}(\bar{V}_1 - \bar{V}_L) = j\omega C_{12}\left(\bar{V}_1 - \frac{\bar{V}_L'}{n}\right)$$

$$= j\omega \frac{C_{12}}{n}(\bar{V}_1 - \bar{V}_L) + j\omega C_{12}\left(\frac{n - 1}{n}\right) \bar{V}_1 \quad \text{A} \qquad (2.136)$$

If the interwinding capacitance current I_{12} entering the secondary terminal is referred to the other side of the ideal transformer, it can be expressed as

$$\frac{\bar{I}_{12}}{n} = j\omega \frac{C_{12}}{n}(\bar{V}_1 - \bar{V}_L) = j\omega \frac{C_{12}}{n}(\bar{V}_1 - \bar{V}_L') - j\omega C_{12}\left(\frac{1 - n}{n^2}\right) \bar{V}_L' \quad \text{A} \qquad (2.137)$$

From these equations, the equivalent capacitances of Fig. 2.36(e) are

$$C_{12}' = \frac{C_{12}}{n} \quad \text{F}$$

$$C_1' = C_1 + C_{12}\left(\frac{n - 1}{n}\right) \quad \text{F}$$

$$C_2' = \frac{C_2}{n^2} + C_{12}\left(\frac{1 - n}{n^2}\right) \quad \text{F} \qquad (2.138)$$

The effects of the winding capacitances on the ratio V_L/E_s are most pronounced in a step-up transformer where $n < 1$. Figure 2.37 shows a typical effect of interwinding capacitance on the frequency response of a transformer. The ratio can be increased substantially above the midfrequency value if the capacitance C_2' in Fig. 2.36(e) is in series resonance with the parallel branches containing L_l and C_{12}'. The potential ratio can be reduced substantially in the higher frequency range of parallel resonance between L_l and C'_{12}.

The equivalent circuit of Fig. 2.36(e) may be used in the analysis of some transient phenomena in transformers; for example, it may be used in an approximate analysis of pulse transformers. The accurate analysis of transient surge potentials in high-potential transformers often requires a more elaborate circuit model in which the distributed nature of the leakage inductance and interwinding capacitances is represented.

Example 2.8   An audio transformer has its primary winding connected to a source and its secondary winding connected to a speaker that has a resistive impedance of 8 ohms over its usable frequency range. The transformer can be represented by the equivalent circuit of Fig. 2.11 with the following parameters:

$$R_1 = 4 \ \Omega \qquad N_1/N_2 = 3.5$$
$$L_{l1} = 0.4 \ \text{mH} \qquad L_{l2} = 45 \ \ \mu\text{H}$$
$$L'_m = 35 \ \text{mH} \qquad R_2 = 0.4 \ \ \Omega$$

a) Determine the source terminal potential difference required to supply a power of 60 watts to the speaker in the midfrequency range.

b) Assume the same source potential difference is maintained over the whole frequency range. Determine the frequencies at which the speaker power will be reduced to 30 watts.

Solution

a) In the midfrequency range, the transformer can be represented by Fig. 2.36(a) with $R_s = 0$. The load power is given by

$$P_L = V_L^2/R_L$$

Thus

$$V_L = (60 \times 8)^{1/2} = 21.9 \quad \text{V}.$$

The source emf is found from Eq. (2.131) as

$$E_s = \left(\frac{n^2 R_L + n^2 R_2 + R_1}{n^2 R_L} \right) n V_L$$

$$= \left(\frac{3.5^2 \times 8 + 3.5^2 \times 0.4 + 4}{3.5^2 \times 8} \right) 3.5 \times 21.9 = 83.6 \ \text{V}$$

b) From Eq. (2.134), the low half-power frequency is $\omega_l = R_q/L'_m$, where

$$R_q = \frac{R_1 \times n^2(R_r + R_L)}{R_1 + n^2(R_2 + R_L)} = \frac{4 \times 3.5^2(0.4 + 8)}{4 + 3.5^2(0.4 + 8)} = 3.85 \quad \Omega$$

Then

$$\omega_l = \frac{3.85}{.035} = 110 \quad \text{rad/s}$$

and

$$f_l = 110/2\pi = 17.5 \quad \text{Hz}$$

From Eq. (2.135), the high half-power frequency is

$$\omega_h = \frac{R_1 + n^2(R_2 + R_L)}{L_{l1} + N^2 L_{l2}}$$

$$= \frac{4 + 3.5^2(0.4 + 8)}{(400 + 3.5^2 \times 45)10^{-6}} = 112{,}400 \text{ rad/s}$$

and

$$f_h = \frac{122{,}400}{2\pi} = 17{,}885 \quad \text{Hz}$$

2.12 TRANSFORMERS FOR POLYPHASE SYSTEMS

Most electrical energy is generated and transmitted using a three-phase system. The 3-phase power may be transformed either by use of suitably connected banks of single-phase transformers or by use of special polyphase transformers. Transformers may also be employed to change the number of phases in a polyphase system. The arrangements for such phase transformations are many and various, and it is possible to mention here only those most commonly encountered.

2.12.1 Connection of Transformers for Three-Phase Systems

Three similar single-phase transformers may be connected to give 3-phase transformation, and since the primary and secondary windings may be connected either in delta or in wye, there are four possible combinations of connections: wye–delta, delta–wye, delta–delta, and wye–wye.

As an example, consider the wye–delta connection shown in Fig. 2.38(a). The transformers are assumed to be ideal. If the primary line-to-line potential difference is V, the primary line-to-neutral potential difference is $V/\sqrt{3}$. If the turns ratio $N_1/N_2 = n$, then the line-to-line potential difference on the secondary side is $V/n\sqrt{3}$. If the winding current on the primary side is I, which is also the line current, then the winding current on the secondary side is nI. The line current on the delta-connected secondary side is therefore $n\sqrt{3}I$. A more convenient method of drawing the circuit is illustrated in Fig. 2.38(b). Other combinations of connections can be similarly represented.

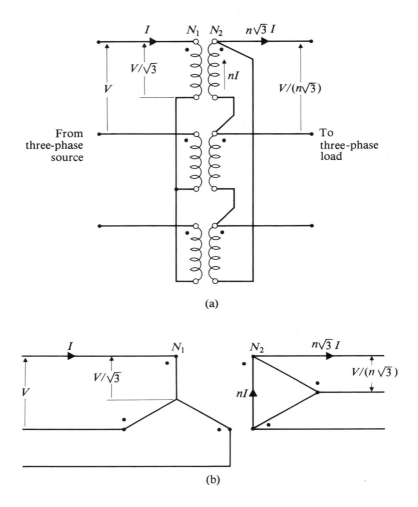

Fig. 2.38 Three-phase wye–delta transformer bank.

If source and load circuit are balanced and the three transformers are for practical purposes identical, then primary and secondary potential differences and currents will be balanced. Only the phase displacements of 120° distinguish the current and potential difference in one phase from those in the others. This means that the variables on the two sides of the transformer bank may be determined by analysis of one phase only. For this purpose it is convenient to consider all sources, transformer windings, and load circuits as if they were wye-connected. A single phase on each side of the bank is then easily isolated.

Figure 2.39(a) shows a delta–wye bank of transformers supplying a delta-connected load circuit. To isolate one phase it is now necessary to replace the real transformers by an equivalent wye–wye bank and the load by an equivalent wye-connected load. The required load circuit is obtained by the well known wye–delta transformation illustrated in Fig. 2.39(b). The secondary potential differences and currents may be determined in terms of the corresponding primary quantities and are shown in Fig. 2.39(a). The turns ratio of the transformers for the equivalent wye–wye transformer bank are now required. These may be determined from the

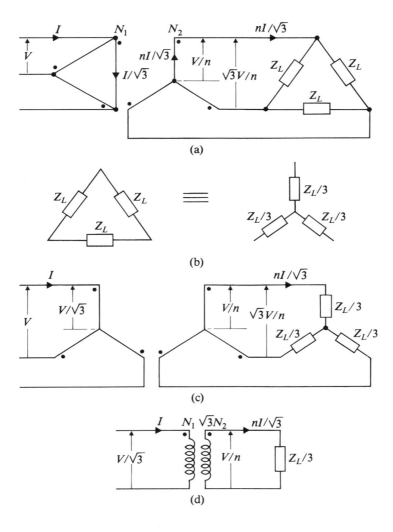

Fig. 2.39 Delta–wye transformer bank. (a) Transformer and load connection. (b) Wye–delta transformation. (c) Equivalent wye–wye bank. (d) Per-phase equivalent circuit.

fictitious circuit in Fig. 2.39(c), where the primary and secondary line currents and line-to-line potential differences must be identical with those in Fig. 2.39(a). From Fig. 2.39(c), the turns ratio n' of the fictitious transformers is seen to be

$$n' = \frac{V/\sqrt{3}}{V/n} = \frac{n}{\sqrt{3}} = \frac{1}{\sqrt{3}}\frac{N_1}{N_2} \tag{2.139}$$

and this is simply the ratio of the line-to-line potential differences on the two sides of the actual transformer bank. One phase of the fictitious circuit may now be isolated for analysis and is shown in Fig. 2.39(d). It should be noted that the phase angle of the secondary potential is not the same as that of the corresponding phase of the primary. This phase angle information is often irrelevant but is of importance in paralleling transformer banks.

One of the advantages of the delta–delta connection is that if one transformer breaks down it may be removed from the circuit and reduced output supplied by the remaining two. This is called the *open-delta connection*. Under these circumstances, the transformer impedances introduce a very slight imbalance into the system. This connection is also useful in a situation in which a growing load is anticipated: a three-phase transformer bank may be designed to deal with the anticipated maximum load, but only two units may be purchased initially. These two would be capable of supplying about 58% of the anticipated maximum load without overheating. Such an arrangement is not economical as a permanent installation, since 67% of the capacity has been installed and is therefore being under-utilized until the three-unit bank is completed.

A wye connection of transformer windings is used when it is desirable to relate all the line potentials to earth potential by grounding the neutral point. This arrangement is used almost universally on the high-potential windings of large power transformers so that the insulation on the transmission line will not be subjected to a steady-state potential difference greater than the line-to-neutral value. The arrangement is also used in many distribution transformers on the low-potential side so that single-phase supplies with one side grounded can be made available.

Example 2.9   Three 10-kVA, 1330:230-V, 60-Hz transformers are connected wye–delta to supply at 230 volts line-to-line a heating load of 2 kW per phase and a three-phase induction-motor load of 21 kVA. The power factor of the induction-motor load is 0.8. Determine the line current supplying the transformers, which may be considered ideal.

Solution   The 3-phase power triangle for the motor load is shown in full line in Fig. 2.40, and has $S_m = 21$ kVA, $\theta_m = \cos^{-1} 0.8$.

The complete power triangle for the heating and motor load combined is shown in broken line in Fig. 2.40, where the heating power component $P_h = 6$ kW has been added. The total active power delivered by the transformers is

$$P = 3 \times 2 + 21 \times 0.8 = 22.8 \quad \text{kW}$$

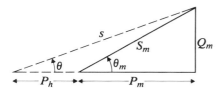

Fig. 2.40 Power triangles for Example 2.9.

The total reactive power is that due to the motor load and is

$$Q_m = 21(1 - 0.8^2)^{1/2} = 12.6 \quad \text{kVAR}$$

The total apparent power delivered by the transformers is therefore

$$S = (P^2 + Q_m^2)^{1/2} = (22.8^2 + 12.6^2)^{1/2} = 26.0 \quad \text{kVA}$$

Since the transformers are considered to be ideal, S is also the input to the primary side of the bank. If the line-to-line potential difference is

$$V = 1330\sqrt{3} \quad \text{V}$$

then

$$\sqrt{3}VI = 26 \times 10^3 \quad \text{VA}$$

from which

$$I = \frac{26 \times 10^3}{3 \times 1330} = 6.52 \quad \text{A}$$

Example 2.10 In the system of the preceding example, the loads are connected to the transformers by means of a common three-phase feeder whose impedance is $0.003 + j0.010 \ \Omega$ per phase. The transformers themselves are supplied from a constant-potential source by means of a three-phase feeder whose impedance is $0.75 + j5.0 \ \Omega$ per phase. The equivalent impedance of one transformer referred to the low-potential side is $0.118 + j0.238 \ \Omega$. Determine the required source potential difference if that at the load is to be 230 V.

Solution Since the secondaries of the transformer bank are connected in delta, it is convenient to refer their equivalent impedances to the primary wye-connected side before determining the equivalent wye–wye bank of ideal transformers. Thus,

$$R'_{eq} + jX'_{eq} = \left[\frac{1330}{230}\right]^2 (0.118 + j0.238) = 3.94 + j7.96 \quad \Omega$$

The required turns ratio of the ideal transformers in the equivalent wye–wye bank is given by the ratio of the line-to-line potential differences of the ideal wye–delta bank:

$$n' = \frac{1330\sqrt{3}}{230} = 10$$

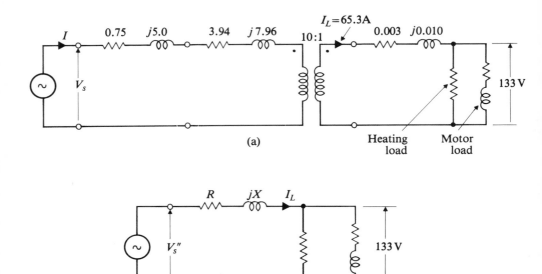

Fig. 2.41 Per-phase equivalent circuit of Example 2.10 circuit.

The per-phase equivalent circuit of the system is therefore that shown in Fig. 2.41(a). The quantities on the primary side of the ideal transformer may now be referred to the secondary and the impedances of the transformer and both feeders combined to give the circuit shown in Fig. 2.41(b), where

$$R = 0.003 + \frac{1}{10^2}(0.75 + 3.94) = 0.050 \quad \Omega$$

$$X = 0.010 + \frac{1}{10^2}(5.0 + 7.96) = 0.140 \quad \Omega$$

Figure 2.42 shows a phasor diagram for the circuit of Fig. 2.41(b), in which the reference phasor is

$$\overline{V}_L = \left(230/\sqrt{3} \right) \underline{|0} = 133 \underline{|0} \quad V$$

From the preceding example, the apparent power per phase delivered to the load is $26/3 = 8.67$ kVA, made up of 7.6 kW of active power and 4.2 kVAR of reactive power. The load current is therefore

$$I_L = \frac{8.67 \times 10^3}{133} = 65.3 \quad A$$

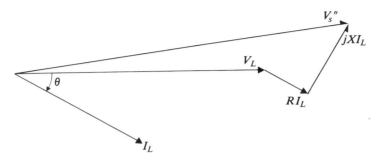

Fig. 2.42 Phasor diagram for circuit of Fig. 2.41.

at angle

$$\theta = -\cos^{-1}\frac{7.6}{8.67} = -28.7°$$

Now

$$\overline{V}''_s = \overline{V}_L + (R + jX)\,\overline{I}_L$$
$$= 133\underline{|0} + (0.050 + j0.140) \times 65.3\underline{|-28.7°}$$
$$= 140\underline{|2.64°}\quad V$$

so that the line-to-neutral potential difference at the source must be

$$10 \times 140 = 1400\quad V$$

and the line-to-line potential difference is

$$1400\sqrt{3} = 1730\quad V$$

Alternative Method of Solution   In the foregoing solution, the equivalent impedances of the transformers were referred to the wye-connected side of the bank so
that the equivalent impedance of one transformer might be shown in Fig. 2.41(a). If
the problem dealt with a delta–delta bank, it would not be possible to do this, and
an alternative method of showing the transformer equivalent impedance in the per-
phase equivalent circuit would have to be employed. This method involves the
wye–delta transformation of Fig. 2.39(b).

The equivalent impedances of the transformers that are connected in delta, may
be transformed to their equivalent wye values. Thus

$$Z''_{eq}\,(Y) = \frac{1}{3}\,Z''_{eq}(\Delta) = \frac{1}{3}\,(0.118 + j0.238)$$
$$= 0.0393 + j0.0793\quad \Omega$$

The per-phase equivalent circuit of the system is therefore that shown in Fig. 2.43,
which is equivalent to Fig. 2.41(a).

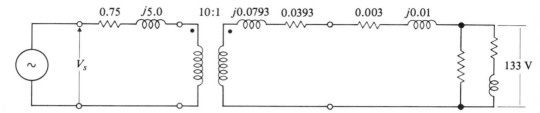

Fig. 2.43 Alternative equivalent circuit to that of Fig. 2.41(a).

*2.12.2 Polyphase Transformers

Instead of using a bank of three single-phase transformers, a polyphase transformer having three sets of primary and secondary windings on a common magnetic structure may be constructed.

   To visualize the development of a three-phase, core-type transformer, it is convenient to start with three single-phase core-type units, as shown in Fig. 2.44(a). Each of these is similar to the transformer of Fig. 2.2, except that windings have been placed on one leg only. For simplicity, only the primary windings have been shown. If the induced emf's in these transformers are sinusoidal and balanced, the fluxes ϕ_a, ϕ_b, and ϕ_c also will be sinusoidal and balanced. Thus, if the three legs

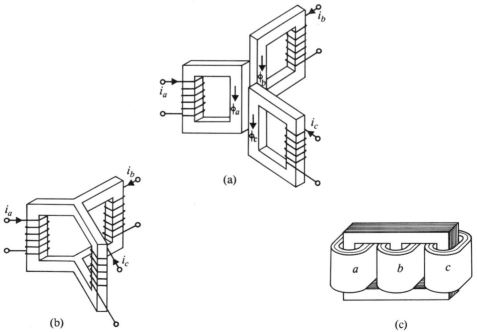

Fig. 2.44 Evolution of the three-phase, core-type transformer. (a) Three single-phase transformers. (b) Return path for $\phi_a + \phi_b + \phi_c$ removed. (c) In-line construction.

carrying these fluxes are merged, the total flux in the combined leg is zero, and this leg can be omitted, as shown in Fig. 2.46(b).

For cores made of stacked laminations, the in-line structure of Fig. 2.44(c) is more convenient. This structure can be evolved from Fig. 2.44(b) by eliminating the yokes of section b and fitting the remainder between a and c. The result does not have complete symmetry, since the magnetic paths of legs a and c are somewhat longer than that of leg b. But the resultant imbalance in the magnetizing currents is seldom significant. A three-phase core of wound grain-oriented strip is illustrated in Fig. 2.45, while Fig. 2.46 shows a three-phase, core-type transformer.

The shell-type, three-phase transformer is evolved by stacking three single-phase, shell-type units, as shown in Fig. 2.47(a). The winding direction of the center unit b is made opposite to that of units a and c. If the system is balanced with

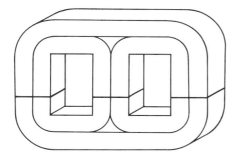

Fig. 2.45 Wound core for three-phase transformer.

Fig. 2.46 Three-phase, 750-kVA transformer. *(Moloney Electric Company of Canada, Limited)*

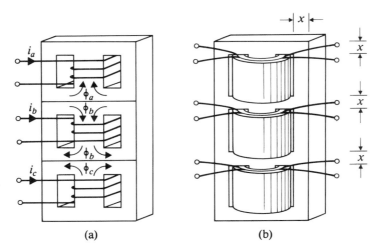

Fig. 2.47 Evolution of the three-phase, shell-type transformer.

phase sequence $a–b–c$, the fluxes are also balanced—that is, $\overline{\Phi}_a = \alpha\overline{\Phi}_b = \alpha^2\overline{\Phi}_c$, where $\alpha = 1\underline{|120°}$. The adjacent yoke sections of units a and b carry a combined flux of

$$\frac{\overline{\Phi}_a}{2} + \frac{\overline{\Phi}_b}{2} = \frac{\overline{\Phi}_a}{2}\left(1\underline{|0} + 1\underline{|240°} \right) = \frac{\overline{\Phi}_a}{2}\underline{|-60°} \quad \text{Wb} \qquad (2.140)$$

Thus the magnitude of this combined flux is equal to the magnitude of each of its components. In this way, the cross-sectional area of the combined yoke sections may be reduced to the same value as that used in the outer legs and in the top and bottom yokes, as shown in Fig. 2.47(b).

The slight imbalance in the magnetic paths among the three phases has very little effect on the performance of the three-phase, shell-type transformer. Its behavior is essentially the same as that of a bank of three single-phase transformers.

The windings of either core- or shell-type, three-phase transformers may be connected in wye or delta as desired.

Three-phase transformers for a given rating weigh less and cost less than a three-transformer bank of the same rating. In addition, the external connections are reduced from twelve to six; and when these are brought into the tank through the elaborate bushings required in high-potential transformers, this alone is a considerable economy. There is the disadvantage that if one phase breaks down, the whole transformer must be removed for repair, and provision of a spare stand-by transformer is more expensive than provision of a single spare transformer for a three-transformer bank.

When testing three-phase transformers to obtain an equivalent circuit, it may be assumed that primary and secondary windings are wye-connected, although this may not be the case.

*2.12.3 Harmonics in Three-Phase Transformer Banks

It has already been explained in Sections 1.10 and 2.5 that the exciting current of a transformer is nonsinusoidal and contains odd harmonics. By far the largest of these harmonics is the third, and in the following discussion it will therefore be assumed that all higher harmonics may be neglected and that the transformer exciting current is made up of a fundamental component and a significant third-harmonic component. The amplitude of the third harmonic depends upon the degree of saturation of the transformer core, but it may reach about 40% of the amplitude of the fundamental component.

Figure 2.48 shows three fundamental-component currents displaced from one another in phase by 120° and also the third harmonics associated with them. The sum of the two components in each case would give a fair approximation to a typical exciting-current waveshape, and it may be assumed that these are the currents required to produce sinusoidal flux variations in the cores of the transformers of a three-phase bank. Note that in Fig. 2.48 the third-harmonic components of current are in phase with each other.

Consider now three similar transformers with the primary windings connected in wye and the secondary windings open-circuited and not connected to each other in any way, as illustrated in Fig. 2.49. The transformers are excited from a three-phase sinusoidal potential source. The line currents in the three-phase system can-

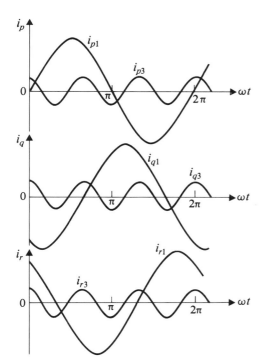

Fig. 2.48 Exciting currents of a three-phase transformer bank.

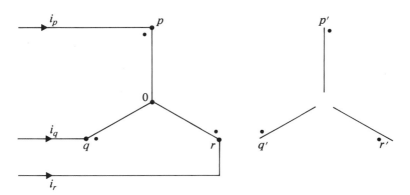

Fig. 2.49 Three-phase transformer bank with secondary windings disconnected.

not contain any third-harmonic components, since these would be in phase and would all have to flow simultaneously toward or away from the star point. Since there is no fourth wire connected to the star point, this is impossible, and only a fundamental component current is able to flow in the transformer windings.

Since the exciting current in each transformer lacks the required third-harmonic component to produce a sinusoidal variation of flux in the core, then the flux must include a third-harmonic component, and this results in a third-harmonic component of induced emf in the transformer. The third-harmonic components of induced emf in the three transformers are in phase and cause the potential of the neutral point 0 to vary at third-harmonic frequency. If the secondary windings of the transformer were connected in wye and were employed to distribute single-phase power by means of a four-wire system, then each single-phase supply would contain unacceptably high third-harmonic components of potential difference. If the star point 0 of the primary windings in Fig. 2.49 were grounded or connected by a fourth wire to the source star point, third-harmonic currents would flow in the primary lines and in the neutral return path. If the impedance of this return path were negligible, the line-to-neutral potential differences of the transformers would be sinusoidal. However, the use of an earth path for conduction of harmonic currents may be objectionable from several points of view, for example, because of interference with adjacent communication cables.

If the secondary windings of the transformers in Fig. 2.49 were connected in delta, however, then the third-harmonic emf's induced in the three secondary windings would drive a third-harmonic current around the secondary delta. This third-harmonic current would supply the missing component of the exciting current and consequently, save for a very small component due to the impedances of the transformers, the third-harmonics of the flux and induced emf would disappear. Similarly, if the primary windings were connected in delta, the required third-harmonic current would flow around the primary delta. In fact a third winding (tertiary) connected in delta may be fitted on the core for the principal purpose of carrying the required third-harmonic component of exciting current when it is desirable to connect both primary and secondary windings in wye.

By using the methods of Section 2.10, an equivalent electric circuit can be produced for the three-phase, core-type transformer. This circuit has a form similar to that of three single-phase transformers connected in delta. The harmonic behavior of this transformer is similar to that of a delta-connected bank in that essentially no third-harmonic components of exciting current flow in the windings under balanced operation, irrespective of the connections. This can be seen by regarding each leg as an equal in-phase source of third harmonics.

*2.12.4 Three-Phase to Two-Phase Transformations

Transformation in the so-called *two-phase* system with two phases 90° from each other may be carried out employing two single-phase transformers as indicated in Fig. 2.50.

Transformation from a three- to a two-phase system (or the converse) may be carried out employing two single-phase transformers in the connection illustrated in Fig. 2.51. The transformer with the center-point tapping is usually referred to as the "main" transformer, while the other is called the "teaser" transformer. Section pn

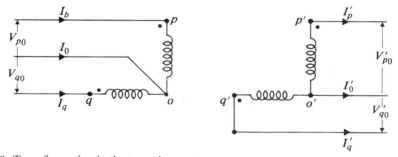

Fig. 2.50 Transformation in the two-phase system.

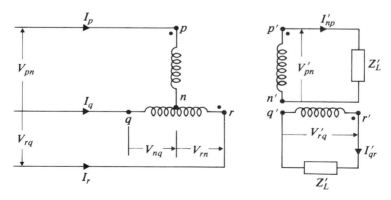

Fig. 2.51 Three-phase to two-phase transformation.

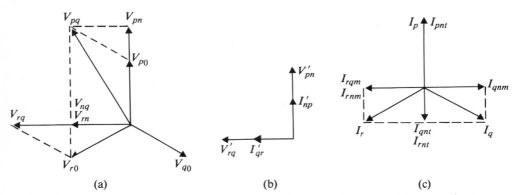

Fig. 2.52 Phasor diagrams for three-phase to two-phase transformation. (a) Three-phase potential phasors. (b) Two-phase diagram. (c) Three-phase current phasors.

of the teaser primary has 0.866 of the potential of the main transformer primary. The transformer arrangement of Fig. 2.51 is known as the Scott connection.

A balanced load on the 2-phase side of this connection will ideally appear as a balanced load on the 3-phase side. It should be noted that the 2-phase secondary circuits may be electrically connected, providing a 2-phase, 3-wire system. The relevant phasor diagrams are shown in Fig. 2.52.

*2.13 SOME SPECIAL TRANSFORMATION DEVICES

The analytical methods discussed earlier in this chapter are sufficient to determine the operating characteristics of most types of transformers. There are, however, a few devices that deserve further comment.

*2.13.1 Autotransformers

The same transformation ratios as are obtained from a 2-winding transformer may be obtained from a single tapped winding on a transformer core. This arrangement, which is illustrated in Fig. 2.53, may be used for a step-up or a step-down transformer. The latter arrangement is shown in Fig. 2.53. If the exciting current is neglected, then I_1 and I_2 are in phase, and the current in the section of winding between the secondary terminals is

$$I_x = I_2 - I_1 \quad \text{A} \tag{2.141}$$

A consequence of this arrangement is that if the current density employed in the windings of an autotransformer is the same as that employed in the windings of a 2-winding transformer of the same kVA rating, then a smaller amount of copper is needed, due to the lower value of I_x. This means that the core window may be smaller, and the weight of the core is therefore reduced.

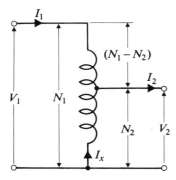

Fig. 2.53 Autotransformer.

The copper ratio for the two types of transformer is approximately as follows:

$$\frac{\text{Autotransformer}}{\text{2-winding transformer}} = \frac{(N_1 - N_2)I_1 + N_2(I_2 - I_1)}{N_1 I_1 + N_2 I_2} = 1 - \frac{N_2}{N_1} \quad (2.142)$$

From this expression it is seen that the economy in material is greatest for transformers with ratio near unity.

The autotransformer has the important disadvantage that electrical isolation of the primary and secondary circuits is not obtained. It has the advantage that if the secondary tapping is replaced by a slider running over a series of studs or the exposed turns of the winding itself, a transformer of virtually continuously variable ratio is obtained.

The autotransformer may be tested in the same way as a normal transformer, and the equivalent circuit of Fig. 2.16 employed to predict its performance. Note, however, that the equivalent-circuit parameters as well as the effective turns ratio depend on the position of the tapping point. If the ratio is near unity, the equivalent resistance R'_{eq} and reactance X'_{eq} will be much lower than for a two-winding transformer.

Example 2.11 The transformer of Example 2.3 has its two windings connected in series to form an autotransformer giving a small reduction to potential difference from a 2300-volt line. Determine

a) The transformation ratio on open circuit.

b) The permissible output of the transformer if the winding currents are not to exceed those for full-load operation as a two-winding transformer.

Solution

a) The connection for operation as an autotransformer is as shown in Fig. 2.53, where $V_1 = 2300$ volts. The original primary winding is now N_2, and the original secondary winding is $(N_1 - N_2)$. The turns ratio of the transformer is thus

$$\frac{V_1}{V_2} = \frac{2300 + 230}{2300} = \frac{2530}{2300} = 1.1$$

b) On open circuit,

$$V_2 = \frac{2300}{2530} \times 2300 = 2090 \quad V$$

Winding $(N_1 - N_2)$ is rated at 43.5 A. Winding N_2 is rated at 4.35 A. Neglecting the exciting current,

$$I_x = I_2 - I_1$$

That is,

$$4.35 = I_2 - 43.5$$

and

$$I_2 = 47.8 \quad A$$

Rating of the autotransformer is thus

$$2090 \times 47.8 = 99.9 \, kVA$$

Compare this with its 10 kVA rating as a two-winding transformer.

*2.13.2 Instrument Transformers

Instrument transformers are of two types: current and potential. The principle of an instrument transformer differs in no fundamental way from that of a power transformer; however, instrument transformers are so conservatively designed that, when employed in accordance with the makers' specifications, they may—with little error—be regarded as having either an ideal potential ratio or an ideal current ratio.

The current transformer serves somewhat the same purpose in an ac circuit as a millivoltmeter shunt serves in a dc circuit. The primary of the transformer is connected in the line and carries the current to be measured. The secondary is connected to an ammeter or the current winding of a wattmeter. A transformer that is rated 100 to 5 amperes is designed to carry up to 100 amperes in its primary winding, and when it does so the current through a meter connected across the secondary terminals is 5 A. Laboratory current transformers have several primary connections for different current ratings. For the largest current ratio, the current-carrying conductor is fed through the center of the transformer core to provide a one-turn primary. In addition to providing a low current in the meter circuit, the current transformer isolates the meter from the line, which may be at high potential.

In the operation of a current transformer, *the secondary terminals should never be open-circuited.* A current transformer is used in such a way that the magnitude of the secondary current has negligible influence on the magnitude of the primary current. Thus, removal of the secondary mmf will not affect the primary mmf, and the net mmf acting on the core will rise to a value many times greater than that for which the core was designed. As a consequence, the core flux density will be switched rapidly between the positive and negative saturation values, causing an excessively high potential difference to appear across the open-circuited secondary terminals. This may endanger the user and damage the transformer insulation. In addition, the excessive core losses may overheat the transformer.

The potential transformer primary is connected across the potential difference to be measured. The secondary is connected to the voltmeter or wattmeter potential winding. In addition to providing a low potential difference in the meter circuit, the potential transformer isolates the meter from the line.

Before an instrument transformer can be used with a wattmeter, its terminal polarity must be known. The polarity of laboratory transformers is indicated by $\pm$ markings on primary and secondary terminals of like polarity. Corresponding markings appear on wattmeter current and potential terminals. The connection for measurement of potential difference, current, and power in a high-potential, high-power circuit is shown in Fig. 2.54.

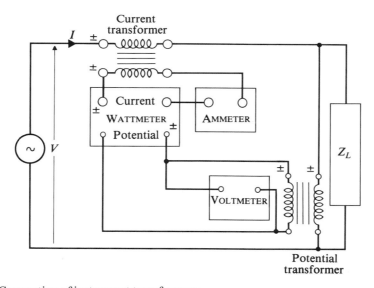

Fig. 2.54 Connection of instrument transformers.

*2.13.3 Measurement of Direct Currents

Measurement of direct current with the meter isolated from the line is sometimes required. This may be accomplished by the system shown, in Fig. 2.55(a), which consists of two identical saturable cores. In the system shown, the primary windings of the two cores consist of the single conductor carrying the unknown current I. The two primary windings are therefore in series. The secondary windings are

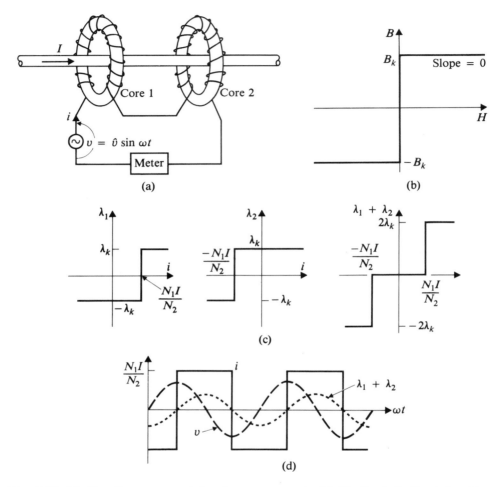

Fig. 2.55 (a) Circuit arrangement for dc measurement. (b) Idealized B–H relationship. (c) Flux linkages λ_1 of core 1, λ_2 of core 2, and $\lambda_1+\lambda_2$. (d) Waveforms of source PD, total flux linkage, and current.

connected in series opposition to a source of alternating potential difference and a measuring instrument. It will be assumed that the B–H relationship of the core material can be represented to a satisfactory approximation by the ideal shape of Fig. 2.55(b) with infinite unsaturated permeability and zero saturated permeability. With a constant direct current I in the N_1 turns of each primary winding (in Fig. 2.55(a), $N_1 = 1$), core 1 is unsaturated only when its secondary current i is

$$i = \frac{N_1 I}{N_2} \quad \text{A} \tag{2.143}$$

and core 2 is unsaturated only when

$$i = -\frac{N_1 I}{N_2} \quad \text{A} \tag{2.144}$$

Figure 2.55(c) shows the idealized relations between the secondary current i and the flux linkage λ_1 of the secondary winding of core 1, the flux linkage λ_2 of core 2, and the total flux linkage $\lambda_1 + \lambda_2$ of the two windings in series. Note that the total flux linkage can change only when the current i has one of the values given by Eqs. (2.143) and (2.144).

Suppose a sinusoidal potential source $\hat{v} \sin \omega t$ is connected as shown. If we neglect the impedance of the meter and the windings, the total flux linkage of the circuit is given by

$$\lambda_1 + \lambda_2 = -\frac{\hat{v}}{\omega} \cos \omega t \quad \text{Wb} \tag{2.145}$$

If the applied potential difference v is limited so that

$$\frac{\hat{v}}{\omega} < 2\lambda_k \quad \text{Wb} \tag{2.146}$$

then current i is equal to $N_1 I/N_2$ when $\lambda_1 + \lambda_2$ is positive and is equal to $-N_1/N_2$ when $\lambda_1 + \lambda_2$ is negative. Thus it has the square waveform shown in Fig. 2.55(d), the amplitude of which is proportional to the current I.

The current i may be measured by an ac instrument, or it may be rectified in a bridge rectifier to produce a direct current that is related to the measured current I by the inverse of the turns ratio.

Inaccuracies in this measurement system arise principally from the finite coercive force required in the cores and from their finite saturated permeability.

*2.13.4 Saturable Reactors

A saturable reactor is a group of saturable magnetic elements connected in such a way as to permit control of the power supplied from a source to a load. Figure 2.56(a) shows two saturable magnetic cores, each with a primary winding of N_p turns and a control winding of N_c turns. The primary windings are connected in series with each other and with the load. A direct current I_c is supplied to the control windings that are connected in series opposition. A large inductance in the control winding ensures that its current is essentially constant and free of ripple.

The operation of this system is based on the same principle as the dc measurement system described in Section 2.13.3. If the control current is I_c, core 1 is unsaturated only when

$$i_L = \frac{N_c}{N_p} I_c \quad \text{A} \tag{2.147}$$

and core 2 is unsaturated only when

$$i_L = -\frac{N_c}{N_p} I_c \quad \text{A} \tag{2.148}$$

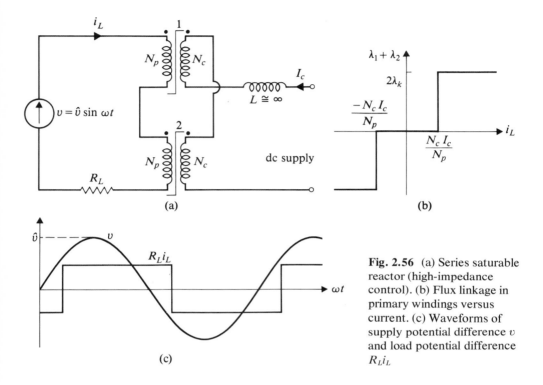

(a)

(b)

(c)

Fig. 2.56 (a) Series saturable reactor (high-impedance control). (b) Flux linkage in primary windings versus current. (c) Waveforms of supply potential difference v and load potential difference $R_L i_L$

By a development similar to that used in Section 2.13.3, the total flux linkage $\lambda_1 + \lambda_2$ of the primary windings is related to the current I_L as shown in Fig. 2.56(b).

When the alternating potential difference of $v = \hat{v} \sin \omega t$ is applied, the current i_L in the load is a square wave having the amplitudes given in Eqs. (2.147) and (2.148). Figure 2.56(c) shows the idealized waveforms of the potential differences v and $R_L I_L$. The saturable reactor acts as a source of constant square-wave current, the magnitude of which is directly proportional to the control current. If ideal cores are assumed, this square waveform of the load current and potential difference is retained over most of the usable range. In practice, the finite saturated permeability of the cores results in some departure from square waveform. In addition, when the control current is zero, the load still carries the small magnetizing current required by the cores.

The large inductance in the control circuit of the saturable reactor of Fig. 2.56(a) causes this circuit to have a large time constant. Following a change in control potential difference, the control current and the load current change slowly. If the inductance is removed to overcome this sluggishness in response, the saturable reactor continues to control the load potential difference, but its action is markedly different.

Figure 2.57 shows a series saturable reactor without control circuit inductance, together with the waveforms of its variables. The total resistance of the control windings is R_c. Let the turns ratio be

$$\frac{N_c}{N_p} = n \qquad (2.149)$$

The resistances of the primary windings are neglected, and the cores are assumed to be ideal saturable elements. Under all conditions

$$v = e_1 + e_2 + v_L \quad \text{V} \qquad (2.150)$$

and

$$V_c = R_c i_c + n e_1 - n e_2 \quad \text{V} \qquad (2.151)$$

At the beginning of a cycle of the applied potential at $\omega t = 0$, assume that core 2 is at negative saturation, making its primary-winding flux linkage $\lambda_2 = -\lambda_k$, and that core 1 has a primary-winding flux linkage of $\lambda_1 = -\lambda_0$. With positive applied potential difference, both flux linkages increase. Since both cores are unsaturated, the currents i_L and i_c must both be zero. As long as this condition obtains, solution of Eqs. (1.313) and (1.314) gives

$$e_1 = \frac{v + V_c/n}{2} \quad \text{V} \qquad (2.152)$$

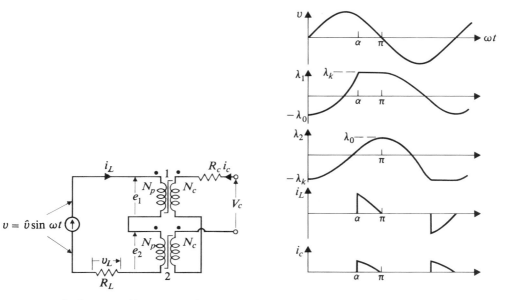

Fig. 2.57 Series saturable reactor with low-impedance control, together with its idealized waveforms.

and

$$e_2 = \frac{v - V_c/n}{2} \quad \text{V} \tag{2.153}$$

At some instant $\omega t = \alpha$, core 1 reaches positive saturation ($\lambda_1 = \lambda_k$) and the core emf e_1 becomes zero.

$$\lambda_{1(\alpha)} = -\lambda_0 + \frac{1}{\omega} \int_0^\alpha \left(\frac{v + V_c/n}{2} \right) d\omega t = \lambda_k \quad \text{Wb} \tag{2.154}$$

Since only core 2 now remains unsaturated, load and control currents may flow according to the relation

$$i_L = n i_c \quad \text{A} \tag{2.155}$$

Rearrangement of Eq. (2.151) with $e_1 = 0$ gives

$$e_2 = \frac{1}{n} (R_c i_c - V_c) = \frac{R_c}{n^2} i_L - \frac{V_c}{n} \quad \text{V} \tag{2.156}$$

Since core 1 started the cycle at flux linkage $-\lambda_0$, core 2 must finish the first half cycle at flux linkage λ_0 if the system is to repeat cyclically. Thus, using Eqs. (2.153) and (2.156),

$$\begin{aligned}
\lambda_{(2\pi)} = -\lambda_k &+ \frac{1}{\omega} \int_0^\alpha \left(\frac{v - V_c/n}{2} \right) d\omega t \\
&+ \frac{1}{\omega} \int_\alpha^\pi \left(\frac{R_c}{n^2} i_L - \frac{V_c}{n} \right) d\omega t = \lambda_0 \quad \text{Wb}
\end{aligned} \tag{2.157}$$

Solution of Eq. (2.154) with Eq. (2.157) gives

$$\frac{V_c \pi}{n} = \frac{R_c}{n^2} \int_\alpha^\pi i_L \, d\omega t \quad \text{V} \tag{2.158}$$

Since the half-cycle average value $i_{L(\text{av})}$ of the load current is

$$i_{L(\text{av})} = \frac{1}{\pi} \int_\alpha^\pi i_L \, d\omega t \quad \text{A} \tag{2.159}$$

Eq. (2.158) can be rewritten as

$$i_{L(\text{av})} = n \frac{V_c}{R_c} \quad \text{A} \tag{2.160}$$

Thus the average load current is directly proportional to the control potential difference. This linear relation applies until essentially the whole of the source potential difference appears across the load, that is,

$$\text{maximum } i_{L(\text{av})} = \frac{2}{\pi} \frac{\hat{v}}{R_L} \quad \text{A} \tag{2.161}$$

As shown in Fig. 2.57, the control current consists of sections of waves, the average value of which is V_c/R_c.

An alternative saturable reactor arrangement is similar to that of Fig. 2.57, but with the primary windings connected in parallel. The control characteristics and the waveform of load potential difference are similar to those for the series connection. Saturable reactors provide an efficient and reliable means of controlling the power delivered to a load from a constant potential ac source.

PROBLEMS

2.1 Refer to the transformer of Fig. 2.58.

   a) Place polarity dots on windings 3 and 4.

   b) State the number of turns of winding 4.   *(Section 2.1)*

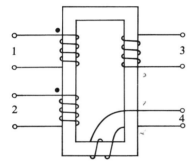

Figure 2.58

2.2 A 2-winding transformer that can be considered ideal has 200 turns on its primary winding and 500 turns on its secondary winding. The primary winding is connected to a 220-V sinusoidal supply, and the secondary winding supplies 10 kVA to a load.

   a) Determine the load potential difference, the secondary current, and the primary current.

   b) Determine the magnitude of the load impedance referred to the primary side.

             (Section 2.1)

2.3 A 5-kVA, 60-Hz, 440:110-V, single-phase transformer that can be considered ideal is drawing a primary current of 10 A at a leading power factor of 0.9. Determine the impedance of the load circuit connected to the low-potential terminals; express it as a phasor quantity.

             (Section 2.1)

2.4 A 5000-Hz ac generator can be modeled as a constant emf of 250 V rms in series with an inductive reactance of 31 Ω. This generator is to be used to supply power to a resistive load of 0.65 Ω through a transformer that may be considered ideal.

   a) What should be the turns ratio of the transformer to achieve maximum power into the load?

   b) What are the required potential and current ratings of the two windings of the transformer?   *(Section 2.1)*

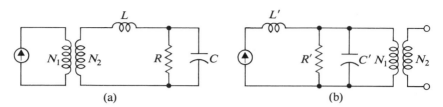

Fig. 2.59 Diagram for Problem 2.6.

2.5 An amplifier is to be connected to a loudspeaker by means of a 2-winding transformer. Consider the loudspeaker a resistive load of 8.0 Ω. The amplifier can be represented as a source of audio-frequency emf in series with a resistance of 1000 Ω. Assuming the transformer to be ideal, determine the turns ratio that will provide maximum power into the loudspeaker.   *(Section 2.1)*

2.6 Refer to Fig. 2.59.

  a) Show that the circuit of Fig. 2.59(a) can be replaced by that of Fig. 2.59(b) for the purpose of determining the source current.

  b) Assuming $L = 0.1$ H, $R = 15$ Ω, $C = 150\ \mu$F, and $N_1/N_2 = 6$, determine L', R', and C'.   *(Section 2.2)*

2.7 A transformer has the following parameters:

$$
\begin{array}{ll}
R_1 = 0.5 \ \ \Omega & R_2 = 30 \ \ \Omega \\
L_{l1} = 3.5 \ \ \text{mH} & L_{l2} = 0.22 \ \ \text{H} \\
L'_m = 5.4 \ \ \text{H} & N_1/N_2 = 0.12
\end{array}
$$

Draw an equivalent circuit of the transformer with all parameters referred to:

  a) The low-potential side.

  b) The high-potential side.   *(Section 2.2)*

*2.8 A 10-kVA, 60-Hz, 2300:115-V, single-phase transformer has the following equivalent-circuit parameters referred to the high-potential side:

$$
\begin{array}{ll}
R_1 = 5.7 \ \ \Omega & R'_2 = 6.20 \ \ \Omega \\
L_{l1} = 32.4 \ \ \text{mH} & L'_{l2} = 32.4 \ \ \text{mH} \\
L'_m = 184 \ \ \text{H} &
\end{array}
$$

Determine the parameters of a coupled-circuit representation of this transformer.

      (Section 2.3)

*2.9 A 2-winding transformer has the following coupled-circuit parameters:

$$
\begin{array}{ll}
L_{11} = \ \ 0.3 \ \ \text{H} & R_1 = 0.02 \ \ \Omega \\
L_{12} = \ \ 3.15 \ \ \text{H} & R_2 = 2.1 \ \ \Omega \\
L_{22} = 35.0 \ \ \text{H} & N_1/N_2 = 0.09
\end{array}
$$

Develop an equivalent circuit for this transformer of the form shown in Fig. 2.11

      (Section 2.3)

2.10 A 100-kVA, 11,000:2,200-V, 60-Hz, single-phase transformer has the following equivalent-circuit parameters referred to the high-potential side:

$$R_1 = 6.1 \ \Omega \qquad R_2' = 7.2 \ \Omega$$
$$X_{l1} = 31.2 \ \Omega \qquad X_{l2}' = 31.2 \ \Omega$$
$$X_m' = 57,300 \ \Omega \qquad R_c' = 124,000 \ \Omega$$

It may be represented by an equivalent circuit similar in form to that of Fig. 2.16. The transformer is supplying at 2200 V a load circuit of 50 Ω impedance and a leading power factor of 0.7. Draw a phasor diagram (not to scale) showing the various phasor magnitudes and angles for this condition of operation. Determine the potential difference and power factor at the high-potential terminals of the transformer. *(Section 2.5)*

2.11 A potential difference $v_1 = 200 \sqrt{2} \sin \omega t$ is applied to the primary terminals of a transformer. The resulting exciting current is

$$i_e = 7.21 \sin(\omega t - 80°) + 2.90 \sin 3(\omega t - 70°) \quad \text{A}$$

a) Determine the rms values of the applied potential difference and the exciting current.
b) Calculate the core loss.
c) Draw a phasor diagram approximately representing the relationships of the applied potential difference and the exciting current and determine the rms value of the magnetizing current.
d) Calculate the approximate reactive volt-amperes input. *(Section 2.5)*

2.12 Assume that the transformer of Problem 2.10 can be represented by the approximate equivalent circuit of Fig. 2.21(b), and draw a phasor diagram (not to scale) showing the various phasor magnitudes and angles for the condition of operation specified in Problem 2.10. Determine the total losses of the transformer:

a) By subtracting output active power from input active power,
b) By calculating the power dissipated in the resistive elements of the approximate equivalent circuit. *(Section 2.6.2)*

2.13 Two single-phase, 10 MVA transformers are connected in parallel on both the primary and secondary sides. Their turns ratios are identical, and their winding resistances are negligible. One transformer has an equivalent reactance of 0.1 per unit while the other has an equivalent reactance of 0.06 per unit.

a) Suppose the transformers are used to supply a 20-MVA load. Determine the load carried by each transformer.
b) Determine the maximum total load that can be supplied without overloading either transformer.
c) If each of the transformers has an equivalent resistance of 0.01 per unit, will the sum of the volt–ampere loads on the two transformers equal the total load supplied by the transformers? *(Section 2.6.2)*

2.14 A 1.5-kVA, 220:110-V, 60-Hz, single-phase transformer gave the following test results:

i) Open-circuit test, low-potential winding excited:

$$V_{oc} = 110 \text{ V}, \qquad I_{oc} = 0.4 \text{ A}, \qquad P_{oc} = 25 \text{ W}, \qquad V_{hp} = 220 \text{ V}$$

ii) Short-circuit test, low-potential winding excited:

$$V_{sc} = 8.25 \text{ V}, \qquad I_{sc} = 13.6 \text{ A}, \qquad P_{sc} = 40 \text{ W}$$

iii) Direct-current winding resistances:

$$R_{1p} = 0.113 \text{ } \Omega, \qquad R_{hp} = 0.413 \text{ } \Omega$$

a) Determine the equivalent circuit of the transformer referred to the low-potential side.
b) Determine the full-load efficiency when the transformer is supplying at 110 V a load circuit with a lagging power factor of 0.8.   *(Section 2.6.3)*

2.15 A 20-kVA, 2200:220-V, 60-Hz, single-phase distribution transformer gave the following test results:

i) Open-circuit test, low-potential winding excited:

$$V_{oc} = 220 \text{ V}, I_{oc} = 1.52 \text{ A}, P_{oc} = 161 \text{ W}$$

ii) Short-circuit test, high-potential winding excited:

$$V_{sc} = 205 \text{ V}, I_{sc} = 9.1 \text{ A}, P_{sc} = 465 \text{ W}$$

iii) Direct-current winding resistances:

$$R_{1p} = 0.0311 \text{ } \Omega, R_{hp} = 2.51 \text{ } \Omega$$

a) Determine the equivalent circuit of the transformer referred to the high-potential side.
b) Determine the all-day efficiency of the transformer on a load cycle supplied at 220 V and approximated by:

> 95% full load for 8 hours at PF = 0.80 lagging
> 60% full load for 8 hours at PF = 0.90 lagging
> 5% full load for 8 hours at PF = 1.0

c) Determine the percentage of rated load at which the transformer has maximum efficiency.   *(Section 2.6.3)*

2.16 A 10-MVA transformer is required to supply a unity-power-factor industrial process. It is to be used at full load for 8 hours per day and at essentially no load for the remainder of the day. Supplier A offers a tranformer having a full-load efficiency of 99% and a no-load loss of 0.5%. Supplier B offers a transformer at the same price having a full-load efficiency of 98.8% and a no-load loss of 0.3%.

a) Which transformer should be chosen?
b) If energy costs 2.5¢ per kWh, what will be the annual difference in energy cost between the two transformers?   *(Section 2.6.3)*

2.17 A 100-kVA, 11,000:2200-V, 60-Hz, single-phase power-system transformer gave the following test results:

i) Open-circuit test, low-potential winding excited:

$$V_{oc} = 2200 \text{ V}, I_{oc} = 1.59 \text{ A}, P_{oc} = 980 \text{ W}$$

ii) Short-circuit test, high-potential winding excited:

$$V_{sc} = 580 \text{ V}, I_{sc} = 9.1 \text{ A}, P_{sc} = 1100 \text{ W}$$

iii) Direct-current winding resistances

$$R_{1p} = 0.288 \ \Omega, \ R_{hp} = 6.10 \ \Omega$$

a) Determine the equivalent circuit of this transformer referred to the low-potential side.

b) Determine the regulation of the transformer when supplying rated load at 2200 V and a lagging PF of 0.8.

c) Determine the modification in turns ratio required to eliminate the transformer regulation for the load in (b). *(Section 2.6.4)*

2.18 A 200-kVA, 2300:230-V, 60-Hz, single-phase transformer has a short-circuit impedance measured on the high-potential side of $0.24 + j \ 1.6 \ \Omega$. The open-circuit admittance measured on the low-potential side is $0.03 - j0.075 \ \mho$ *. The rated nameplate quantities are to be used as base values. Determine each of the following:

a) The base voltage, current, impedance, and power for both the high- and low-potential sides of the transformer.

b) The per-unit value of the equivalent resistance and equivalent leakage reactance.

c) The per-unit value of the loss in the transformer operated on open circuit at rated potential difference.

d) The per-unit value of the exciting current on open circuit at rated potential difference.

e) The per-unit value of the loss in the transformer operated at rated load at rated potential difference. *(Section 2.6.5)*

2.19 Tabulate the per-unit values of the exciting currents and equivalent impedances of the transformers of Problems 2.14, 2.15, and 2.17. *(Section 2.6.5)*

2.20 A transformer has an equivalent impedance of $0.01 + j \ 0.05$ per unit and a no-load power loss of 0.01 per unit, the base for all unit quantities being the rating of the transformer.

a) Determine the efficiency of the transformer when supplying rated load at rated potential difference and 0.8 lagging power factor.

b) Determine the transformer regulation for the condition in (a). *(Section 2.6.5)*

*2.21 A transformer has a core with a magnetic path length of 1.8 m and a cross-section area of $0.025 \ m^2$. The core material may be considered to require zero magnetic field intensity up to a flux density of 1.8 T. Beyond 1.8 T the B–H relation can be approximated by a straight line of slope 5.5 μ_0. The primary winding has 200 turns, negligible resistance, and a leakage inductance of 65 μH.

a) Draw an equivalent circuit for the transformer in the form of Fig. 2.23(b).

b) Assuming no residual flux density in the core, estimate the peak value of the inrush current when a potential difference of 2200 V (rms) at 60 Hz is applied to the primary terminals at the instant of zero voltage crossover.

c) Assume a maximum residual flux density of 0.9 T, and repeat (b) to estimate the peak current. *(Section 2.7)*

*2.22 Suppose that the transformer in Problem 2.15 is of the core type shown in Fig. 2.26(d), with both primary coils and both secondary coils connected in parallel. The user of this transformer is informed that the power supply is to be changed to 4400 V.

a) How should the transformer windings be reconnected?

b) What will be the equivalent resistance and reactance of the reconnected transformer referred to the 4400-volt side?

---

*The unit mho ($\mho$) is also known as siemens (S).

c) When the transformer is reconnected, will its regulation and efficiency be increased, decreased, or unchanged?   *(Section 2.8)*

*2.23   The transformer shown in Fig. 2.60 is to operate with 115 V rms at 60 Hz applied to its primary winding. The secondary winding is to produce an output at 500 V rms. All core dimensions are given in mm. The core material is to operate at peak flux density of 1.4 T. The core stacking factor k_i is 0.95. The windings can be operated at a current density of 2 A rms per mm². The winding space factor k_w is estimated at 0.45.

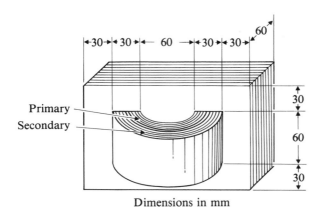

Dimensions in mm

Figure 2.60

a) Determine the required numbers of turns on the primary and secondary windings.
b) Determine the kVA rating of the transformer.
c) At the operating temperature the resistivity of the copper is about $2 \times 10^{-8} \ \Omega \cdot m$. Assuming that the projecting ends of the coil are semicircular in shape, estimate the power loss in the windings when operating at rated current.
d) If the core laminations are made of M-36 sheet steel with a density of 7800 kg/m³, what is the power loss in the core?   *(Section 2.9)*

*2.24   Two single-phase transformers designated as A and B have the same physical proportions. The linear dimensions of all parts of transformer A are k times the corresponding dimensions of transformer B. The core laminations are of the same thickness in both transformers. The operating value of magnetic flux density, conductor current density, and frequency are the same for the two transformers.
   Determine the value of the following quantities for transformer A relative to the corresponding quantities for transformer B: rated potential difference, rated current, rated volt amperes, mass, output per unit of mass, winding loss, winding resistances, core loss, heat loss per unit of surface area, total loss in per-unit of rating, leakage reactance in ohms, and per-unit magnetizing reactance.   *(Section 2.9)*

*2.25   Design a 5-kVA, 2200:220-V, 60-Hz single-phase distribution transformer of the configuration shown in Fig. 2.26(d). A core of the type shown in Fig. 1.51 may be used, and its relative dimensions may be as given in Eq. (1.105). The peak flux density should be limited to 1.45 T, and the rms current density in the copper conductor should not exceed 3.5 A/mm².

Provision should be made for the low-potential winding to be connected as two 110-V windings in series. Assume that the winding space factor is 0.55.   *(Section 2.9)*

*2.26 Estimate the losses in the transformer of Problem 2.25 when operated at rated load, assuming that the material has the core loss of Fig. 1.53, a mass density of 7640 kg/m^3, and a stacking factor of 0.95. Also estimate the exposed surface area of the transformer available for cooling, and determine the heat loss in W/m^2 from this surface at rated load.

(Section 2.9)

*2.27 The magnetic system of Fig. 2.61 has a coil of 50 turns on its middle leg. The magnetic material may be assumed to have a constant relative permeability of 4000. Leakage flux may be neglected.

    a) Derive a magnetic equivalent circuit for this system showing the values of all parameters.

    b) Determine the total inductance of the 50-turn coil.

    c) Suppose a 100-turn coil is now placed on the right-hand leg. Insert an appropriate ideal transformer into the electric equivalent circuit.

    d) If a 10-V alternating potential difference is applied to the 50-turn coil, what will be the open-circuit potential difference at the terminals of the 100-turn coil? *(Section 2.10.1)*

*2.28 A magnetic system having two coils is shown in Fig. 2.62. The reluctance of the ferromagnetic material may be neglected. At each air gap, the magnetic field is not entirely confined to the gap volume. The fringing flux around the gap edges may be taken into account in the calculation of reluctance by an appropriate increase in the cross-section area of the gap. A useful approximation is to assume that the region of uniform flux density extends outward from each of the gap edges by a distance equal to half the gap length.

    a) Develop an electric equivalent circuit for the system including the values of all parameters. The coil resistances may be neglected.

    b) If the frequency is 180 Hz, what is the reactance of the system with the two coils connected in series, terminal *b* connected to terminal *d*?

    c) Repeat part (b) with terminal *b* connected to terminal *c*. *(Section 2.10.2)*

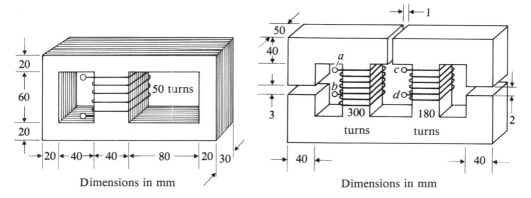

Fig. 2.61 Diagram for Problem 2.27.   **Fig. 2.62** Diagram for Problem 2.28.

*2.29 In electric welding a potential difference of 50–70 V is required to strike an arc. After the arc is struck, an essentially constant current supply is desired. Figure 2.63 shows a 2-winding transformer designed for use in a welder. To limit the load current, a low-reluctance path is provided for leakage flux between the primary and secondary windings.

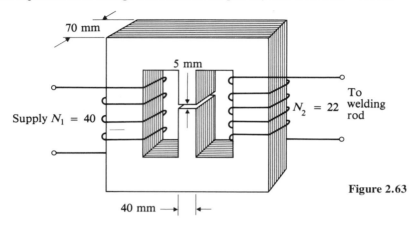

Figure 2.63

a) Derive an electric equivalent circuit for the transformer. As a first approximation, the core material may be regarded as perfect, the leakage path around each of the windings may be ignored, and the winding resistances may be neglected.

b) The supply is 115 V at 60 Hz. Determine the open-circuit secondary potential difference and the short-circuit current in the secondary winding.

c) If the arc can be regarded as a variable resistance dependent on the arc length, what is the maximum power that can be delivered to the arc?

d) Would the power determined in part (c) have been increased if the effects of the leakage paths around the windings had been included?   *(Section 2.10.2)*

*2.30 A magnetic system comprises two toroidal ferromagnetic cores. Core A has a cross-section area of 300 mm^2 and an air gap 1 mm in length. Core B has a cross-section area of 500 mm^2 and an air gap 0.5 mm in length. A primary coil of 240 turns links both cores, while a secondary coil of 400 turns links only core B. Each of the coils has a resistance of 0.04 Ω/ turn. On the assumption that the ferromagnetic material has infinite permeability, develop an electric equivalent circuit for the system showing the numerical values of all parameters.   *(Section 2.10.2)*

*2.31 The magnetic material of the device shown in Fig. 2.64 may be regarded as perfect. The core depth is 50 mm.

a) Develop an equivalent electric circuit for the system as seen from the resistive load circuit.

b) Determine the power delivered to the load.   *(Section 2.10.2)*

*2.32 Figure 2.65 shows the core and winding arrangement of a transformer that is supposed to maintain a constant ratio of output to input potential difference, regardless of load. The supply frequency is 400 Hz.

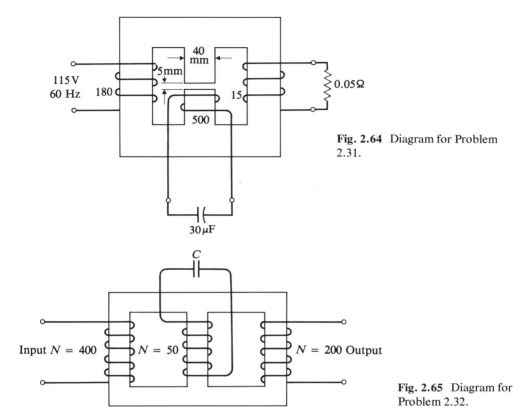

Fig. 2.64 Diagram for Problem 2.31.

Fig. 2.65 Diagram for Problem 2.32.

a) Develop an equivalent electric circuit for the system. The winding resistances and the magnetizing current may be considered negligible. The leakage path around each winding may be assigned a reluctance of 10^7 A/Wb. (Mutual leakage fluxes may be neglected.)

b) Determine the value of capacitance C required to achieve the desired condition of zero regulation.

c) Show that the system of Fig. 2.65 is identical in operation to a normal 2-winding transformer with a capacitor connected in series with one of its windings.   *(Section 2.10.2)*

*2.33 A transformer for use at audio frequencies can be represented by the equivalent circuit of Fig. 2.66 on page 174. An ideal source of variable frequency supplies the input potential difference V_1. The output potential difference V_2 is applied to a resistive load of 10 Ω.

a) Derive and sketch the ratio of the magnitude of V_2 to the magnitude of V_1 as a function of frequency over the reasonably useful frequency range.

b) Assuming the magnitude of the supply potential difference is held constant, determine the values of frequency for which the load power is half that obtained in the mid-frequency range.   *(Section 2.11)*

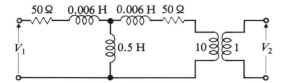

Fig. 2.66 Diagram for Problem 2.33.

*2.34 Suppose that the transformer of Problem 2.33 has a nearly closed loop of metal foil between its windings and that the foil is connected to one terminal of each of the two windings. This makes the interwinding capacitance C_{12} of Fig. 2.36(d) equal to zero. Assume the capacitance C_2 in Fig. 2.36(d) is 1.0 μF. Repeat part (a) of Problem 2.33 with this capacitance included.   *(Section 2.11)*

2.35 Three similar 2-kVA, 60-Hz, 220:110-V, single-phase transformers are connected in turn in wye–wye, delta–wye, wye–delta, and delta–delta. Determine the rated values of the primary and secondary line-to-line potential differences, the rated primary and secondary line currents, the ratio of line-to-line potential differences, and the ratio of line currents for each of the three-phase transformer banks so formed. (The transformers may be considered ideal.)   *(Section 2.12.1)*

2.36 Three similar 10-kVA, 60-Hz, 2300:115-V, single-phase transformers are connected in delta–wye to supply on their low-potential side a balanced three-phase load of 24 kVA. Determine the primary and secondary winding and line-to-line potential differences, winding currents, and line currents for this arrangement. (The transformers may be considered ideal.)   *(Section 2.12.1)*

2.37 A 3-phase, 60-Hz synchronous generator produces an output of 100 MVA at 15.8 kV line to line. This output is to be fed into a transmission line at 345 kV line to line. The transformer bank is to consist of three 2-winding transformers connected in delta on the low-potential side and in wye on the high-potential side.

   a) Specify the required rated values for the winding potential differences and currents in each transformer.
   b) Suppose that each transformer has a leakage reactance of 0.15 per unit, based on its rating. Determine the required generator line-to-line potential difference to provide rated balanced output to the transmission line at 0.8 lagging power factor and 345 kV.
   c) Draw a per-phase equivalent circuit of the transformer bank.   *(Section 2.12.1)*

2.38 The transformers of Problem 2.36 have an equivalent impedance referred to the primary side of $\overline{Z}'_{eq} = 11.9 + j\,24.4\,\Omega$. On their low-potential side, they supply, through a three-phase feeder of line impedance $\overline{Z}_F = 0.0315 + j\,0.104\,\Omega$, a balanced three-phase load of 24 kVA at a power factor of 0.85 lagging. Draw an equivalent circuit and phasor diagram (not to scale) for this system, and determine the line-to-line potential difference required at the primary terminals of the transformers in order that the line-to-line potential difference at the load shall be 200 V.   *(Section 2.12.1)*

2.39 Three similar 20-kVA, 60-Hz, 2200:220-V, single-phase transformers are connected in delta–wye to supply from their low-potential terminals a balanced 3-phase load of power factor 0.8 lagging. The equivalent impedance of one of the transformers referred to the high-

potential side is $\overline{Z}'_{eq} = 5.62 + j\,21.8\ \Omega$. The load is connected to the transformers by a 3-phase feeder of line impedance $\overline{Z}_{F2} = 0.025 + j\,0.700\ \Omega$. The transformer primary terminals are connected to a balanced 2200-V, 3-phase source by a 3-phase feeder of line impedance $\overline{Z}_{F1} = 1.0 + j3.0\ \Omega$. Draw an equivalent circuit and phasor diagram (not to scale) of this system, and determine the potential difference at the terminals of the load when the transformers are carrying rated current.
(*Section 2.12.1*)

2.40 Figure 2.67 shows the open–delta connection of two transformers which may be used to connect a 3-phase load to a 3-phase supply. Assuming the transformers are similar and each rated at 10 MVA, determine the load that may be supplied by this arrangement without overloading either of the transformers. The transformers may be considered ideal.
(*Section 2.12.1*)

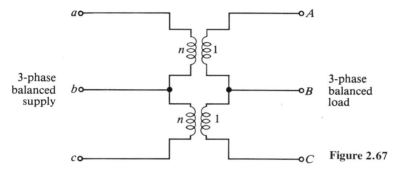

Figure 2.67

2.41 Consider a load that is supplied from a source by means of two 3-phase transmission lines connected in parallel electrically but following different geographical routes. Suppose the shorter line is overloaded while the longer line is underloaded. The current distribution in the two lines can be altered by inserting a phase-shifting transformer into one of the lines. Figure 2.68 shows a simple means of obtaining a small phase shift.

a) Assume that each transformer has a turns ratio of $n : 1$ and may be considered ideal. Show that the phase shift of the output potential differences with respect to those of the input is given by $\theta = \tan^{-1}\left(\sqrt{3}/n\right)$.

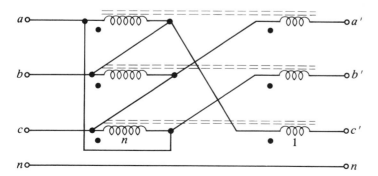

Figure 2.68

b) If the turns ratio of each transformer is 5 : 1, what is the ratio of the phasors $\overline{V}_{a'b'}$ and $\overline{V}_{ab}$?   *(Section 2.12.1)*

*2.42 Suppose that the three-phase, core-type transformer shown in Fig. 2.44(c) has windings of N_1 and N_2 turns on each leg, the N_1-turn winding being on the inside.

   a) Develop a magnetic equivalent circuit for the transformer. Assume that the core material has infinite permeability. Include reluctance elements to represent the leakage paths between the pair of windings on each leg. Also include a reluctance element to represent the flux paths in air from the top yoke to the bottom yoke outside the windings.

   b) Derive an electric equivalent circuit for the transformer with all reactances referred to the N_1-turn windings.

   c) Suppose the three N_1-turn windings are connected in parallel. Show that the impedance of the resultant 2-terminal system is $R_1/3 + j(X_L/3 + X_o/9)$, where R_1 is the resistance of each winding, X_L is its leakage reactance, and X_o is the reactance representing the yoke-to-yoke path.

   d) Is it possible to use the windings on only one leg as a single-phase transformer? Explain.

   e) Is it practical to connect the three N_1-turn windings in parallel and use the resultant system as a single-phase transformer? Explain.   *(Section 2.12.2)*

*2.43 Figure 2.47(b) shows a 3-phase, shell-type transformer with windings N_1 and N_2 turns in each window of the core.

   a) Develop a magnetic equivalent circuit for this transformer assuming infinite permeability of the core material. Include reluctance elements to represent the leakage paths between the pair of windings in each window.

   b) Derive an electric equivalent circuit for the transformer.

   c) Is it practical to connect the three N_1-turn windings in parallel and the three N_2-turn windings in series and to use the resultant system as a single-phase transformer? Explain.   *(Section 2.12.2)*

*2.44 A 400-Hz, 2-phase control motor requires 115 V rms on each phase. The only 400-Hz supply available is a 3-wire, 3-phase system with a line-to-line potential difference of 240 V rms. Magnetic cores having a cross-section area of 400 mm² and a peak operating flux density of 1.4 T are available. Specify the required numbers of turns on the windings of two transformers that use these cores and are connected in the Scott connection as in Fig. 2.51.

(Section 2.12.4)

*2.45 An industrial process requires a single-phase supply of 500 kVA at 2400 V. The available utility supply is 4000 V. An autotransformer is to be used.

   a) Determine the required current ratings of the two sections of the transformer winding.

   b) Suppose the two sections of the autotransformer winding are disconnected. Determine the kVA rating of the resultant two-winding transformer.   *(Section 2.13.1)*

*2.46 Figure 2.69 shows a variable ratio autotransformer. A sliding brush makes contact with a bared part of the top of the single-layer winding. Determine the required number of turns N and the dimension d for an autotransformer with an output of 0 to 115 V rms at 60 Hz and an output current of 8.7 A rms. Square wire having a 0.25-mm insulation thickness may be used with a current density of 1 A rms/mm² in the copper. The toroidal core may be operated at a peak flux density of 1.5 T.   *(Section 2.13.1)*

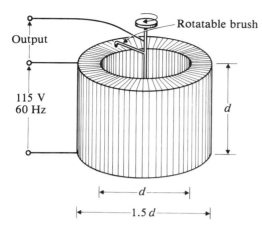

Rotatable brush

Output

115 V
60 Hz

d

d

1.5 d

Fig. 2.69 Diagram for Problem 2.46.

*2.47 A toroidal current transformer has a core with a mean diameter of 60 mm and a cross-section area of 100 mm². The material has a relative permeability of 5000 for small values of flux density. The primary consists of a single bus bar passing through the aperture of the core and carrying a 60 Hz current.

a) How many turns are required on the core to give a nominal current ratio of 800 : 5?

b) Determine the magnetizing reactance as seen from the secondary side.

c) Suppose the secondary winding is connected by means of a 2-conductor cable to a 5-A ammeter. Assume that the combined impedance of the meter, the cable, the winding resistance, and leakage reactance is $0.07 + j\,0.06$ Ω. Determine the magnitude error of the current measuring system.

d) If the current transformer supplies the current coil of a wattmeter rather than an ammeter, the error in phase of the current is significant. For the combined impedance given in (c), determine the phase error.

e) Would the phase error of (d) cause the wattmeter to read too high or too low if the short-circuit loss of a transformer were being measured?   *(Section 2.13.2)*

*2.48 Figure 2.55(a) shows an arrangement of two cores that may be used for the measurement of the direct current in a bus bar. Suppose each core has a cross-section area of 100 mm², a mean diameter of 100 mm, and 4000 turns. The core material has a residual flux density of 1.5 T and negligible coercive force.

a) Determine the largest rms value of 60-Hz sinusoidal potential difference that should be applied to the windings connected in series.

b) If the meter indicates the full-wave rectified average value, what will be the meter reading when the bus bar current is 2000 A?

c) If the coercive force of the material is 20 A/m and the hysteresis loop is rectangular, what is the meter reading with no current in the bus bar?

d) Does the magnetizing current produce an error in the meter reading? (Assume that the core has an ideal rectangular hysteresis loop.)   *(Section 2.13.3)*

*2.49 A series saturable reactor is required to control the power supplied to a 25-Ω resistive load from a 220-V, 400-Hz sinusoidal source. Toroidal cores having a cross-section area of

200 mm² and a mean diameter of 60 mm are available. The core material has a rectangular hysteresis loop with a residual flux density of 1.6 T and negligible coercive force.

a) What is the minimum number of primary turns required on each of the cores to produce zero output potential difference when the control current is zero?

b) If the control current is not to exceed 0.5 A, how many control turns are required on each core to produce maximum load potential difference?

c) When the control current is zero, what is the maximum value of the emf induced in each of the control windings?   *(Section 2.13.4)*

3 / Basic Principles of Electric Machines

The function of electromagnetic machines is to transmit energy or to convert it from one form to another. The transformer transmits electrical energy, changing only the potential difference and current at which it is available. It also converts a small amount of electrical energy to heat, but this is an undesired feature that is minimized by the designer. A translational or rotational electromagnetic machine converts energy from electrical to mechanical form, or the converse; that is, it "motors" or "generates." It also converts some electrical or mechanical energy to heat, but this again is undesired.

Electric motors and generators of all kinds may be classed as *electromechanical energy converters.* Such devices embody three essential parts: (1) an electric system, (2) a mechanical system, and (3) a coupling field. The device illustrated in Fig. 3.1 will act as an electric motor, since if a sufficiently high current is passed through

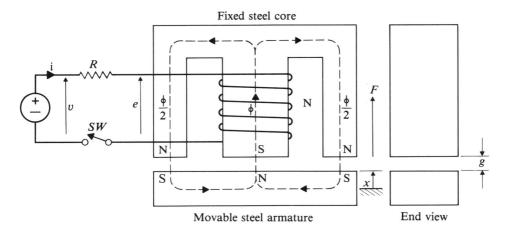

Fixed steel core

Movable steel armature

End view

Fig. 3.1 Actuator of vertical-lift contactor ($x = 0$ is "DOWN" position).

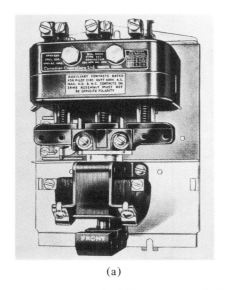

(a)

(b)

Fig. 3.2 (a) Vertical-lift contactor. (b) Contactor actuator.

the coil, the movable armature will be attracted upwards to the fixed core. The device is in fact the actuator of a vertical-lift contactor. The contactor and an actuator mounted for instructional purposes are shown in Fig. 3.2(a) and 3.2(b) respectively.

The motion of this actuator is translational and restricted in extent. Nevertheless this simple device possesses the three essential parts of an electromechanical energy converter, and analysis of its operation can illustrate very clearly the basic principles underlying the operation of the most complicated rotating machine.

3.1 ELECTROMECHANICAL ENERGY CONVERSION

In engineering systems in which the conversion of energy to mass or vice versa is not a practical consideration, the principle of the conservation of energy applies. Thus, within an isolated system, energy may be converted from one kind to another and transferred from an energy source to an energy sink, but it can be neither created nor destroyed. The total energy in the system is therefore constant. It is possible to write an equation describing the energy conversion and redistribution that have taken place during any time interval. If radiation of energy can be ignored (and in most energy converters this is the case), then for motor action, this equation is

$$\begin{array}{ccccccc} \text{Electrical} & & \text{Mechanical} & & \text{Increase} & & \text{Energy} \\ \text{energy} & = & \text{energy} & + & \text{of field} & + & \text{converted} \\ \text{from source} & & \text{to load} & & \text{energy} & & \text{to heat} \end{array} \qquad (3.1)$$

For generator operation, the first and second terms in Eq. (3.1) would have negative values.

The actuator illustrated in Fig. 3.1 is designed to act as a motor, and the energy sink, or load, is the movable armature whose potential energy in this case is increased due to its being raised against the force of gravity. Although under some conditions, this actuator may momentarily act as a generator, Eq. (3.1) still applies. As will be seen when machines are analyzed in detail, the main energy-conversion process in any electromagnetic machine can be reversed—that is, any machine can either motor or generate, but many are better suited to one purpose than the other.

The last term on the right-hand side of Eq. (3.1) includes losses of all kinds— that is, energy that escapes the main electromechanical conversion process. These losses may be divided into three parts, each one associated with one of the principal parts of the system:

$$\begin{matrix} \text{Energy} \\ \text{converted} \\ \text{to heat} \end{matrix} = \begin{matrix} \text{Resistance} \\ \text{losses} \end{matrix} + \begin{matrix} \text{Friction and} \\ \text{windage} \\ \text{losses} \end{matrix} + \begin{matrix} \text{Field} \\ \text{losses} \end{matrix} \qquad (3.2)$$

The first two terms on the right-hand side of Eq. (3.2) require no comment. The losses in the third term are made up almost entirely of hysteresis and eddy-current losses in the magnetic system. The actuator of Fig. 3.2 is laminated to reduce such losses. However, a small amount of energy is also dissipated in dielectric loss due to the electric field in the insulation material. Also, some energy may be radiated electromagnetically when the electric or magnetic fields change rapidly. If dielectric and radiation losses are neglected, the term "field losses" may be changed to read "losses in the magnetic coupling field." If the term "increase of field energy" in Eq. (3.1) is similarly interpreted as applying to the magnetic coupling field only, then substitution from Eq. (3.2) in Eq. (3.1) yields

$$\begin{matrix} \text{Electrical energy} \\ \text{from source} \\ \text{minus resistance} \\ \text{losses} \end{matrix} = \begin{matrix} \text{Mechanical energy} \\ \text{to load plus} \\ \text{friction and} \\ \text{windage losses} \end{matrix} + \begin{matrix} \text{Increase of} \\ \text{magnetic coupling} \\ \text{field energy plus} \\ \text{core losses} \end{matrix} \qquad (3.3)$$

If the electric field energy is negligible, the same current flows in all parts of the coil, and the effect of loss in the coil conductor may be modeled by a lumped resistance R, as shown in Fig. 3.1.

At any instant, the emf e induced in the coil by the change in flux linkage λ is

$$e = \frac{d\lambda}{dt} \quad \text{V} \qquad (3.4)$$

Consider now a differential time interval dt, during which the current in the coil is changing and the armature is moving. The differential energy transferred in time dt from the electric source to the coupling field is given by the energy output of the source minus the resistance loss; that is,

$$dW_e = v_t i \, dt - Ri^2 dt = (v_t - Ri)i \, dt = ei \, dt \quad \text{J} \qquad (3.5)$$

The coupling field forms an energy storage to which energy is supplied by the electric system. At the same time, energy is released from the coupling field to the mechanical system and also dissipated in the form of heat produced by field losses. The rate of release of energy *from* the field is not necessarily equal at any instant to the rate of supply of energy *to* the field, so that the amount of energy stored in the coupling field may vary. In time dt, let dW_f be the energy supplied to the field and either stored or dissipated. Let dW_m be the energy converted to mechanical form, useful or as loss, in the same time, dt. Then, by the principle of conservation of energy, the following energy-balance equation may be written *for the field:*

$$dW_e = dW_m + dW_f \quad \text{J} \tag{3.6}$$

Equation (3.6) is simply Eq. (3.3) expressed in symbols. It provides a basis for the analysis of the operation of machines.

3.1.1 Field Energy

The electrical energy input to the field dW_e can be found from the terminal quantities v and i, as shown in Eq. (3.5). If an expression can be derived for the energy supplied to the coupling field dW_f, then the mechanical energy output can be found from Eq. (3.6). The term dW_f includes core losses due to the changing magnetic field. Since these losses are usually small, a first step in modeling the system will be to neglect them. From this point on, therefore, W_f signifies the energy *stored* in the coupling field and recoverable from it.

To obtain an expression for dW_f of Eq. (3.6) in terms of the system variables, it is first necessary to find an expression for the energy stored in the magnetic field for any position of the armature. The armature will therefore be clamped at some value of air-gap length g so that no mechanical output can be produced. That is,

$$dW_m = 0 \tag{3.7}$$

If switch SW in Fig. 3.1 is now closed, the current will rise to a value v_t/R, and flux will be established in the magnetic system.

Let the relationship between coil flux linkage λ and current i for the chosen air-gap length be that shown in Fig. 3.3. Since core loss is being neglected, this will be a single-valued curve passing through the origin. In the absence of any mechanical output energy, all of the electric input energy must be stored in the magnetic field:

$$dW_e = dW_f \quad \text{J} \tag{3.8}$$

Substitution from Eqs. (3.4) and (3.8) in Eq. (3.5) yields

$$dW_f = dW_e = iedt = id\lambda \quad \text{J} \tag{3.9}$$

If now v_t is changed, resulting in a change in current from i_1 to i_2, there will be a corresponding change in flux linkage from λ_1 to λ_2. The states at the beginning and end of this change will correspond to two points such as a and b on the character-

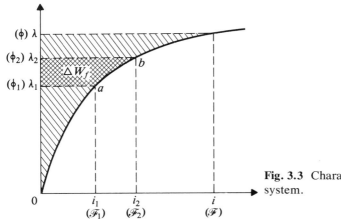

Fig. 3.3 Characteristic of the magnetic system.

istic in Fig. 3.3. The increase in energy stored during the transition between these two states is

$$\Delta W_f = \int_{\lambda_1}^{\lambda_2} i d\lambda \quad \text{J} \tag{3.10}$$

This is shown as the cross-hatched area in Fig. 3.3. When the coil current and the flux linkage are zero, the field energy is zero. When the flux linkage is increased from zero to λ, the total energy stored in the field is

$$W_f = \int_{0}^{\lambda} i d\lambda \quad \text{J} \tag{3.11}$$

This integral represents the area between the λ–i characteristic and the λ-axis, the entire shaded area of Fig. 3.3.

The expression of Eq. (3.11) for the energy stored in a magnetic field applies generally to any lossless magnetic system. Several other useful expressions for stored energy can be derived by introducing further assumptions. If it is assumed that there is no leakage flux, so that all flux ϕ in the magnetic system links all N turns of the coil, then

$$\lambda = N\phi \quad \text{Wb} \tag{3.12}$$

From Eqs. (3.9) and (3.12),

$$dW_f = i d\lambda = Ni d\phi = \mathscr{F} d\phi \quad \text{J} \tag{3.13}$$

where

$$\mathscr{F} = Ni \quad \text{A} \tag{3.14}$$

If the characteristic of Fig. 3.3 is rescaled to represent the relationship between ϕ and $\mathscr{F}$, the shaded area again represents the stored energy.

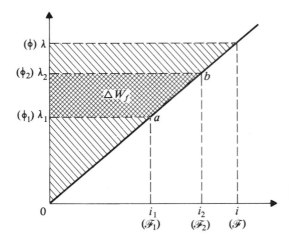

Fig. 3.4 Characteristic of the idealized magnetic system.

If the reluctance of the air gap forms a large part of the total reluctance of the magnetic system, then that of the steel may be neglected and the λ–i characteristic becomes the straight line through the origin shown in Fig. 3.4. For this idealized system,

$$\lambda = Li \quad \text{Wb} \tag{3.15}$$

where L is the inductance of the coil. Substitution in Eq. (3.11) gives the energy W_f in several useful forms:

$$W_f = \int_0^\lambda \frac{\lambda}{L}\, d\lambda = \frac{\lambda^2}{2L} = \frac{Li^2}{2} = \frac{i\lambda}{2} \quad \text{J} \tag{3.16}$$

If the reluctance of the magnetic system (that is, of the air gap) as seen from the coil is $\mathscr{R}$, then $\mathscr{F} = \mathscr{R}\phi$, and from Eq. (3.13),

$$W_f = \int_0^\phi \mathscr{F}\, d\phi = \frac{\mathscr{R}\phi^2}{2} = \frac{\mathscr{F}^2}{2\mathscr{R}} \quad \text{J} \tag{3.17}$$

As a further step in modeling, assume that no fringing takes place at the air gaps. Under these circumstances the entire field energy is distributed uniformly throughout the volume of the air gaps. If A is the cross-section area of the center leg of the core, $A/2$ is the cross-section area of the outer legs, and $l = 2g$ is the total length of air gap in a flux path, then from Eq. (3.16),

$$W_f = \frac{i\lambda}{2} = \frac{\mathscr{F}\phi}{2} = \frac{1}{2} HBlA \quad \text{J} \tag{3.18}$$

where B is the flux density in the air gaps. Since $B/H = \mu_0$ and lA is the total gap volume, it follows from Eq. (3.18) that the energy density in the air gaps is

$$w_f = \frac{W_f}{lA} = \frac{1}{2} BH = \frac{1}{2} \mu_0 H^2 = \frac{1}{2} \frac{B^2}{\mu_0} \quad \text{J/m}^3 \tag{3.19}$$

Equations (3.16), (3.17), and (3.19) represent three different ways of expressing the field energy, each of which is useful at different stages of machine analysis.

Example 3.1   The core and armature dimensions of the actuator of Fig. 3.1 are shown in Fig. 3.5. Both parts are made of 29-gauge M-36 sheet steel, whose magnetization curve is given in Fig. 1.7. The stacking factor is 0.95. The coil has 2000 turns. Leakage flux and fringing at the air gaps may be neglected. The armature is fixed, so that the length of the air gaps, $g = 10$ mm, and a direct current is passed through the coil, producing a flux density of 1.2 T in the air gap.

a) Determine the required coil current.

b) Determine the energy stored in the air gap.

c) Determine the energy stored in the steel.

d) Determine the total field energy.

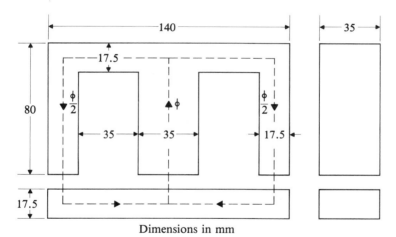

Dimensions in mm

Fig. 3.5   Diagram for Examples 3.1 and 3.2.

Solution

a) Flux density in the steel is

$$B_s = \frac{1.2}{0.95} = 1.26 \quad T$$

From Fig. 1.7, magnetic field intensity in the steel is

$$H_s = 320 \quad A/m$$

Length of flux path in the steel is

$$l_s \simeq 2 \times 80 + 2(70 - 17.5/2) = 282.5 \quad mm$$

The mmf required by the steel is

$$\mathscr{F}_s = 320 \times 0.2825 = 90.4 \quad \text{A}$$

Total flux in the magnetic system is

$$\phi = 35^2 \times 10^{-6} \times 1.2 = 1.470 \times 10^{-3} \quad \text{Wb}$$

Reluctance of air gaps in the magnetic path is

$$\mathscr{R}_a = \frac{2 \times 10 \times 10^{-3}}{4\pi \times 10^{-7} \times 35^2 \times 10^{-6}} = 12.99 \times 10^6 \quad \text{A/Wb}$$

The mmf required by air gaps is

$$\mathscr{F}_a = 1.470 \times 10^{-3} \times 12.99 \times 10^6 = 19.09 \times 10^3 \quad \text{A}$$

Total mmf required is

$$\mathscr{F} = \mathscr{F}_a + \mathscr{F}_s = 19.09 \times 10^3 + 90.4 = 19.18 \times 10^3 \quad \text{A}$$

Coil current is

$$I = \frac{19.18 \times 10^3}{2000} = 9.59 \quad \text{A}$$

b) From Eq. (3.19), energy density in the air gaps is

$$w_a = \frac{1}{2} \times \frac{1.2^2}{4\pi \times 10^{-7}} = 0.573 \times 10^6 \quad \text{J/m}^3$$

Volume of air gaps is

$$V_a = 2 \times 10 \times 35^2 \times 10^{-9} = 24.5 \times 10^{-6} \quad \text{m}^3$$

Energy stored in the air gap is

$$W_a = 24.5 \times 0.573 = 14.0 \quad \text{J}$$

c) Energy density in the steel is given by the area enclosed between the characteristic and the B axis in Fig. 1.7 up to a value of 1.26 T. By employing a straight-line approximation, this area is

$$w_s = \frac{1.26 \times 300}{2} = 189 \quad \text{J/m}^3$$

Volume of steel is

$$V_s = 35 \times 0.95(140 \times 97.5 - 2 \times 35 \times 62.5) \times 10^{-9} = 0.308 \times 10^{-3} \quad \text{m}^3$$

Energy stored in the steel is

$$W_s = 0.308 \times 10^{-3} \times 189 = 0.0582 \quad \text{J}$$

d) Total field energy is

$$W_f = W_a + W_s = 14.0 + 0.0582 \approx 14.06 \quad \text{J}$$

The proportion of field energy stored in the steel is, therefore, seen to be negligibly small.

Example 3.2 For the actuator of Example 3.1, determine the stored field energy, assuming that the steel of the core and armature has infinite permeability.

Solution From Example 3.1,
$$\mathcal{R}_a = 12.99 \times 10^6$$
$$\phi = 1.470 \times 10^{-3}$$
From Eq. (3.17),
$$W_f = \frac{12.99 \times 10^6 \times 1.470^2 \times 10^{-6}}{2} = 14.0 \quad \text{J}$$

3.1.2 Mechanical Energy in a Linear System

Electromechanical energy conversion is possible when the amount of energy stored in the coupling field depends upon the relative positions of the parts of the mechanical system. The truth of this general principle must be demonstrated. The ideal linear magnetic system introduced in Section 3.1.1 will first be examined.

It will be assumed that the armature of the actuator in Fig. 3.1 may move within certain limits, but that stops ensure that air gaps still exist when it has reached the upper limit of its movement. The $\lambda{-}i$ characteristics for the two extreme positions of the armature may be assumed to be the two straight lines passing through the origin of the diagram in Fig. 3.6. Further, the source potential difference has the constant value V_t. The following operations are now carried out:

i) The armature is held in the "Down" position, and switch SW is closed.

ii) The current in the coil rises to the steady value

$$I = \frac{V_t}{R} \quad \text{A} \tag{3.20}$$

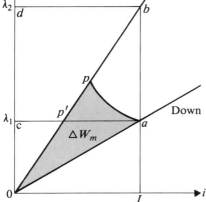

Fig. 3.6 Ideal system: $\lambda{-}i$ characteristics for extreme positions of the armature.

iii) The armature is allowed to move upwards against the action of a restraining force, which may possibly be only that of gravity. The reluctance of the magnetic system decreases; hence the flux linkage increases, and an emf e is induced in the coil, reducing the current to the value

$$i = \frac{V_t - e}{R} \quad \text{A} \tag{3.21}$$

iv) The armature reaches the "Up" position, at which instant current and flux linkage have the values corresponding to point p in Fig. 3.6.

v) Current and flux linkage now increase with time until e falls to zero, and once more the steady value of current I is established. The flux linkage has now risen to λ_2. This state is represented by point b in Fig. 3.6.

The shape of the λ–i curve between points a and b depends upon the speed at which the armature has been permitted to move. If it has moved very slowly, e will be negligibly small, and i will remain virtually constant at value I. If it has moved quickly, e will have reduced the current appreciably, and the operating point on the λ–i diagram in Fig. 3.6 will have followed a curve such as apb. The limiting case is $ap'b$, in which the armature motion is so rapid that it is completed before any significant change in flux linkage has occurred.

The energy stored in the magnetic field for any given operating point has been shown to be equal to the area between the λ–i characteristic on which the point lies and the λ-axis. Thus, the stored field energy for state a is

$$W_{fa} = \text{area } 0ac0 \quad \text{J} \tag{3.22}$$

The stored field energy for state b is

$$W_{fb} = \text{area } 0bd0 \quad \text{J} \tag{3.23}$$

Thus the change in stored field energy resulting from the movement of the armature to the top limit of its travel is

$$\Delta W_f = \text{area } 0bd0 - \text{area } 0ac0 \quad \text{J} \tag{3.24}$$

The energy supplied to the field from the source in a differential time dt during the movement of the armature is

$$dW_e = iedt = i\frac{d\lambda}{dt} \cdot dt = id\lambda \quad \text{J} \tag{3.25}$$

Thus the total energy supplied to the field by the source during the time in which the field energy is increased by ΔW_f is

$$\Delta W_e = \int_{\lambda_1}^{\lambda_2} id\lambda = \text{area } capbdc \quad \text{J} \tag{3.26}$$

If now ΔW_m is the mechanical work done in raising the armature through the limited distance between the stops, then by conservation of energy,

$$\Delta W_e = \Delta W_m + \Delta W_f \quad \text{J} \tag{3.27}$$

Thus, from Eqs. (3.24), (3.26), and (3.27),

$$\text{area } capbdc = \Delta W_m + \text{area } 0bd0 - \text{area } 0ac0 \quad \text{J} \tag{3.28}$$

from which

$$\Delta W_m = \text{area } 0ap0 \quad \text{J} \tag{3.29}$$

The mechanical work done, or the energy released by the field to the mechanical system in raising the armature is given by the shaded area in Fig. 3.6. If the magnetization characteristic, and hence the energy in the coupling field, were not affected by the movement of the armature, then ΔW_m would be zero. This proves the "General Principle."

A more complete understanding of the energy conversion process may be obtained by considering a movement of the operating point along a differential segment of curve ap of the λ–i diagram in Fig. 3.6. This movement of the operating point corresponds to a differential armature displacement dx. During the analysis of the consequences of this movement, useful expressions for calculating the force exerted on the armature will appear.

Figure 3.7 illustrates the differential movement of the operating point in Fig. 3.6 resulting from a displacement dx. The state at position x is represented by point a'; that at position $x + dx$, by point b'. By analogy with Eq. (3.29), the differential mechanical energy dW_m associated with movement dx is given by the shaded area $0a'b'0$ in Fig. 3.7. As this differential mechanical energy is equal to force times displacement, the force on the armature at position X is

$$F = \frac{dW_m}{dx} \quad \text{N} \tag{3.30}$$

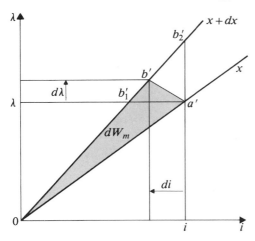

Fig. 3.7 Ideal system: λ–i characteristics for a differential displacement of the armature.

As has been explained in relation to Fig. 3.6, the slope of the segment $a'b'$ in Fig. 3.7 depends upon the speed at which the armature is permitted to move. For very rapid armature movement, segment $a'b'_1$, is traced out, and

$$dW_m = \text{area } 0a'b'_1 0 \quad \text{J} \tag{3.31}$$

For very slow movement, segment $a'b'_2$ is traced out, and

$$dW_m = \text{area } 0a'b'_2 0 \quad \text{J} \tag{3.32}$$

While area $0a'b'0$ in Fig. 3.7 is a differential quantity, areas $a'b'_1 b'a'$ and $a'b'b'_2 a'$ are second-order differential quantities, so that in the limit, as $dx \to 0$,

$$\text{area } 0a'b'0 = \text{area } 0a'b'_1 0 = \text{area } 0a'b'_2 0 \quad \text{J} \tag{3.33}$$

and dW_m is of the same magnitude, no matter whether the armature movement is fast or slow. It follows that the force F exerted on the armature in any position x is independent of the manner in which the armature is moving.

Consider first a very rapid differential armature displacement dx. It may be assumed that it takes place at essentially constant flux linkage, as illustrated in Fig. 3.8(a). Since λ does not change, no emf is induced in the coil, and dW_e is zero. The differential mechanical energy is given by the reduction in the area to the left of the λ-i characteristic—that is,

$$dW_m = -dW_f \quad \text{J} \tag{3.34}$$

and from Eq. (3.30),

$$F = -\left.\frac{\partial W_f}{\partial x}\right|_{\lambda = \text{constant}} \quad \text{N} \tag{3.35}$$

The mechanical output energy has been supplied entirely by the coupling field. From Eq. (3.16),

$$W_f = \frac{\lambda^2}{2L} \quad \text{J} \tag{3.36}$$

so that substitution in Eq. (3.35) from Eqs. (3.36) and (3.15) yields:

$$F = \frac{\lambda^2}{2L^2}\frac{dL}{dx} = \frac{i^2}{2}\frac{dL}{dx} \quad \text{N} \tag{3.37}$$

As i^2 is always positive, Eq. (3.37) shows that the force on the armature in the x direction is positive if inductance increases with motion in the x direction. In other words, the force acts in a direction to increase the inductance of the system.

If λ is constant, then the flux ϕ in the magnetic system is constant; and since, from Eq. (3.17),

$$W_f = \frac{\mathscr{R}\phi^2}{2} \quad \text{J} \tag{3.38}$$

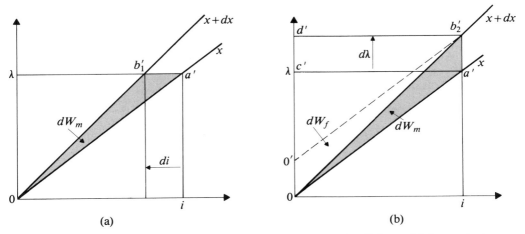

Fig. 3.8 Ideal system: Differential displacement. (a) At constant flux linkage. (b) At constant current.

it follows from Eq. (3.35) that

$$F = -\frac{\phi^2}{2}\frac{d\mathcal{R}}{dx} \quad J \tag{3.39}$$

Thus the force acts to decrease the reluctance of the magnetic system.

Consider now a very slow differential armature displacement dx. It may be assumed that it takes place at essentially constant current as illustrated in Fig. 3.8(b). By analogy with Eq. (3.29), the energy transferred to the mechanical system is

$$dW_m = \text{area } 0a'b_2'0 \quad J \tag{3.40}$$

The change in field energy during this displacement is, as before, given by the change in area to the left of the λ–i characteristic. This area is increased. Adding the construction line $0'b_2'$ parallel to $0a'$ to Fig. 3.8(b) shows that area $0b_2'0'0 =$ area $0a'b_2'0$, and area $0'b_2'd'0' = $ area $0a'c'0$. Since the field energy for position $x + dx$ is given by area $0b_2'$ $d'0$, then it is seen that $dW_f = dW_m$. Since there has been a change in flux linkage, dw_e is not zero. This again may be confirmed from Fig. 3.8(b) on geometrical principles, since from Eq. (3.25),

$$dW_e = id\lambda = \text{area } a'b_2' \, d'c' \quad J \tag{3.41}$$

Since area $a'b_2' \, d'c'a' = $ area $0a'b_2' \, 0'0$, then it follows that in this ideal linear system the source is delivering equal increments of energy to the mechanical system and to the stored field energy. This may be contrasted with the conclusion reached for very rapid armature movement, expressed in Eq. (3.34).

From Eq. (3.6),

$$dW_m = dW_e - dW_f \quad J \tag{3.42}$$

Substitution in Eq. (3.30) yields

$$F = \frac{dW_e}{dx} - \frac{dW_f}{dx} \quad \text{N}$$ (3.43)

and it therefore follows that for the present condition of slow motion at constant current

$$F = \frac{\partial W_e}{\partial x}\bigg|_{i=\text{constant}} - \frac{\partial W_f}{\partial x}\bigg|_{i=\text{constant}} \quad \text{N}$$ (3.44)

Substitution in Eq. (3.44) from Eqs. (3.15), (3.16), and (3.25) gives

$$F = \frac{i \, d\lambda}{dx} - \frac{i}{2}\frac{d\lambda}{dx} \quad \text{N}$$ (3.45)

and substitution in Eq. (3.45) from Eq. (3.15) gives

$$F = \frac{i^2}{2}\frac{dL}{dx} \quad \text{N}$$ (3.46)

This is the same conclusion as was reached in Eq. (3.37), confirming that the force F exerted on the armature in any position x is dependent on the position x and the current i, but is independent of the manner in which the armature is moving.

As has already been remarked, the force acts in the x direction to increase the inductance of the system. This statement may be generalized to apply to all deformable electromagnetic systems: "If an electromagnetic system is such that a deformation of some kind will increase the inductance of the winding or windings, then, when current flows, a force will arise that tends to produce that deformation." In other words, the system tends to assume a configuration with increased inductance or reduced reluctance.

From Fig. 3.1, it is seen that a positive displacement dx will correspond to a reduction dg in the air-gap length. Thus

$$dx = -dg \quad \text{m}$$ (3.47)

Substitution in Eq. (3.39) yields

$$F = -\frac{\phi^2}{2}\frac{d\mathcal{R}}{dx} = \frac{\phi^2}{2}\frac{d\mathcal{R}}{dg} \quad \text{N}$$ (3.48)

Since

$$\mathcal{R} = \frac{2g}{\mu_0 A} \quad \text{A/Wb}$$ (3.49)

then

$$F = \frac{\phi^2}{2}\left(\frac{2}{\mu_0 A}\right) = \frac{B^2}{2\mu_0}(2A) \quad \text{N}$$ (3.50)

The total cross-section area of the air gaps is $2A$. Thus the force per unit area of air gap is

$$F_A = \frac{B^2}{2\mu_0} \quad \text{N/m}^2 \tag{3.51}$$

If the coil of the actuator of Fig. 3.1 is excited with alternating current, a fluctuating force is developed that has an average value equal and opposite to the restraining force exerted on the armature.

Example 3.3   For the actuator of Example 3.1, determine the force acting on the armature for the given armature position and the conditions determined in Example 3.1. Assume the system is linear.

Solution   The total cross-section area of the air gaps is

$$2A = 2 \times 35^2 \times 10^{-6} \quad \text{m}^2$$

From Eq. (3.51), the force per unit area is

$$F_A = \frac{1.2^2}{2 \times 4\pi \times 10^{-7}} \quad \text{P}$$

Total force exerted is

$$F = 2A F_A = \frac{2 \times 35^2 \times 10^{-6} \times 1.2^2}{2 \times 4\pi \times 10^{-7}} = 1400 \quad \text{N}$$

*3.1.3 Mechanical Energy in a Saturable System

Figure 3.9 shows a diagram illustrating the λ–i characteristics for the magnetic system of the actuator in Fig. 3.1 when the effect of the ferromagnetic material is taken into account. In the "down" position, where the air gap is large, the characteristic is almost a straight line. In the "up" position, where the air gap is very small, the characteristic is nearly linear for small values of flux linkage, but as flux linkage is increased, the curvature of the characteristic due to saturation of the steel is pronounced. The areas of this diagram may be interpreted in exactly the same way as were those of Fig. 3.6 for the ideal linear system. Field energy is still given by Eq. (3.11), and the energy supplied is still given by Eq. (3.26). The shaded area of Fig. 3.9 represents the mechanical energy.

If an analytical expression is available that gives the coil current as a function of λ and x, then the force on the armature for a given value of x can readily be determined. Figure 3.10(a) illustrates a differential movement of the operating point in the λ–i diagram corresponding to a differential displacement dx of the armature made at high speed—that is, at constant flux linkage. By integrating to obtain an

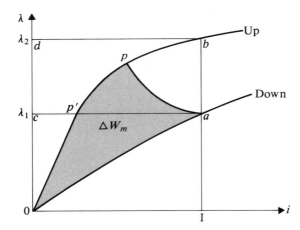

Fig. 3.9 Saturating system: $\lambda-i$ characteristics for extreme positions of the armature.

expression for the area between the $\lambda-i$ curve for any x and the λ-axis, W_f is obtained as a function of λ and x and can be written as

$$W_f = W_f(\lambda, x) \quad \text{J} \tag{3.52}$$

For the movement, the electrical energy input is zero, since λ does not change and the emf is zero. Consequently,

$$dW_m = dW_f(\lambda, x) \tag{3.53}$$

and

$$F = \frac{dW_m}{dx} = -\frac{\partial W_f}{\partial x}(\lambda, x)\bigg|_{\lambda=\text{constant}} \quad \text{N} \tag{3.54}$$

This corresponds to the expression for a linear system in Eq. (3.35).

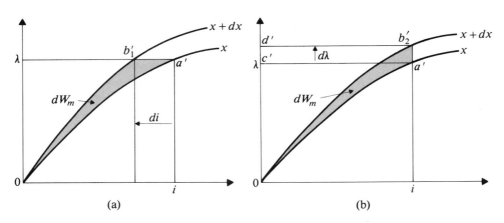

Fig. 3.10 Saturating system: Differential displacement. (a) At constant flux linkage. (b) At constant current.

More usually, however, it is convenient to express λ as a function of x and i and to employ a different approach. Figure 3.10(b) illustrates a differential movement of the operating point in the λ–i diagram corresponding to a differential displacement dx of the armature made at low speed—that is, at constant current. For this displacement, during which the flux linkage changes, the emf is not zero, and therefore dW_e is not zero. From Eq. (3.6),

$$dW_m = dW_e - dW_f = d(W_e - W_f) \quad \text{J} \tag{3.55}$$

Thus for the displacement illustrated in Fig. 3.10(b),

$$F = \frac{\partial}{\partial x}(W_e - W_f)\Big|_{i=\text{constant}} \quad \text{N} \tag{3.56}$$

The energy dW_m is given by the area $0a'b'_20$ in Fig. 3.10(b), and it can be seen that this is equal to the increase in area below the λ–i characteristic. This last area has the dimensions of energy. The quantity of energy that it represents does not exist anywhere inside (or outside) the magnetic system; nevertheless, it is a useful quantity and is given the name *coenergy* and symbol W'. By integrating the given relationship between λ, x, and i to obtain an expression for the area between the λ–i curve and the i-axis for any x, W' is obtained as a function of i and x. Thus, for any armature position,

$$W'(i, x) = \int_0^i \lambda di \quad \text{J} \tag{3.57}$$

The force on the armature is then

$$F = \frac{\partial W'}{\partial x}(i, x)\Big|_{i=\text{constant}} \quad \text{N} \tag{3.58}$$

Also, since $\lambda = N\phi$, and $i = \mathscr{F}/N$, substitution in Eq. (3.57) yields the coenergy as a function of the mmf and displacement

$$W'(\mathscr{F}, x) = \int_0^{\mathscr{F}} \phi d\mathscr{F} \quad \text{J} \tag{3.59}$$

and

$$F = \frac{W'(\mathscr{F}, x)}{x}\Big|_{\mathscr{F}=\text{constant}} \quad \text{N} \tag{3.60}$$

For a linear system in which the λ-i characteristic is a straight line, it is clear that at all times the coenergy is equal in magnitude to the stored field energy. But, for a saturable system, the coenergy is greater than the stored energy.

Example 3.4 The flux-linkage–current relationship for an actuator can be expressed approximately by

$$\lambda = \frac{0.08i^{1/2}}{g}$$

between the limits $0 < i < 5$ A and $0.02 < g < 0.10$ m. If the current is maintained at 4 A, what is the force on the armature for $g = 0.06$ m?

Solution   The λ–i relationship is nonlinear, and thus the force must be determined using Eq. (3.54) or (3.60). Since the flux linkage is given as a function of the current, the latter equation is convenient. The coenergy is

$$W' = \int_0^i \frac{0.08 i^{1/2}}{g} \, di = \frac{0.08}{g} \left(\frac{2}{3}\right) i^{3/2} \quad \text{J}$$

From Eq. (3.60), where, since i is constant, $\mathscr{F}$ is constant,

$$F = \left. \frac{\partial W'}{\partial g} \right|_{i \, = \, 4} = -\frac{0.08}{g^2} \left(\frac{2}{3}\right) i^{3/2}$$

$$= -\frac{0.08}{0.06^2} \times \frac{2}{3} \times 4^{3/2} = -119 \quad \text{N}$$

The negative sign indicates a force tending to decrease the air-gap length g.

3.1.4 Rotating Actuator and Reluctance Motor

The diagram in Fig. 3.11 illustrates the essential parts of an actuator that will give a rotational motion, instead of the translational motion produced by the actuator of the vertical-lift contactor. The fixed part of the magnetic system is now called the *stator,* and the moving part is called the *rotor.* The latter is mounted on a shaft whose axis is normal to the plane of the diagram, so that it may rotate freely

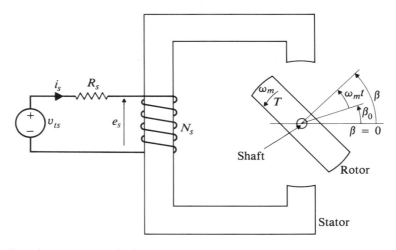

Fig. 3.11 Rotating actuator and reluctance motor.

between the poles of the stator. The assumptions and approximations employed in modeling this actuator are identical with those employed in modeling the translational actuator.

If a current is passed through the coil of the actuator in Fig. 3.11 and the rotor is free to move, then the rotor will be attracted into alignment with the pole pieces of the stator. If, on the other hand, the rotor is displaced from this aligned position and a restraining torque is applied to the shaft, then—when the current is passed through the coil—the rotor will move toward a position in which the torque developed by the actuator balances the restraining torque. The differential work done as the rotor moves through a differential angle $d\beta$ radians is

$$dW_m = T d\beta \quad \text{J} \tag{3.61}$$

where T is the torque developed by the actuator in the direction of β. By a procedure exactly analogous to that employed to determine an expression for the force developed by the translational actuator, it may be shown that for a linear system,

$$T = \frac{1}{2} i_s^2 \frac{dL}{d\beta} = -\frac{1}{2} \phi^2 \frac{d\mathcal{R}}{d\beta} \quad \text{N} \cdot \text{m} \tag{3.62}$$

Thus if L or $\mathcal{R}$ is known as a function of β, the torque for any position of the rotor may be determined. The torque always acts in such a direction that the resulting rotation increases the inductance, or decreases the reluctance.

From Eq. (3.62), it is seen that there are two relationships that may be employed in predicting the behavior of this actuator. The one chosen depends upon the type of electrical excitation applied, i.e., whether the current i or the flux ϕ is known. If the permeability of the steel parts of the magnetic system is assumed infinite, then both $dL/d\beta$ and $d\mathcal{R}/d\beta$ may be determined as functions of β from the dimensions of the device.

If excitation is from a source of constant direct potential difference V_{ts}, then in the steady-state condition,

$$i_s = I_s = \frac{V_{ts}}{R_s} \quad \text{A} \tag{3.63}$$

If now an external torque T_L is applied to the shaft of the rotor, the rotor will be deflected from alignment with the pole pieces of the stator to an angle β at which the internal torque T balances the external torque. Under these conditions,

$$T = T_L = \frac{1}{2} I_s^2 \frac{dL}{d\beta} \quad \text{N} \cdot \text{m} \tag{3.64}$$

The deflection β may be obtained by solution of Eq. (3.64).

On the other hand, if excitation is from a source of constant-amplitude, sinusoidally alternating potential difference v_{ts}, and if the effect of the winding resistance may be neglected, then, since

$$v_{ts} = N_s \frac{d\phi}{dt} \quad \text{V} \tag{3.65}$$

the amplitude and time variation of ϕ in the magnetic system is known and is unaffected by the angle β of the rotor. The sinusoidal flux variation will have a peak value of $\hat{\phi}$; and for any angle β, T will vary over the range

$$0 \geqslant T \geqslant -\frac{1}{2}\hat{\phi}^2 \frac{d\mathcal{R}}{d\beta} \quad \text{N} \cdot \text{m} \tag{3.66}$$

If now an external torque T_L is applied to the shaft of the rotor, the rotor will be deflected from alignment with the pole pieces of the stator and will oscillate through an angle determined by the relation

$$J \frac{d^2\beta}{dt^2} = T - T_L \quad \text{N} \cdot \text{m} \tag{3.67}$$

where J is the rotational inertia of the rotor and connected system in which friction is assumed to be negligible. If the frequency of the alternating source is high, and the inertia of the system is large, the rotor will take up an essentially constant angle β, such that the average torque T_{av} developed by the actuator is equal to the load torque T_L. The relationship in Eq. (3.62) then becomes

$$T_{av} = T_L = -\frac{1}{4}\hat{\phi}^2 \frac{d\mathcal{R}}{d\beta} \quad \text{N} \cdot \text{m} \tag{3.68}$$

Again, Equation (3.68) may be solved for angle β.

When the actuator is excited from an alternating source it may also operate as a continuously rotating motor. Under these circumstances, $d\mathcal{R}/d\beta$ in Eq. (3.62) becomes a periodic function of time t. Let $\mathcal{R}_d$ be the reluctance of the magnetic system of Fig. 3.11 when the rotor is in the direct-axis position ($\beta = 0$) and $\mathcal{R}_q$ be the reluctance when the rotor is in the quadrature-axis position ($\beta = \pi/2$). Then the mean reluctance is

$$\mathcal{R}_a = \frac{\mathcal{R}_q + \mathcal{R}_d}{2} \quad \text{A/Wb} \tag{3.69}$$

It will be assumed that the shape of the stator and rotor poles is such that the reluctance varies sinusoidally about the mean value as illustrated in Fig. 3.12. The diagram shows that there are two cycles of reluctance for each revolution of the rotor. Thus, if

$$\frac{\mathcal{R}_q - \mathcal{R}_d}{2} = \mathcal{R}_b \quad \text{A/Wb} \tag{3.70}$$

then for any angle β, the reluctance of the magnetic system is

$$\mathcal{R} = \mathcal{R}_a - \mathcal{R}_b \cos 2\beta \quad \text{A/Wb} \tag{3.71}$$

Substitution from Eq. (3.71) in Eq. (3.62) yields

$$T = -\phi^2 \mathcal{R}_b \sin 2\beta \quad \text{N} \cdot \text{m} \tag{3.72}$$

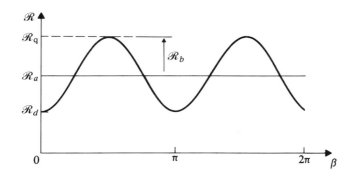

Fig. 3.12   Variation of reluctance with rotor position.

Thus, if the rotor is displaced from the direct axis position, then T will tend to move it into a position where $\sin 2\beta = 0$ — that is, where $\beta = 0$ or π radians.

Consider that the rotor is rotating at a constant angular velocity $\omega_m = d\beta/dt$, and the coil is excited from a source of alternating potential difference v_{ts} of angular frequency ω_s. If $R_s i_s << v_{ts}$, then the emf e_s induced in the coil must be effectively sinusoidal, and since $e_s = N_s d\phi/dt$, it follows that the flux in the magnetic system must also vary sinusoidally. The coil current will be whatever is necessary to maintain the sinusoidal flux. For convenience, the origin of the time scale will be chosen such that at $t = 0$, $\phi = \hat\phi$, and

$$\phi = \hat\phi \cos \omega_s t \quad \text{Wb} \tag{3.73}$$

Also, at $t = 0$, let the rotor have some angular position β_0, so that

$$\beta = \omega_m t + \beta_0 \quad \text{rad} \tag{3.74}$$

Substitution from Eqs. (3.73) and (3.74) in Eq. (3.72) yields

$$T = -\hat\phi^2 [\cos^2 \omega_s t] \mathcal{R}_b \sin 2(\omega_m t + \beta_0)$$

$$= -\frac{\hat\phi^2}{2} [1 + \cos 2\omega_s t] \mathcal{R}_b \sin 2(\omega_m t + \beta_0)$$

$$= -\frac{\mathcal{R}_b \hat\phi^2}{2} \left\{ \sin 2(\omega_m t + \beta_0) + \frac{1}{2} \sin 2 [(\omega_m + \omega_s)t + \beta_0] \right.$$

$$\left. + \frac{1}{2} \sin 2 [(\omega_m - \omega_s)t + \beta_0] \right\} \quad \text{N·m} \tag{3.75}$$

The three sine terms on the right-hand side of Eq. (3.75) are functions of time whose average value is zero unless, in one of them, the coefficient of t is zero. Since $\omega_m \neq 0$, then the necessary condition for nonzero average torque is

$$|\omega_m| = |\omega_s| \quad \text{rad/s} \tag{3.76}$$

under which circumstances

$$T_{\text{av}} = -\frac{\mathcal{R}_b \hat{\phi}^2}{4} \sin 2\beta_0 \quad \text{N} \cdot \text{m} \tag{3.77}$$

Equation (3.76) shows that an average torque will be produced for either direction of rotation. This torque will tend to maintain the speed of the rotor against friction, windage, and any external load torque applied to the rotor shaft.

It follows from Eq. (3.76) that, at $\omega_s t = 2n\pi$, where n is an integer, $\phi = \hat{\phi}$, and $\beta = \beta_0$; that is, the rotor always reaches a particular position when the flux reaches its peak value. Rotation and flux variation may thus be said to be synchronized. In general, any machine whose speed of rotation in radians per second is equal to the angular frequency of the source (or indeed to any integral submultiple of that angular frequency) is said to rotate at *synchronous speed*. Since the torque in this particular form of electromechanical energy converter is due to the variation of reluctance with rotor position, it is called a *synchronous reluctance machine*.

From Eq. (3.77) a curve of T_{av} as a function of β_0 may be drawn, as shown in Fig. 3.13. For $\beta_0 < 0$, $T_{\text{av}} > 0$; that is, the developed torque acts in the direction of rotation, and the machine motors. Since β_0 is negative, flux reaches its peak value as the poles of the rotor are approaching those of the stator. The maximum motoring torque occurs when $\beta_0 = -\pi/4$. If the combination of torque due to friction and windage plus any external torque applied to the shaft exceeds this maximum value, then the machine is pulled out of synchronism and brought to a standstill.

For $\beta_0 > 0$, $T_{\text{av}} < 0$; that is, the torque developed opposes the rotation, and an external driving torque must be applied to the rotor shaft to maintain the rotor at synchronous speed. This means that mechanical energy must be supplied to the system. If this energy exceeds that dissipated in friction and windage, the excess will be converted to electrical energy, and the machine generates. This can only occur, however, if the coil is already connected to an ac source, which becomes a sink as the driving torque is applied, and the machine begins to generate. Since under generating conditions β_0 is positive, the flux reaches its peak value as the

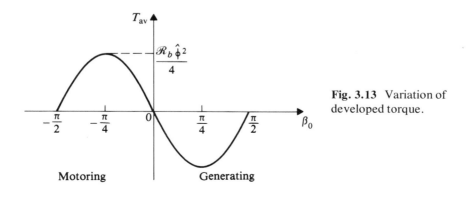

Fig. 3.13 Variation of developed torque.

poles of the rotor are receding from those of the stator. The maximum generating torque occurs when $\beta_0 = \pi/4$. If the driving torque applied to the shaft exceeds the combination of developed torque, plus that due to friction and windage, then the machine is driven above synchronous speed, and may run away unless the prime mover speed is limited. Under such conditions, continuous energy conversion ceases. Thus, there is a limit to the rate of energy conversion or to the power that the machine can develop.

The emf induced in the stator coil when the machine is motoring or generating is

$$e_s = N_s \frac{d\phi}{dt} = -N_s \omega_s \hat{\phi} \sin \omega_s t \quad \text{V} \tag{3.78}$$

and the rms magnitude of this emf is

$$E_s = \frac{N_s \omega_s \hat{\phi}}{\sqrt{2}} \quad \text{V} \tag{3.79}$$

The current in the coil may be determined by substituting from Eqs. (3.71), (3.73), and (3.74) in the expression $\phi = N_s i_s / \mathcal{R}$ for the condition $\omega_s = \omega_m$. This yields

$$i = \frac{\hat{\phi}}{N_s} \cos \omega_s t \, [\mathcal{R}_a - \mathcal{R}_b \cos 2(\omega_s t + \beta_0)]$$

$$= \frac{\hat{\phi}\mathcal{R}_a}{N_s} \cos \omega_s t - \frac{\hat{\phi}\mathcal{R}_b}{2N_s} [\cos(\omega_s t + 2\beta_0) + \cos (3\omega_s t + 2\beta_0)] \quad \text{A} \tag{3.80}$$

From Eq. (3.80) it is seen that the coil current contains a third-harmonic component. This is an undesirable feature of such machines, which renders them useless as practical generators and limits their size as motors. They may be used, however, to drive electric clocks, record players, or other small mechanisms requiring accurate and constant speed.

Example 3.5 A reluctance motor of the type illustrated in Fig. 3.11 has a stator and rotor of cross section 25 mm × 25 mm and a rotor 50 mm in length. The air gap is approximately 4 mm in length, so that the magnetic circuit reluctances are

Direct-axis reluctance: $\mathcal{R}_d = 10 \times 10^6$ A/Wb
Quadrature-axis reluctance: $\mathcal{R}_q = 40 \times 10^6$ A/Wb

It may be assumed that reluctance varies with rotor position in the sinusoidal manner illustrated in Fig. 3.12 and expressed in Eq. (3.71). The stator coil has 3000 turns and is excited from a 115 V, 25 Hz supply. The coil resistance is negligibly small.

a) Determine the maximum value of average mechanical power which this machine can develop.

b) Determine the rms value of the coil current.

Solution

a) From Eqs. (3.69), (3.70), and (3.71),

$$\mathscr{R} = (25 - 15 \cos 2\beta) \times 10^6 \quad \text{A/Wb}$$

from which

$$\frac{d\mathscr{R}}{d\beta} = 30 \times 10^6 \sin 2\beta$$

Since the resistance of the coil is negligible, from Eq. (3.79),

$$E_s = V_{ts} = \frac{N_s \omega_s \hat{\phi}}{\sqrt{2}}$$

from which

$$\hat{\phi} = \frac{115\sqrt{2}}{3000 \times 50\pi} = 0.345 \times 10^{-3} \quad \text{Wb}$$

From Eq. (3.77), the maximum average motoring torque is developed when $\beta_0 = -45°$ and

$$T_{av} = \frac{\mathscr{R}_b \hat{\phi}^2}{4} = \frac{15 \times 10^6 \times 0.345^2 \times 10^{-6}}{4} = 0.447 \quad \text{N·m}$$

Necessarily, for the machine to develop torque

$$|\omega_m| = |\omega_s| = 50\pi \quad \text{rad/s}$$

Thus the power developed is

$$P = T_{av} \times \omega_m = 0.447 \times 50\pi = 70 \quad \text{W}$$

b) For $\beta_0 = -45°$, from Eq. (3.80),

$$i = \frac{0.345 \times 10^{-3} \times 25 \times 10^6}{3000} \cos \omega_s t$$

$$+ \frac{0.345 \times 10^{-3} \times 15 \times 10^6}{2 \times 3000} [\sin \omega_s t + \sin 3\omega_s t]$$

$$= 2.88 \cos \omega_s t + 0.863 \sin \omega_s t + 0.863 \sin 3\omega_s t$$

$$= 3.01 \sin (\omega_s t + 73.3°) + 0.863 \sin 3\omega_s t$$

The rms value of this current is

$$I = \left[\frac{3.01^2}{2} + \frac{0.863^2}{2} \right]^{1/2} = 2.21 \quad \text{A}$$

3.1.5 Doubly Fed Machine

The synchronous reluctance machine would work equally well with the exciting coil on the rotor instead of on the stator. This fact prompts the thought that some advantage may be gained in fitting such coils on both members of the machine. Such an arrangement is shown in Fig. 3.14. The rotor coil is connected to its source by means of slip rings.

In order to analyze the behavior of this two-winding device, it is convenient to make the same assumptions and approximations as were made in the case of the single-winding rotary actuator and then to carry out a procedure analogous to that followed in analyzing the behavior of the ideal linear translational actuator.

For the system of Fig. 3.14 also, the electrical input energy from the sources minus the resistance losses will be equal to the mechanical output energy (including mechanical losses) plus the increase in the stored magnetic field energy. An expression for the field energy can be derived by locking the rotor at some arbitrary angle β, so that no mechanical work can be done, and then establishing the magnetic field by increasing the currents i_s and i_r.

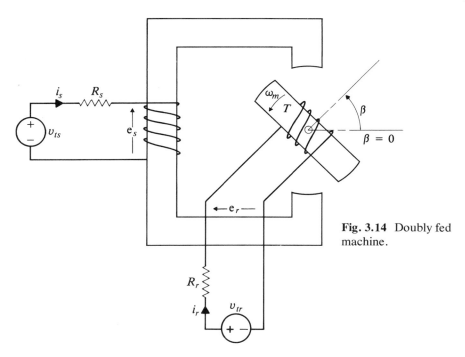

Fig. 3.14 Doubly fed machine.

The differential electrical energy input to the field resulting from the mmf's of the two windings is

$$dW_e = e_s i_s dt + e_r i_r dt \quad \text{J} \tag{3.81}$$

In the absence of mechanical output,

$$dW_e = dW_f \quad \text{J} \tag{3.82}$$

If λ_s and λ_r are the flux linkages of the stator and rotor windings respectively, then the emf's induced in the windings are

$$e_s = \frac{d\lambda_s}{dt} \qquad e_r = \frac{d\lambda_r}{dt} \quad \text{V} \tag{3.83}$$

The flux linkages can be expressed as functions of the coil currents using the self- and mutual inductances of the coils, which are constant for a fixed rotor position. Thus

$$\lambda_s = L_{ss}i_s + L_{sr}i_r \quad \text{Wb} \tag{3.84}$$

$$\lambda_r = L_{sr}i_s + L_{rr}i_r \quad \text{Wb} \tag{3.85}$$

Now consider a two-stage process, in which the first stage consists of increasing the stator current from zero while the rotor winding is open-circuited, so that $i_r = 0$. From Eqs. (3.83) and (3.84),

$$e_s = \frac{d\lambda_s}{dt} = L_{ss}\frac{di_s}{dt} \quad \text{V} \tag{3.86}$$

Substitution in Eq. (3.81) gives

$$dW_{e(1)} = L_{ss}i_s di_s \quad \text{J} \tag{3.87}$$

Integrating this expression from zero up to a final current i_s gives the electrical energy input during the first stage as

$$W_{e(1)} = \int_0^{i_s} L_{ss}i_s di_s = \frac{1}{2}L_{ss}i_s^2 \quad \text{J} \tag{3.88}$$

Note that since $i_r = 0$, no electrical energy enters the rotor winding during the first stage, even though there is an emf induced in it.

The second stage consists of holding i_s constant while i_r is increased from zero. During this operation, from Eqs. (3.84) and (3.85),

$$e_s = \frac{d(L_{ss}i_s + L_{sr}i_r)}{dt} = L_{sr}\frac{di_r}{dt} \quad \text{V} \tag{3.89}$$

$$e_r = \frac{d(L_{sr}i_s + L_{rr}i_r)}{dt} = L_{rr}\frac{di_r}{dt} \quad \text{V} \tag{3.90}$$

Substitution from Eqs. (3.89) and (3.90) in Eq. (3.81) gives

$$dW_{e(2)} = L_{sr}i_s di_r + L_{rr}i_r di_r \quad \text{J} \tag{3.91}$$

Integrating Eq. (3.91) from zero up to a final current i_r gives the electrical energy input during the second stage:

$$W_{e(2)} = \int_0^{i_r} L_{sr}i_s di_r + \int_0^{i_r} L_{rr}i_r di_r = L_{sr}i_s i_r + \frac{1}{2}L_{rr}i_r^2 \quad \text{J} \tag{3.92}$$

In the absence of any mechanical energy, all of the net electrical energy input during stages 1 and 2 must have been stored in the magnetic field. Thus, the stored energy with currents i_s and i_r is,

$$W_f = W_{e(1)} + W_{e(2)} = \frac{1}{2} L_{ss} i_s^2 + L_{sr} i_s i_r + \frac{1}{2} L_{rr} i_r^2 \quad \text{J} \qquad (3.93)$$

An expression for the torque developed in the doubly-fed machine may now be derived. Suppose that the rotor is free to move so that the inductances L_{ss}, L_{rr}, and L_{sr} are all functions of β. At any instant, the potential differences of the sources in Fig. 3.14 are

$$v_{ts} = R_s i_s + \frac{d\lambda_s}{dt} \quad \text{V} \qquad (3.94)$$

$$v_{tr} = R_r i_r + \frac{d\lambda_r}{dt} \quad \text{V} \qquad (3.95)$$

These can be converted to power equations by multiplying Eq. (3.94) by i_s and Eq. (3.95) by i_r, and then substituting from Eqs. (3.84) and (3.85). Thus

$$v_{ts} i_s = R_s i_s^2 + i_s \frac{d}{dt} (L_{ss} i_s + L_{sr} i_r)$$

$$= R_s i_s^2 + L_{ss} i_s \frac{di_s}{dt} + i_s^2 \frac{dL_{ss}}{dt} + L_{sr} i_s \frac{di_r}{dt} + i_s i_r \frac{dL_{sr}}{dt} \quad \text{W} \qquad (3.96)$$

$$v_{tr} i_r = R_r i_r^2 + i_r \frac{d}{dt} (L_{sr} i_s + L_{rr} i_r)$$

$$= R_r i_r^2 + L_{sr} i_r \frac{di_s}{dt} + i_s i_r \frac{dL_{sr}}{dt} + L_{rr} i_r \frac{di_r}{dt} + i_r^2 \frac{dL_{rr}}{dt} \quad \text{W} \qquad (3.97)$$

The total electrical power input from the sources is the sum of the left-hand sides of Eqs. (3.96) and (3.97). The copper losses are $R_s i_s^2 + R_r i_r^2$. The sum of the remaining terms must therefore be the electrical power input to the field, which is equal to the mechanical power released from the field plus the power into storage in the field. The power, or rate of flow of energy, into storage is given from Eq. (3.93) as

$$\frac{dW_f}{dt} = \frac{d}{dt} \left(\frac{1}{2} L_{ss} i_s^2 + L_{sr} i_s i_r + \frac{1}{2} L_{rr} i_r^2 \right)$$

$$= L_{ss} i_s \frac{di_s}{dt} + \frac{i_s^2}{2} \frac{dL_{ss}}{dt} + L_{sr} i_s \frac{di_r}{dt} + L_{sr} i_r \frac{di_s}{dt}$$

$$\quad + i_s i_r \frac{dL_{sr}}{dt} + L_{rr} i_r \frac{di_r}{dt} + \frac{i_r^2}{2} \frac{dL_{rr}}{dt} \quad \text{W} \qquad (3.98)$$

The rate of energy flow into the mechanical system is therefore

$$\frac{dW_m}{dt} = [\text{RHS of (3.96)} - R_s i_s^2] + [\text{RHS of (3.97)} - R_r i_r^2]$$

$$- \text{RHS of (3.98)} \quad \text{W} \qquad (3.99)$$

where RHS signifies the right-hand side of equations whose numbers are given. From Eq. (3.99), the mechanical power is

$$p_{\text{mech}} = \frac{i_s^2}{2} \frac{dL_{ss}}{dt} + i_s i_r \frac{dL_{sr}}{dt} + \frac{i_r^2}{2} \frac{dL_{rr}}{dt} \quad \text{W} \tag{3.100}$$

Since

$$p_{\text{mech}} = T\omega_m = T \frac{d\beta}{dt} \quad \text{W} \tag{3.101}$$

the torque is then

$$T = \frac{i_s^2}{2} \frac{dL_{ss}}{d\beta} + i_s i_r \frac{dL_{sr}}{d\beta} + \frac{i_r^2}{2} \frac{dL_{rr}}{d\beta} \quad \text{N} \cdot \text{m} \tag{3.102}$$

The terms in Eq. (3.102) involving angular rate of change of self-inductance are reluctance torques similar to the torque arising in the rotational actuator discussed in Section 3.1.3. The term involving angular rate of change of mutual inductance expresses the torque that has been caused by the interaction of fields produced by the stator and rotor currents. It is this mutual inductance torque that is most commonly exploited in practical rotating machines.

3.2 CYLINDRICAL MACHINES

In the torque expression of Eq. (3.102), the first and last terms on the right-hand side describe reluctance torques. It has been explained that a reluctance motor or generator must run at synchronous speed for it to develop an average torque and be capable of energy conversion. If the stator and rotor sources were of different frequencies, then the machine would have two different synchronous speeds due to the reluctance torques. Only one of these could be satisfied at a time, and the other torque term would be oscillatory and would produce unwanted speed oscillation. It is therefore desirable to eliminate one or both of the reluctance–torque terms and concentrate on the mutual inductance term that, as will be seen, may permit variable-speed operation of the machine.

The reluctance–torque term due to the rotor excitation would disappear from Eq. (3.102) if the space derivative of the rotor self-inductance $dL_{rr}/d\beta$ were identically zero. This can be achieved by mounting the rotor within a stator which consists of a hollow cylinder coaxial with the rotor. This configuration is illustrated in Fig. 3.15, where L_{rr} has the same magnitude no matter what the position of the rotor.

The question now arises as to how the winding should be placed on the cylindrical stator. The manner in which this may be done is illustrated in Fig. 3.16. Conductors are embedded in slots on the inner periphery of the stator, and the ends of these conductors are joined by end connectors so that conductors and connectors form a coil mounted inside the stator. The end connectors shown in full line are at

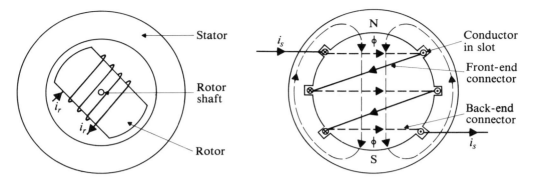

Fig. 3.15 Elimination of reluctance torque due to rotor excitation.

Fig. 3.16 Mounting of winding on a cylindrical stator.

the end of the stator nearest to the reader. The end connectors shown in broken line are at the end of the stator remote from the reader. In practice, both front and back connectors are led around the stator periphery; otherwise they would prevent the introduction of the rotor into the stator. In addition, several conductors are placed in each slot, so that the six-slot stator of Fig. 3.16 would have more than the three turns illustrated.

Direction of the current in the conductors of Fig. 3.16 is indicated by the conventional signs: a cross indicating a current down into the plane of the diagram, and a dot indicating a current up out of the plane of the diagram. The path of the flux ϕ due to the stator winding is also shown. The stator mmf produces a north magnetic pole on the upper half of the stator and a south pole on the lower half. The N and S marking in Fig. 3.16 simply indicates the central axis of the flux density distribution due to the stator mmf.

The complete machine is illustrated in Fig. 3.17, where the rotor winding is shown in the form of two coils that are connected in series and mounted one on each of the *salient*, or projecting, rotor pole pieces. Connection of the source to

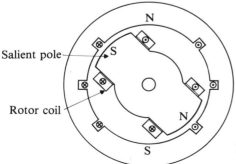

Fig. 3.17 Salient-pole rotor in a cylindrical stator.

these rotor coils is made by means of slip rings. The end connectors of the stator conductors are omitted to simplify the diagram. For this machine, the torque expression in Eq. (3.102) reduces to

$$T = \frac{i_s^2}{2} \frac{dL_{ss}}{d\beta} + i_s i_r \frac{dL_{sr}}{d\beta} \quad \text{N·m} \tag{3.103}$$

This machine has a synchronous speed determined by the reluctance torque expressed by the first term on the right-hand side of Eq. (3.103)—that is, the speed in radians per second is equal to the angular frequency of the stator source. The second term in Eq. (3.103) is simultaneously useful only if it produces average torque at this same synchronous speed.

However, many purposes require machines that will run at continuously variable speed, so that it is desirable to eliminate the remaining reluctance–torque term from Eq. (3.103). This would leave only the torque due to the interaction of the two winding mmf's, which is not subject to any speed constraint imposed by variation in the reluctance of the magnetic system.

In order to reduce $dL_{ss}/d\beta$ to zero, the rotor configuration must provide a path of constant reluctance for the flux produced by the stator winding, no matter what the rotor position. This is achieved by making the rotor cylindrical and embedding the rotor winding in axial slots in its surface, as illustrated in Fig. 3.18. The conductors in the rotor slots are connected into a coil by end connectors in the same manner as was employed for the stator coil of Fig. 3.16. Once again, several conductors would normally be laid in each slot to increase the number of turns in the rotor coil. For this machine, the reluctance-torque terms are effectively zero, and

$$T = i_s i_r \frac{dL_{sr}}{d\beta} \quad \text{N·m} \tag{3.104}$$

If the stator and rotor windings of the machine of Fig. 3.18 are to be excited from sinusoidal sources of potential difference, then it is desirable that all variables

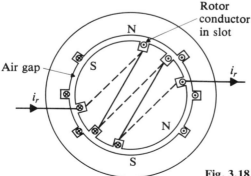

Fig. 3.18 Cylindrical rotor in a cylindrical stator.

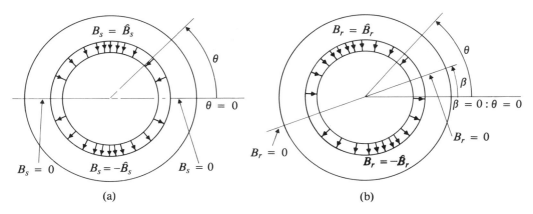

Fig. 3.19 Sinusoidal distribution of flux density in the air gap. (a) Due to stator mmf. (b) Due to rotor mmf.

in the entire system, including L_{sr}, the mutual inductance between the stationary stator winding and the rotating rotor winding, should be sinusoidal functions of time. This can only be the case if current i_s and i_r acting singly or at the same time produce a sinusoidal distribution of flux density around the air gap of the machine. As illustrated in Fig. 3.19(a), the space variation of flux density around the air gap due to stator current i_s acting alone must be described by the relation

$$B_s = \hat{B}_s \sin \theta \quad \text{T} \tag{3.105}$$

Similarly, as illustrated in Fig. 3.19(b), the space variation of the flux density around the air gap due to rotor current i_r acting alone with the rotor at angle β must be described by the relation

$$B_r = \hat{B}_r \sin (\theta - \beta) \quad \text{T} \tag{3.106}$$

The resultant of these two distributions for any angle β and any magnitudes of $\hat{B}_s$ and $\hat{B}_r$, depending upon the magnitudes of i_s and i_r, would then also be a sinusoidal distribution of flux density around the air gap.

Production of the sinusoidal flux–density distribution described in Eq. (3.105) requires that the mmf acting across the air gap due to the stator conductors also be sinusoidally distributed. This could be accomplished if the stator conductors were distributed in such a way that the conductor density, in conductors per radian of air-gap arc at position θ became proportional to $\cos \theta$. The required conductor distribution is therefore expressed by

$$n_s = \frac{N_s}{2} \cos \theta \text{ conductors/rad} \tag{3.107}$$

where N_s is the total number of *turns* on the stator. That is, N_s is equal to half the number of conductors in all the stator slots, since each turn embodies two conductors. The required flux–density distribution illustrated in Fig. 3.19(a) is achieved when current i_s flows in all turns of the stator winding connected in series.

A distribution of conductors on the rotor similar to that described in the preceding paragraph for the stator would achieve the flux–density distribution illustrated in Fig. 3.19(b) and described by Eq. (3.106).

Since it is not possible to distribute a finite number of stator conductors in slots in the ideal manner described in Eq. (3.107), it is therefore necessary to see what can be done in practice to produce at least a good approximation to the sinusoidal conductor distribution. Such practical measures could then be applied equally to the rotor winding.

3.2.1 Concentric or Spiral Windings

Figure 3.20(a) shows the end view of the stator of a cylindrical machine that has twelve uniformly distributed slots. Coils are to be laid in these slots in such a way that the distribution of conductors around the stator periphery approximates as closely as possible the ideal sinusoidal distribution described in Eq. (3.107). In general, if there are s slots in the stator, then the number of conductors in slot q is given by gathering together the ideal conductor distribution of Eq. (3.107) over an arc of $2\pi/s$ radians—that is,

$$N_q = \frac{N_s}{2} \int_{\theta \,=\, 2\pi(q-1)/s}^{\theta \,=\, 2\pi q/s} \cos \theta \; d\theta \tag{3.108}$$

For the twelve-slot stator illustrated in Fig. 3.20, two sets of three coils each are required. In the upper set, the outermost coil occupies slots a and f and has 0.25 N_s turns. The middle coil occupies slots b and e and has 0.183 N_s turns. The innermost coil occupies slots c and d and has 0.067 N_s turns. The total number of

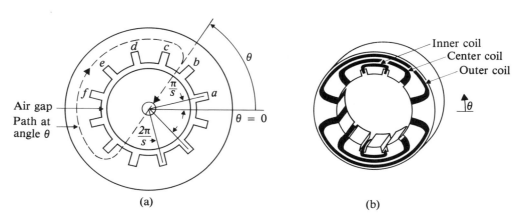

(a)   (b)

Fig. 3.20 Concentric winding for a 12-slot stator.

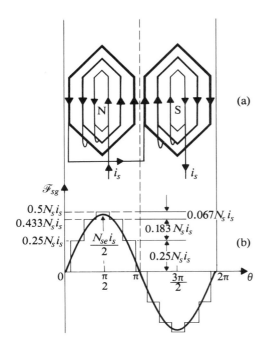

(a)

(b)

Fig. 3.21 Developed winding and mmf distribution for a 12-slot stator.

turns in the upper set of coils is therefore 0.5 N_s. Figure 3.20(b) shows the coils in the slots.

Figure 3.21(a) shows a development or linear representation of the stator winding of Fig. 3.20, and illustrates how the coils may be connected to give a *concentric* or *spiral distributed winding*.

The space distribution of mmf acting across the air gap may be determined by considering the path at angle θ shown in broken line in Fig. 3.20(a). The mmf acting around the path is, by Ampere's circuital law, given by the *net* current enclosed by the path. Because of the symmetry of the winding, the mmf across each of the two air gaps in the path will be equal to half the total path mmf. The space distribution of the air gap mmf at angle θ is shown approximately by the stepped waveform in Fig. 3.21. This waveform may be described by the Fourier series

$$\mathscr{F}_{sg} = N_s i_s [0.494 \sin \theta - 0.0449 \sin 11\theta - 0.382 \sin 13\theta \cdots] \quad \text{A} \quad (3.109)$$

The higher harmonic components are small in comparison with the fundamental component and those below the eleventh are absent. The fundamental-component waveform is shown superimposed on the stepped mmf waveform in Fig. 3.21(b). However, the slots have finite width, so that the vertical segments of the stepped waveform should actually be sloped over the angle subtended by one slot. If, more-

Fig. 3.22 Concentric stator winding. (*The Harland Engineering Company of Canada Ltd.*)

over, allowance is made for the fringing of the magnetic flux, which would result in a flux–density wave without the sharp corners of the mmf wave, then the actual flux–density wave produced by the winding of Figs. 3.20 and 3.21 would be very close to sinusoidal. A stator with a concentric winding is illustrated in Fig. 3.22.

The maximum amplitude of the stepped wave in Fig. 3.21(b) is $N_s i_s/2$. That of the fundamental component is $a_1 N_s i_s/2$. The fundamental component may be considered to be produced by an *effective* number of turns N_{se}, ideally distributed in accordance with Eq. (3.107), and giving a sinusoidal mmf wave of amplitude $N_{se} i_s/2$. Thus the effective number of sinusoidally distributed turns is

$$N_{se} = a_1 N_s = 0.989\, N_s \qquad (3.110)$$

Other types of windings and the calculation of their effective numbers of turns are discussed further in Chapter 5.

From this point on, the analysis of ac rotating machines will be carried out in terms of N_{se}, the effective turns of a stator winding that gives a sinusoidal mmf wave, and N_{re}, the corresponding quantity for a rotor. While the real numbers of turns N_s and N_r are necessarily integral, N_{se} and N_{re} need not be.

3.2.2 Torque Production in the Cylindrical Machine

A complete cylindrical machine with N_{se} effective turns on the stator and N_{re} effective turns on the rotor is illustrated in Fig. 3.23, where only a small number of conductors is shown to represent each winding. The magnetic poles due to the mmf of the stator winding acting alone are shown on the inner periphery of the stator.

Once more, let it be emphasized that these pole markings simply indicate the points on the stator periphery where flux density would be a maximum if only the stator winding were excited.

The magnetic poles due to the mmf of the rotor winding acting alone are shown at the surface of the rotor. If both windings were excited simultaneously, each would produce a sinusoidally distributed mmf, and the result of the summation of the two mmf's would be a sinusoidally distributed air gap mmf and flux density. The position of the maximum flux density in the air gap would be dependent on the rotor position and the stator and rotor currents.

A consequence of the effectively sinusoidal conductor distribution on the rotor and stator is that the mutual inductance between the two windings of Fig. 3.23 may be expressed by the relation

$$L_{sr} = \hat{L}_{sr} \cos \beta \quad \text{H} \tag{3.111}$$

where β is the rotor position and is therefore the angle between the stator and rotor winding axes.

Let the currents in the two windings be

$$i_s = \hat{i}_s \cos \omega_s t \quad \text{A} \tag{3.112}$$

$$i_r = \hat{i}_r \cos (\omega_r t + \alpha) \quad \text{A} \tag{3.113}$$

and let the rotor rotate at angular velocity

$$\omega_m = \frac{d\beta}{dt} \quad \text{rad/s} \tag{3.114}$$

The position of the rotor at instant t is

$$\beta = \omega_m t + \beta_0 \quad \text{rad} \tag{3.115}$$

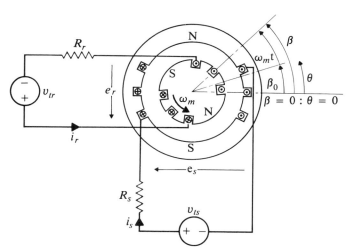

Fig. 3.23 Torque production in a cylindrical machine.

where β_0 is the rotor position at $t = 0$. Then the instantaneous torque developed in the machine may be expressed by substitution from Eqs. (3.111), (3.112), (3.113), and (3.115) in Eq. (3.104). This yields

$$T = -\hat{i}_s\hat{i}_r\hat{L}_{sr}\cos\omega_s t\,\cos(\omega_r t + \alpha)\sin(\omega_m t + \beta_0) \quad \text{N}\cdot\text{m} \tag{3.116}$$

The product of three trigonometric terms in Eq. (3.116) may be expanded to give

$$
\begin{aligned}
T = \frac{-\hat{i}_s\hat{i}_r\hat{L}_{sr}}{4}\{&\sin[(\omega_m + (\omega_s + \omega_r))t + \alpha + \beta_0] \\
+&\sin[(\omega_m - (\omega_s + \omega_r))t - \alpha + \beta_0] \\
+&\sin[(\omega_m + (\omega_s - \omega_r))t - \alpha + \beta_0] \\
+&\sin[(\omega_m - (\omega_s - \omega_r))t + \alpha + \beta_0]\} \quad \text{N}\cdot\text{m} \tag{3.117}
\end{aligned}
$$

The average value of each of the sinusoidal terms in Eq. (3.117) is zero, unless the coefficient of t is zero in that term, that is, the average torque T_{av} developed by the machine is zero unless

$$\omega_m = \pm(\omega_s \pm \omega_r) \quad \text{rad/s} \tag{3.118}$$

The $\pm$ sign outside the brackets in Eq. (3.118) shows that the direction of rotation of the machine is immaterial, and the necessary condition for the development of an average torque in the machine may be stated as

$$|\omega_m| = |\omega_s \pm \omega_r| \quad \text{rad/s} \tag{3.119}$$

The various combinations of electrical and mechanical excitations that satisfy Eq. (3.119) determine the various types of machine that can be built. Some particular cases are of special interest.

Let $\omega_s = \omega_r = 0$; that is, the excitations are direct currents I_s and I_r, and consequently the phase angle α is zero. Also, let $\omega_m = 0$, so that Eq. (3.119) is satisfied. Substitution of these values in Eq. (3.117) then yields

$$T = -I_s I_r \hat{L}_{sr}\sin\beta_0 \quad \text{N}\cdot\text{m} \tag{3.120}$$

Under these conditions, the machine operates as a dc rotary actuator, developing a constant torque against any displacement β_0 produced by an external torque applied to the rotor shaft.

Let $\omega_s = \omega_r$; that is, both excitations are alternating currents of the same frequency. Also let $\omega_m = 0$, so that Eq. (3.119) is satisfied. Substitution of these values in Eq. (3.117) yields a torque made up of two alternating and two constant terms. The machine operates as an ac rotary actuator. The developed torque is fluctuating and has an average value

$$T_{av} = \frac{-\hat{i}_s\hat{i}_r\hat{L}_{sr}}{2}\sin\beta_0\cos\alpha \quad \text{N}\cdot\text{m} \tag{3.121}$$

If the two windings were connected in series, α would be zero, and the $\cos\alpha$ term in Eq. (3.121) would become unity.

Let $\omega_r = 0$, $\omega_s \neq 0$; that is, the rotor excitation is direct current I_r, and consequently the phase angle α is zero. Also let $\omega_m = \omega_s$, so that Eq. (3.119) is satisfied. Substitution of these values in Eq. (3.117) yields an average torque

$$T_{av} = \frac{-\hat{i}_s \hat{i}_r \hat{L}_{sr}}{2} \sin \beta_0 \quad \text{N} \cdot \text{m} \tag{3.122}$$

Thus, if the rotor were brought up to a speed of ω_s rad/s, an average unidirectional torque would be established, and continuous energy conversion would take place at synchronous speed.

Let ω_s and ω_r be two different angular frequencies, and consider the condition

$$\omega_m = \omega_s - \omega_r \quad \text{rad/s} \tag{3.123}$$

that satisfies Eq. (3.119). Substitution of these values in Eq. (3.117) yields an average torque

$$T_{av} = \frac{-\hat{i}_s \hat{i}_r \hat{L}_{sr}}{4} \sin (\alpha + \beta_0) \quad \text{N} \cdot \text{m} \tag{3.124}$$

Thus, if the rotor were brought up to a speed of $(\omega_s - \omega_r)$ rad/s, an average unidirectional torque would be established, and continuous energy conversion would take place at asynchronous speed ω_m.

It should be noted that neither of the rotating machines whose average torque expressions are given in Eqs. (3.122) and (3.124) would accelerate from standstill with constant values of supply frequencies ω_s and ω_r, since neither develops an average unidirectional torque at $\omega_m = 0$, a condition that would not satisfy Eq. (3.119) for the specified electrical excitations.

Each of the possible conditions satisfying Eq. (3.119) results in the production of an average unidirectional torque and hence in the possibility of energy conversion. But apart from the case of the dc actuator, the instantaneous torque in each case is pulsating, due to the sinusoidal terms remaining when the specified excitations are substituted in Eq. (3.117). This pulsating torque is an undesirable feature in a motor or generator, since it may result in speed fluctuation, vibration, noise, and waste of energy. Only in some small machines is such a pulsating torque acceptable.

Since torque pulsates while speed is relatively constant, the power developed by an energy converter of the type illustrated in Fig. 3.23 must also pulsate. Pulsating power is a feature of systems excited by single-phase alternating sources, and this indicates that an improvement may be found in employing polyphase sources, which develop constant power in a balanced system.

Example 3.6 A device such as that shown in Fig. 3.23 has similar windings on stator and rotor. The electrical parameters are

Stator winding resistance $R_s = 0.4 \ \Omega$
Stator winding self-inductance $L_{ss} = 45$ mH

Rotor winding resistance R_r $\qquad = 0.4\ \Omega$
Rotor winding self-inductance L_{rr} $= 45\ \text{mH}$
Mutual inductance $\hat{L}_{sr}$ $\qquad\qquad = 40\ \text{mH}$

a) Determine the average torque produced at a rotor angle of 45° when the machine is employed as an actuator with its two windings in series, as indicated in Fig. 3.24, and excited from

   i) a dc source of 10 volt

   ii) an ac source of 115 volt, 60 Hz.

b) If the stator were excited from a 60-Hz source, and the rotor from a 25-Hz source, at what speed or speeds would the machine be capable of energy conversion?

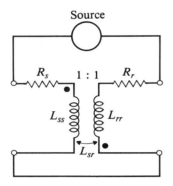

Figure 3.24

Solution

a) The average torque is calculated as follows:

   i) With dc excitation, the current in the windings is limited only by the winding resistances when the actuator is not moving, so that

$$I = \frac{10}{0.8} = 12.5 \quad \text{A}$$

$$L_{sr} = 40 \times 10^{-3} \quad \cos\beta$$

$$\frac{dL_{sr}}{d\beta} = -40 \times 10^{-3} \quad \sin\beta$$

From Eq. (3.104) for $\beta = 45°$, the instantaneous torque is

$$T = -12.5^2 \times 40 \times 10^{-3} \times 0.707 = -4.42 \quad \text{N}\cdot\text{m}$$

Since I_s and I_r are constant direct currents, this is also the average torque produced. The minus sign shows that the torque opposes the displacement through angle β.

ii) With ac excitation, from the circuit of Fig. 3.24,

$$\overline{V} = (R_s + R_r)\,\overline{I} + j\omega\,(L_{ss} + L_{rr} + 2\hat{L}_{sr}\cos 45)\overline{I}$$

that is,

$$115 = 0.8\,\overline{I} + j\,120\pi \times 0.1466\,\overline{I} \quad \text{V}$$

from which

$$I = 2.08 \quad \text{A}$$

The machine is operating as an ac actuator, so that from Eq. (3.121), in which $\alpha = 0$,

$$T_{av} = -\frac{\hat{i}_s\hat{i}_r\hat{L}_{sr}}{2}\sin\beta_0$$

$$= -\frac{(\sqrt{2} \times 2.08)^2 \times 40 \times 10^{-3} \times 0.707}{2}$$

$$= -0.122 \quad \text{N} \cdot \text{m}$$

b) The necessary condition for development of an average torque in the machine, and hence for energy conversion, is given by Eq. (3.119) as

$$|\omega_m| = |\omega_s \pm \omega_r|$$

In this case,

$$\omega_s = 120\pi \quad \text{rad/s}$$
$$\omega_r = 50\pi \quad \text{rad/s}$$

Thus,

$$\omega_m = 170\pi \text{ or } 70\,\pi \quad \text{rad/s}$$

in either direction.

3.3 CONDITIONS FOR CONSTANT TORQUE

The necessary conditions for the development of constant torque in a cylindrical machine may be determined by examining an expression for the torque existing at any instant. Figure 3.23 shows a diagram representing a cylindrical machine with effectively sinusoidally distributed windings on both rotor and stator. The position of the rotor is expressed by angle β, while angle θ expresses the position of any point in the air gap. Current directions in the windings are shown for a particular instant, as also are the centers of the magnetic poles that would be produced on the rotor and stator peripheries if the corresponding winding only were excited with current flowing in the direction indicated. The distribution of the fundamental component of mmf for the stator winding may be expressed from Eqs. (3.109) and (3.110) as

$$\mathscr{F}_{sg1} = \frac{N_{se}i_s}{2} \sin \theta \quad \text{A} \tag{3.125}$$

The distribution of the fundamental component of mmf for the rotor winding when the rotor is in the position shown in Fig. 3.23 may correspondingly be expressed as

$$\mathscr{F}_{rg1} = \frac{N_{re}i_r}{2} \sin (\theta - \beta) \quad \text{A} \tag{3.126}$$

As was stated earlier, the higher harmonic components of mmf may, for most purposes, be neglected. Equations (3.125) and (3.126) will therefore be taken as expressing the entire mmf due to each of the two windings, and the subscripts indicating that they are the fundamental components will be omitted in the following. The two mmf distributions are illustrated in the developed diagram of Fig. 3.25.

The amplitudes of the mmf waves for the instant illustrated in Fig. 3.25 are

$$\hat{\mathscr{F}}_{sg} = \frac{N_{se}i_s}{2} : \hat{\mathscr{F}}_{rg} = \frac{N_{re}i_r}{2} \quad \text{A} \tag{3.127}$$

Substitution from Eqs. (3.111) and (3.127) in Eq. (3.104) yields the following expression for instantaneous torque:

$$T = - K \, \hat{\mathscr{F}}_{sg} \, \hat{\mathscr{F}}_{rg} \sin \beta \quad \text{N} \cdot \text{m} \tag{3.128}$$

where

$$K = \frac{4\hat{L}_{sr}}{N_{se}N_{re}} \quad \text{H} \tag{3.129}$$

Equation (3.129) shows that K is a constant determined by the parameters of the machine. Equation (3.128) shows that torque is proportional to the peak values of the stator and rotor mmf's and also to the sine of the angle between the axes of symmetry of the mmf distributions produced by the two windings. Since for positive winding currents $T < 0$ for $\beta > 0$, the torque acts in such a direction as to tend to bring the mmf axes into alignment.

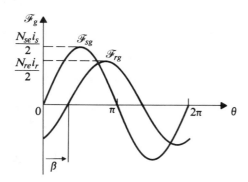

Fig. 3.25 Developed diagram of stator and rotor mmf's.

It is clear from Eq. (3.128) that constant torque, varying neither with time nor with rotor position, would be obtained provided that two mmf waves of constant amplitude and constant angular displacement from one another could be produced.

Perhaps the first question prompted by the statement in the preceding paragraph is, "If one winding is stationary, and the other is rotating, how is it possible to maintain a constant angle between the mmf axes of rotor and stator?" There are three possible answers to this question:

i) If the stator mmf is fixed in space, the rotor mmf axis must be fixed in space, even when the rotor winding is rotating.

ii) Alternatively, if the rotor mmf is fixed relative to the rotor, the stator mmf axis must rotate at rotor speed relative to the fixed stator windings.

iii) Alternatively, the two mmf axes must rotate at such speeds relative to their windings that they remain stationary relative to one another.

In putting any one of these three answers into effect, $\hat{\mathscr{F}}_{sg}$ and $\hat{\mathscr{F}}_{rg}$ must also be constant, so that a constant torque is obtained.

The angle β in Eq. (3.128) defines the position of the rotor as well as the angle between the mmf axis of the rotor and stator windings. If the rotor mmf axis is caused to rotate relative to the rotor windings, then the rotor position will no longer define the angle between the mmf axes. The symbol δ will therefore be employed in the following to signify this angle, while β will be reserved for rotor position. Equation (3.128) thus becomes

$$T = - K \, \hat{\mathscr{F}}_{sg} \, \hat{\mathscr{F}}_{rg} \sin \delta \quad \text{N} \cdot \text{m} \tag{3.130}$$

where δ is the angle between the mmf axes of the two machine members.

3.3.1 Commutator Action

The necessary condition for energy conversion was expressed in Eq. (3.118) by

$$\omega_m = \pm(\omega_s \pm \omega_r) \quad \text{rad/s} \tag{3.131}$$

Let $\omega_s = 0$, so that the stator is excited with dc. The mmf axis of the stator is stationary and $\hat{\mathscr{F}}_{sg}$ is constant, provided that the stator current is constant. Constant torque will be produced if it is possible (a) to hold the rotor mmf axis stationary while the rotor is rotating, and (b) to maintain $\hat{\mathscr{F}}_{rg}$ at constant magnitude.

Consider now the condition of operation

$$\omega_m = + \omega_r \quad \text{rad/s} \tag{3.132}$$

If f is the frequency of the rotor current in hertz, and n_r is the speed of rotation of the rotor in revolutions per second, then

$$f = n_r \quad \text{Hz} \tag{3.133}$$

$\hat{\mathscr{F}}_{rg}$ may be made constant by exciting the rotor also with a constant direct current, but in that case $f \neq n_r$. Equation (3.133) may, however, be satisfied by the use of a *commutator*.

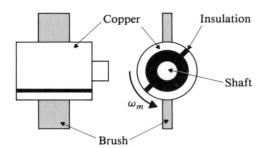

Fig. 3.26 Two-segment commutator.

A commutator is a rotary switch, consisting of a cylinder of copper with brushes of carbon—or of some composition of carbon and other materials—held against the outside of the cylinder. The commutator is mounted on the rotor shaft, and hence rotates at speed ω_m or n_r. The copper cylinder is split into a number of segments, and the ends of the coils on the rotor are connected to these segments. First consider the simplest case of a two-segment commutator with a single-turn rotor coil connected to it. Figure 3.26 shows the physical arrangement of such a commutator and its brushes.

Figure 3.27 shows the electrical connections to the brushes and the commutator segments. The direction of current in the coil for the instants illustrated and the resulting rotor mmf axis positions are also shown. As the mmf axis passes through the vertical position, the direction of the current in the coil is reversed, since the insulation strips between the segments pass beneath the brushes. This reversal of current reverses the direction of the rotor mmf. The conditions immediately before and immediately after reversal are illustrated in Fig. 3.27(a) and (b), respectively.

Reversal of current in the rotor coil takes place every half revolution, so that Eq. (3.133) is satisfied. The time variation of current in the coil is not sinusoidal, but is a rectangular wave. This is not important, since the machine is not excited from a sinusoidal ac source. It should be noted that the rotor position at which switching takes place depends upon the position of the brushes on the commutator. The mmf axis of the rotor in Fig. 3.27 is not stationary, but the S pole is always on the right and the N pole on the left, so that the poles rotate at speed ω_m for one-half revolution and are then abruptly interchanged. Thus, while $\hat{\mathscr{F}}_{rg}$ is of constant magnitude, the position of the mmf axis varies through π radians.

The variation in the angular position of the mmf axis may be halved by adding a second coil to the rotor. For the moment, this second coil may be considered to be supplied through a second commutator as illustrated in Fig. 3.28. The rotor mmf is now the resultant of the two coil mmf's, and since all the brushes are on the vertical axis, the switching of each coil occurs as it reaches the horizontal position. At each switching of one coil, the mmf axis is abruptly moved $\pi/2$ radians in a direction opposite to the rotor rotation. The situation immediately before and immediately after current reversal in one coil is indicated in Fig. 3.28(a) and (b), respectively.

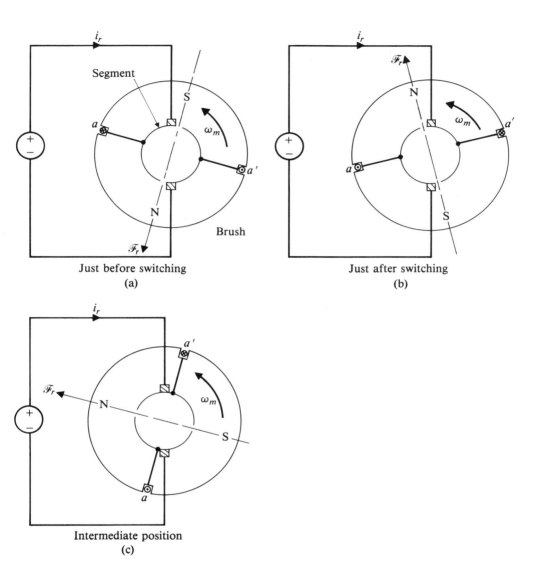

Fig. 3.27 Switching effect of a two-segment commutator.

As the number of commutated coils on the rotor is increased, the angle of oscil-lation of the rotor mmf axis is reduced, until with a large number of coils on the rotor the mmf axis becomes virtually stationary. If all the coils carry the same current (and this is constant in magnitude), then $\hat{\mathscr{F}}_{rg}$ is constant, and the condi-tions for constant torque are established.

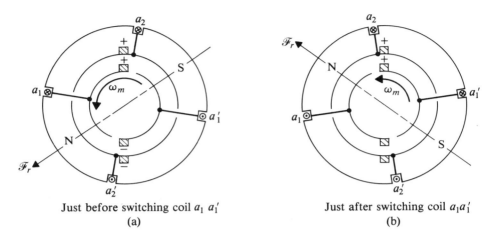

Just before switching coil $a_1\,a_1'$ Just after switching coil $a_1 a_1'$

(a)   (b)

Fig. 3.28 Effect of switching two coils.

The alternative condition to that expressed in Eq. (3.132), that is $\omega_m = -\omega_r$, simply corresponds to rotation in the opposite direction to that considered in the foregoing.

3.3.2 Direct-Current Machine

A machine that is excited from dc sources only, or that itself acts as a source of dc, is termed a *dc machine*.

The rotor of a dc machine carries a large number of coils that are fitted into axial slots in the rotor periphery. Usually each coil has more than a single turn and two coil sides are fitted into each slot. As an example that is simple to illustrate, a machine having a rotor carrying only twelve single-turn coils will be considered. Twelve slots in the rotor will carry the coils, provided that two coil sides lie in each slot. The resulting arrangement of coil sides or conductors is shown in Fig. 3.29,

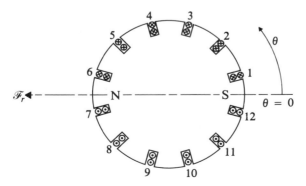

Fig. 3.29 Twelve-slot rotor for a dc machine, as seen from the commutator end.

where the slots are numbered, and the current directions required to maintain the rotor mmf axis in an approximately horizontal position are shown.

In practice it is possible to arrange for the necessary switching of the coil currents without providing a separate two-segment commutator for each coil. How this may be done is illustrated in the developed diagram of the rotor winding shown in Fig. 3.30. The current directions in this diagram correspond to those in Fig. 3.29. At the bottom of the winding diagram is a development of the required 12-segment commutator with two brushes. The N and S poles shown in Fig. 3.30 should be ignored for the present discussion.

Examination of the winding in Fig. 3.30 shows that there are two paths through the winding by means of which the current passes from the positive brush to the negative brush. These two paths are therefore in parallel, and each one includes half the coils making up the winding. As the rotor turns and as the conductors and commutator segments move past the stationary brushes, coils are transferred two at a time from one parallel path to the other, thus maintaining the rotor mmf axis in an approximately stationary position. When the rotor has moved one slot pitch (that

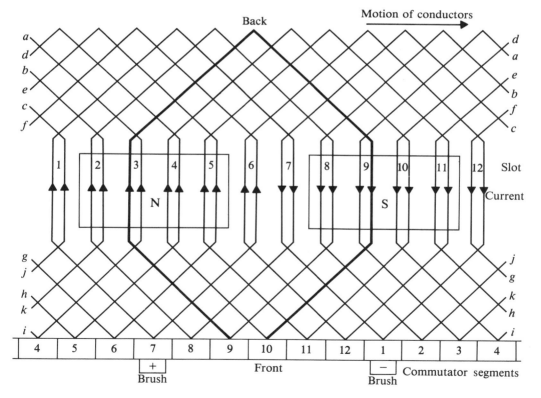

Fig. 3.30 Twelve-slot rotor winding for a dc machine (current directions shown for motor operation).

is, the distance between two adjacent slots) then commutator segments 6 and 12 are in contact with the brushes, and the current in the coils lying in slots 6 and 12 has been reversed.

The variation of rotor mmf around the rotor periphery is illustrated in the developed diagram of Fig. 3.31. If the conductors in slots 4–9 are considered as forming three concentric coils (the actual end connections of the conductors are unimportant) and those in slots 10–3 as forming three more concentric coils, then the mmf of these coils may be represented by the stepped wave in Fig. 3.31. With a large number of conductors on the rotor, this stepped mmf wave becomes virtually triangular, and such a rotor mmf wave is shown in the diagram.

The adoption of a commutated dc source to excite the rotor results in nonsinusoidal distribution of mmf around the rotor periphery. As has already been remarked, this is not important, as the machine is not to be connected to an alternating current system. For the same reason, it is not necessary for the distribution of stator mmf around the air gap to be sinusoidal. The stator may thus be constructed in a form that is simpler to manufacture. It is therefore normal practice to build the stators of dc machines with salient pole pieces upon which the stator windings or field coils may be conveniently mounted. This arrangement is illustrated in Fig. 3.32.

The brushes are placed on the commutator so that the coils are switched at the

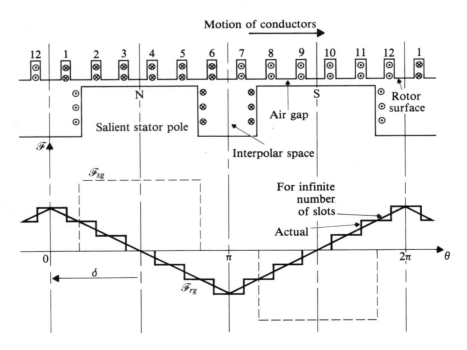

Fig. 3.31 Magnetomotive force distribution for the machine of Figs. 3.29 and 3.30.

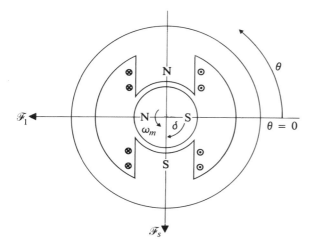

Fig. 3.32 Salient-pole structure for a dc machine.

instant at which their conductors are passing through the interpolar space of the stator, and the coils have virtually zero emf induced in them at that instant. This arrangement also ensures that the rotor and stator mmf axes are at all times at right angles to one another. The field poles and the wave of stator mmf are also shown in Fig. 3.31.

The expression for the torque given in Eq. (3.130) is applicable to the dc machine, provided the constant K is adjusted for the nonsinusoidal mmf distributions and sin δ is made equal to unity. The conditions for constant torque are fulfilled, since $\hat{\mathscr{F}}_{sg}$, $\hat{\mathscr{F}}_{rg}$, and δ are all constant. Moreover, the condition for maximum torque and hence for maximum energy conversion is also fulfilled, since $\delta = -\pi/2$ radians. Since $\delta < 0$, $T > 0$, and the machine acts as a motor driving in the positive direction. This also can be seen by considering the polarities shown on the rotor and stator in Fig. 3.32. Reversal of the current in either the stator or rotor system causes the machine to act as a generator.

The reasons why dc machines are built in their normally accepted form should now be sufficiently obvious. The so-called "armature winding" must be on the rotor so that the commutator can produce in it alternating currents that satisfy the necessary condition for energy conversion expressed in Eq. (3.131).

3.3.3 Rotating Magnetic Fields

The first of the possible methods of producing constant torque suggested at the end of Section 3.3 led to the development of the direct-current machine. The second and third of the possible methods of producing constant torque required that the stator mmf axis should rotate relative to its fixed windings.

Since a single-phase ac source of supply can only develop fluctuating power and hence a fluctuating torque in a rotating machine, a possible means of achieving constant torque would be to employ a polyphase ac source. Then, if the system is balanced, constant power can be developed. If polyphase ac stator and rotor

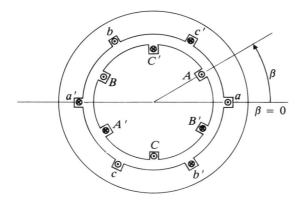

Fig. 3.33 Machine with polyphase stator and rotor windings.

sources are to be employed and the system is to be balanced, then polyphase windings must be fitted to both the stator and the rotor of the cylindrical machine, and these windings must be distributed on the machine in some symmetrical fashion.

One version of the required machine is illustrated in Fig. 3.33, where, to simplify the diagram, only the center turn of each phase winding has been shown. Each phase winding, however, must be considered as sinusoidally distributed about the entire stator or rotor periphery, as the case may be. Three-phase windings have been shown on both stator and rotor, but literally any two polyphase sources of equal or different numbers of phases could be employed for the two excitations.

If now the mutual inductances between each pair of stator and rotor windings are determined as functions of β, the angular position of the rotor, then an expression for the torque developed due to each pair of windings may be obtained. Since each rotor phase will interact with each stator phase, there will be nine components of instantaneous torque, and if these are added it will be found that a constant torque may be developed. However, there is a less cumbersome and more enlightening method of arriving at this same conclusion.

It is a commonplace of mechanics that two simple harmonic motions of equal frequency and amplitude may be combined to produce a circular motion. This same principle is illustrated if two alternating potential differences of equal magnitude and frequency but displaced $\pi/2$ radians in phase from one another are applied to the vertical and horizontal inputs of an oscilloscope. The result is a pattern on the screen formed by the spot of light traveling in a circle. This principle may also be applied to the sinusoidally distributed time-varying mmf's produced by coils excited with alternating current.

3.3.4 Two-Winding Stator

Figure 3.34 is a diagram of a stator carrying two similar windings at right angles to one another. Again only the center conductor of each sinusoidally distributed winding is shown. When carrying current, each of these windings produces a sinusoidal distribution of mmf around the stator periphery. Positive directions of current in the

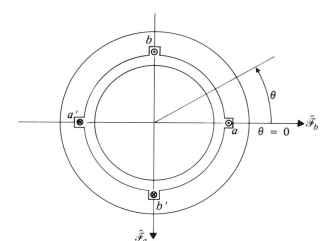

Fig. 3.34 Machine with two-winding stator.

coils are defined in the diagram. The resulting positive peak values of mmf act along the axes of the two coils in the direction indicated.

By "peak" mmf is meant the maximum value of mmf occurring across the air gap at any instant. This is a variable quantity depending upon the magnitude and direction of the current in the winding. As an example, the mmf distribution due to a positive current in winding aa' at a particular instant is shown in broken line in the developed diagram in Fig. 3.35. $\hat{\mathscr{F}}_{ag}$ is the maximum value of mmf across the air gap due to winding aa' at that instant. When the current in winding aa' has its positive maximum value, then the mmf distribution will be that (shown in full line in Fig. 3.35) for which $\hat{\mathscr{F}}_{ag} = \hat{\mathscr{F}}_{g\max}$.

Let the currents in the two windings be

$$i_a = \hat{i}_s \cos \omega_s t \quad \text{A} \tag{3.134}$$

$$i_b = \hat{i}_s \cos\left(\omega_s t - \frac{\pi}{2}\right) = \hat{i}_s \sin \omega_s t \quad \text{A} \tag{3.135}$$

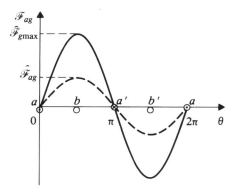

Fig. 3.35 Magnetomotive force due to winding aa' in Fig. 3.34.

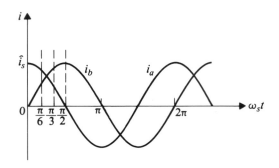

Fig. 3.36 Time variation of currents in the windings of Fig. 3.34.

The time variations of the currents in the two windings are shown in Fig. 3.36. Since for

$$i_a = \hat{i}_s, \qquad \hat{\mathscr{F}}_{ag} = \hat{\mathscr{F}}_{g\max} \quad \text{A} \tag{3.136}$$

and for

$$i_b = \hat{i}_s, \qquad \hat{\mathscr{F}}_{bg} = \hat{\mathscr{F}}_{g\max} \quad \text{A} \tag{3.137}$$

then

$$\hat{\mathscr{F}}_{ag} = \hat{\mathscr{F}}_{g\max} \cos \omega_s t \quad \text{A} \tag{3.138}$$

$$\hat{\mathscr{F}}_{bg} = \hat{\mathscr{F}}_{g\max} \sin \omega_s t \quad \text{A} \tag{3.139}$$

When each winding is excited separately, an mmf wave is produced that is sinusoidally distributed in space around the air gap; when both windings are excited simultaneously they produce together a resultant sinusoidal mmf wave around the air gap. By considering a series of instantaneous values of the two winding mmf's, the magnitude and position of the peak value of the resultant mmf wave may be determined for each instant of the series. Four convenient instants shown in Fig. 3.36 are for the series of values $\omega_s t = 0, \pi/6, \pi/3, \pi/2$.

At $\omega_s t = 0$, $\hat{\mathscr{F}}_{ag} = \hat{\mathscr{F}}_{g\max}$ and $\hat{\mathscr{F}}_{bg} = 0$. The positive peak value of the resultant of the two mmf's at this instant is therefore $\hat{\mathscr{F}}_{g\max}$ and this peak value of the resultant mmf wave may be represented by the vector shown in Fig. 3.37(a).

At $\omega_s t = \pi/6$, $\hat{\mathscr{F}}_{ag} = (\sqrt{3}/2) \hat{\mathscr{F}}_{g\max}$, and $\hat{\mathscr{F}}_{bg} = (1/2) \hat{\mathscr{F}}_{g\max}$. The positive peak values of the waves due to the two windings may be represented by the vectors shown in Fig. 3.37(b). The resultant of these two vectors has the magnitude $\hat{\mathscr{F}}_{g\max}$ and is rotated in a positive direction by $\pi/6$ radians from its initial position, shown in Fig. 3.37(a).

At $\omega_s t = \pi/3$, $\hat{\mathscr{F}}_{ag} = (1/2) \hat{\mathscr{F}}_{g\max}$, and $\hat{\mathscr{F}}_{bg} = (\sqrt{3}/2) \hat{\mathscr{F}}_{g\max}$. The positive peak values of the waves due to the two windings may be represented by the vectors shown in Fig. 3.37(c). The resultant of these two vectors has the magnitude $\hat{\mathscr{F}}_{g\max}$ and is rotated in a positive direction by $\pi/3$ radians from its initial position, shown in Fig. 3.37(a).

At $\omega_s t = \pi/2$, $\hat{\mathscr{F}}_{ag} = 0$, and $\hat{\mathscr{F}}_{bg} = \hat{\mathscr{F}}_{g\max}$. The positive peak value of the resultant of the two mmf's at this instant is therefore $\hat{\mathscr{F}}_{g\max}$, and this peak value

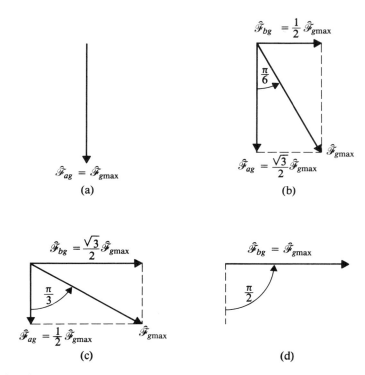

Fig. 3.37 Direction of resultant mmf in the stator of Fig. 3.34 for instants $\omega t = 0,\ \pi/6,\ \pi/3,\ \pi/2$.

of the resultant mmf wave may be represented by the vector shown in Fig. 3.37(d), which is rotated by $\pi/2$ radians from the initial position in Fig. 3.37(a).

Thus a resultant mmf wave of constant amplitude rotating at ω_s rad/s has been produced by the stator windings. This rotating mmf wave may be envisaged as producing a pair of poles on the inner surface of the stator that travel round it at ω_s rad/s, as illustrated in Fig. 3.38.

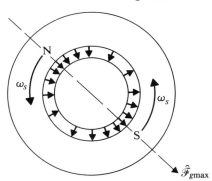

Fig. 3.38 Rotating stator poles.

Note: The peak value of the resultant stator mmf wave coincides with the axis of a particular winding when that winding is carrying its peak current.

3.3.5 Three-Phase Stator

Figure 3.39 shows a single sinusoidally distributed winding pp' on a stator. Only the center turn of the winding is shown, and it is at angle γ to the chosen angular datum line on the stator periphery. If the winding has N_{se} effective turns and is carrying current i_p, then it will produce a peak mmf $N_{se}i_p$ acting along the axis of the winding perpendicular to the plane of the center turn. The peak mmf applied to the air gap will be $N_{se}i_p/2$, and this will act in a positive direction (that is, from stator to rotor) at angle $\theta = \gamma + \pi/2$, and in a negative direction (that is, from rotor to stator) at angle $\theta = \gamma - \pi/2$.

The total mmf due to winding pp' acting around a path in the magnetic circuit at angle θ will be, from Fig. 3.39,

$$\mathscr{F}_{\text{path}} = N_{se}\,i_p \cos\left[\frac{\pi}{2} - (\theta - \gamma)\right] \quad \text{A} \qquad (3.140)$$

Thus, the mmf acting across the air gap at any angle θ is

$$\mathscr{F}_{sg} = \frac{N_{se}\,i_p}{2}\cos\left[\frac{\pi}{2} - (\theta - \gamma)\right] = \frac{N_{se}\,i_p}{2}\sin(\theta - \gamma) \quad \text{A} \qquad (3.141)$$

Figure 3.40 represents a stator carrying three phase windings, each sinusoidally distributed, each having N_{se} effective turns, and each displaced by $2\pi/3$ radians from its neighbor around the periphery in the sequence $a\ b\ c$. As in previous diagrams only the center turn of each winding is shown.

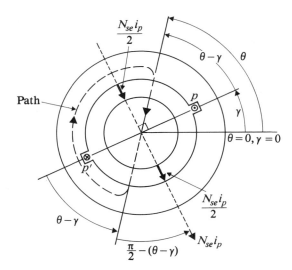

Fig. 3.39 Magnetomotive force due to a single sinusoidally distributed stator winding.

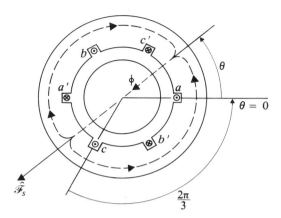

Fig. 3.40 Three-phase stator.

If the winding currents are i_a, i_b, i_c, then the resultant mmf produced by all of the windings across the air gap at angle θ may be found by determining for each winding an expression such as that in Eq. (3.141) and by adding all such expressions. Thus, for the three-phase stator,

$$\mathscr{F}_{sg} = \frac{N_{se}\,i_a}{2} \sin\theta + \frac{N_{se}\,i_b}{2} \sin\left(\theta - \frac{2\pi}{3}\right) + \frac{N_{se}\,i_c}{2} \sin\left(\theta - \frac{4\pi}{3}\right) \quad \text{A} \quad (3.142)$$

Let the winding currents be a balanced three-phase set of peak amplitude $\hat{i}_s$, sequence $a\,b\,c$, and frequency ω_s. That is,

$$i_a = \hat{i}_s \sin(\omega_s t + \alpha_s) \quad \text{A}$$

$$i_b = \hat{i}_s \sin\left(\omega_s t + \alpha_s - \frac{2\pi}{3}\right) \quad \text{A} \quad\quad (3.143)$$

$$i_c = \hat{i}_s \sin\left(\omega_s t + \alpha_s - \frac{4\pi}{3}\right) \quad \text{A}$$

Substitution from Eq. (3.143) in Eq. (3.142) yields

$$\mathscr{F}_{sg} = \frac{N_{se}\hat{i}_s}{2} \Bigg[\sin(\omega_s t + \alpha_s) \sin\theta + \sin\left(\omega_s t + \alpha_s - \frac{2\pi}{3}\right) \sin\left(\theta - \frac{2\pi}{3}\right)$$

$$+ \sin\left(\omega_s t + \alpha_s - \frac{4\pi}{3}\right) \sin\left(\theta - \frac{4\pi}{3}\right) \Bigg] \quad \text{A} \quad\quad (3.144)$$

Expansion of the sinusoidal terms in Eq. (3.144) yields

$$\mathscr{F}_{sg} = \frac{N_{se}\hat{i}_s}{4} \Bigg[3\cos(\omega_s t + \alpha_s - \theta) - \cos(\omega_s t + \alpha_s + \theta)$$

$$-\cos\left(\omega_s t + \alpha_s + \theta - \frac{4\pi}{3}\right) - \cos\left(\omega_s t + \alpha_s + \theta - \frac{8\pi}{3}\right) \Bigg] \quad \text{A} \quad (3.145)$$

The last three terms in the square brackets of Eq. (3.145) sum to zero, so that

$$\mathscr{F}_{sg} = \hat{\mathscr{F}}_g \cos (\omega_s t + \alpha_s - \theta) \quad \text{A} \qquad (3.146)$$

where

$$\hat{\mathscr{F}}_g = \frac{3 N_{se} \hat{i}_s}{4} \quad \text{A} \qquad (3.147)$$

The expression for $\mathscr{F}_{sg}$ in Eq. (3.146) describes a sinusoidally distributed rotating magnetic field. At any particular point on the stator periphery (where θ is a constant) the mmf acting across the air gap is a sinusoidal function of time. At any instant in time (when t is a constant) the mmf acting across the air gap is a sinusoidal function of θ, and the peak positive value of that mmf occurs where $\theta = \omega_s t + \alpha_s$. As t increases, the angle at which this peak positive mmf occurs must also increase, so that the mmf distribution advances as a space wave of unchanging shape and amplitude around the air gap. As with the two-winding stator, an N and S pole may be envisaged as appearing on the inner periphery of the stator and traveling around it at ω_s rad/s as illustrated in Fig. 3.38. The direction of rotation of the mmf wave may be reversed by reversing the phase sequence of the stator currents. It should also be noted that when $i_a = \hat{i}_s$, that is, when $\omega_s t + \alpha_s = \pi/2$, then from Eq. (3.146) the peak positive value of mmf acting across the air gap will occur at $\theta = \pi/2$, along the axis of winding a. *In general, when the current in any winding is at its peak magnitude, then the mmf on the axis of that winding is $\hat{\mathscr{F}}_{g\max}$.*

Equations (3.144) and (3.147) describe the rotating field produced by a three-phase winding. Other numbers of phases may also be used to produce a rotating field. For m_s phases, the windings are displaced from each other by $2\pi/m_s$ rad and the air-gap mmf is as given in Eqs. (3.146) and (3.147), except that the factor 3 is replaced by m_s.

Example 3.7 A machine with a 4-phase stator is excited from a 4-phase, 5-wire, 60-Hz system of 115 volts line-to-neutral. The machine parameters are

$$N_{se} = 300, \quad L_{ss} = 0.07 \text{ H}, \quad L_{ab} = 0, \quad \text{air-gap length } g = 1.5 \text{ mm}$$

The four windings are sinusoidally distributed and displaced $\pi/2$ radians from one another. The winding resistances are negligible. The circuit diagram is shown in Fig. 3.41.

Assuming negligible reluctance of the ferromagnetic part of the magnetic circuit, obtain an expression for the flux density produced in the air gap when terminals c and d in Fig. 3.41 are disconnected, and determine the peak flux density produced in the air gap.

Solution

$$X_s = \omega L_{ss} = 120\pi \times 0.07 = 26.4 \quad \Omega$$

$$I_a = I_b = \frac{115}{26.4} = 4.36 \quad \text{A}$$

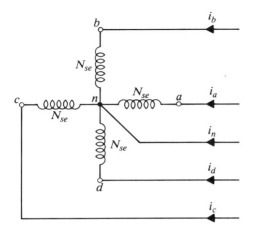

Fig. 3.41 Diagram for Example 3.7.

Thus

$$i_a = 4.36 \sqrt{2} \sin \omega_s t \quad \text{A}$$

$$i_b = 4.36 \sqrt{2} \sin \left(\omega_s t - \frac{2\pi}{4} \right) \quad \text{A}$$

$$i_c = 0$$

$$i_d = 0$$

For each phase winding, from Eq. (3.141),

$$\mathcal{F}_{sg} = \frac{N_{se}\, i}{2} \sin (\theta - \gamma)$$

when

$$\gamma_a = 0, \quad \gamma_b = \frac{\pi}{2}, \quad \gamma_c = \pi, \quad \gamma_d = \frac{3\pi}{2} \quad \text{rad}$$

By superposition, the total mmf acting across the air gap at angle θ will be

$$\mathcal{F}_{sg} = \frac{N_{se} i_a}{2} \sin \theta + \frac{N_{se} i_b}{2} \sin \left(\theta - \frac{\pi}{2} \right)$$

$$+ \frac{N_{se} i_c}{2} \sin (\theta - \pi) + \frac{N_{se} i_d}{2} \sin \left(\theta - \frac{3\pi}{2} \right)$$

$$= \frac{300 \times 4.36 \sqrt{2}}{2} \left[\sin \omega_s t \sin \theta + \sin \left(\omega_s t - \frac{\pi}{2} \right) \sin \left(\theta - \frac{\pi}{2} \right) \right]$$

$$= \frac{300 \times 4.36 \sqrt{2}}{2} \left[\frac{1}{2} \cos (\omega_s t - \theta) - \frac{1}{2} \cos (\omega_s t + \theta) \right.$$

$$\left. + \frac{1}{2} \cos (\omega_s t - \theta) - \frac{1}{2} \cos (\omega_s t + \theta - \pi) \right]$$

$$= \frac{300 \times 4.36 \sqrt{2}}{2} \cos (\omega_s t - \theta) \quad \text{A}$$

$$B_{sg} = \mu_0 H_{sg} = \mu_0 \frac{\mathscr{F}_{sg}}{g} = \frac{4\pi \times 10^{-7}}{1.5 \times 10^{-3}} \cdot \frac{300 \times 4.36 \sqrt{2}}{2} \cos{(\omega_s t - \theta)}$$

$$= 0.78 \cos{(\omega_s t - \theta)} \quad \text{T}$$

So the peak flux density produced in the air gap is 0.78 T.

Thus, excitation of only two of the phases nevertheless produces a rotating wave. A machine with only two of four phases excited (and in fact with only two windings on the stator) is often employed and supplied from a so-called "two-phase" source.

3.4 POLYPHASE ALTERNATING-CURRENT MACHINES

A machine that receives its sole or principal excitation from an ac source, or that, itself, acts as a source of ac, is termed an *ac machine*.

The conditions required to produce constant torque in a cylindrical machine were investigated in Section 3.3 and related to the equation

$$T = -K \hat{\mathscr{F}}_{sg} \hat{\mathscr{F}}_{rg} \sin{\delta} \quad \text{N} \cdot \text{m} \tag{3.148}$$

This equation showed that constant torque would be produced provided that two mmf waves of constant amplitude and constant angular displacement from one another could be produced.

Constant angular displacement was achieved in the dc machine by means of the commutator, which brought the rotor mmf axis to a standstill and thus exemplified the first of the three possible methods suggested at the end of Section 3.3. The other two possible methods required that the stator mmf axis should rotate relative to the fixed stator windings. The discussion in Sections 3.3.3–3.3.5 has shown how that may be done by employing polyphase stator windings.

The essential condition for energy conversion in the cylindrical machine with single stator and rotor windings was expressed by the equation

$$\omega_m = \pm \, (\omega_s \pm \omega_r) \quad \text{rad/s} \tag{3.149}$$

where the $\pm$ outside the parentheses indicated that in such single-winding machines the direction of rotation was immaterial. In a machine with a polyphase stator winding this can no longer be the case, since the rotor mmf axis must rotate in the same direction as that of the stator in order to satisfy the conditions for constant torque deduced from Eq. (3.130). For machines with polyphase stators, Eq. (3.149) must be reduced, therefore, to

$$\omega_m = \omega_s \pm \omega_r \quad \text{rad/s} \tag{3.150}$$

The second and third methods of achieving constant torque suggested at the end of Section 3.3 may therefore be investigated in the light of Eqs. (3.148) and (3.150).

3.4.1 Synchronous Machine

The second possible method of producing constant torque was seen to be that of causing the stator mmf axis to rotate at rotor speed relative to the fixed stator windings. To express this in the reverse order, the rotor must rotate at the same speed as the stator mmf axis. In a machine with a polyphase stator, the speed of rotation of the stator mmf axis has been shown to be ω_s, the angular frequency of the polyphase stator source. Thus if $\omega_m = \omega_s$, then Eq. (3.150) can only be satisfied if $\omega_r = 0$; that is, the rotor is excited with direct current.

If a rotor carrying a single sinusoidally distributed winding excited with constant direct current is fitted within a polyphase stator and rotated at ω_s rad/s, this rotor will produce a sinusoidally distributed mmf wave of constant amplitude in the air gap, and its mmf axis will remain at a constant angle to that of the stator. The conditions for energy conversion due to the development of a constant torque will then be satisfied. The resulting physical system is illustrated in Fig. 3.42.

A polyphase machine in which the rotor rotates in synchronism with the rotating mmf wave produced by the stator is called a *synchronous machine*.

In Fig. 3.42,

$$T = -K\,\hat{\mathscr{F}}_{sg}\,\hat{\mathscr{F}}_{rg}\sin\delta \quad \text{N}\cdot\text{m} \tag{3.151}$$

For $\delta < 0$, $T > 0$—that is, the machine is motoring since the developed torque is positive and acts in the direction of rotation. If the machine were driven by a prime mover, so that δ became positive, then the torque would become negative, and the energy conversion process would be reversed with the machine functioning as a generator.

The torque produced in the machine may be envisaged in terms of the stator and rotor "poles" shown in Fig. 3.42, which attract or repel one another. These "poles" correspond to peak values of gap mmf components produced by the rotor and stator windings. The flux density distribution in the air gap is produced by the

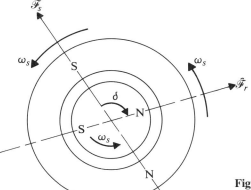

Fig. 3.42 Magnetomotive-force axes of a synchronous machine.

resultant mmf. Maximum flux density in the air gap therefore occurs at some angle between the mmf axes. Since the rotor is rotating at the same speed as the resultant flux wave produced in the air gap, no variation in flux linkage takes place in the rotor winding, and hence no emf is induced in it.

It should be noted that at standstill, when $\omega_m = 0$, no average unidirectional torque is developed by the synchronous machine—that is, it has no starting torque.

A synchronous machine could also be built with a polyphase winding on the rotor and a dc winding on the stator, but for practical reasons, which will be discussed later, such an arrangement is not normally adopted.

3.4.2 Induction Machine

The third possible method of producing constant torque was seen to be that of causing the mmf axes of stator and rotor to rotate at such speeds relative to their windings that they remain stationary relative to one another. The mmf axes of both stator and rotor may be caused to rotate relative to their windings if the windings are polyphase and carry polyphase alternating currents. Such a machine would therefore have polyphase stator excitation at angular frequency ω_s, polyphase rotor excitation of angular frequency ω_r, and rotor speed ω_m as defined by Eq. (3.150). First consider the rotor speed

$$\omega_m = \omega_s - \omega_r \quad \text{rad/s} \tag{3.152}$$

If a polyphase stator is excited from a source of angular frequency ω_s, then a rotating magnetic field of constant amplitude is produced, and this field rotates at ω_s rad/s relative to the stator. Similarly, if a polyphase rotor is excited from a source of angular frequency ω_r, then a rotating magnetic field of constant amplitude will be produced, and this field will rotate at ω_r rads/s relative to the rotor. The situation with the stationary rotor is illustrated in Fig. 3.43, where the phase sequences of the sources are such that both magnetic fields rotate in a positive direction.

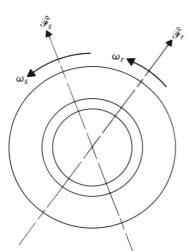

Fig. 3.43 Machine with separate stator and rotor excitations.

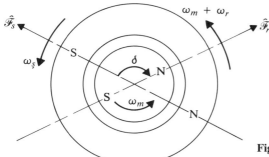

Fig. 3.44 Magnetomotive-force axes of an induction machine.

If the rotor is now given a positive speed of rotation ω_m, the speed of rotation of the rotor magnetic field relative to the stator is $\omega_m + \omega_r$. If

$$\omega_m + \omega_r = \omega_s \quad \text{rad/s} \tag{3.153}$$

then the conditions for energy conversion at constant torque are satisfied. This situation is illustrated in Fig. 3.44. The machine under these circumstances is operating as a doubly-fed polyphase machine. It remains to show how the rotor windings may be conveniently excited at such a frequency that Eq. (3.153) is satisfied. Normally, in a machine with polyphase stator and rotor windings, only a stator source is employed, and the rotor excitation is induced from the stator winding. It is for this reason that the machine is called an *induction machine*.

Figure 3.45 is a diagram of a machine with a two-winding stator similar to that in Fig. 3.34, which was discussed in Section 3.3.4. Two windings similar to those on the stator are also placed on the rotor, each sinusoidally distributed winding, as usual, represented by its center turn only. As before, let the stator currents be

$$i_a = \hat{\imath}_s \cos \omega_s t \quad \text{A} \tag{3.154}$$

$$i_b = \hat{\imath}_s \cos\left(\omega_s t - \frac{\pi}{2}\right) = \hat{\imath}_s \sin \omega_s t \quad \text{A} \tag{3.155}$$

so that the resulting field rotates in a positive direction.

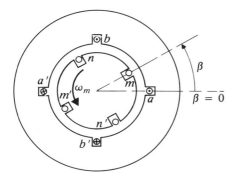

Fig. 3.45 Induction of rotor emf's.

At $t = 0$, let the rotor have some angular position β_0, so that if the rotor is rotating at speed $\omega_m = d\beta/dt$ in a positive direction, then

$$\beta = \omega_m t + \beta_0 \quad \text{rad} \tag{3.156}$$

The mutual inductances between the various pairs of windings are

$$L_{ma} = \hat{L} \cos \beta \qquad\qquad\qquad \text{H}$$

$$L_{mb} = \hat{L} \cos \left(\beta - \frac{\pi}{2} \right) = \hat{L} \sin \beta \qquad \text{H}$$

$$L_{na} = \hat{L} \cos \left(\beta + \frac{\pi}{2} \right) = - \hat{L} \sin \beta \quad \text{H} \tag{3.157}$$

$$L_{nb} = \hat{L} \cos \beta \qquad\qquad\qquad \text{H}$$

$$L_{ab} = L_{mn} = 0 \qquad\qquad\qquad \text{H}$$

If the rotor windings are open-circuited, the emf induced in winding mm' due to the rotating field produced by the stator excitation will be

$$e_m = \frac{d\lambda_m}{dt} \quad \text{V} \tag{3.158}$$

where λ_m is the flux linkage in winding mm' due to the rotating field. This flux linkage will be made up of two components, λ_{ma}, which is the flux linkage in winding mm' due to i_a, and λ_{mb}, the flux linkage due to i_b. Thus

$$\lambda_m = \lambda_{ma} + \lambda_{mb} = L_{ma} i_a + L_{mb} i_b \quad \text{Wb} \tag{3.159}$$

Substitution from Eq. (3.159) in Eq. (3.158) then yields

$$e_m = L_{ma} \frac{di_a}{dt} + i_a \frac{dL_{ma}}{dt} + L_{mb} \frac{di_b}{dt} + i_b \frac{dL_{mb}}{dt} \quad \text{V} \tag{3.160}$$

Substitution from Eqs. (3.154), (3.155), and (3.157) in Eq. (3.160) yields

$$e_m = \hat{L} \cos \beta \left(-\omega_s \hat{i}_s \sin \omega_s t \right) + \hat{i}_s \cos \omega_s t (-\hat{L} \sin \beta) \frac{d\beta}{dt}$$

$$+ \hat{L} \sin \beta \left(\omega_s \hat{i}_s \cos \omega_s t \right) + \hat{i}_s \sin \omega_s t (\hat{L} \cos \beta) \frac{d\beta}{dt}$$

$$= (\omega_s - \omega_m) \hat{L} \hat{i}_s \left(\cos \omega_s t \sin \beta - \sin \omega_s t \cos \beta \right)$$

$$= -(\omega_s - \omega_m) \hat{L} \hat{i}_s \sin \left(\omega_s t - \beta \right) \quad \text{V} \tag{3.161}$$

Substitution from Eq. (3.156) in Eq. (3.161) then yields

$$e_m = - (\omega_s - \omega_m) \hat{L} \hat{i}_s \sin \left[(\omega_s - \omega_m) t - \beta_0 \right] \quad \text{V} \tag{3.162}$$

Similarly, it may be shown that

$$e_n = (\omega_s - \omega_m) \hat{L} \hat{i}_s \cos \left[(\omega_s - \omega_m) t - \beta_0 \right] \quad \text{V} \tag{3.163}$$

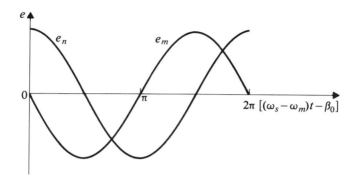

Fig. 3.46 Electromotive forces induced in rotor of Fig. 3.45.

Thus emf's of frequency $(\omega_s - \omega_m)$, displaced in phase by $\pi/2$ radians from one another, are induced in the rotor windings. The time variation of these emf's is shown in Fig. 3.46 from which it will be seen that the phase sequence is *mn*.

If passive external circuits of equal impedance are connected to the rotor windings, or if the windings are simply short-circuited, then currents will be produced in the rotor windings by emf's e_m and e_n. These currents will produce a rotating field that, due to the phase sequence *mn*, will rotate in a positive direction relative to the rotor. Thus the necessary conditions have been fulfilled for energy conversion by means of a constant unidirectional torque developed by the machine. Equation (3.153) is satisfied, and the machine operates as illustrated in Fig. 3.44.

At standstill, $\omega_m = 0$, the frequency of the induced rotor currents is $\omega_r = \omega_s$, and a starting torque is developed that accelerates the rotor in the direction of rotation of the stator field. Since this torque has developed as a result of electrical excitation only, it is clear that the machine is operating as a motor under conditions where $\omega_m < \omega_s$.

The amplitudes of the induced rotor emf's are proportional to $(\omega_s - \omega_m)$. Thus if the stator is excited and the rotor is driven at speed $\omega_m = \omega_s$, that is, at synchronous speed, no rotor currents flow and torque is zero. Under these conditions no useful energy conversion takes place, and the machine simply "floats" on the electrical system.

If the rotor were driven at such a speed that $\omega_m > \omega_s$, then e_n would be reversed. This means that the phase sequence of the rotor currents would be reversed, and the rotating field produced by the rotor windings would reverse its direction relative to the rotor. The speed of the rotor field relative to the stator would then be

$$\omega_m - \omega_r = \omega_s \quad \text{rad/s} \tag{3.164}$$

This situation is illustrated in Fig. 3.47, where it may be seen that the torque developed in the machine opposes the rotation. Mechanical energy must thus be supplied

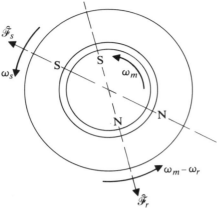

Fig. 3.47 Induction machine operation for $\omega_m > \omega_s$.

Fig. 3.48 Rotor squirrel cage.

to the machine in order to maintain the rotor at a speed greater than that of the stator rotating field. Under these conditions the machine operates as an induction generator.

If the operation of a machine with a two-winding stator, but with a three-phase rotor winding were analyzed, it would be found that balanced three-phase emf's would be induced in the rotor windings, the necessary rotor rotating field would be developed, and energy conversion would take place. Conversely, a machine with a three-phase stator and a two-winding rotor would also operate satisfactorily. In general, a rotor with any number of phases will develop torque in a stator of the same or any other number of phases, single-winding stators only excepted.

One kind of rotor winding consists simply of a number of conductor bars connected to end rings, the result being a "squirrel cage" of conductor as illustrated in Fig. 3.48. The emf's induced in the conductors produce currents that circulate through the bars and the end rings, effectively forming a rotor winding with a large number of phases. The limiting case of this type of rotor is a "winding" consisting of a cylinder of conductor enclosing the rotor magnetic core. Such machines in small sizes are frequently employed in servomechanisms.

3.4.3 Starting of Synchronous Motors

Section 3.4.1 pointed out that the synchronous machine has no starting torque. This is inconvenient if the machine is to be used as a motor. This problem is solved, as illustrated in Fig. 3.49, by the addition of a starting winding that consists of a partial squirrel cage of conductors embedded in the faces of the salient poles of the synchronous motor. This pole-face winding permits starting the synchronous motor as an induction motor. When the rotor reaches synchronous speed, no emf's are induced in the squirrel cage, and currents cease to flow in it.

The pole-face winding also has the function of damping out speed oscillations when the machine is operating at synchronous speed. When the load torque imposed on a synchronous motor changes, the axis of the rotor must take up a new angle relative to the axis of the rotating mmf wave due to the stator windings. Before the machine settles down to a new steady-state operating condition, the rotor is liable to oscillate about the new angle. For the half cycle of an oscillation in which the rotor speed is less than synchronous, the pole-face winding produces a positive induction-motor torque. For the half cycle in which the speed is more than synchronous, the pole-face winding produces a negative torque. This action damps out the oscillations.

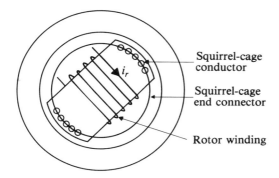

Fig. 3.49 Synchronous machine starting winding.

3.5 MACHINES WITH MORE THAN TWO POLES

For synchronous machines, a required condition of operation is $\omega_m = \omega_s = 2\pi f$, where f is the stator source frequency, which in large machines is normally 60 Hz. Thus $\omega_m = 120\pi$ rad/s, and the speed of such a machine in revolutions per minute is

$$n = 120\pi \times \frac{60}{2\pi} = 3600 \quad \text{r/min} \tag{3.165}$$

It is not always convenient to operate machines at so high a speed. In order to avoid the necessity of doing so, additional windings are mounted on both the stator and the rotor.

Figure 3.50(a) illustrates a stator carrying a single sinusoidally distributed winding, represented as usual by the center turn only. The development diagram of Fig. 3.50(b) shows the distribution of mmf around the air gap due to this single winding when it is carrying current i_a. Two poles are produced on the inner periphery of the stator.

Figure 3.51(a) illustrates a stator carrying two identical windings $a_1 a_1'$ and $a_2 a_2'$, each extending over only π radians of the air-gap periphery. These windings also are sinusoidally distributed, but the conductor density is a function of 2θ. The development diagram of Fig. 3.51(b) shows the distribution of mmf around the air

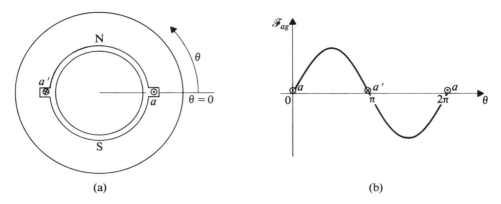

Fig. 3.50 Two-pole stator mmf distribution.

gap due to these two windings when they are carrying currents of the same magnitude. The two windings may be considered to be connected either in series or in parallel to a single alternating-emf source, so that at all times the current in the two is the same. Four poles are now produced on the inner periphery of the stator.

If two more windings are now fitted to the stator with their center conductors halfway between those of the first two in the positions indicated by the unmarked conductors in Fig. 3.51(a), and if these windings carry current i_b, then the distribution of mmf around the air gap will be the resultant of the excitations of the two sets of windings. If i_a and i_b are alternating currents with a phase displacement of $\pi/2$ radians, then a 4-pole rotating field will be produced in the air gap.

In analyzing the operation of machines with more than two poles, it is convenient to recognize that the physical conditions existing in the neighborhood of any two poles of the machine are reproduced exactly in the neighborhood of any other

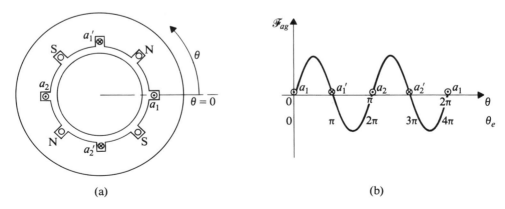

Fig. 3.51 Four-pole stator mmf distribution.

two poles. For example, in the 4-pole stator of Fig. 3.51, conditions in the air gap over the domain $0 < \theta < \pi$ are identical with those over the domain $\pi < \theta < 2\pi$. For this reason, it is sufficient to analyze the physical conditions in the neighborhood of one pair of poles only—that is, for one complete cycle of the resultant space mmf wave.

The scale of the horizontal axis in developed diagrams of machines may therefore be calibrated in "electrical radians," such that 2π electrical radians encompass one complete cycle of the resultant mmf wave in the air gap. A scale in electrical radians for what may be called the "electrical angle" θ_e is, therefore, shown in Fig. 3.51(b). The original calibration may be called the "mechanical angle" θ. For the 4-pole machine, $\theta_e = 2\theta$, and in general, for a machine with p poles

$$\theta_e = \frac{p}{2}\,\theta \text{ electrical radians} \tag{3.166}$$

Increasing the number of poles on the stator has an important practical result. If the excitations employed to produce a 2-pole rotating field are applied to the 4-pole stator of Fig. 3.51(a), that is,

$$i_a = \hat{i}_s \cos \omega_s t \quad \text{A} \tag{3.167}$$
$$i_b = \hat{i}_s \sin \omega_s t \quad \text{A} \tag{3.168}$$

then the rotating magnetic field produced in the air gap will traverse only 2π electrical radians during one cycle of i_a and i_b. Thus the speed of the rotating field produced by a 4-pole stator is half of that of the field produced by a 2-pole stator. Since the rotor of a synchronous machine must keep in synchronism with the rotating field, it follows that for a p-pole machine

$$\omega_m = \frac{2}{p}\,\omega_s \quad \text{rad/s} \tag{3.169}$$

Thus, by increasing the number of poles, the speed of the machine may be reduced to some integral submultiple of 3600 r/min, provided that the number of poles on the rotor also is increased to the same number as that of the stator. If different numbers of stator and rotor poles were employed, the net torque developed in the machine would be zero.

In the case of the induction machine, the rotor does not rotate at synchronous speed. Nevertheless, the provision of multipole stator and rotor windings reduces the operating speed of the machine in the same ratio as in the case of the synchronous machine. For the induction machine, the equation corresponding to Eq. (3.169) is

$$\omega_m = \frac{2}{p}\,(\omega_s - \omega_r) \quad \text{rad/s} \tag{3.170}$$

The desirability of multipole construction of direct-current machines arises from considerations other than that of operating speed alone, as will be seen when these machines are discussed in more detail in Chapter 4.

Example 3.8 A 3-phase, 60-Hz induction motor has a full-load speed of 840 r/min, and a no-load speed of nearly 900 r/min.

a) Determine the number of poles of the motor;

b) Determine the rotor frequency at full load;

c) Determine the speed of rotation of the rotor field at full load
   i) with respect to the rotor,
   ii) with respect to the stator.

Solution.

a) On no load the motor is required to develop only an extremely small torque to overcome its own losses. It therefore runs at virtually synchronous speed. Thus, from Eq. (3.169),

$$\omega_m = \omega_s = \frac{2\pi}{60} \times 900 = \frac{2}{p} \times 120\pi \quad \text{rad/s}$$

Thus

$$p = 8$$

b) For the 2-pole machine, Eqs. (3.162) and (3.163) show that the rotor frequency is $\omega_s - \omega_m$. For a p-pole machine, it is, from Equation (3.170),

$$\omega_r = \omega_s - \frac{p}{2} \omega_m \quad \text{rad/s}$$

On full load

$$\omega_m = \frac{2\pi}{60} \times 840$$

Thus

$$\omega_r = 120\pi - \frac{8}{2} \times \frac{2\pi}{60} \times 840 = 8\pi \quad \text{rad/s}$$

$$f_r = \frac{8\pi}{2\pi} = 4 \text{ Hz}$$

c) The speed of the rotor field is determined as follows:
   i) With respect to the rotor, it is

$$\frac{2}{p} \omega_r = \frac{2}{8} \times 8\pi = 2\pi \quad \text{rad/s}$$

$$= 60 \quad \text{r/min}$$

ii) With respect to the stator, it is

$$\omega_m + \frac{2\omega_r}{p} = \frac{2\pi}{60} \times 840 + 2\pi = \frac{2\pi}{60} \times 900 = \text{synchronous speed}$$

This must be the case if the rotor field is to be stationary with respect to the stator field.

3.6 CONSTRUCTION OF ELECTROMAGNETIC MACHINERY

At this stage, the mental pictures conjured of machines that are possible to construct may not be very realistic. It is therefore important to gain some concept of what an electric machine may look like. For this purpose Figs. 3.52 through 3.62 are provided.

Figure 3.52 shows a 500-kilowatt, 220-volt, 1000-r/min, dc generator, and shows the commutator and brush gear. From the length of the commutator and the number of brushes it may be seen that this machine has a high armature current. The stator of a similar machine is shown in Fig. 3.53. This stator has six main poles and six interpoles (see Chapter 4). Figure 3.54 shows the commutator of a smaller relatively low-current dc machine.

Fig. 3.52   500-kW dc generator. *(The Harland Engineering Company of Canada Ltd.)*

Fig. 3.53 Direct-current machine stator with interpoles. (*The Harland Engineering Company of Canada Ltd.*)

Fig. 3.54 Direct-current machine commutator. (*Canadian Westinghouse Company Limited*)

Fig. 3.55 Polyphase ac stator winding. *(The Harland Engineering Company of Canada Ltd.)*

The stator windings of polyphase ac synchronous and induction machines in no way differ from one another, and an example of one such stator winding is shown in Fig. 3.55. The construction of the rotor of a salient-pole synchronous machine is shown in Fig. 3.56. This is a 10-pole machine that would run at 720 r/min if connected to a 60-Hz electric system.

The machine in Fig. 3.57 is an ac synchronous generator designed to be driven at 3600 r/min by a steam turbine. Because of this high speed, the rotor must be long and of small diameter. The machine in Fig. 3.58, on the other hand, is an ac synchronous generator designed to be driven at 94.7 r/min by a water turbine. Although much bigger than the machine of Fig. 3.57, it only develops about half the power of the high-speed machine.

The rotor shown in Fig. 3.59 is for a wound-rotor induction motor, and would be fitted inside a stator with a winding arrangement similar to that in Fig. 3.55. This type of rotor is employed in induction motors when variable speed is required. The much simpler rotor winding shown in Fig. 3.60 would also be fitted into a stator similar to that in Fig. 3.55, but this type of rotor would be employed when an essentially constant-speed motor was required.

Fig. 3.56 Rotor for a salient-pole synchronous machine. (*The Harland Engineering Company of Canada Ltd.*)

Fig. 3.57 High-speed ac synchronous generator. (*Canadian General Electric Company Limited*)

Fig. 3.58 Low-speed ac synchronous generator. (*Canadian General Electric Company Limited*)

Fig. 3.59 Wound rotor with slip rings and brush gear for an induction motor. (*Canadian General Electric Company Limited*)

Fig. 3.60 Large bar-wound squirrel-cage rotor with end ring to be brazed to the conductors. *(The Harland Engineering Company of Canada Ltd.)*

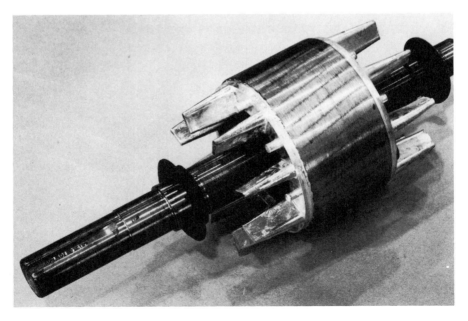

Fig. 3.61 Die-cast squirrel-cage rotor. *(Canadian Westinghouse Company Limited)*

Fig. 3.62 Squirrel-cage induction motor. *(Canadian Westinghouse Company Limited)*

Figure 3.61 shows a smaller squirrel-cage rotor, whose construction is much simplified by die-casting the complete aluminum cage with end rings directly into the stack of steel laminations forming the core of the rotor. "Flingers" are cast on the end rings to ventilate the machine. A complete squirrel-cage induction motor with one end cover unshipped is shown in Fig. 3.62. This relatively small motor has wire stator conductors rather than the rectangular copper conductor used in the stator of Fig. 3.55.

PROBLEMS

3.1 A solenoid coil consists of a single layer of 250 circular turns each of 0.02 m radius. The length of the coil is 0.3 m, and it is self-supporting, so that it contains only air.

a) Determine the inductance of the coil, assuming the magnetic field intensity to be uniform inside the coil and to be zero elsewhere.

b) Find the stored energy in the magnetic field when the coil current is 18 A.

(Section 3.1.1)

3.2 Figure 3.63 shows a pair of parallel conducting plates. A cloud of ionized gas, or plasma, is established between the plates at a mean distance z from the ends, to which a current source of 1000 A is connected.

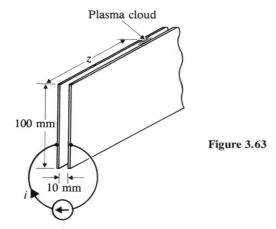

Figure 3.63

a) Determine the inductance of the conducting loop, assuming the magnetic field intensity to be uniform between the plates and zero elsewhere.
b) Determine the force acting on the plasma cloud.
c) Assuming that the cloud is bounded by the edges of the plates, determine the pressure applied to the cloud. *(Section 3.1.2)*

*3.3 A coil with a ferromagnetic core has a flux linkage–current relationship that can be approximated by

$$\lambda = 0.72 i^{1/3} \quad \text{Wb}$$

Determine the energy stored in the magnetic system when the current is 2.1 A.

(Section 3.1.3)

3.4 Determine the force tending to change the length of the solenoid coil in Problem 3.1 when the coil current is 18 A. *(Section 3.1.2)*

3.5 The magnetic holding device shown in Fig. 3.64(a) consists of two ferrite magnets, each having a length of 8 mm in the direction of magnetization and a cross-section area of

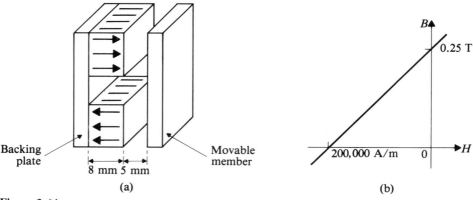

(a)

(b)

Figure 3.64

700 mm². The magnets are fixed to a high-permeability iron backing plate and faced by a high-permeability movable member. The ferrite material is described by the linear $B–H$ relation shown in Fig. 3.64(b). Fringing of flux around the air gaps may be ignored. Determine the force on the movable member when it is 5 mm from the magnets. *(Section 3.1.2)*

3.6 A 13.2-kV transmission line consists of two conductors, each 20 mm in diameter and spaced 1.5 m apart. The span between supporting poles is 100 m.

a) Derive an expression for the inductance of the line in terms of its length, conductor diameter, and spacing.

b) Derive an expression for the force acting on the conductor as a function of the line current.

c) Determine the average force acting on each conductor when a sinusoidal current of 8000 A rms flows in the line.

d) Confirm the answer to (c) by determining the flux density produced by one conductor in the region of the other and then using the relationship $F = Bli$ to evaluate the force.

e) Suppose the transmission line is shorted by a round bar placed on top of the conductors and free to roll along them. If a current of 8000 A is passed along the line and through the bar, what is the force on the bar? *(Section 3.1.2)*

*3.7 The relationship between flux linkage λ, current i, and displacement x of the moving part of a ferromagnetic actuator can be expressed approximately by the equation

$$\lambda = \frac{0.04\sqrt{i}}{x} \quad \text{Wb}$$

for the ranges $0 < i < 5$ A, $0.01 < x < 0.05$ m.

If the current is maintained constant at 4 A, what is the force produced by the actuator when the displacement is 0.03 m? *(Section 3.1.3)*

*3.8 Over the intended operating range, the relationship between flux linkage λ, current i, and displacement x of the moving member of a ferromagnetic actuator may be expressed approximately by

$$\lambda = \frac{4.3\,i^{1/3}}{2} \quad \text{Wb}$$

Determine the force in the x direction when $i = 0.6$ A and $x = 0.02$ m. *(Section 3.1.3)*

3.9 The magnetic system shown in Fig. 3.65 has a square cross section 30 mm × 30 mm. When the two sections of the core are fitted together, air gaps, each of length $x = 1$ mm, separate them. The coil has 250 turns and a resistance of 7.5 Ω. The magnetic field intensity required by the core material is negligible.

a) If a potential difference of 40 volts is applied to the coil terminals, what is the total force holding the two sections of the core together? (Fringing at the air gaps may be neglected.)

b) If an alternating potential difference of 100 V rms at 60 Hz is applied to the coil terminals, what is the average value of the force holding the two sections together? (Fringing at the air gaps may be neglected.)

c) The effect of fringing may be allowed for by assuming that the cross section of the air gap is $(30 + x) \times (30 + x)$ and that the flux density is uniform over that cross-section

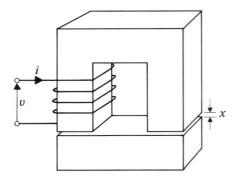

Figure 3.65

area for the gap length of x. Repeat parts (a) and (b), making this correction for fringing.   *(Section 3.1.2)*

3.10 The actuator shown in Fig. 3.66 is to be used to raise a mass m through a distance y. The coil has 500 turns and can carry a current of 2 A without overheating. The magnetic material can support a flux density of 1.5 T with negligible field intensity. (Fringing of flux at the air gaps may be neglected.)

a) Determine the maximum air gap y for which a flux density of 1.5 T can be established with a current of 2 A.

b) With the air gap determined in part (a), what is the force exerted by the actuator?

c) The mass density of the material is 7800 kg/m³. Determine the approximate value of the net mass m of the load that can be lifted against the force of gravity by the actuator at the air gap determined in (a).

d) What current is required in the coil to lift the unloaded actuator at the air gap determined in (a)?

e) What is the initial acceleration of the unloaded actuator if it is released at the air gap determined in (a) when the coil current is 0.3 A?   *(Section 3.1.2)*

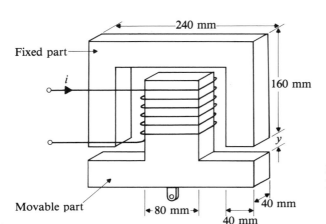

Fig. 3.66 Diagram for Problems 3.10, 3.18, and 3.19.

3.11 In the relay mechanism constructed of steel parts and illustrated in Fig. 3.67, the two poles are 30 mm × 30 mm in cross section and, when they meet, the center line of the armature is vertical. The coil has 1000 turns. Assume the permeability of the steel is infinite. Determine the force on the armature when the coil carries a direct current of 0.5 A and the armature is 2° from the vertical position. (Fringing and the effect of the air gap at the hinge may be neglected.)   *(Section 3.1.2)*

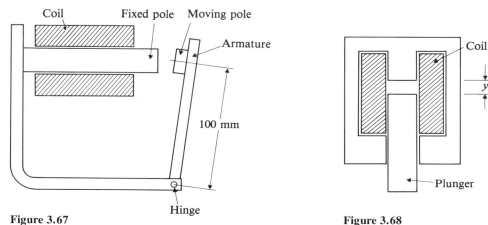

Figure 3.67   **Figure 3.68**

3.12 Figure 3.68 shows a cross section of a cylindrical magnetic actuator. The plunger, of cross-section area .0015 m², is free to slide vertically through a circular hole in the outer magnetic casing, the air gap between the two being negligible. The coil has 3000 turns and a resistance of 8 Ω. A potential difference of 12 V is applied to the coil terminals. The magnetic material may be assumed perfect up to its saturation flux density of 1.6 T. (Fringing at the air gap may be neglected.)

a) Determine the static force on the plunger as a function of the air gap length y.
b) Over what range of gap length y will the force on the plunger be essentially constant because saturation flux density has been reached?
c) Suppose that the plunger is constrained to move slowly from a gap of 10 mm to the fully closed position. What will be the mechanical energy produced?
d) Suppose that the plunger is allowed to close so quickly from an initial gap of 10 mm that the flux linkage of the coil does not change appreciably during the motion. How much mechanical energy will be produced?   *(Section 3.1.2)*

3.13 Figure 3.69 illustrates the movement of a moving-iron ammeter in which a curved ferromagnetic rod is drawn into a curved solenoid against the torque of a restraining spring. The inductance of the coil is $L = 5 + 20\theta$ μH, where θ is the deflection angle in radians. Its resistance is 0.01 Ω. The spring constant is 7×10^{-4} N·m/rad.

a) Show that the instrument measures the root-mean-square value of the coil current.
b) What will be the full-scale deflection if the rated current is 10 A?
c) What will be the potential difference at the coil terminals when the current is 5 A rms at a frequency of 180 Hz?   *(Section 3.1.4)*

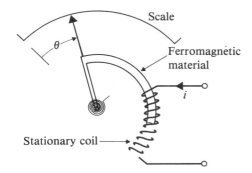

Scale

Ferromagnetic material

i

Stationary coil

Fig. 3.69 Diagram for Problem 3.13.

3.14 A rotating actuator of the form shown in Fig. 3.70 has the following dimensions: $g = 1$ mm, $r = 20$ mm, $z = 40$ mm. The coil has 400 turns. The magnetic material may be considered perfect up to 1.2 T.

   a) Determine the maximum coil current if the flux density is to be limited to 1.2 T in the material bounding the air gap.

   b) Assume the current found in part (a), and determine the torque produced at angle θ, where θ is the angle of overlap of the stator and rotor poles. (Fringing at the air gaps may be neglected.)

   c) What work will be done as the shaft moves from $\theta = \pi/2$ to $\theta = 0$?   (*Section 3.1.4*)

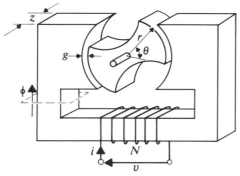

Fig. 3.70 Diagram for Problem 3.14.

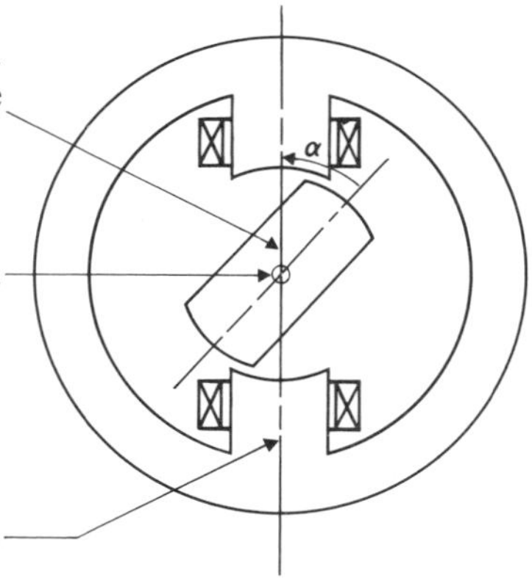

Steel armature

Shaft

Field pole

α

Fig. 3.71 Diagram for Problem 3.15.

3.15 Figure 3.71 illustrates the end view of a rotational actuator. The mean diameter of the air gap between the armature and the two field poles is 50 mm, and both curved surfaces subtend an angle of 60° at the axis of rotation of the armature. The axial length of the armature and field poles is 80 mm, and the air-gap length is 2 mm. The two field coils are connected in series, and each has 500 turns.

   a) Derive an expression for the torque that must be applied to the armature shaft in order to maintain the axis of the armature poles at an angle α to that of the field poles (for $0 < \alpha < 60°$) when a direct current of 7.5 A is flowing in the field coils.

b) Calculate the magnitude of the shaft torque required for $\alpha = 15°$, $30°$, and $45°$. (Fringing at the air gaps may be neglected, and the permeability of the steel may be assumed infinite.)
(Section 3.1.4)

3.16 In some control applications, a rotating actuator with reversible torque is required. Figure 3.72 illustrates a device that has this feature. In addition, it provides a torque that is linearly proportional to the product of its two coil currents. Each of the four stator poles subtends an angle of 45° at the axis of rotation of the armature. Each of the rotor poles subtends an angle of 90°. The axial length of the rotor and stator poles (into the plane of the diagram) is z. Each stator pole has two N-turn coils, one carrying a current i_1, the other carrying a current i_2. The coils are connected so that the mmf on each of the horizontal poles is $N(i_1 + i_2)$ while that on each of the vertical poles is $N(i_1 - i_2)$. (Fringing at the air gaps may be neglected, and the permeability of the steel may be assumed infinite.)

Note that, because of the symmetry of the magnetic system, mmf's on the horizontal poles do not produce flux in the vertical poles.

a) Derive an expression for the torque due to the mmf's on the horizontal poles. Derive a similar expression for the torque due to the mmf's on the vertical poles. Combine these two expressions and show that

$$T = \frac{4N^2\mu_0 z r i_1 i_2}{g} \quad \text{N} \cdot \text{m}$$

over the range of $0 < \theta < \pi/4$, where θ is the angle of overlap illustrated in Fig. 3.72.

b) Let $r = 25$ mm, $z = 50$ mm, $g = 1$ mm, and $N = 1000$ turns. If the air-gap flux density is not to exceed 1.5 T, and the currents i_1 and i_2 are equal, what is the maximum permissible value of each of these currents? If the current i_1 is maintained constant at this value, what is the torque produced per ampere of current i_2? *(Section 3.1.4)*

3.17 A synchronous reluctance machine of the type illustrated in Fig. 3.11 has a stator and rotor cross section of 25 mm × 25 mm. The pole faces are flat, as shown in Fig. 3.73. The

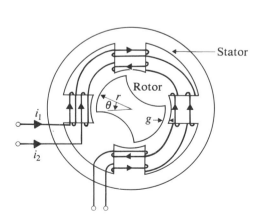

Fig. 3.72 Diagram for Problem 3.16.

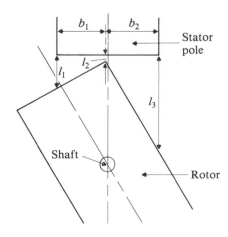

Fig. 3.73 Diagram for Problem 3.17.

length of the rotor between the pole faces is 50 mm, and the distance between the stator pole faces is 60 mm. The reluctance of each air gap may be approximated by considering it to consist of two reluctances in parallel. These are, from Fig. 3.72,

$$\mathscr{R}_1 = \frac{1}{\mu_o} \frac{l_1 + l_2}{2} \cdot \frac{1}{b_1 \times 0.025} \qquad \mathscr{R}_2 = \frac{1}{\mu_o} \frac{l_2 + l_3}{2} \cdot \frac{1}{b_2 \times 0.025}$$

where all dimensions are expressed in meters. When the corner of the rotor pole is no longer beneath the stator pole, there will be only one such reluctance to calculate for each rotor position.

a) By determining the reluctance of the magnetic circuit at 10° intervals, plot a curve of $\mathscr{R}$ versus θ. The permeability of the steel may be assumed infinite.

b) Determine the approximate maximum average torque that this machine will develop if the stator coil has 1500 turns and is excited from a 220-V, 60-Hz source.

(Section 3.1.4)

3.18 A reluctance motor of the form shown in Fig. 3.70 has a magnetic path whose reluctance can be approximately expressed as

$$\mathscr{R} = 5.06 \times 10^1 \, (2.5 + 1.5 \cos 2\theta) \quad \text{A/Wb}$$

The coil has 15 turns of negligible resistance. If a sinusoidal potential difference of 110 V rms at 60 Hz is applied to the coil terminals,

a) What is the magnetic flux in the machine?

b) At what angular velocity of rotation does the machine develop an average unidirectional torque?

c) What is the maximum value of average torque that this motor can produce?

d) What is the mechanical power output in part (c)?   *(Section 3.1.4)*

3.19 A rotating machine of the form shown in Fig. 3.70 has a coil inductance that can be expressed approximately by

$$L = 0.01 - 0.03 \cos 2\theta - 0.02 \cos 4\theta \quad \text{H}$$

A current of 5 A rms at 50 Hz is passed through the coil, and the rotor is driven at a controllable speed of ω_m rad/s.

a) At what values of speed will the machine develop useful torque?

b) Determine the maximum torque at each of the speeds obtained in (a).

c) Determine the maximum mechanical power output at each of the synchronous speeds obtained in (a).   *(Section 3.1.4)*

3.20 Figure 3.74(a) illustrates the cross-section of a reluctance motor with four rotor poles. These poles are shaped so that the reluctance of the magnetic system is a sinusoidally varying function of β, as illustrated in Fig. 3.74(b). The coil has 150 turns and negligible resistance. An alternating potential difference of 115 V rms at 60 Hz is applied to the coil terminals.

Determine the synchronous speed of the rotor and the maximum average torque that the machine can develop.   *(Section 3.1.4)*

3.21 A small coil having 500 tightly packed turns and a diameter of 10 mm is placed inside a solenoid consisting of a single layer of 250 circular turns each of 40 mm diameter. The length

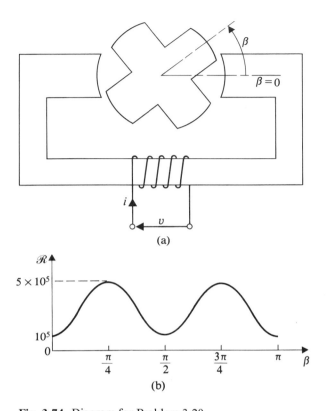

Fig. 3.74 Diagram for Problem 3.20.

of the solenoid is 300 mm. The coil is free to rotate about a diameter, so that its axis is at an angle θ to that of the solenoid. It is also free to move along the axis of the solenoid.

   a) Determine the mutual inductance between the small coil and the solenoid coil in terms of θ.

   b) Assuming that the coil currents are 0.5 A and 80 A respectively, find the maximum torque on the small coil and the angle θ at which it will occur.

   c) Assume the currents of (b). What will be the force tending to move the coil along the solenoid axis?   *(Section 3.1.5)*

3.22 A dynamometer instrument has one stationary coil and one moving coil. The moving coil is mounted within the stationary coil and is free to rotate about a common diameter of the two coils. The self- and mutual inductances of the coils are $L_{11} = 0.01$ H, $L_{22} = 0.004$ H, and $L_{12} = 0.003 \cos \theta$ H, where θ is the angle between the axes of the two coils.

   a) Suppose the two coils are connected in series so that the instrument is a dynamometer ammeter. Assuming that an alternating current of 0.5 A rms is passed through the coils, determine the average value of torque between them as a function of angle θ.

   b) Suppose the helical restraining spring on the moving coil is adjusted to give a rest position for zero current at $\theta = 90°$. What should the spring constant be if a full-scale deflection of $60°$ is to occur with a current of 0.5 A?

c) What current is required to produce half-scale deflection? *(Section 3.1.5)*

3.23 A small single-phase induction motor is to have a single winding of 400 turns. A concentric winding of the type shown in Figs. 3.20 and 3.21 is to be accommodated in 16 slots. Determine the required number of turns in each of the slots. *(Section 3.2.1)*

3.24 For a machine such as that illustrated in Fig. 3.23, the mutual inductance between the effectively sinusoidally distributed windings is

$$L_{sr} = 37.9 \times 10^{-3} \cos \beta \quad \text{H}$$

a) What average torque would the machine develop at standstill with $\beta = 30°$?

b) At what speeds, if any, would the machine operate as an energy converter?

c) What maximum average torque, if any, would the rotating machine develop for the following excitations:

  i) A direct current of 5 A in both windings.

  ii) An alternating current of 5 A rms in both windings connected in series.

  iii) Stator current $i_s = 5\sqrt{2} \cos(120\pi t + 30°)$ A,
  Rotor current $i_r = 5\sqrt{2} \cos(50\pi t + 60°)$ A.

  iv) Stator current $I_s = 5$ A dc,
  Rotor current $i_r = 5\sqrt{2} \cos(50\pi t + 60°)$ A. *(Section 3.2.2)*

3.25 A device has one stationary coil with a self-inductance of 0.1 H and one rotatable coil with a self-inductance of 0.04 H. Mutual inductance between the two coils is $0.05 \cos \theta$ H, where θ is the angle between the axes of the coils.

a) Suppose the rotatable coil is rotated at 200 rad/s. If a current of $10 \sin 200t$ A is passed through one of the coils, what is the peak value of the induced emf in the other coil?

b) Suppose a current of $10 \sin 200t$ A is passed through both coils in series. At what speeds will this device develop an average torque?

c) What is the maximum value of average torque that can be obtained in (b)? *(Section 3.2.2)*

3.26 The machine shown in Fig. 3.75 has three *concentrated* eight-turn stator coils. The coils are arranged symmetrically around the stator. The poles of the rotor are so shaped that

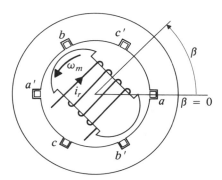

Figure 3.75

the flux density distribution around the stator periphery due to the rotor mmf is sinusoidal. The total flux produced by rotor current i_r is 0.40 Wb, and the rotor is driven at 3000 r/min.

  a) Determine the frequency of the emf's induced in the stator coils.
  b) If the stator terminals a', b', c' are connected to form a neutral point 0, what is the rms potential difference between any two of the terminals a, b, c?
  c) For the direction of rotation shown in Fig. 3.75, write expressions for the emf's e_{a0}, e_{b0}, e_{c0} induced in the stator coils.   *(Section 3.2.2)*

3.27 Figure 3.76 shows a cross section of a simple electric brake. The rotor winding consists of two shorted loops, each having a resistance of 0.005 Ω. Each of the stator poles covers a 90° arc of the rotor. The axial length of the rotor is 50 mm. (The magnetic material in the stator and rotor may be considered ideal; fringing may be neglected; and the magnetic field produced by the shorted loops may be ignored.)

  a) Derive expressions for the emf and for the current in each of the shorted loops during the interval in which the loop sides are under the poles.
  b) Show that the total power dissipated in the two loops is essentially constant throughout each revolution, provided the rotor speed ω_m and the field current i_f are constant.
  c) Derive an expression for the electromagnetic torque acting on the rotor.
  d) What field current is required to provide a braking power of 200 W when the speed is 400 rad/s?   *(Section 3.2.2)*

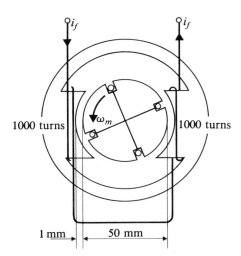

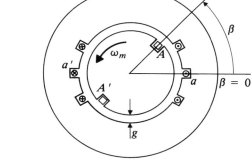

Fig. 3.76 Diagram for Problem 3.27.   **Fig. 3.77** Diagram for Problem 3.28.

3.28 Figure 3.77 shows a machine with a single effectively sinusoidally distributed winding of N_{se} turns on the stator and a single concentrated winding of N_r turns on the rotor.

  a) Determine an expression for the flux linkage of the open-circuited rotor winding in any position β when the stator excitation current is $i_s = \hat{i}_s \sin(\omega_s t + \alpha)$, and thence show that the mutual inductance between the two windings is expressed by

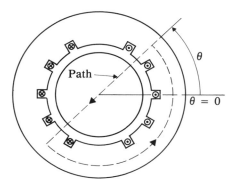

Fig. 3.78 Diagram for Problem 3.30.

$$L_{sr} = \frac{\mu_0 N_{se} N_r l r}{g} \cos \beta \quad \text{H}$$

where l = effective axial length of the rotor, r = air-gap radius, and g = air-gap length.

b) Determine an expression for the emf induced in the rotor winding when the rotor is driven at speed $\omega_m = \omega_s$.   *(Section 3.2.2)*

3.29 By means of a series of diagrams for $\Delta\omega t = 30°$ intervals, demonstrate the production of a rotating magnetic field by a three-phase stator winding, employing the technique applied in Section 3.3.4 to the two-winding stator.   *(Section 3.3.4)*

3.30 The sinusoidally distributed winding on the stator of Fig. 3.78 has $N_{se} = 120$ turns. The stator has an axial length of 100 mm and an air-gap radius of 50 mm. The effective length of the air gap (making allowance for the stator slots) is 1.0 mm. A current of 10 sin ωt A is passed through the winding.

a) Determine the mmf acting around a closed path at angle θ.
b) Assuming ideal magnetic material, determine the magnetic flux density in the air gap as a function of angle θ and time t. Sketch a graph of flux density versus angle θ for $\omega t = 0$, $\pi/6$, $\pi/2$, and $2\pi/3$.
c) Determine the flux linkage of the winding as a function of time.
d) Determine the self-inductance of the winding.
e) Suppose two other similar windings are placed in the stator slots to make a symmetrical three-phase winding. Determine the mutual inductance between a pair of these windings.
f) Suppose the three-phase windings carry currents

$$i_a = 10 \sin \omega_s t \quad \text{A}$$
$$i_b = 10 \sin (\omega_s t - 2\pi/3) \quad \text{A}$$
$$i_c = 10 \sin (\omega_s t - 4\pi/3) \quad \text{A}$$

Sketch a graph of the air gap flux density at $\omega t = 0$, $\pi/6$, $\pi/2$, and $2\pi/3$. Determine the peak value of the flux density.
g) For the condition of part (f) determine the maximum induced emf in each of the windings if the frequency is 60 Hz.   *(Section 3.3.5)*

3.31 For a synchronous machine

a) Write an expression for the speed of rotation in r/min in terms of the number of poles $p = 2, 4, 6, 8, \cdots$, etc. when the machine is to operate at
  i) 60 Hz,
  ii) 50 Hz.
b) Determine the maximum possible speed for a 50 to 60 Hz frequency conversion set using the expressions obtained in (a). *(Section 3.4.1)*

3.32 If a 60-Hz induction motor is driven at its synchronous speed with the rotor terminals open-circuited, what will be the frequency of the potential differences appearing at the rotor terminals when:

a) The stator windings are excited from a 3-phase, 25-Hz source (two answers).
b) Two of the stator terminals are supplied from a dc source. *(Section 3.4.2)*

3.33 A 4-pole, 60-Hz induction motor is mechanically coupled to an 8-pole, 60-Hz synchronous motor. The stator terminals of both machines are connected to the three-phase 60-Hz supply, but the rotor terminals of the induction motor are open-circuited. If this set is driven by the synchronous motor, determine:

a) The two possible frequencies of the potential differences appearing at the rotor terminals of the induction motor.
b) The ratio of the magnitudes of the rotor emf's for these two conditions of operation. *(Section 3.5)*

3.34 For machines with more than two poles,

a) Modify Eq. (3.163) to apply to a p-pole machine. Is β_0 affected?
b) Assume that a 60-Hz induction motor runs at 1790 r/min on no load and at 1700 r/min on full load. Determine the frequencies of the rotor currents at these two speeds and the ratio of the induced rotor emf's.
c) For the motor in (b), determine the ratio of the induced rotor emf's at full load to those at starting. *(Section 3.5)*

4 / Direct-Current Machines

The basic principles of direct-current machines have already been discussed in Sections 3.3.1 and 3.3.2. In this chapter their behavior will be analyzed in more detail, and equivalent-circuit models for prediction of their steady-state performance will be developed.

4.1 MAGNETIC SYSTEM OF A DIRECT-CURRENT MACHINE

Figure 4.1(a) shows the magnetic system of a simple two-pole machine. The system is magnetized by one or more field coils encircling each pole core. The magnetic flux passes along the pole core and is distributed over an arc of the rotor periphery by the pole shoe. It then crosses the air gap and passes through the *teeth* formed in the rotor surface between the slots that contain the rotor winding. After passing through the rotor core, the teeth, the gap, and the other pole, the magnetic flux returns along the two magnetically parallel paths through the yoke of the machine stator.

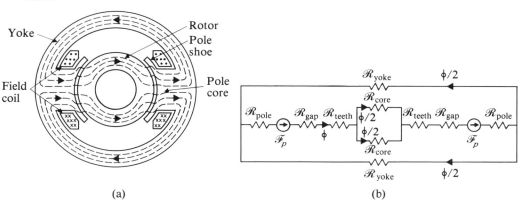

(a)   (b)

Fig. 4.1 (a) Magnetic system of two-pole machine. (b) Magnetic equivalent circuit.

To determine the performance of the machine, it is necessary to know the relation between the magnetic flux ϕ that crosses the air gap under each pole and the magnetomotive force $\mathscr{F}_p$ of the coils on each pole. Following the methods of Section 1.8, the equivalent magnetic circuit of Fig. 4.1(b) is obtained. In this circuit, a separate reluctance has been assigned to each section of the magnetic system that has a reasonably uniform flux density.

At low values of flux ϕ, the magnetic material in the machine may be considered as ideal, possessing infinite permeability. If all the corresponding reluctances are set to zero, the magnetic flux in each pole is then approximated by

$$\phi = \frac{2\mathscr{F}_p}{2\mathscr{R}_g} = \frac{\mathscr{F}_p}{\mathscr{R}_g} \quad \text{Wb} \tag{4.1}$$

The reluctance $\mathscr{R}_g$ is inversely proportional to the pole area and directly proportional to the effective air-gap length g_e, taking into account the effect of rotor slotting.

As the magnetic flux ϕ is increased, saturation will occur in various parts of the magnetic system, particularly in the rotor teeth. The resultant relation between the flux, ϕ, in each air gap and the mmf per pole, $\mathscr{F}_p$, is shown in Fig. 4.2(a). The curve of Fig. 4.2(a) may, to a good approximation, be described by the relationship of Eq. (4.1) until saturation becomes significant.

Although magnetic hysteresis may be neglected in most analyses, its effects are significant in certain types of machine. Figure 4.2(b) shows, in somewhat accentuated form, the multivalued relationship between flux and mmf, which may be obtained experimentally.

In a 2-pole machine, such as that illustrated in Fig. 4.1(a), the interpolar space is large, and it is difficult to make efficient use of the available rotor surface. By using a larger number of poles, as shown in Fig. 4.3, a more compact design, requir-

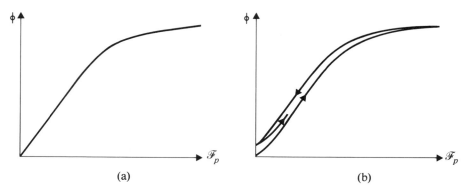

(a)   (b)

Fig. 4.2 Relation between air-gap flux ϕ and magnetomotive force $\mathscr{F}_p$ per pole. (a) Neglecting hysteresis. (b) Including hysteresis.

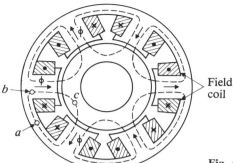

Fig. 4.3 Magnetic system of a six-pole machine.

ing less magnetic material per unit of output power, may be achieved. The mmf's on adjacent poles act in opposite directions. Because of the symmetry of the system, the relation between the flux per pole, ϕ, in each air gap and the mmf per pole, $\mathscr{F}_p$, may be determined by considering only one flux path, such as abc in Fig. 4.3.

*4.1.1 Determination of Magnetization Characteristic

The curve relating flux and mmf may be predicted from the dimensions of the machine and the magnetic properties of the various materials forming the parts of the magnetic system. Since the reluctance $\mathscr{R}_g$ is dominant, its calculation deserves first consideration. If the armature surface were smooth and the air-gap length constant, the reluctance could be approximated by

$$\mathscr{R}_g = \frac{g}{\mu_0 A_p} \quad \text{A/Wb} \tag{4.2}$$

where A_p is the surface area of the face of the pole shoe and g is the gap length. With a slotted rotor, however, the effective area of the flux path is substantially reduced. If the air-gap length is very small, all the flux can be considered to cross the gap sections above the teeth. The effective area A_g of the air-gap path is thus reduced to

$$A_g = A_p \frac{t}{t + d} \quad \text{m}^2 \tag{4.3}$$

where t and d are the tooth and slot widths at the rotor surface, as illustrated in Fig. 4.4(a). There is, however, some fringing of flux around the tooth edges, and this flux penetrates down the slot as shown in the field pattern of Fig. 4.4(b). The effective air-gap area is therefore between the value given in Eq. (4.3) and the area A_p of the pole surface. In an empirical approximation, which is adequate for most cases, the effective air-gap area may be expressed as

$$A_g = A_p \frac{(t + kd)}{(t + d)} \quad \text{m}^2 \tag{4.4}$$

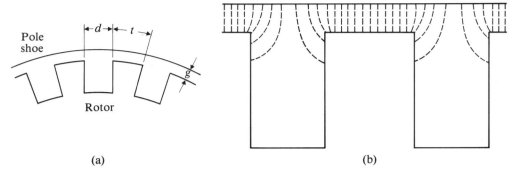

Fig. 4.4 The determination of air-gap reluctance.

where

$$k = \frac{1}{1 + d/5g} \tag{4.5}$$

Alternatively, the reciprocal of the correction factor in Eq. 4.4 may be applied to the air-gap length, giving an effective air-gap length

$$g_e = g \frac{(t + d)}{(t + kd)} \quad \text{m} \tag{4.6}$$

Thus, in Eq. (4.2) either A_p may be replaced by A_g or g replaced by g_e for increased accuracy.

As the flux in the magnetic system is increased, the first region to experience magnetic saturation is generally the teeth. The slots are usually made with parallel sides to accommodate coils of rectangular cross section, and the teeth are therefore tapered, as shown in Fig. 4.4(a). For a given value of flux ϕ, the mmf absorbed in, or the magnetic potential across, the tooth reluctance may be found by (1) determining the flux in one tooth, (2) determining the flux density B at various points along the path through the tooth, (3) obtaining corresponding values of magnetic field intensity H from an appropriate B–H curve for the material, and (4) integrating H numerically along the tooth path to obtain the magnetic potential difference.

For a given value of flux ϕ, the calculation of the magnetic potentials across the remaining reluctances may be made by using appropriate B–H curves for the various materials. The magnetic potentials across the reluctances of a closed path in Fig. 4.1(b) may then be summed to give the required mmf of the field windings. Figure 4.2(a) shows the resultant relation between the flux ϕ in each air-gap and the mmf per pole, $\mathscr{F}_p$. If the machine is designed so that all parts of the magnetic system enter the saturated region of their B–H curves more or less simultaneously, the ϕ–$\mathscr{F}_p$ relation of Fig. 4.2(a) enters saturation rapidly. But if different parts enter saturation at different values of the flux ϕ, the change of slope of the $\phi-\mathscr{F}_p$ characteristic is gradual.

There is a leakage flux across the air path between the tips of adjacent poles. As this flux does not enter the rotor, it does not contribute to the energy-conversion process. But it does increase the flux linkage of the field coil beyond the value that would be predicted by the equivalent magnetic circuit of Fig. 4.1(b). It also contributes to the magnetic saturation of the pole cores and the yoke sections.

4.2 ARMATURE WINDINGS

The winding on the rotor of a dc machine is called the *armature* winding; that on the stator is called the *field* winding.

In practice, the dc machine discussed in Section 3.3.2 is modified in a number of respects. First, a large number of multiturn coils are fitted on the rotor, with all the conductors forming one side of one coil lying in one slot, while those forming the other side of the coil lie in another slot. The arrangement of conductors and end connectors for a two-turn coil is illustrated in Fig. 4.5. Second, the slots sometimes do not run axially along the rotor surface, but are skewed one slot pitch, thus forming helices. This arrangement avoids an abrupt change in the reluctance of the magnetic system as a slot passes under one edge of a field pole, thus eliminating the vibration and noise that would otherwise occur. Third, in larger machines, more than two field poles are fitted, so that the maximum possible number of armature conductors will lie in a region of high flux density.

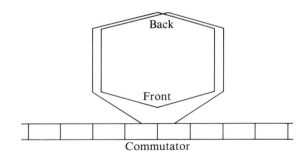

Fig. 4.5 Two-turn armature coil for lap winding.

The flux paths in the magnetic system of a 4-pole machine are illustrated in Fig. 4.6(a). The existence of four poles on the stator calls for an armature winding that will produce four poles on the rotor. The developed diagram of air-gap mmf $\mathscr{F}_{sg}$ versus θ for the 4-pole machine in Fig. 4.6(b) shows that there are two complete cycles of mmf variation around the air-gap. Since, as explained in Section 3.5, it is sufficient to analyze the physical conditions in the neighborhood of one pair of poles only, then the scale of the horizontal axis in Fig. 4.6(b) is most conveniently calibrated in electrical radians.

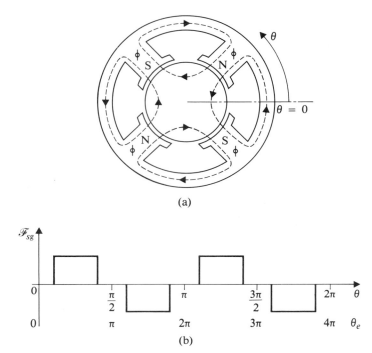

(a)

(b)

Fig. 4.6 Flux paths and mmf distribution in a four-pole machine.

Figure 4.7 shows the method of developing the commutator-end view of a dc machine to yield a convenient diagram of the armature winding. The direction in which the armature winding is viewed is shown in Fig. 4.7(c). The eye of the observer may be considered to be located at the axis of the rotor, seeing the stator poles *through* the winding. Positive rotor rotation ω_m thus yields a positive linear motion of the conductors from left to right in Fig. 4.8 and similar diagrams. This arrangement also ensures that the first half cycle of the mmf wave, already shown in Fig. 4.6(b), is positive.

Figure 4.8 shows the armature winding for a 4-pole machine in which an odd number of rotor slots is used. For simplicity, the number of coils is made unrealistically small, and single-turn coils are employed. There are two conductors per slot, giving a double-layer winding. This is called a *lap winding* since the path of current through the winding regularly turns back and overlaps its earlier part. Note also that adjacent commutator segments are shorted together momentarily by the brushes as the segments pass under them. (Owing to the small number of rotor slots in this illustration it has been necessary to exaggerate the size of the arc of commutator covered by the brush to illustrate this point.) Thus a coil connected to two such segments will be momentarily short-circuited by the brush. For the time being, the currents in these short-circuited coils will be ignored, but their effects will be discussed later.

The current paths existing between the brushes may be traced through the armature winding. Starting from the positive brush on segments 11 and 12 in Fig. 4.7, it is possible to trace a path to each of the two negative brushes. Similarly, starting from the positive brush on segments 1 and 2 it is possible to trace a path to each of the negative brushes. Since the positive brushes are connected to each other, as also are the negative brushes, there are four parallel paths from the posi-

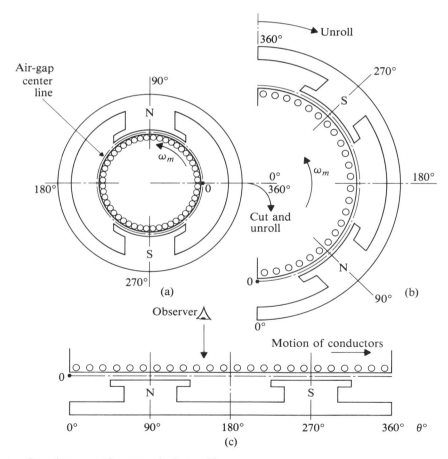

Fig. 4.7 Development of a two-pole dc machine.

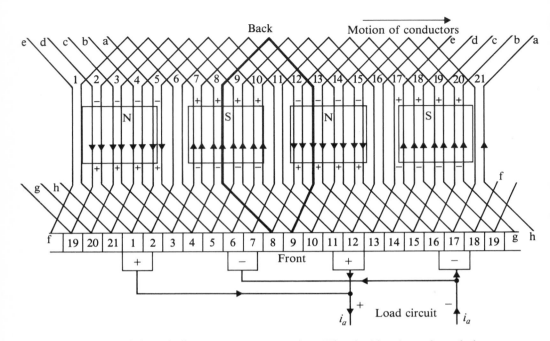

Fig. 4.8 Four-pole lap winding—generator operation. (View looking down through the unrolled winding as illustrated in Fig. 4.7.)

tive to the negative armature terminal. *In a lap winding, the number of parallel paths is always equal to the number of poles and to the number of brushes.*

Figure 4.9 shows another method of connecting the coils of the armature winding for a machine with more than two poles. The arrangement of conductors and end connectors for a two-turn coil of this winding is illustrated in Fig. 4.10, and a comparison with Fig. 4.5 shows that the essential difference between the two types of winding lies in the way in which the ends of any individual coil are connected to the commutator segments. The winding in Fig. 4.9 is called a *wave winding*. The reason for this name becomes clear when the current path is traced through the winding. The path for one coil is shown in heavy line in Fig. 4.9 and is seen to form one cycle of a "wave."

If the paths of the current through the wave winding from a positive to a negative brush are traced, only two parallel paths from the positive to the negative armature terminal will be found. The two positive brushes are connected to essentially the same point in the winding, since they make contact with segments that have only one coil between them, and such coils are therefore shorted by the brushes. The same is true of the two negative brushes. Between the positive brushes and the

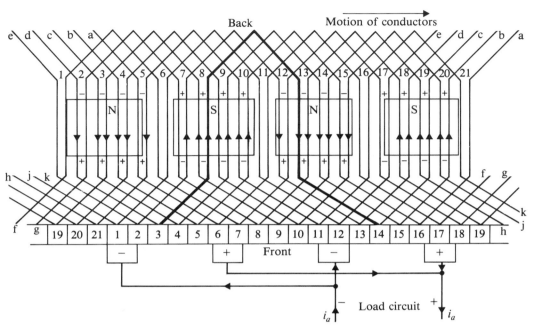

Fig. 4.9 Four-pole wave winding—generator operation.

negative brushes, however, the current is carried by a large number of coils connected in series.

In a wave winding, all except one set of brushes could be omitted, and this is often done in small machines. In large machines it is usual to fit more than two brushes, since this permits the area of contact between brushes and commutator to be increased without lengthening the commutator. *In a wave winding, the number of parallel paths is always two. There may be two or more brushes.*

Figure 4.11 shows the completed armature of a dc machine. The winding consists of a heavy rectangular conductor occupying the whole width of one slot. In the machine of Fig. 4.12, on the other hand, the coils are formed from four parallel conductors formed of thin copper strap. These four conductors are connected to four adjacent commutator segments; thus the advantage of a large number of coils is obtained without the need for an equal number of slots in the rotor. Figure 4.13 shows a single rotor lamination. The small nicks in the teeth near the tops of the slots hold long wedges, driven in after the coils have been placed in the slots. These wedges prevent the windings from rising out of the slots due to centrifugal force as the rotor revolves.

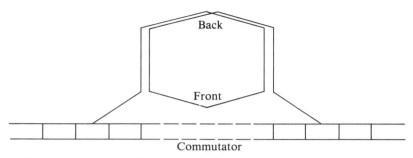

Fig. 4.10 Two-turn armature coil for wave winding.

Fig. 4.11 Large bar-wound armature. *(The Harland Engineering Company of Canada Ltd.)*

4.2.1 EMF Induced in the Armature Winding

For the purpose of the discussion in this section, it will be assumed that the magnetic material is ideal and that Eq. (4.1) applies.

Figure 4.14 shows a sector of a multipole dc machine. The armature winding could be either the lap winding of Fig. 4.8 or the wave winding of Fig. 4.9. The field coils, which are not shown, are excited from a dc source, and the rotor is being

Fig. 4.12 Partly wound armature. *(Canadian Westinghouse Company Limited)*

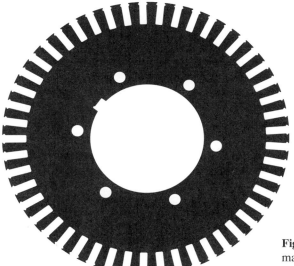

Fig. 4.13 Rotor lamination for a dc machine. *(Canadian Westinghouse Company Limited)*

driven by some prime mover at constant speed ω_m rad/s. The distribution of field mmf $\mathscr{F}_{sg}$ around the rotor periphery will be in the form of a rectangular wave, as in Fig. 4.6(b). The resulting distribution of flux density B in the air gap will be assumed to be similar to that of the stator mmf $\mathscr{F}_{sg}$, since the flux will pass almost directly across the air gap between the field poles and the rotor. The flux linking

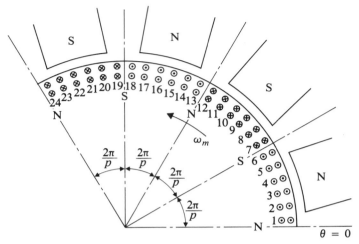

Fig. 4.14 Sector of a multipole dc machine—generator operation.

each coil in the armature winding will vary as the rotor rotates, and this variation will induce an emf in each coil. The directions of the induced emf's in the conductors are shown by the plus and minus signs at the ends of the conductors in Figs. 4.8 and 4.9. Similarly, in Fig. 4.14, the conductors passing under the N-poles have emf's directed toward the reader. Zero emf will be induced in coils whose conductors are in the interpolar space, since there is no variation in the flux linking a coil whose conductors are moving through that space. On the other hand, a constant emf will be induced in coils whose conductors are beneath the field poles, since the rate of change of flux linking a coil whose conductors are moving beneath the field poles is constant. It therefore follows that the time variation of coil emf will be a rectangular wave similar to that presented by the space variation of flux density around the rotor periphery. One cycle of this emf variation, corresponding to a rotor movement of two pole pitches, is shown in Fig. 4.15(a), where the induced emf e_{coil} is shown as a function of time.

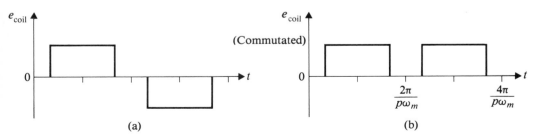

Fig. 4.15 Time variation of emf induced in an armature coil.

In Figs. 4.8 and 4.9, the brushes are shown in the *neutral* position; that is, they are resting on commutator segments connected to conductors that are moving through the interpolar space, so that the coils that they form have no emf induced in them. The discussion of motor action in Section 3.3.1 showed that this brush position produced maximum torque, and hence resulted in maximum rate of energy conversion. Placing the brushes in the neutral position also results in the maximum induced emf that can be obtained in the armature winding.

The emf induced in any parallel path in the armature winding may be obtained by adding the instantaneous values of the coil emf's in that path. If one conductor enters the region under a pole at the same time as another conductor leaves the region under the pole, the total emf in the path will be an integral multiple of the peak value of coil emf shown in Fig. 4.15(a). However, if the number of conductors under the pole is continually changing by ±1, there will be a ripple in the total emf of the path. This is normally called *commutator ripple*, and may usually be ignored in analysis of machine performance. However, it may cause interference in communication circuits unless steps are taken to filter it from circuits connected to the machine.

The action of the commutator is to transfer a coil from one path to another as the commutator segments to which the ends of the coil are connected pass beneath the brushes. The effect of this action is to rectify the coil emf so that its contribution to the emf between the brushes is unidirectional, as shown in Fig. 4.15(b). In the analysis of machine performance, it is the average value of this rectified coil emf that is of importance, since this average value multiplied by the number of coils in one parallel path gives the magnitude of the average emf induced in the armature winding.

Consider a single-turn coil, such as that connected to commutator segments 11 and 12 in Fig. 4.8. In the position shown, this coil links the total flux ϕ leaving the N field pole. When the coil is moved to the right, to a similar position in relation to the S pole, it is then linked by the same quantity of flux ϕ, but acting in the opposite direction through the coil. Thus, for this movement of the single-turn coil, the change in flux linkage is 2ϕ.

The motion just described is equivalent to a rotation of π electrical radians or $2\pi/p$ mechanical radians in Fig. 4.14. The speed of rotation in electrical radians per second is

$$\omega_{me} = \frac{p}{2}\,\omega_m \quad \text{rad/s} \tag{4.7}$$

The time taken for the coil to move π electrical radians is π/ω_{me}, so that the average rate of change of flux linkage in the coil during this movement is

$$\left[\frac{d\lambda}{dt}\right]_{av} = \frac{2\phi}{\pi/\omega_{me}} = \frac{p}{\pi}\,\omega_m\phi = [e_{\text{turn}}]_{av} \quad \text{V} \tag{4.8}$$

Let N = total number of turns in the armature winding, and a = number of parallel paths in the winding. Then the number of turns in series in each path is

N/a, and the average emf induced per path, that is, the average emf induced in the complete armature winding, is

$$e_a = \frac{N}{a}[e_{\text{turn}}]_{\text{av}} = \frac{N}{a}\frac{p}{\pi}\omega_m\phi \quad \text{V} \tag{4.9}$$

For any particular machine,

$$e_a = k\omega_m\phi \quad \text{V} \tag{4.10}$$

where

$$k = \frac{Np}{a\pi} \quad \text{V} \tag{4.11}$$

and is a constant *for that machine.*

Note in passing that a coil spanning only one pole arc would have the same average emf induced in it as would a full-pitch coil—that is, one spanning π electrical radians. Thus an armature winding may be made up of coils of less than full pitch without reducing the emf of the armature. In this way, a considerable saving in copper may be made due to the shortening of the end connections of the coils.

The no-load saturation curve for a dc machine is a curve of induced armature emf e_a versus field current i_f for the machine running at rated speed on no load. This curve can be determined directly from the pole-flux mmf curves of Fig. 4.2, by multiplying the ordinate ϕ by $k\omega_m$ to give e_a, as in Eq. (4.10), and dividing the abscissa $\mathscr{F}_p$ by the number of turns per field coil, N_f, to give i_f.

For an existing dc machine, the no-load saturation curve can be measured. If the field winding of a dc machine is excited by means of a constant source of potential difference V through a potentiometer, as indicated in Fig. 4.16(a), and the rotor is driven at mechanical speed Ω_0, then the emf induced in the armature winding will appear as a potential difference at the open-circuited armature terminals.

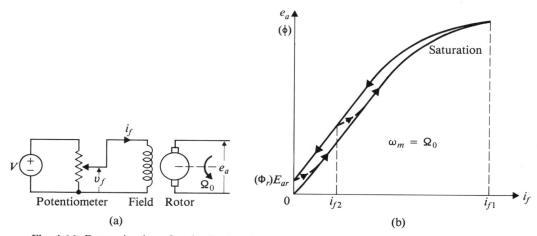

Fig. 4.16 Determination of no-load saturation curve.

Since $e_a = k\omega_m\phi$, if k is known for the machine under test, then e_a gives a measure of ϕ, the flux per pole. If the field poles are initially unmagnetized, and the field current i_f is gradually increased from zero, then ϕ and hence e_a will also increase from zero as shown in the lower limb of the double-valued curve in Fig. 4.16(b). Until saturation appears, the relationship between ϕ and i_f will be virtually linear. When saturation commences, the slope of the curve decreases.

If, after reaching a maximum value i_{f1} as shown in Fig. 4.16(b), the field current is reduced once more to zero, hysteresis in the ferromagnetic parts of the magnetic system will result in the upper limb of the double-valued curve in Fig. 4.16(b). When i_f has been reduced to zero, residual magnetism will result in a small residual flux per pole Φ_r, and this will induce a small emf:

$$E_{ar} = k\Omega_0\Phi_r \quad V \tag{4.12}$$

If i_f were again increased from zero, e_a would increase from E_{ar} as indicated by the broken line in Fig. 4.16(b). If, in tracing out the upper limb of the curve, the field current were reduced from i_{f1} only to the value i_{f2} and were then once more increased, the broken line star ing from the upper limb at $i_f = i_{f2}$ would give the relationship between e_a and i_f for i_f increasing.

As a consequence of hysteresis, it is never easy to predict what precise value of e_a will result from a given value of i_f. The operating point may lie anywhere on or between the two limbs of the curve of e_a versus i_f.

At the cost of a slight inaccuracy in predicting performance, the no-load saturation curve may be approximated by a single-valued curve giving the mean value of e_a for the range of field current to be employed in the machine. Such a curve is shown in Fig. 4.17. A straight line drawn through the origin, coinciding as far as possible with the straight-line part of this single-valued curve, is called the *air-gap line*. This line corresponds to a fictitious air gap of greater length than the real air gap, the increase of length allowing for the reluctance of the ferromagnetic part of the magnetic system in the linear unsaturated range.

If a resistive circuit is connected between the brushes of the machine in the developed diagrams of Figs. 4.8 and 4.9, then current will flow through that circuit

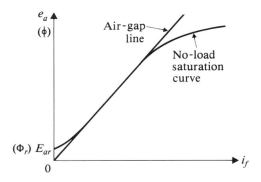

Fig. 4.17 No-load saturation curve and air-gap line.

from the positive to the negative brush and through the winding from the negative to the positive brush. The directions of current in the individual conductors of Figs. 4.8 and 4.9 are shown by arrowheads. (No currents are indicated in coils short-circuited by the brushes.) Under these conditions, the machine is acting as a generator. The sector end view of a multipole machine shown in Fig. 4.14 also has the current directions marked on the conductors for generator operation.

In the machine of Fig. 4.14, there is an mmf acting in the magnetic system due to the armature conductor currents. This mmf has a maximum value at the center of the space between the stator poles, as shown by the positions of the rotor pole markings. The armature mmf produces very little flux between the field poles because of the long air path. It produces some flux on one side of a field pole but also produces an equal and oppositely directed flux on the opposite side of the field pole. If the magnetic system is linear, the net effect of the rotor mmf on the average induced armature emf is zero, since the flux per pole ϕ in Eq. (4.9) remains unchanged.

A further discussion of the effect of the rotor mmf, usually called *armature reaction,* is given in Section 4.10.

Example 4.1   A small 4-pole, wave-wound dc machine is employed as a tachometer, being driven from the shaft of which the speed is to be measured. A constant current is supplied to the field coils, and a voltmeter of extremely high impedance is connected to its armature terminals. The machine has 800 conductors on the rotor, and the flux per pole produced by the field current is 0.3×10^{-3} Wb. What speed in radians per second will be indicated by each one-volt division on the voltmeter scale?

Solution   Since the voltmeter has an extremely high impedance, negligible current will flow in the armature winding, and the potential difference appearing at the armature terminals will be equal to the emf induced in the armature winding.

From Eq. (4.10):

$$\frac{\omega_m}{e_a} = \frac{1}{k\phi} \qquad \text{rad} \cdot \text{V/s}$$

Since the machine is wave-wound, there are only two parallel paths in the armature winding, so that

$$k = \frac{Np}{a\pi} = \frac{800}{2 \times 2} \cdot \frac{4}{\pi} = \frac{800}{\pi}$$

Thus

$$\frac{\omega_m}{e_a} = \frac{\pi}{800} \times \frac{10^3}{0.3} = 13.1 \qquad \text{rads} \cdot \text{V/s}$$

4.3 INTERNAL TORQUE

The developed diagrams of Figs. 4.8 and 4.9 show a condition where electric power is leaving the armature terminals. Examination of the directions of the conductor currents and the field–pole fluxes indicate that there is a force on the conductors to the left—that is, against the direction of motion. Correspondingly, in Fig. 4.14 the positions of the rotor poles, relative to the field poles, show that there is a torque in a clockwise direction—that is, against the direction of rotation. All three diagrams have been drawn with current directions appropriate for the operation of the machine as a generator. Reversal of the current direction would reverse the torque direction. Such reversal could be brought about by replacing the resistive load circuit connected to the brushes by a source of potential difference opposing and exceeding the emf induced in the armature winding.

In steady-state operation, the flux per pole ϕ is constant, so that the total energy stored in the magnetic field of the machine is also constant. Thus, if the machine is operating as a generator, the rate at which energy is supplied to the field by the mechanical system must be equal to the rate at which it is withdrawn from the field by the two electrical systems: one connected to the field winding, the other to the armature. Conversely, if the machine is operating as a motor, the rate at which energy is supplied to the magnetic field by the electrical systems must be equal to the rate at which it is withdrawn from the field by the mechanical system. Thus

$$T\omega_m = e_a i_a + e_f i_f \quad \text{W} \tag{4.13}$$

where T is the internal or electromagnetic torque developed across the air gap of the machine, e_a and e_f are the emf's induced in the armature and field windings respectively, while i_a and i_f are their currents. However, since ϕ is constant, e_f is zero, and Eq. (4.13) becomes

$$T\omega_m = e_a i_a \quad \text{W} \tag{4.14}$$

Substitution from Eq. (4.10) then yields

$$T = k\phi i_a \quad \text{N} \cdot \text{m} \tag{4.15}$$

where k is the machine constant specified in Eq. (4.11).

Example 4.2 In the speed-measuring system of Example 4.1, the high-impedance voltmeter is replaced by a meter of 1000 ohms resistance. If the resistance of the armature circuit of the machine is 400 ohms and the field current is the same as in the previous example, what are the internal torque and power developed by the machine when driven at 3000 r/min?

Solution Since k and ϕ are already known, it simply remains to determine i_a, and substitute in Eq. (4.15). The armature current i_a will be produced by the induced armature emf e_a acting in a circuit of total resistance, $R = 1000 + 400 = 1400 \ \Omega$.

$$e_a = k\omega_m\phi = \frac{800}{\pi} \times 3000 \times \frac{2\pi}{60} \times 0.3 \times 10^{-3} = 24 \quad V$$

$$i_a = \frac{e_a}{R} = \frac{24}{1400} = 17.2 \times 10^{-3} \quad A$$

Thus

$$T = \frac{800}{\pi} \times 0.3 \times 10^{-3} \times 17.2 \times 10^{-3} = 1.31 \times 10^{-3} \quad N \cdot m$$

Internal power developed $= e_a i_a = 24 \times 17.2 \times 10^{-3} = 0.412 \quad W$

$$= T\omega_m = 1.31 \times 10^{-3} \times 3000 \times \frac{2\pi}{60} = 0.412 \quad W$$

4.4 METHODS OF EXCITATION

So far the dc machine has been discussed in terms of armature or rotor mmf and field or stator mmf, but no particular attention has been given to the manner of providing the excitations producing these mmf's. The excitation of a motor is purely electrical, and it is often convenient to employ a single source of direct emf to supply both the armature and the field windings. Different ways in which this may be done are illustrated in Fig. 4.18.

Figure 4.18(a) shows the circuit of a *shunt-wound* motor, so called because the field circuit branch is in shunt, or parallel, with that of the armature. The field circuit branch represents all field coils in series. These coils have a large number of turns and take only a small current, of the order of 5% or less of the line current. A rheostat may be included in the field circuit to control the field current to some extent and thus to vary the field mmf.

Figure 4.18(b) shows the circuit of a *series-wound* motor. The field coil shown represents all field coils in series. These coils have a small number of turns and carry the same current as the armature. The ratio of field mmf to armature mmf is fixed in this machine.

Figure 4.18(c) represents a combination of (a) and (b), and shows the circuit of a *compound-wound* motor. The field mmf in this machine is supplied by two different sets of coils on the field poles. One set of coils constitutes the series field, and these have a small number of turns and carry the line current. The other set of coils constitutes the shunt field similar to that of the shunt-wound motor. The particular compound-wound arrangement shown in Fig. 4.18(c) represents a *short-shunt* machine, so called because the shunt field is in parallel with the armature only. If the shunt field were connected to the motor supply terminals, so that it was in parallel with a branch consisting of the series field and the armature, the result would be a *long-shunt* machine. There is little practical difference between these two arrangements.

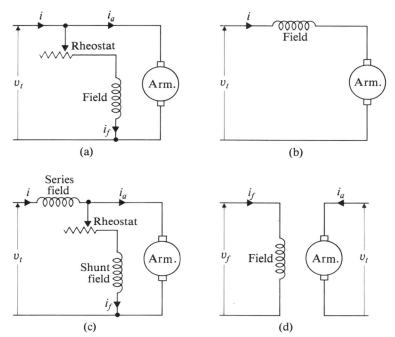

Fig. 4.18 Methods of excitation.

When a dc motor is an element of a variable-speed drive or a servomechanism, it is often convenient to employ separate sources to excite the field and the armature. Such a machine is said to be *separately excited,* and its circuit is illustrated in Fig. 4.18(d). Field excitation may also be provided by permanent magnets. Machines of this type are discussed in Section 4.12.

The energy input to a generator is principally mechanical, and is provided by a driving torque applied to the rotor shaft. An electrical excitation for the field winding is also required, however. The field may be excited from an independent electrical source, in which case the circuit is that shown in Fig. 4.18(d), and the generator is separately excited. On the other hand, when certain conditions are fulfilled, the generator's own armature winding may be employed as a source of field excitation. When this is done, any one of the circuit arrangements shown in Fig. 4.18(a), (b), or (c) may be employed with i and i_a reversed, depending upon the generator characteristic desired. Such generators are said to be *self-excited.*

Figure 4.19 shows a shunt–field coil formed of many turns of small cross-section conductor. Before being fitted on the stator pole, the coil would be wrapped in insulation and impregnated with an insulating compound. Figure 4.20 shows a coil for a compound machine in which a small number of turns of large cross-section conductor, forming the series coil, are wound on top of the shunt coil.

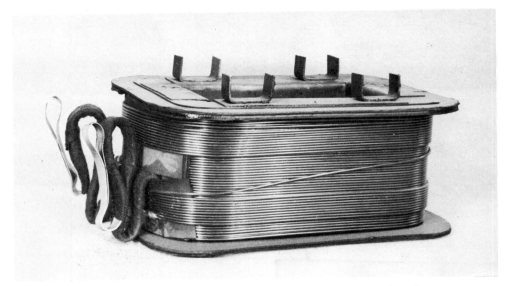

Fig. 4.19   Shunt field coil. *(The Harland Engineering Company of Canada Ltd.)*

Fig. 4.20   Series over shunt field coil. *(The Harland Engineering Company of Canada Ltd.)*

4.5 EQUIVALENT CIRCUIT AND GENERAL EQUATIONS

For the purposes of analysis, it is convenient to visualize a dc machine as an ideal energy converter in combination with parameters that represent its various imperfections. It is therefore first of all necessary to examine the properties of the ideal machine.

Suppose a machine is driven at ω_m rad/s by a mechanical prime mover (gasoline engine, steam turbine, etc.). The emf induced in the armature winding will be, from Eq. (4.10),

$$e_a = k\omega_m\phi \quad \text{V} \tag{4.16}$$

If a load circuit is connected to the armature terminals, a current i_a flows. The electric power generated in the machine is given by

$$P_{\text{elec}} = e_a i_a = k\phi\omega_m i_a \quad \text{W} \tag{4.17}$$

From Eq. (4.15), the armature current i_a reacts with the flux to produce a torque

$$T = k\phi i_a \quad \text{N·m} \tag{4.18}$$

in such a direction as to oppose the rotation. The mechanical power converted in the interaction between rotor and stator is then given by

$$P_{\text{mech}} = T\omega_m = k\phi i_a \omega_m = P_{\text{elec}} \quad \text{W} \tag{4.19}$$

Thus, internally, the machine behaves as an ideal energy converter whose properties are expressed by Eqs. (4.16) and (4.18).

Figure 4.21 shows an equivalent circuit for a dc machine that includes this ideal machine; the remainder of the equivalent circuit introduces the imperfections of the real machine.

When current flows in the armature winding, there is a power loss in the armature coils. The resistance R_a of the a parallel groups of armature coils is

$$R_a = \rho \frac{N}{a^2} \frac{l}{A} \quad \Omega \tag{4.20}$$

where

N = total number of armature turns,
l = length of one turn,
A = conductor cross-section area,
ρ = conductor resistivity at the operating temperature.

This resistance may be measured between the commutator segments under adjacent brushes when the rotor is stationary.

There is also an electric-power loss in the brushes that make contact with the commutator. These brushes are normally made of carbon in either hard or graphite form and sometimes contain metal such as copper in solution. The permissible current density in a brush varies from about 0.05 to 0.25 A/mm², and a potential differ-

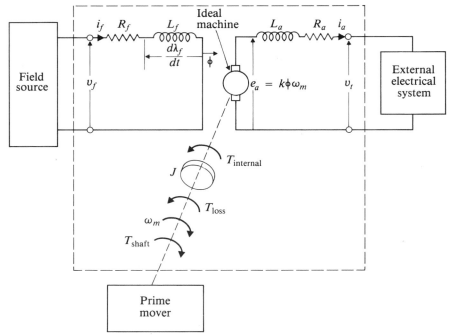

Fig. 4.21 Equivalent circuit of a dc machine.

ence occurs between the commutator surface and the brush material close to the brush contact surface. This potential difference varies nonlinearly with current density, as shown for a typical brush in Fig. 4.22. Since the total potential difference across the two sets of brushes by which the current enters and leaves the armature winding may be as high as two volts, the brush loss is significant in low-potential

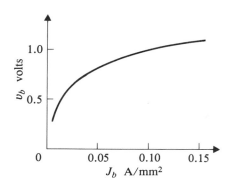

Fig. 4.22 Potential difference across a typical brush-commutator contact.

machines. In high-potential machines, this loss may often be neglected in approximate analyses.

The armature circuit also has an inductance L_a representing the flux linkage of the armature circuit per ampere of armature current i_a. This armature circuit may include not only the armature winding but also additional stator windings, to be discussed later. The armature inductance may generally be regarded as a constant.

With a constant field current i_f, all the power input to the field coils represents a power loss. The field-circuit resistance R_f is p times the resistance of each coil, if the coils are connected in series. Parallel and series-parallel connections of the field coils also may be made.

The inductance L_f of a field circuit of p field coils, each having N_f turns connected in series, may be expressed as

$$L_f = \frac{pN_f\phi_f}{i_f} \quad \text{H} \tag{4.21}$$

The pole flux ϕ_f consists of the air-gap flux per pole ϕ plus the leakage flux that crosses the air-gap path between adjacent pole tips. The relationship between the air-gap flux per pole ϕ and the field current i_f takes the graphical form of Fig. 4.17. In the linear range, the inductance of a series-connected field circuit may be expressed as

$$L_f = \frac{\lambda_f}{i_f} = \frac{pN_f\phi_f}{i_f} \simeq 1.15 \frac{pN_f\phi}{i_f} \quad \text{H} \tag{4.22}$$

The shaft torque of the machine differs from the electromagnetic or internal torque—Eq. (4.18)—by the sum of the torque required to overcome the mechanical losses and the torque required by the inertia of the machine. Components of loss torque arise from bearing friction, brush friction, and friction with the air (windage). It is usually convenient to include with these components the torque that arises from hysteresis and eddy-current loss in the rotor as it rotates through the pole fluxes. The total loss torque is nonlinearly dependent on the speed ω_m. But since the loss torque is generally a small fraction of the total torque, a linear relation with speed may often be assumed.

The kinetic energy stored in the rotor of the machine at speed ω_m is

$$W_{\text{kin}} = \frac{1}{2} J\omega_m^2 \quad \text{J} \tag{4.23}$$

where J is the polar moment of inertia in kg·m². A change in rotor speed is therefore accompanied by a change in the stored energy. The rate of change with time of the stored energy W_{kin} is equal to the accelerating torque multiplied by the speed. Thus,

$$T_{\text{accel}} = \frac{1}{\omega_m} \frac{d}{dt} W_{\text{kin}} = J \frac{d\omega_m}{dt} \quad \text{N·m} \tag{4.24}$$

The equivalent circuit of Fig. 4.21 shows that the machine has six terminal variables: v_f and i_f for the field circuit, v_t and i_a for the armature circuit, ω_m and T_{shaft} for the mechanical system. These variables are related by three equations:

$$v_f = R_f i_f + L_f \frac{di_f}{dt} \quad \text{V} \tag{4.25}$$

$$v_t = k\phi\omega_m - L_a \frac{di_a}{dt} - R_a i_a \quad \text{V} \tag{4.26}$$

$$T_{\text{shaft}} = k\phi i_a + J\frac{d\omega_m}{dt} + T_{\text{loss}} \quad \text{N}\cdot\text{m} \tag{4.27}$$

If any of the variables i_f, i_a, or ω_m is constant, the term in the equations involving its derivative with time becomes zero. If all three variables are constant, the resultant equations are:

$$v_f = R_f i_f \quad \text{V} \tag{4.28}$$

$$v_t = k\phi\omega_m - R_a i_a \quad \text{V} \tag{4.29}$$

$$T_{\text{shaft}} = k\phi i_a + T_{\text{loss}} \quad \text{N}\cdot\text{m} \tag{4.30}$$

These last three equations apply to the steady-state performance of the machine.

Equations (4.25)–(4.30) include a seventh variable, ϕ, related to one of the terminal variables by

$$\phi = f(i_f) \quad \text{Wb} \tag{4.31}$$

For steady-state operation, the relationship is most conveniently expressed in the form

$$e_a = k\Omega_0\phi = F(i_f) \quad \text{V} \tag{4.32}$$

Equation (4.32) is the relationship describing the no-load saturation curve of the machine taken at speed Ω_0 as shown in Fig. 4.17.

If the steady-state performance of the machine is to be determined, Eqs. (4.28)–(4.30) and Eq. (4.32) must be combined with three other equations to solve for the seven system variables. These additional equations describe the terminal properties of the field source, the electrical system connected to the armature, and the mechanical system coupled to the shaft. They depend upon the application of the machine.

4.6 PERFORMANCE OF GENERATORS

Because of the wide variety of applications of dc machines, it is convenient to classify them as generators or motors, depending upon the dominant direction of energy flow. Generators are further classified by the means used to provide excitation for the field windings, as discussed in Section 4.4.

In many applications a generator is merely required to be driven at constant speed and to supply a constant or slowly varying load current. Under such circumstances, the problem facing the applications engineer is simply that of determining from data supplied by the machine builder what the terminal potential difference of the machine will be for a particular speed, field current, and load current. In other words, he or she is required to determine a curve of armature terminal potential difference, v_t, versus load current, i_L. This curve is known as the *external characteristic* of the machine.

4.6.1 Separately Excited Generator

As explained in Section 4.4, a separately excited generator is one in which the source of the field current is external to the machine. This source may be another dc generator, a controlled or diode rectifier, or it may be a battery, as shown in Fig. 4.23. The relationship between the source terminal potential difference and the field current is described by the equation

$$i_f = \frac{V_{fs}}{R_e + R_f} \quad \text{A} \tag{4.33}$$

The value of the steady-state field current in this system may be controlled by adjustment of source, V_{fs}, or the field-rheostat resistance, R_e.

The armature emf induced in the steady state is expressed from Eqs. (4.16) and (4.32) by

$$e_a = k\omega_m\phi = F(i_f)|_{\omega_m = \Omega_0} \quad \text{V} \tag{4.34}$$

The resistance of the external load circuit is

$$R_L = \frac{v_t}{i_L} \quad \Omega \tag{4.35}$$

With i_f, ω_m, and i_L all constant, from Fig. 4.23:

$$v_t = e_a - R_a i_L \quad \text{V} \tag{4.36}$$

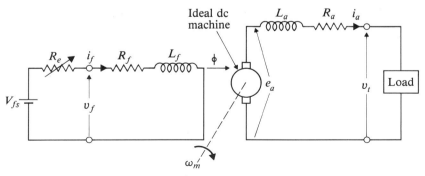

Fig. 4.23 Connections for separately excited generator.

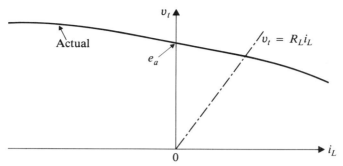

Fig. 4.24 External characteristic of a separately excited generator.

Thus

$$i_L = \frac{k\omega_m\phi}{R_a + R_L} \quad \text{A} \tag{4.37}$$

The external characteristic of the generator is shown in Fig. 4.24. If the load characteristic described by Eq. (4.35) is shown on the same diagram, then the point of its intersection with the external characteristic gives the operating values of v_t and i_L.

Equation (4.36) indicates that the terminal potential difference of a separately excited generator drops linearly with increase of load current because of armature-circuit resistance. At high values of armature current, the demagnetizing effect of armature reaction reduces e_a and causes a divergence from the linear relationship. This effect can usually be neglected for armature currents below the rated value. The drop in terminal potential difference can be counteracted by an increase of field current i_f, and a constant value of v_t can be maintained over the whole range of i_L if the field current is varied as indicated in Fig. 4.25.

Under certain circumstances, the load may become an electrical source causing the machine to operate as a motor. Figure 4.24 shows that the external character-

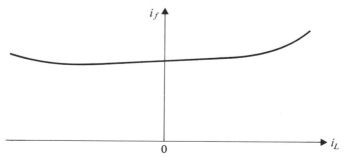

Fig. 4.25 Variation of field current to maintain constant terminal potential difference.

istic with reversed current i_L is represented adequately by Eq. (4.36) for the major part of the current range. At high values of current, the effect of armature reaction again becomes apparent.

The analysis of the generator performance under transient conditions involves the solution of differential equations similar to Eqs. (4.25)–(4.27).

Example 4.3   A 4½-kW, 125-V, 1150-r/min, separately excited dc generator has an armature–circuit resistance of 0.37 Ω. When the machine is driven at rated speed, the no-load saturation curve obtained is that shown in Fig. 4.26.

If the field rheostat is adjusted to give a field current of 2 A, and the machine is driven at 1000 r/min, what will be the terminal potential difference when the load current is at the rated value? (The effects of armature reaction and brush contact resistance may be neglected.)

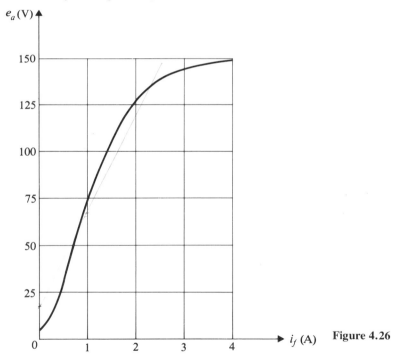

Figure 4.26

Solution   The machine is being driven at reduced speed, and since $e_a = k\omega_m \phi$, the induced armature emf at any value of ϕ or i_f will be proportional to speed. From the no-load saturation curve at rated speed for $i_f = 2$ A, $e_a = 126$ V. Thus at 1000 r/min,

$$e_a = \frac{1000}{1150} \times 126 = 109 \quad V$$

Rated current is

$$I_L = \frac{4.5 \times 10^3}{125} = 36 \quad \text{A}$$

From Eq. (4.36),

$$v_t = 109 - 0.37 \times 36 = 96 \quad \text{V}$$

*4.6.2 Shunt Generator

When certain conditions are fulfilled, the generator's own armature winding may be employed as a source of field excitation. The required conditions for this mode of operation will now be investigated.

The circuit for a shunt-wound generator on no-load is shown in Fig. 4.27. If the machine is to operate as a self-excited generator, some residual magnetism must exist in the magnetic system of the stator. If none is present, as in a newly built machine, then some must be created by passing a current through the field winding from some convenient dc source, such as a battery.

When the machine is driven on open circuit with switch SW in Fig. 4.27 open, a small potential difference equal to E_{ar}, the induced armature emf due to residual magnetism, appears at the terminals. Since no current is flowing,

$$e_a = v_t = E_{ar} \quad \text{V} \tag{4.38}$$

If switch SW is now closed, e_a drives a small current through the field winding, and

$$e_a = (R_a + R_e + R_f)i_f + (L_a + L_f)\frac{di_f}{dt} \quad \text{V} \tag{4.39}$$

Also,

$$v_t = (R_e + R_f)i_f + L_f\frac{di_f}{dt} \quad \text{V} \tag{4.40}$$

The field mmf due to the small current i_f will either (a) reduce the flux per pole below the residual value Φ_r, or (b) increase the flux per pole.

In case (a), $e_a < E_{ar}$, and both i_f and v_t fall to very low values. One of two things may then be done to reverse the field current due to residual magnetism and

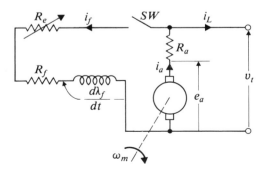

Fig. 4.27 Connections for shunt generator.

produce case (b): either the direction of drive may be reversed (if the machine builder's specification does not prohibit this), thus reversing e_a and i_f; or, more simply, the field connections (or armature connections) may be interchanged. When one or other is done, case (b) is obtained, and i_f produces flux in the same direction as the residual flux. The increase in flux per pole increases e_a, which further increases i_f, and a cumulative buildup of these two variables takes place. The question is, Where does the cumulative buildup of e_a stop?

Normally, $(R_e + R_f) >> R_a$, so that when a steady-state condition has been reached, and $di_f/dt = 0$, then

$$v_t \simeq e_a \quad \text{V} \tag{4.41}$$

Since Eq. (4.41) holds for any value of i_f, the curve of v_t versus i_f on no-load is indistinguishable from the curve of e_a versus i_f, which is the no-load saturation curve of the machine, described by

$$e_a = F(i_f) \quad \text{V} \tag{4.42}$$

In addition, for any steady-state condition,

$$v_t = (R_e + R_f)i_f \quad \text{V} \tag{4.43}$$

and the straight line described by Eq. (4.43) is shown with the no-load saturation curve in Fig. 4.28.

From Eqs. (4.41) and (4.42),

$$v_t \simeq F(i_f) \quad \text{V} \tag{4.44}$$

The two simultaneous equations, Eqs. (4.43) and (4.44), have their steady-state solution at the point of intersection of the two lines in Fig. 4.28, where $v_t \simeq e_{a(ss)}$. From this diagram, the process of buildup may also be understood, since the vertical distance between the two lines is equal to $(L_a + L_f)di_f/dt$, and when the field current reaches its steady-state value, this distance is necessarily zero.

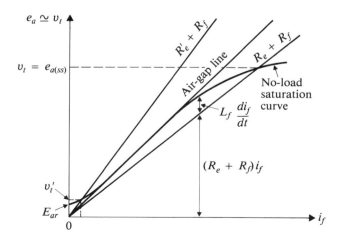

Fig. 4.28 Buildup of e_a and i_f in a shunt generator.

When the machine has built up to a terminal potential difference, such as $v_t = e_{a(ss)}$, shown in Fig. 4.28, then it may be employed as a useful energy source. However, if the field rheostat were set to a value R'_e, yielding the field circuit line designated $R'_e + R_f$, the terminal potential difference would only build up to the value v'_t. The machine then would be virtually useless as an energy source. There is a critical value of rheostat resistance for which the resistance line coincides with the air-gap line. For values of rheostat resistance higher than this critical value, v_t does not build up to a useful magnitude.

Provided that the field rheostat resistance is less than the critical value and that conditions are correct for the cumulative buildup of e_a and i_f, the shunt generator is capable of delivering a load current i_L to a load circuit of resistance R_L, which may be connected to its terminals. The equations that describe a steady-state condition of operation on load are then:

$$\phi = f(i_f) \quad \text{Wb} \tag{4.45}$$

$$e_a = k\omega_m\phi = F(i_f) \quad \text{V} \tag{4.46}$$

$$v_t = (R_e + R_f)i_f \quad \text{V} \tag{4.47}$$

$$v_t = R_L i_L \quad \text{V} \tag{4.48}$$

$$v_t = e_a - R_a i_a \quad \text{V} \tag{4.49}$$

$$i_a = i_L + i_f \quad \text{A} \tag{4.50}$$

If the generator is operating in a steady-state condition, delivering load current i_L, a reduction in R_L will cause an increase in i_a and hence a reduction in v_t. This reduction in v_t will in turn cause a decrease in i_f, and consequently e_a will be reduced, further reducing v_t until a new steady-state condition is reached.

From the foregoing description of the effect of a reduction in R_L, it can be seen that Eq. (4.49) does not represent a linear relationship between v_t and i_a, since e_a is a nonlinear function of i_a. The regulation of a shunt-wound generator is therefore greater than that of a separately excited generator.

The determination of a point on the external characteristic of v_t versus i_L requires the solution of two simultaneous equations relating e_a and i_f. One of these is Eq. (4.46), describing the no-load saturation curve. The other is obtained from Eqs. (4.47) and (4.49) and is

$$e_a = R_a i_a + (R_e + R_f)i_f \quad \text{V} \tag{4.51}$$

For any given values of R_e and i_a, a straight line representing Eq. (4.51) may be drawn on the same axes as the no-load saturation curve of the generator. Two such lines are shown in Fig. 4.29 for a single value of R_e less than the critical value, and two values of i_a, where $i_{a2} > i_{a1}$.

For armature current i_{a1}, the line described by Eq. (4.51) has a small intercept $R_a i_{a1}$ on the e_a axis, and slope $R_e + R_f$. It intersects the magnetization characteristic at point (i_{f1}, e_{a1}), representing the steady-state operation of the machine. The

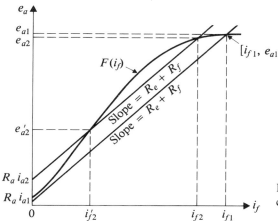

Fig. 4.29 Determination of external characteristic of a shunt generator.

corresponding values i_{L1} and v_{t1} may now be obtained from Eqs. (4.50) and (4.47) or (4.48). This gives one point on the external characteristic.

For a fixed value of R_e and a series of values of i_a, an external characteristic of v_t versus i_L may be drawn by repetition of the foregoing procedure and is illustrated in Fig. 4.30. For part of the range of i_L, the curve may be double-valued. Figure 4.29 shows that the line for armature current i_{a2} yields two possible conditions of operation, and the corresponding points on the external characteristic are shown in Fig. 4.30. The point at which the machine operates is determined by the value of the load-circuit resistance. A line corresponding to Eq. (4.48) of slope R_L may be drawn on the same axes as the external characteristic. The intersection of this straight line and the external characteristic defines the operating condition of the machine.

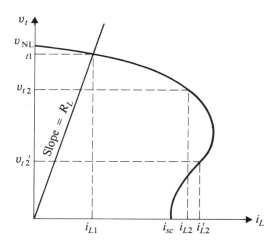

Fig. 4.30 External characteristic of a shunt generator.

If the machine were driven at constant speed at a fixed field-rheostat setting, and R_L were varied from infinity (open circuit) to zero (short circuit) the whole external characteristic would be traced out, from the no-load terminal potential difference v_{NL} to the short-circuit current i_{SC}. If such a procedure were followed in practice, there would be a danger that the current delivered by the machine might greatly exceed the rated value, causing the machine to overheat and damaging the commutator by sparking due to excessive brush current density. There would also be a danger of flashover between adjacent commutator segments caused by heavy overload.

Example 4.4   The machine of Example 4.3 is driven as a shunt generator at rated speed, and the field rheostat is adjusted to give a no-load terminal potential difference of 135 V.

What will be the terminal potential difference of the machine when it is delivering its rated current of 36 A?

Solution   To determine the steady-state condition of operation, it is necessary to know the armature current i_a. However, since the terminal potential difference and consequently the field current are not known, it is not possible to say what the armature current will be when the load current is 36 A. Since i_f is small, and

$$i_a = i_f + i_L \quad A$$

it is best to assume two values of i_a close to 36 A, determine the corresponding points on the external characteristic, and then determine the required point by interpolation.

From the magnetization characteristic of Fig. 4.26, when $e_a = 135$ V, $i_f = 2.42$ A. The slope of the field-resistance line shown in Fig. 4.31 is thus

$$R_f = \frac{135}{2.42} = 55.8 \quad \Omega$$

1. Assume $i_a = 36$ A. From Eq. (4.51),

$$e_{a1} = 0.37 \times 36 + 55.8 \, i_{f1} = 13.3 + 55.8 \, i_{f1} \quad V$$

A line drawn in Fig. 4.31 with an intercept of 13.3 V and a slope of 55.8 Ω gives an operating point

$$e_{a1} = 127 \quad V, \qquad i_{f1} = 2.05 \quad A$$

This line also gives another much lower operating point, but it will be assumed that the machine is operating normally on the upper part of the external characteristic. Then, from Eq. (4.49),

$$v_{t1} = 127 - 0.37 \times 36 = 114 \quad V$$

$$i_{L1} = 36 - 2.05 = 34 \quad A$$

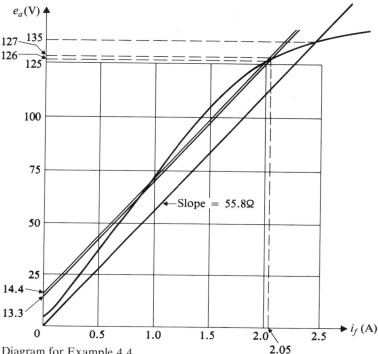

Fig. 4.31 Diagram for Example 4.4.

which gives one point on the external characteristic.

2. Assume $i_a = 39$ A. From Eq. (4.51),

$$e_{a2} = 0.37 \times 39 + 55.8 \, i_{f2} = 14.4 + 55.8 \, i_{f2} \quad \text{V}$$

A line drawn in Fig. 4.31 with an intercept of 14.4 V and a slope of 55.8 Ω gives an operating point

$$e_{a2} = 126 \quad \text{V}, \qquad i_{f2} = 2.0 \quad \text{A}$$

Thus

$$v_{t2} = 126 - 0.37 \times 39 = 112 \quad \text{V}$$
$$i_{L2} = 39 - 2.0 = 37 \quad \text{A}$$

From these two results, it may be seen that for $i_L = 36$ A, v_t will be 113 V.

Another procedure, which eliminates the need for a graph, is to carry out calculations for a series of values of i_f below the no-load value that yield the results shown in Table 4.1.

Without drawing the external characteristic, this table shows an operating condition of $v_t = 112$ V for $i_L = 35.8$ A, which closely approximates the result from the graph.

Table 4.1

Formula	Value of i_f				
	2.4	2.2	2.0	1.8	1.6
$v_t = R_f i_f$	134.0	123.0	112.0	100.0	89.0
e_a from Fig. 4.26	135.0	131.0	126.0	120.0	110.0
$R_a i_a = e_a - v_t$	1.0	8.0	14.0	20.0	21.0
$i_a = (e_a - v_t)/R_a$	2.7	21.6	37.8	54.1	56.8
$i_L = i_a - i_f$	0.3	19.4	35.8	52.3	55.2

*4.6.3 Series Generator

Like the shunt generator, the series generator is a self-excited machine. The circuit diagram is shown in Fig. 4.32, and from this it may be seen that the field circuit is not complete unless a load circuit is connected to the terminals. For this machine

$$i_a = i_f = i_L \quad \text{A} \tag{4.52}$$

so that the no-load saturation curve may be described by

$$e_a = F(i_L) \quad \text{V} \tag{4.53}$$

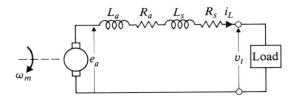

Fig. 4.32 Series generator circuit.

A typical no-load saturation curve is shown in Fig. 4.33. In the steady state, the terminal potential difference may be expressed as

$$v_t = e_a - (R_a + R_s)i_L \quad \text{V} \tag{4.54}$$

The external characteristic for the machine may be obtained from the no-load saturation curve of Fig. 4.33—that is, Eq. (4.53)—together with Eq. (4.54), and is shown in Fig. 4.34. Since the ratio of field to armature mmf is fixed in this machine, it has only one possible external characteristic. However, the no-load saturation curve and the external characteristic may be changed by connecting a diverter resistor in parallel with the series field winding.

If the load circuit in Fig. 4.32 is a resistor R_L, the series generator will operate at the point of intersection of the external characteristic and the potential–current characteristic of the load circuit, as shown in Fig. 4.34. If R_L is too large, the ter-

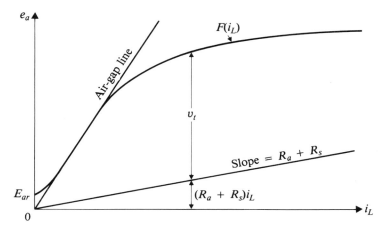

Fig. 4.33 Determination of external characteristics of a series generator.

minal potential difference of the generator will not build up. Over the near-linear portion of the external characteristic, the series generator acts as a negative resistance of value equal to the slope of that portion. This feature is sometimes exploited in using series generators as potential boosters in dc transmission lines to compensate for the potential drop in the transmission-line resistance.

Series machines are frequently used in traction applications, where they may be operated in a braking, or *regenerative,* mode. In such a situation, the ''load'' circuit is the power-supply system, and the current generated by the series machine may be controlled by a dc–to–dc converter or chopper. Under these circumstances, the generator may be operated at any point on the external characteristic of Fig. 4.34.

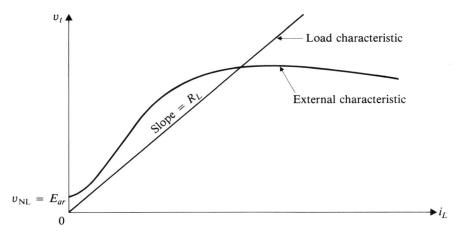

Fig. 4.34 External characteristic of a series generator.

Example 4.5   A 600-V dc source is provided from a 440-V, 3-phase, ac system by means of a bank of rectifiers. The dc load is supplied via a feeder of resistance 0.41 Ω per conductor. In order to maintain approximately constant potential difference at the load end of the feeder, a series booster generator is fitted in one side of the feeder, as shown in Fig. 4.35.

The booster generator is rated at 1 kW, 25 V. Its no-load saturation curve, obtained at rated speed by separately exciting the series field winding, is shown in Fig. 4.36. The resistance of the booster armature circuit is 0.08 Ω and that of the series field is 0.05 Ω.

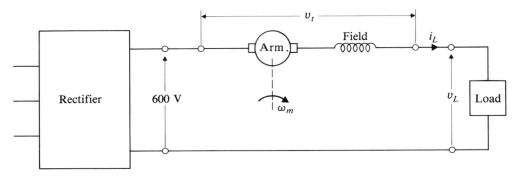

Fig. 4.35   Diagram I for Example 4.5.

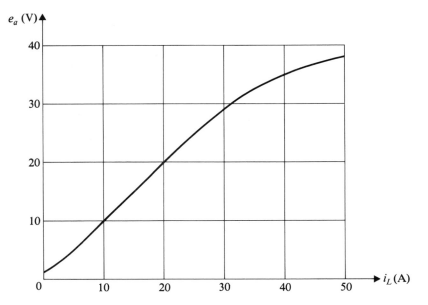

Fig. 4.36   Diagram II for Example 4.5.

Determine the potential difference v_L at the load end of the feeder when the rectifier is supplying (a) 20 A, (b) 40 A direct current; and determine the terminal potential difference of the booster generator in each case.

Solution The steady-state equivalent circuit of the system is shown in Fig. 4.37. From that circuit,

$$v_L = v_R + e_a - (2R_F + R_a + R_s)i_L \tag{1}$$

$$v_t = e_a - (R_a + R_s)i_L \tag{2}$$

a) $i_L = 20$ A.

From Fig. 4.36, $e_a = 20$ V. Substitution in Eq. (1) gives

$$v_L = 600 + 20 - 20(2 \times 0.41 + 0.08 + 0.05) = 601 \quad \text{V}$$

Substitution in Eq. (2) gives

$$v_t = 20 - 20(0.08 + 0.05) = 17.4 \quad \text{V}$$

b) $i_L = 40$ A.

From Fig. 4.36, $e_a = 35$ V.

$$v_L = 600 + 35 - 40(2 \times 0.41 + 0.08 + 0.05) = 597 \quad \text{V}$$
$$v_t = 35 - 40(0.08 + 0.05) = 29.8 \quad \text{V}$$

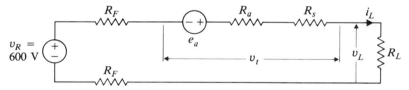

Fig. 4.37 Diagram III for Example 4.5.

*4.6.4 Compound Generator

If a generator is to run untended, the external characteristic of the shunt machine may be very unsatisfactory, and that of a series machine even more so, since a source of constant potential difference supplying a varying load current is usually required. The situation is even less satisfactory if the load is supplied via a feeder with appreciable resistance, since this will introduce an additional fall in potential at the load end of the feeder. What is required is a generator with a rising external characteristic, since this would counteract the effect of feeder resistance. Such a characteristic may be obtained from a compound generator.

An increase of armature induced emf with increase of load current may be obtained by arranging that field mmf increases as load current increases. To achieve this, the generator is fitted with both a shunt and a series winding, as illustrated in

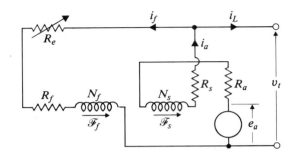

Fig. 4.38 Compound generator circuit.

Fig. 4.38, where the armature current passes through the series winding in such a direction as to produce an mmf aiding the mmf due to the shunt winding. The generator is then said to be *cumulatively compounded*. As may be seen from Fig. 4.38, the compound generator will build up a no-load terminal potential difference in exactly the same way as a shunt generator.

If the number of series turns, N_s, is made large enough, then v_t may be made to increase as i_L increases. For a given increase in load current Δi_L that results in a corresponding increase in armature current Δi_a, it can be arranged that, for the connection of Fig. 4.38,

$$\Delta e_a > (R_a + R_s)\Delta i_a \quad \text{V} \tag{4.55}$$

Eventually, however, due to saturation of the magnetic system, e_a rises more slowly with increase of i_L, and as a consequence, v_t begins to fall.

If the windings are suitably designed, it may be arranged that, at the rated load current for the machine, the terminal potential difference is the same as that for no load. The machine is then said to be *flat-compounded,* and provides a nearly constant potential source. Characteristics above and below the flat-compounded char-

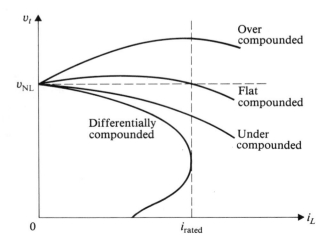

Fig. 4.39 Compound generator characteristics.

acteristic are said to be *over-compounded* and *under-compounded* respectively. Typical characteristics are illustrated in Fig. 4.39.

If the series-field mmf is arranged to oppose the shunt-field mmf, the machine is *differentially compounded*. The result of this arrangement is that e_a, and consequently v_t, fall rapidly as i_L is increased. A differentially compounded generator acts approximately as a current source, and is useful in such applications as welders, when the generator may be frequently short-circuited.

The external characteristics of separately excited, shunt, and flat-compounded generators are shown in Fig. 4.40, where in each case the field rheostat has been adjusted to give rated terminal potential difference at rated load current.

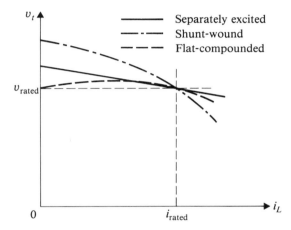

Fig. 4.40 Comparison of generator characteristics.

4.7 PERFORMANCE OF MOTORS

The engineer wishing to apply a dc motor is principally concerned with the relation between torque and the speed of the machine. The extensive use of dc motors derives from the wide range of accurate speed and torque control they provide. In contrast, unless elaborate control equipment is employed, the synchronous and induction motors are essentially constant-speed machines.

4.7.1 Separately Excited Motor

Figure 4.41 illustrates the equivalent circuit of a dc machine employed as a motor. The positive direction of armature current i_a has been chosen so that when v_t and i_a are both positive, power is entering the armature winding. The circuit layout has been rearranged (compare with Fig. 4.21) so that the direction of energy flow for motor operation is from left to right. In addition, the torque directions on the shaft have been assigned so that the internal electromagnetic torque T produced by the motor opposes the mechanical loss and load–torque components.

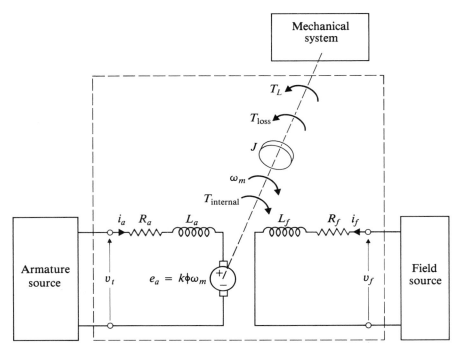

Fig. 4.41 Equivalent circuit of a dc motor.

As a consequence of these changes in datum directions, the equations relating the terminal variables of the machine are:

$$v_f = R_f i_f + L_f \frac{di_f}{dt} \quad \text{V} \tag{4.56}$$

$$v_t = k\phi\omega_m + L_a \frac{di_a}{dt} + R_a i_a \quad \text{V} \tag{4.57}$$

and

$$T_L = k\phi i_a - J \frac{d\omega_m}{dt} - T_{\text{loss}} \quad \text{N} \cdot \text{m} \tag{4.58}$$

These may be compared with Eqs. (4.25)–(4.27), which are written for generator operation.

If any of the variables i_f, i_a, or ω_m is constant, the term in Eqs. (4.56)–(4.58) involving its time derivative becomes zero. If all three variables are constant, the resultant equations are:

$$v_f = R_f i_f \quad \text{V} \tag{4.59}$$

$$v_t = k\phi\omega_m + R_a i_a \quad \text{V} \tag{4.60}$$

$$T_L = k\phi i_a - T_{\text{loss}} \quad \text{N·m} \tag{4.61}$$

These last three equations apply to the steady-state performance of the machine.

As before, Eqs. (4.56)–(4.61) include the seventh variable, ϕ. This is again most conveniently included in the form

$$e_a = k\omega_m\phi = F(i_f)|_{\omega_m=\Omega_0} \quad \text{V} \tag{4.62}$$

The steady-state performance of the machine is then described by Eqs. (4.59)–(4.62) and three additional equations that describe the terminal properties of the field and armature sources and the mechanical system coupled to the motor shaft.

The internal torque of the machine has been shown to be expressed by

$$T = k\phi i_a \quad \text{N·m} \tag{4.63}$$

Substitution in Eq. (4.60) yields

$$\omega_m = \frac{v_t}{k\phi} - \frac{R_a T}{(k\phi)^2} \quad \text{rad/s} \tag{4.64}$$

From Eq. (4.61),

$$T = T_{\text{loss}} + T_L \quad \text{N·m} \tag{4.65}$$

Several possible methods of speed control can be visualized from Eq. (4.64). The steady-state speed can be controlled directly by controlling the potential difference v_t applied to the armature terminals. If the armature resistance R_a is small, the speed is seen to be essentially independent of load torque. The speed may also be controlled by varying the field current i_f, and therefore the flux per pole ϕ. These two methods of speed control are frequently combined in systems employing separately excited motors, to give a very wide speed range.

As a third possibility, the resistance R_a of the armature circuit may be augmented by a series resistor, thus altering the speed–torque relationship and giving the machine a large speed regulation for a small increase of load torque. This method of control is not normally employed with separately excited motors and indeed is generally undesirable, since it is wasteful of energy.

If the field current of the motor is adjusted to the maximum value for which the machine is designed, so that ϕ is large, then Eq. (4.64) shows that for a constant value of v_t the speed–torque relationship can be represented by a straight line of small negative slope with an intercept on the speed axis. Such a straight line is shown, marked v_{t1}, in Fig. 4.42. If then, for the motor and the driven mechanical system, the relationship between $(T_{\text{loss}} + T_L)$ and speed can be represented by the line shown in Figure 4.42, the system will operate at point p_1, given by the intersection of the two lines. On the other hand, uncoupling of the mechanical system will give the line marked T_{loss} and the no-load operating point p_0.

If the armature is supplied from a controllable source of direct potential difference, then the speed may be controlled from zero up to a value at which v_t equals

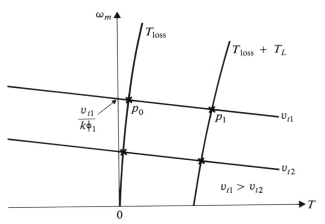

Fig. 4.42 Speed control by variation of v_t.

the maximum potential difference for which the machine has been designed. Any fixed value of v_t, such as v_{t2}, will yield a characteristic parallel to that for v_{t1}.

If the mechanical system driven by the motor has considerable inertia, and if a reduction in speed is commanded by means of a reduction in v_t, then the motor will momentarily be driven by the inertia of the load at more than the command speed. Under these circumstances, the motor will regenerate, supplying energy to the armature source. The negative torque developed due to this regeneration will cause the motor to operate momentarily in the second quadrant of Fig. 4.42, until the load has been decelerated to the command speed.

Reversal of the polarity of v_t gives a pattern of characteristics and operating points in the third and fourth quadrants similar to that in Fig. 4.42, but rotated through 180°. This would provide for operation of the motor in the reversed direction.

When v_t has been brought to its maximum permissible value and ϕ is at the maximum value ϕ_1 for which the machine is designed, the machine is said to be operating at *base speed*. Increase of speed above this value can be obtained by *field weakening*—that is, by a reduction in v_f and consequent reduction in i_f and ϕ. Equation (4.64) shows that the resulting characteristic will have an increased intercept on the axis of ω_m and an increased negative slope, as illustrated in Fig. 4.43. Speed control by field weakening is limited by the mechanical design of the motor and maximum speed may be from three to six times the base speed.

The full-load operating conditions of the motor are given on its nameplate for operation at base speed ω_b. The quantities specified are v_t, i_a, i_f, the power developed at the coupling P_L, and the permissible speed range by field control ω_b to ω_{max}. Consider now a system in which v_t may be controlled from zero up to the nameplate value. If armature-circuit resistance is negligible, control of motor speed from zero

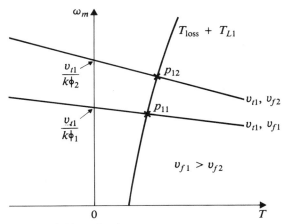

Fig. 4.43 Speed control by field weakening.

up to base speed ω_b is obtained with rated field current. Over this speed range, the armature current must normally be limited to its rated value, and hence the developed torque will also be limited, since

$$T = k\phi i_a \quad \text{N·m} \tag{4.66}$$

Power is expressed by

$$P = T\omega_m \quad \text{W} \tag{4.67}$$

and is a linear function of ω_m, having a maximum value at $\omega_m = \omega_b$.

Since the terminal potential difference v_t may not normally be increased much above the nameplate value, both because of the commutator insulation and because of the limit on the available supply, any further speed increase requires field weakening. If armature-circuit resistance is negligible, Eq. (4.64) shows that for constant v_t,

$$\omega_m \propto \frac{1}{\phi} \quad \text{rad/s} \tag{4.68}$$

Since again i_a must normally be limited to its nameplate value, the torque available is

$$T = k\phi i_a \propto \frac{1}{\omega_m} \quad \text{N·m} \tag{4.69}$$

and

$$P = T\omega_m = \text{constant} \quad \text{W} \tag{4.70}$$

Thus, up to base speed, the machine is limited to operating continuously at nameplate current and, hence, at constant torque; above base speed, it is limited to oper-

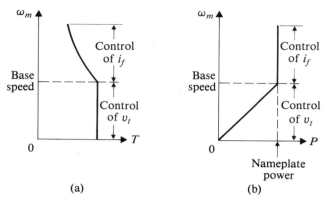

Fig. 4.44 Speed control of an ideal separately excited motor.

ating at constant input power. These limiting relationships are illustrated in Fig. 4.44. It may be assumed that the torque and power developed at the coupling will vary with ω_m in a manner similar to the quantities shown in Fig. 4.44, but that both will be reduced by the machine's losses. Thus, if operating point p_{11} in Fig. 4.43 represented machine operation at nameplate conditions, it would not be permissible to increase speed by field weakening without reducing T_L, since operation at point p_{12} would call for an unacceptably high value of i_a.

Separately excited motors are used in many industrial applications where speed or torque must be controlled over a wide range. Examples are rolling mills, paper mills, and machine tools.

Example 4.6 The nameplate data on a dc machine that is to be used as a separately excited motor in a control system are as follows: $v_t = 230$ V, $P_L = 18.5$ kW, $i_a = 93$ A, $i_f = 2.5$ A, $R_a = 0.185$ Ω, speed = 1200 to 3600 r/min. When the motor is driven at base speed on no load, the power input to the motor armature circuit is 1145 W. It may be assumed that the torque required to overcome mechanical losses is expressed by

$$T_{\text{loss}} = B\omega_m \quad \text{N} \cdot \text{m}$$

where B is a constant.

For this motor, draw the diagrams corresponding to Fig. 4.44(a) and (b) and superimpose upon them curves for T_L and P_L.

Solution When the machine is operating at nameplate conditions, the input power is

$$P = v_t i_a = 230 \times 93 = 21{,}390 \quad \text{W}$$

Base speed is

$$\omega_b = \frac{2\pi}{60} \times 1200 = 125.7 \quad \text{rad/s}$$

When losses are neglected, the developed torque is

$$T = \frac{P}{\omega_b} = \frac{21{,}390}{125.7} = 170.2 \quad \text{N} \cdot \text{m}$$

This gives a point on each of the curves of Fig. 4.44. Above the base speed, the curve of Fig. 4.44(a) is a rectangular hyperbola, since the product $T\omega_m$ is constant at the value 21,390 W.

At 3600 r/min, the maximum permitted speed,

$$T = \frac{21{,}390}{3 \times 125.7} = 56.7 \quad \text{N} \cdot \text{m}$$

The ideal curves of T and P versus ω_m are shown below in Fig. 4.45.

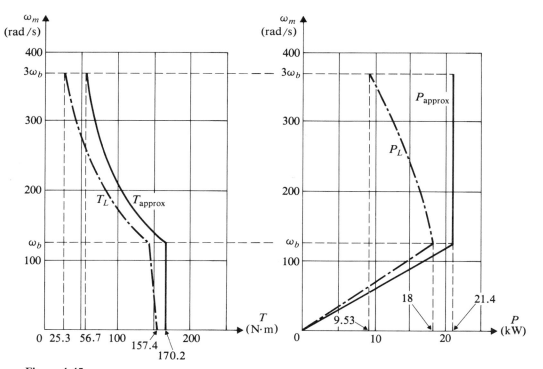

Figure 4.45

To determine the actual curves of T_L and P_L versus ω_m, it is necessary to express T_{loss} as a function of ω_m—that is, to determine the constant B.

For the no-load test,

$$i_a = \frac{1145}{230} = 4.978 \quad A$$

Armature copper loss is

$$P_c = 0.185 \times 4.978^2 = 4.6 \quad W$$

Thus, power dissipated in mechanical losses is,

$$P_{\text{loss}} = 1145 - 4.6 = 1140 \quad W$$

$$P_{\text{loss}} = T_{\text{loss}}\omega_b = B\omega_b^2$$

so that

$$B = \frac{1140}{125.7^2} = 0.07215$$

First the output torque and power on full load at base speed (nameplate conditions) will be determined.

$$P_L = 230 \times 93 - 0.185 \times 93^2 - 1140 = 18,650 \quad W$$

$$T_L = \frac{P_L}{\omega_b} = \frac{18,650}{125.7} = 148.4 \quad N \cdot m$$

At $\omega_m = 3\omega_b$,

$$P_{\text{loss}} = \frac{1140}{125.7^2} \times (3 \times 125.7)^2 = 10,260 \quad W$$

$$P_L = 230 \times 93 - 0.185 \times 93^2 - 10,260 = 9,530 \quad W$$

Note that the assumption here is that mechanical losses may increase without a compensating reduction in copper losses. Since the increased mechanical losses arise, to some extent, from driving additional cooling air through the machine; and, because the field winding losses are being reduced, this assumption is usually acceptable.

$$T_L = \frac{9530}{125.7 \times 3} = 25.3 \quad N \cdot m$$

By the same procedure, at $\omega_m = 2\omega_b$,

$$P_L = 15,230 \text{ W}, \quad \text{and} \quad T_L = 60.6 \quad N \cdot m$$

At $\omega_m = \omega_b$, the emf induced in the armature winding will be

$$e_a = k\phi\omega_b = v_t - R_a i_a = 230 - 0.185 \times 93 = 212.8 \quad V$$

and

$$k\phi = \frac{e_a}{\omega_b} = \frac{212.8}{125.7} = 1.693$$

The internal torque developed by the motor will be

$$T = k\phi i_a = 1.693 \times 93 = 157.4 \quad \text{N} \cdot \text{m}$$

This value of internal torque will apply over the speed range $0 < \omega_m < \omega_b$. At $\omega_m = 0.5 \, \omega_b$,

$$T_{\text{loss}} = \frac{P_{\text{loss}}}{0.5\omega_b} = \frac{1140}{125.7^2} \times 0.5 \times 125.7 = 4.53 \quad \text{N} \cdot \text{m}$$

$$T_L = T - T_{\text{loss}} = 157.4 - 4.53 = 152.9 \quad \text{N} \cdot \text{m}$$

$$P_L = 0.5\omega_b T_L = 0.5 \times 125.7 \times 152.9 = 9610 \quad \text{W}$$

At $\omega_m = 0$,

$$T_{\text{loss}} = 0$$

$$T_L = T = 157.4 \quad \text{N} \cdot \text{m}$$

$$P_L = 0$$

The curves of P_L and T_L obtained are shown in Fig. 4.45.

4.7.2 Shunt Motor

A dc motor is rarely installed in a situation where it is required to run at constant speed under constant load, since an ac induction motor performs such duties satisfactorily, costs only a fraction of the price of a dc machine of equal power and speed, and requires minimal maintenance.

Many simple variable-speed systems, including their driving motors, are inherently stable in operation, so that the steady-state behavior of a dc motor is frequently all that an engineer need take into consideration. For such simple systems, a dc shunt motor excited from a single source is often satisfactory, and provides a reasonable range of adjustable speed and torque.

The steady-state equivalent circuit for a shunt motor is shown in Fig. 4.46, where the common source for armature and field provides a constant terminal potential difference V_t. Control of field current is now obtained by the inclusion of a rheostat R_e in the field circuit. The motor behaves like a separately excited motor in which speed control is obtained by variation of the field current. Thus, from Eqs. (4.59)–(4.65),

$$V_t = (R_e + R_f)i_f \quad \text{V} \tag{4.71}$$

$$V_t = k\phi\omega_m + R_a i_a \quad \text{V} \tag{4.72}$$

$$T_L = k\phi i_a - T_{\text{loss}} \quad \text{N} \cdot \text{m} \tag{4.73}$$

$$e_a = k\omega_m\phi = F(i_f)\big|_{\omega_m = \Omega_0} \quad \text{V} \tag{4.74}$$

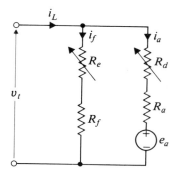

Fig. 4.46 Equivalent circuit of a shunt motor.

$$T = k\phi i_a \quad \text{N·m} \tag{4.75}$$

$$\omega_m = \frac{V_t}{k\phi} - \frac{R_a T}{(k\phi)^2} \quad \text{rad/s} \tag{4.76}$$

$$T = T_{\text{loss}} + T_L \quad \text{N·m} \tag{4.77}$$

Figure 4.47 illustrates speed control by variation of the field rheostat setting.

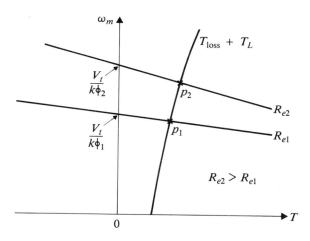

Fig. 4.47 Speed control of a shunt motor by field rheostat.

As has been noted in connection with the separately excited motor, speed control may also be achieved by including some external resistance, R_d, in the armature circuit. Under these conditions, Eq. (4.76) becomes

$$\omega_m = \frac{V_t}{k\phi} - \frac{(R_a + R_d)T}{(k\phi)^2} \quad \text{rad/s} \tag{4.78}$$

This method of speed control is effective only if the motor is loaded, since it simply increases the motor's speed regulation. It has the unsatisfactory feature that if the

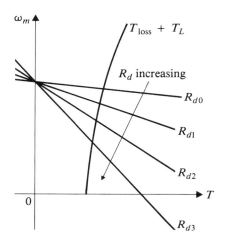

Fig. 4.48 Speed control of a shunt motor by armature circuit resistance.

load torque changes, then speed changes considerably for a constant value of R_d. This effect and the result of varying R_d are illustrated in Fig. 4.48.

Speed control by variation of armature-circuit resistance is seldom employed continuously with shunt motors, since much energy is wasted in the external resistance introduced into the armature circuit. However, this is the standard method of starting a dc motor excited from a source of constant potential difference, since at standstill, from Eq. (4.72),

$$V_t = R_a i_a \quad \text{V} \tag{4.79}$$

Since R_a is small, and V_t is fixed, i_a would be extremely high if no additional resistance were included in the armature circuit. In general, i_a must not exceed about twice the rated value for the machine. If it does so, the commutator is liable to be damaged due to excessive sparking at the brushes.

A rapid start can be achieved without excessive armature current if the resistance R_d is reduced gradually or in steps as the speed increases. Various types of manual and automatic starters provide this capability. The manual starter of Fig. 4.49 has a spring-loaded contact arm that is rotated clockwise, providing the initial connection of the field circuit and the gradual reduction of the armature-circuit resistance. The hold-on coil holds the arm in the zero-resistance position as long as adequate terminal potential difference is available. The field rheostat should be set to give maximum field current during starting, so that a high flux per pole ϕ is produced, giving a high starting torque per ampere of armature current, and minimizing the effect of armature mmf on flux-density distribution under the poles.

Automatic starters perform essentially the same functions as the starter of Fig. 4.49 with electromagnetic relays that short out sections of the starting resistance, either in a predetermined time sequence, or when the armature current has dropped to a predetermined value.

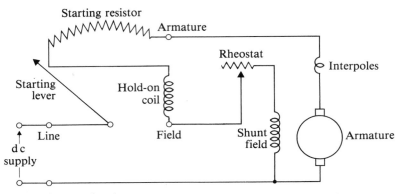

Fig. 4.49 Manual starter for shunt motor.

The external characteristic of a shunt motor is a curve of speed versus load torque—that is, the torque applied or appearing at the coupling. If this characteristic is to be predicted with any accuracy, then saturation of the magnetic system of the machine must be taken into account. This is achieved by employing the no-load saturation curve of the motor, described by

$$e_a = F(i_f)|_{\omega_m = \Omega_0} \quad \text{V} \tag{4.80}$$

Allowance must be made for friction and windage losses within the machine as well as for core losses. These may all be lumped together to give a composite quantity called *rotational losses*. For a motor that is operated over only a small speed range, it is common practice to consider rotational losses as constant and equal to the power input to the armature when the motor is running at nameplate speed on no load. Thus

$$P_{\text{loss}} = V_t i_{a\text{NL}} \quad \text{W} \tag{4.81}$$

In the case of a machine that is operated over a wide speed range, the rotational losses may be determined approximately by measuring the no-load input to the armature over the desired speed range.

For any particular machine, the nameplate data will be known. The values of the armature-circuit resistance R_a, the field-winding resistance R_f (not including the field rheostat), and the "constant" rotational loss P_{loss}, as well as a no-load saturation curve at some known speed Ω_0 will be supplied by the machine builder. If these data are not available, they can be obtained by means of simple tests.

The no-load saturation curve, provided or measured, may be described by

$$e_a = k\Omega_0\phi = F(i_f) \quad \text{V} \tag{4.82}$$

from which

$$\frac{e_a}{\Omega_0} = k\phi \quad \text{V} \cdot \text{s/rad} \tag{4.83}$$

The vertical axis of the no-load saturation curve may therefore be calibrated in units of $k\phi$. The resulting curve of $k\phi$ versus i_f does not change with the motor speed.

When the data specified in the preceding paragraph are available, an external characteristic of the motor may be determined, taking as a starting point a particular speed on no load. The first point on the characteristic is then $(0, \omega_{NL})$, as shown in Fig. 4.50. On no load, $e_{aNL} \gg R_a i_{aNL}$, so that from Fig. 4.46,

$$V_t \simeq e_{aNL} = k\phi\omega_{NL} \quad V \qquad (4.84)$$

from which

$$k\phi = \frac{e_{aNL}}{\omega_{NL}} \simeq \frac{V_t}{\omega_{NL}} \quad V\cdot s/rad \qquad (4.85)$$

From the curve of $k\phi$ versus i_f obtained from the no-load saturation curve and illustrated in Fig. 4.51, the value of i_f corresponding to this no-load speed may be determined. The field rheostat must then be adjusted so that

$$R_e + R_f = \frac{V_t}{i_f} \quad \Omega \qquad (4.86)$$

This field-rheostat setting must be maintained for any one external characteristic of the motor.

Assume that a load torque is applied such that the motor draws line current i_{L1}. A useful value to start with would be the nameplate current of the motor. The speed ω_{m1} and coupling torque T_{L1} corresponding to this line current are to be determined. From the circuit of Fig. 4.46,

$$i_{a1} = i_{L1} - i_f \quad A \qquad (4.87)$$

so that

$$e_{a1} = V_t - R_a i_{a1} \quad V \qquad (4.88)$$

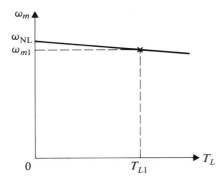

Fig. 4.50 External characteristic of a shunt motor.

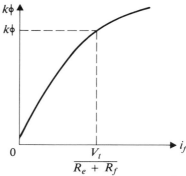

Fig. 4.51 Curve of $k\phi$ versus i_f for a shunt motor.

and

$$e_{a1} = k\phi\omega_{m1} \quad \text{V} \tag{4.89}$$

Speed ω_{m1} is now known. The output power at this speed is

$$P_{L1} = \text{(input power)} - \text{(resistance losses)} - \text{(rotational losses)} \quad \text{W} \tag{4.90}$$

The whole of the power delivered to the motor field circuit is dissipated as resistance loss. Part of the power delivered to the armature is also dissipated as loss in the armature winding. From Eq. (4.88),

$$V_t = e_{a1} + R_a i_{a1} \quad \text{V} \tag{4.91}$$

and the power delivered to the armature circuit is

$$V_t i_{a1} = e_{a1} i_{a1} + R_a i_{a1}^2 \quad \text{W} \tag{4.92}$$

The second term on the right-hand side of Eq. (4.92) represents armature resistance loss. The first term expresses the power converted to mechanical form. Thus the power output at the coupling is

$$P_{L1} = e_{a1} i_{a1} - P_{\text{loss}} \quad \text{W} \tag{4.93}$$

and

$$T_{L1} = \frac{P_{L1}}{\omega_{m1}} \quad \text{N·m} \tag{4.94}$$

This gives an additional point on the external characteristic, as shown in Fig. 4.50. Other points may be similarly determined and a curve drawn.

Example 4.7 The nameplate data for a small dc motor are 125 V, 36 A, 1150 r/min. The armature-circuit resistance R_a is given as 0.370 Ω, and the armature power input on no load at rated speed is 325 W. The no-load saturation curve for the motor taken at rated speed is shown in Fig. 4.26. The motor is to be started by means of the starter of Fig. 4.49, with the field current i_f set at 2.5 A.

Assuming that the current during starting will be allowed to vary over the range $40 < i_L < 75$ A, determine

a) The total starter resistance required.

b) The speed at which the first reduction in starter resistance takes place.

c) The magnitude of the first reduction in starter resistance.

d) The speed when all starter resistance has been cut out, and $i_L = 36$ A.

e) The no-load speed if the mechanical load is uncoupled.

Solution

a) At standstill,

$$i_a = i_L - i_f = 75 - 2.5 = 72.5 \quad \text{A}$$

$$R_a + R_d = \frac{125}{72.5} = 1.724 \quad \Omega$$

$$R_d = 1.724 - 0.370 = 1.354 \quad \Omega$$

b) The machine accelerates until

$$i_a = 40 - 2.5 = 37.5 \quad A$$

From Eqs. (4.72) and (4.74),

$$e_a = V_t - (R_a + R_d)i_a = 125 - 1.724 \times 37.5 = 60.35 \quad V$$

From Fig. 4.26, for $i_f = 2.5$ A, $e_a = 137$ V. This curve was taken at speed

$$\Omega_0 = \frac{2\pi}{60} \times 1150 = 120.4 \quad rad/s$$

Thus, at $i_f = 2.5$ A,

$$k\phi = \frac{137}{120.4} = 1.138$$

Thus, for $i_a = 37.5$ A, and $e_a = 60.35$ V,

$$\omega_m = \frac{60.35}{1.138} = 53.03 \text{ rad/s} = 506.4 \text{ r/min}$$

c) When the resistance step is removed, the armature current may rise to 72.5 A. Thus,

$$R_a + R_{d1} = \frac{V_t - e_a}{i_a}$$

and

$$R_{d1} = \frac{125 - 60.35}{72.5} - 0.370 = 0.5217 \quad \Omega$$

The reduction in R_d has therefore been

$$\Delta R_d = 1.354 - 0.5217 = 0.832 \quad \Omega$$

d) When all starting resistance has been cut out,

$$e_a = V_t - R_a i_a = 125 - 0.370 \times (36 - 2.5) = 112.6 \quad V$$

$$\omega_m = \frac{e_a}{k\phi} = \frac{112.6}{1.138} = 98.95 \text{ rad/s} = 944.9 \text{ r/min}$$

e) On no load, the power input to the armature is (approximately):

$$V_t i_a = 325 \quad W$$

$$i_a = \frac{325}{125} = 2.6 \quad A$$

$$e_a = V_t - R_a i_a = 125 - 0.370 \times 2.6 = 124.0 \quad V$$

$$\text{Speed} = \frac{124.0}{1.138} = 109.0 \text{ rad/s} = 1041 \text{ r/min}$$

*4.7.3 Series Motor

The steady-state equivalent circuit of a series motor is shown in Fig. 4.52, where R_s represents the resistance of the series field winding and R_d the added series resistance for starting and speed control. Since the field, armature, and line current of a series motor are one and the same, the no-load saturation curve may be described by

$$e_a = F(i_L)|_{\omega_m = \Omega_0} \quad V \tag{4.95}$$

From Eqs. (4.60)–(4.65),

$$v_t = k\phi\omega_m + (R_a + R_s + R_d)i_L \quad V \tag{4.96}$$

$$T_L = k\phi i_L - T_{\text{loss}} \quad N \cdot m \tag{4.97}$$

$$e_a = k\omega_m\phi \quad V \tag{4.98}$$

$$T = k\phi i_L \quad N \cdot m \tag{4.99}$$

$$\omega_m = \frac{v_t}{k\phi} - \frac{(R_a + R_s + R_d)T}{(k\phi)^2} \quad \text{rad/s} \tag{4.100}$$

$$T = T_{\text{loss}} + T_L \quad N \cdot m \tag{4.101}$$

When the motor current is within the linear range of the no-load saturation curve (usually below rated value):

$$k\phi = k_f i_L \tag{4.102}$$

and

$$T = k_f i_L^2 \quad N \cdot m \tag{4.103}$$

Thus the motor current increases only as the square root of the torque, in contrast with the shunt machine, in which motor current and torque are directly proportional. From Eqs. (4.102) and (4.103),

$$k\phi = (k_f T)^{1/2} \tag{4.104}$$

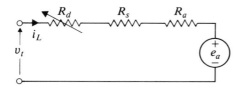

Fig. 4.52 Equivalent circuit of a series motor.

Substitution in Eq. (4.100) yields

$$\omega_m = \frac{V_t}{(k_f T)^{1/2}} - \frac{(R_a + R_s + R_d)}{k_f} \quad \text{rad/s} \tag{4.105}$$

If the resistance of the machine and external resistor is neglected, the speed is inversely proportional to the square root of the torque. An increase in torque is accompanied by an increase in motor current, which in turn causes an increase in flux per pole. Figure 4.53 shows a typical torque–speed relation for a series motor with series resistance control.

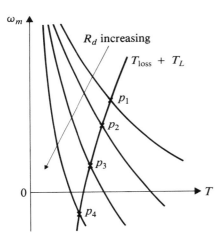

Fig. 4.53 Speed control of a series motor by series resistance.

The significant decrease in speed with increase in load torque is desirable for many types of mechanical load. For example, in traction applications, a low speed when a vehicle is climbing or accelerating at high torque, and a high speed when it is running on the level at low torque allows more nearly constant power to be taken from the supply lines than would be the case with a shunt motor.

It may be seen from Eq. (4.105) that as T approaches zero, ω_m approaches infinity. While T can never reach zero, since its lower limit is T_{loss}, the speed of the motor may reach a destructively high value. For this reason, series motors should not be run without a mechanical load coupled to the shaft.

The normal method of controlling the speed of a series motor operating at a constant supply potential difference is to introduce series resistance R_d. A supplementary method is that of connecting a *diverter resistor* in parallel with the series field. This changes the ratio of field current to armature current and, in effect, gives the machine a different speed–torque characteristic.

A family of motor characteristics for different values of added series resistance is illustrated in Fig. 4.53. The intersections of these characteristics with that of $(T_{\text{loss}} + T_L)$ versus speed represent various operating conditions for the machine.

Point p_4 shows that if the motor were being employed as a hoist drive (a common application), the introduction of too much series resistance would result in the load's driving the motor backward, and descending instead of rising.

Example 4.8   A 230V 15 kW (20 hp), 75 A, 915 r/min dc series motor has an armature circuit resistance of 0.080 Ω and a field-winding resistance of 0.055 Ω. When the motor is driven at rated speed as a separately excited generator, the following results are obtained:

i_f(A)	0	25	50	75	100	125	150	175	200	225	250
e_a (V)	12	77	159	219	255	274	285	293	298	302	304
$k\phi$ (Wb)	0.13	0.82	1.69	2.32	2.71	2.91	3.02	3.11	3.16	3.20	3.23

The motor is to start against full-load torque, and during acceleration the current may vary between 200 A and 100 A. The rotational losses of the motor may be considered constant at 750 W.

Design the starting resistor, and determine the final operating speed.

Solution   The curve of $k\phi$ vs. i_L is shown in Fig. 4.54. This curve is independent of motor speed.

At standstill, point 1 on Fig. 4.55(a),

$$i_L = \frac{V_t}{R_a + R_s + R_d} \quad \text{A}$$

and the current is at the maximum permitted value. Thus,

$$R_1 = 0.080 + 0.055 + R_{d1} = \frac{230}{200} = 1.15 \quad \Omega$$

from which

$$R_{d1} = 1.015 \quad \Omega$$

At

$$i_L = 200 \text{ A}, \quad k\phi = 3.16, \quad \text{and} \quad T = 3.16 \times 200 = 632 \quad \text{N} \cdot \text{m}$$

The motor accelerates on the characteristic corresponding to R_{d1} until the current falls to 100 A at point 2.

At point 2 in Fig. 4.55,

$$i_L = 100 \text{ A}, \qquad k\phi = 2.71, \qquad T = 2.71 \times 100 = 271 \text{ N} \cdot \text{m}$$

$$e_a = V_t - R_1 i_L = 230 - 1.15 \times 100 = 115 \quad \text{V}$$

$$\omega_{m1} = \frac{e_a}{k\phi} = \frac{115}{2.71} = 42.5 \quad \text{rad/s}$$

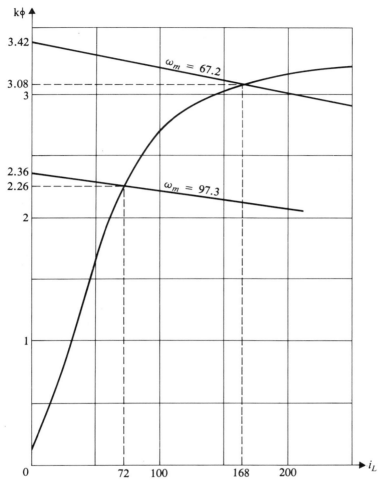

Fig. 4.54 Diagram I for Example 4.8.

R_d is now reduced to R_{d2}, so that i_L rises instantaneously to 200 A, and T rises to 632 N·m.

At point 3,

$$i_L = 200 \text{ A}, \qquad k\phi = 3.16, \qquad T = 632 \text{ N·m}, \qquad \omega_m = 42.5 \text{ rad/s}$$

$$e_a = k\phi\omega_m = 3.16 \times 42.5 = 134 \quad \text{V}$$

$$e_a = V_t - R_2 i_L = 134 = 230 - R_2 \times 200$$

from which

$$R_2 = 0.480 \quad \Omega$$

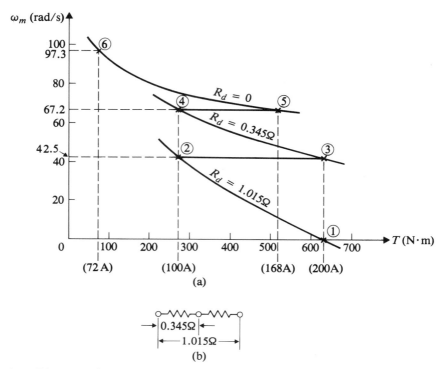

Fig. 4.55 Diagram II for Example 4.8.

and

$$R_{d2} = 0.480 - 0.080 - 0.055 = 0.345 \quad \Omega$$

The motor accelerates on the characteristic corresponding to R_{d2} until the current again falls to 100 A at point 4.

At point 4,

$$i_L = 100 \text{ A}, \qquad k\phi = 2.71, \qquad T = 271 \text{ N·m}$$

$$e_a = V_t - R_2 i_L = 230 - 0.480 \times 100 = 182 \text{ V}$$

$$\omega_{m2} = \frac{e_a}{k\phi} = \frac{182}{2.71} = 67.2 \quad \text{rad/s}$$

R_d is now reduced to R_{d3}, so that i_L rises to 200 A, and T rises to 632 N·m.

At point 5,

$$i_L = 200 \text{ A}, \qquad k\phi = 3.16, \qquad T = 632 \text{ N·m}, \qquad \omega_m = 67.2 \text{ rad/s}$$

$$e_a = k\phi\omega_m = 3.16 \times 67.2 = 212 \quad \text{V}$$

$$e_a = V_t - R_3 i_L = 212 = 230 - R_3 \times 200$$

from which

$$R_3 = 0.090 \quad \Omega$$

But this is less than $R_a + R_s = 0.135 \, \Omega$. So all that can be done is to remove all external resistance and let $R_3 = 0.135 \, \Omega$:

$$R_{d3} = 0.$$

The arrangement of the starting resistor is shown in Fig. 4.55(b). The real point 5, for which the current will be less than 200 A, must now be found.

A relationship between $k\phi$ and i_L for the motor at speed 67.2 rad/s permits a graphical solution. From Eq. (4.96)

$$230 = 67.2k\phi + 135i_L$$

from which

$$k\phi = 3.42 - 2.01 \times 10^{-3} \, i_L$$

This line, drawn on the curve of Fig. 4.54, yields an intersection (168, 3.08), which corresponds to point 5 in Fig. 4.55(a).

The motor now accelerates to its final steady-state operating condition. At point 6, speed and armature current are unknown, but the torque at the coupling must be the rated value. Nameplate speed is

$$\Omega_0 = \frac{2\pi}{60} \times 915 = 95.8 \quad \text{rad/s}$$

Rated output torque is then

$$T_L = \frac{15,000}{95.8} = 157 \quad \text{N} \cdot \text{m}$$

The corresponding armature current will be approximately equal to the nameplate value, so two values, one on either side of the nameplate value, may be assumed, and the required current obtained by assuming that the short section of the $\omega_m - T_L$ curve between them is a straight line and interpolating from Table 4.2.

Table 4.2

Formula	Values of i_a	
	70	80
From Fig. 4.54: $k\phi$	2.22	2.92
$(R_a + R_s)i_a$	9.5	10.8
$V_t - (R_a + R_s)i_a = e_a$	220.5	219.2
$e_a/k\phi = \omega_m$	99.5	90.5
$P_{\text{mech}} = e_a i_a$	15,460	17,520
$P_{\text{mech}} - P_{\text{loss}} = e_a i_a - 750 = P_L$	14,710	16,770
$P_L/\omega_m = T_L$	1,480	1,850

Interpolation between points (148, 99.5) and (185, 90.5) on the speed–torque curve gives the final speed for $T_L = 157$ N·m as

$$97.3 \text{ rad/s} = 928 \text{ r/min}$$

This is somewhat greater than the nameplate speed. Once again, i_a at this speed may be determined graphically. From Eq. (4.96),

$$230 = 97.3k\phi + 0.135i_a$$

from which

$$k\phi = 2.36 - 1.39 \times 10^{-3}I_a$$

This line, drawn on the curve of Fig. 4.54, yields an intersection (72, 2.26). The steady-state operating current is therefore 72 A, which is somewhat less than the nameplate value.

*4.7.4 Universal Motors

If a dc series motor were connected to an ac supply, it would run, since both the field flux and the armature current would reverse simultaneously each half cycle. A unidirectional torque pulsating at double the source frequency would therefore be developed. The motor would not operate satisfactorily, however, for a number of reasons, of which the chief are

a) Excessive core losses in the stator magnetic system, due to the alternating flux.

b) Large potential difference across the terminals of the field winding due to its high inductance. This would result in low armature terminal potential difference.

c) Poor commutation (see Section 4.11) due to emf's induced in short-circuited coils by transformer action between the armature and field coils.

It is convenient for such applications as domestic appliances, hand tools, and so forth, to be able to operate from either ac or dc sources. For such purposes, universal motors have been developed with the disadvantages listed above reduced to an acceptable level by the following design and operating features:

a) Eddy-current losses in the field structure are reduced by laminating the stator poles and yoke.

b) Armature terminal potential difference can be raised by increasing the ratio of armature to field turns. There is a limit to this process, however, as the increased armature mmf introduces great distortion of the flux-density distribution in the air gap and consequent reduction in rms flux density due to magnetic saturation (see Section 4.10). This can be overcome by fitting a compensating winding to the stator (see Section 4.11.2), but the result is an expensive machine, and few compensated universal motors are manufactured. A more practical step is to operate the motor at high speed, causing high induced armature emf. Nevertheless, due to the high field-winding reactance, the speed of a universal motor is usually lower on ac than on dc for equal load torques.

The typical universal motor is a series unidirectional machine of less than one kilowatt output power that operates at full-load speeds of 4,000 to 10,000 r/min. The laminated stator is built with only two salient poles carrying concentrated field windings, while the armature is similar to that of a dc series motor. The series speed–torque characteristic is appropriate to purposes where a reduction of speed with increase of load is desirable, and the high speed of its operation results in a small, light motor for a given power output.

*4.7.5 Compound Motors

For some special purposes a compound motor fitted with both series and shunt field windings is required. If the two windings are connected so that their mmf's act in the same direction in the magnetic system, the motor is said to be *cumulatively compounded*. If the two windings are connected so that their mmf's oppose one another, the motor is *differentially compounded*. Such motors are usually unstable, being liable to accelerate suddenly to dangerously high speeds when a load torque is applied; they are therefore not normally used.

The speed–torque characteristic of a cumulatively compounded motor is intermediate to that of a series motor and a shunt motor. Depending upon the ratio of series to shunt-field turns, the characteristic will approach that of one or the other type (see Fig. 4.56).

Two special examples of compounding are noteworthy. The simpler is a series motor on which a small shunt field is fitted to ensure that the speed–torque characteristic of the motor has an intercept on the speed axis that limits operation to less than destructive speeds. If the motor should be switched on inadvertently without a mechanical load coupled to it, no damage will result.

The second example concerns the operation of a shunt motor and shows what would occur if a motor were differentially compounded. The shunt-motor characteristic shown in Fig. 4.50 has negative slope throughout its length. This indicates

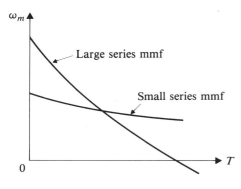

Fig. 4.56 Compound motor characteristics.

stable operation: if the load torque is increased, machine speed falls. Conditions may arise, however, when this stability may be lost due to the demagnetizing effect of armature mmf (see Section 4.9).

Under any operating conditions, $e_a >> R_a i_a$, so that $e_a \simeq v_t$. As v_t is normally constant, $k\phi\omega_m$ is constant and, to a fair approximation,

$$\omega_m \propto \frac{1}{\phi} \quad \text{rad/s} \tag{4.106}$$

As load on the motor is increased, ϕ may be reduced due to the demagnetizing effect of armature mmf. This means, in effect, that the whole speed–torque characteristic of the motor rises. Since load torque invariably increases with speed, and $T = k\phi i_a$, the armature current increases, causing a further reduction in ϕ so that the speed–torque characteristic rises further. If this process is allowed to continue, the machine will drive the load up to a destructive speed and draw an excessive armature current that may damage the commutator and cause flashover between the brushes. This instability may be guarded against by including a circuit breaker of appropriate rating in the armature circuit. Instability may be prevented by fitting a small series winding that gives cumulative compounding sufficient to overcome the demagnetizing effect of armature mmf at normal armature currents. Such a compound machine is known as a *stabilized shunt motor*.

*4.8 SOLID-STATE DRIVES FOR DIRECT-CURRENT MOTORS

The controllability of the dc machine, always great, has been much increased in recent years by the development of power semiconductor devices and the evolution of flexible and efficient converters which eliminate the need for control resistors, in which much energy was wasted. For dc machines these converters are of two principal types:

a) *DC-to-DC converters,* or *choppers,* which may be employed when a dc source of suitable and constant potential difference is already available; and

b) *Phase-controlled rectifiers,* which may be employed when only an ac source is available.

*4.8.1 Chopper Drives

The basic principle of operation of a drive employing a dc-to-dc converter is illustrated in Fig. 4.57. The motor is separately excited, and the field circuit is omitted from Fig. 4.57(a), since it is in no way unusual. The chopper may be considered as an ideal electronic switch that applies the source potential difference V to the motor armature terminals in a series of pulses, as illustrated in Fig. 4.57(b). Thus v_t consists of a series of rectangular pulses of amplitude V, and the average value of v_t lies within the range $0 \leq v_{t\,\text{av}} \leq V$. This average value may be varied either by

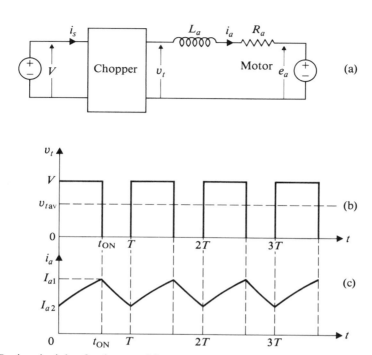

Fig. 4.57 Basic principle of a chopper drive.

varying the pulse width or by varying the pulse frequency, or even by a combination of both of these methods. For the sake of simplicity, only *pulse-width modulation* will be considered here, so that the pulse duration varies within range $0 \leqslant t_{ON} \leqslant T$, with T maintained constant.

To understand the waveform of the armature current resulting from the pulsed v_t, it is necessary to consider the basic chopper power circuit shown in Fig. 4.58. This circuit incorporates two major semiconductor devices, a *diode, D,* and a *thyristor, Q*. In a low-power chopper, Q might well be replaced by a *power transistor;* but the symbol enclosed in a circle in Fig. 4.58 represents a thyristor, which can be turned on into the conducting state *(gated)* and turned off into the blocking state

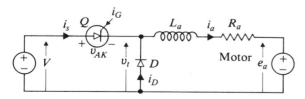

Fig. 4.58 Basic chopper power circuit.

(commutated) by auxiliary circuitry. A pulse of gating current i_G permits Q to conduct source current i_s, provided that the anode-to-cathode potential difference v_{AK} is positive. In order to restore to Q its forward blocking capability, i_G must be removed and v_{AK} driven negative for an interval of the order of 20 μs. The thyristor must therefore be gated at the commencement and commutated at the end of each pulse.

When Q is commutated and i_s no longer flows, the armature-circuit inductance L_a maintains current i_a, which then flows through the *free-wheeling diode D* until Q is again turned on. The waveform of i_a therefore consists of sectors of alternately increasing and decreasing exponential curves. If t_{ON} is made small, i_a may fall to zero during the interval $t_{ON} < t < T$, so that the motor armature current is discontinuous. Under conditions approaching full load, however, i_a will be continuous, as shown in Fig. 4.57(c), since the *chopping frequency* will be sufficiently high to maintain the current and hence the motor torque sensibly constant.

In the chopper circuit of Fig. 4.58, energy flow can take place in one direction only—that is, from source to motor armature—since i_a cannot reverse. This means that the motor is unable to regenerate. It has been seen in Section 4.7 that braking by regeneration is a very desirable feature in high-inertia systems, such as subway trains, where the kinetic energy of the train would otherwise have to be dissipated as heat in braking resistors or friction brakes. The slightly elaborated chopper circuit of Fig. 4.59(a) permits regeneration to take place.

In the two-quadrant chopper of Fig. 4.59, Q_1 and D_1 perform the functions of Q and D in Fig. 4.58. The thyristor gating-current waveforms i_{G1} and i_{G2} for a relatively low-load condition of operation are shown in Fig. 4.59(b). Note that the thyristors are turned on alternately. If the motor is running at constant speed and constant field current, then the armature induced emf e_a will be constant, and for some part of the period $t_{ON} < t < T$ when $v_t = 0$, i_a may reverse and flow through Q_2. When Q_2 is commutated, the armature inductance L_a maintains the reverse direc-

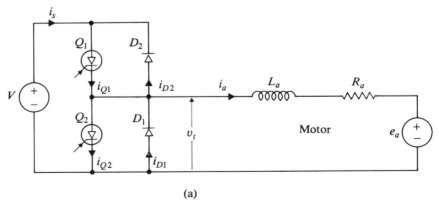

(a)

Fig. 4.59 Two-quadrant chopper drive.

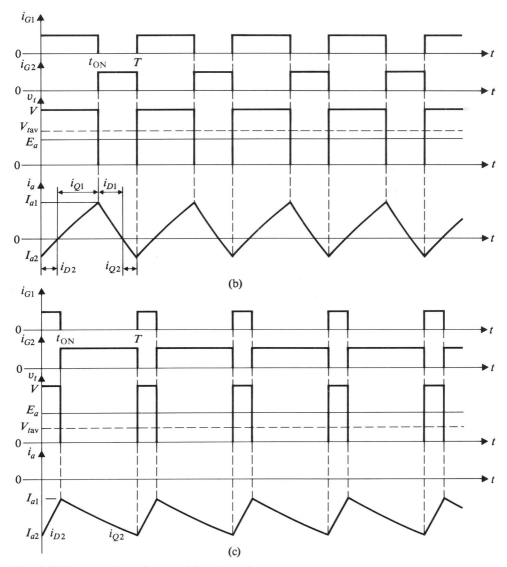

Fig. 4.59 Two-quadrant chopper drive   (cont.)

tion of i_a, which then flows through D_2 and delivers energy to source V. Thus, for part of each cycle of v_t, regeneration may be taking place. The devices conducting i_a during various parts of the cycle are indicated on the waveform of i_a. The waveform of source current i_s consists of those segments of the waveform of i_a during which conduction by Q_1 or D_2 is taking place.

If the motor load were increased and t_{ON} were also increased to maintain the speed, then the whole waveform of i_a would rise and resemble that in Fig. 4.57. If, on the other hand, a reduction in speed were required and t_{ON} were therefore reduced to the low value shown in Fig. 4.59(c), then Q_1 and D_1 would temporarily cease to conduct. Thus i_a would be continuously negative during the early part of the deceleration. Regenerative braking would then be taking place.

The current waveforms shown in Fig. 4.57 and 4.59 are such as would be anticipated from a relatively small motor with a low armature-circuit time constant, resulting in wide variation of i_a. In large machines, and particularly near full load, i_a is continuously positive, and its variation is much less than is illustrated in Fig. 4.57(c). In fact, for preliminary system design calculations, it is usually permissible to make the assumption that L_a is great enough to maintain i_a sensibly constant, so that

$$i_a = \frac{1}{R_a} \left[\frac{t_{ON}}{T} V - e_a \right] \quad \text{A} \tag{4.107}$$

If this assumption is made, then the analysis of Section 4.7.1 may be applied to the chopper-driven, separately excited motor.

*4.8.2 Controlled Rectifier Drives

When the available power source is ac, a phase-controlled rectifier may be employed to drive the dc machine. If the motor output power is 1 kW or less, a single-phase rectifier is satisfactory. The power circuit of such a rectifier drive is illustrated in Fig. 4.60. For greater power, a 3-phase source is necessary, and the power circuit of such a drive is illustrated in Fig. 4.61. Since the problems of single-phase drives are greater than those of polyphase drives, this discussion will initially concentrate on the single-phase case. As in the chopper drive, the field is separately excited by means of a separate, controlled or diode rectifier.

The average current in the motor armature circuit, and hence the average torque developed by the motor, is changed by varying the point in the cycle of the ac

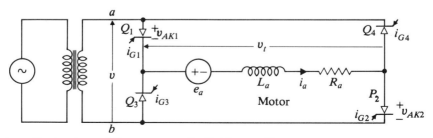

Fig. 4.60 Single-phase, full-wave controlled bridge rectifier.

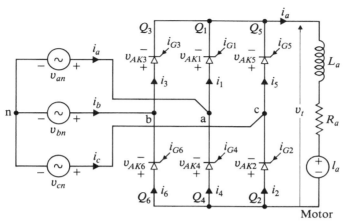

Fig. 4.61 Three-phase, full-wave controlled bridge rectifier.

source at which the gating currents are applied to the thyristors. The secondary terminal potential difference of the transformer in Fig. 4.60 may be described by

$$v = \sqrt{2}\ V \sin \omega t \quad \text{V} \tag{4.108}$$

During the positive half-cycle of v, positive armature current will tend to flow if thyristors Q_1 and Q_2 are turned on. During the negative half-cycle, positive armature current will tend to flow if thyristors Q_3 and Q_4 are turned on. Thus for each cycle of the ac source, two pulses of current will flow in the armature circuit.

While it is necessary to turn on the thyristors by gating currents in a controlled rectifier, it is not necessary to commutate them. The current will naturally fall to zero at some point in the cycle when, if the gating currents have been removed, the thyristors turn off; or, alternatively, thyristors Q_1 and Q_2 will be commutated when Q_3 and Q_4 are gated during the negative half-cycle of v, since this will make v_{AK1} and v_{AK2} negative. At any instant in the cycle when Q_1 and Q_2 are conducting, $v_t = v_{ab}$. At any instant when Q_3 and Q_4 are conducting, $v_t = v_{ba}$. If at any instant in the cycle $i_a = 0$, then $v_t = e_a$. The result is that if the motor is small or lightly loaded and i_a is therefore discontinuous, as illustrated in Fig. 4.62, the waveform of v_t will be that also shown in Fig. 4.62.

The *delay angle* α is the point in the cycle at which the reference thyristors Q_1 and Q_2 are turned on. When α is reduced and the motor is so fully loaded that the armature current is continuous, then the horizontal sector of the waveform of v_t in Fig. 4.62 is eliminated, and the waveform is defined by

$$\begin{aligned} v_t &= v_{ab} \quad \text{V} \qquad \alpha < \omega t < \alpha + \pi \quad \text{rad} \\ v_t &= v_{ba} \quad \text{V} \qquad \alpha + \pi < \omega t < \alpha + 2\pi \quad \text{rad} \end{aligned} \tag{4.109}$$

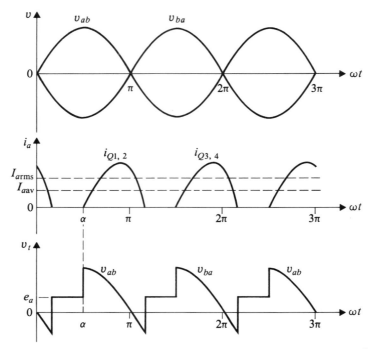

Fig. 4.62 Waveforms of i_a and v_t in a motor supplied by a single-phase controlled rectifier.

Under these circumstances, the average value of v_t is seen to be

$$v_{tav} = \frac{1}{\pi} \int_{\alpha}^{\alpha+\pi} \sqrt{2}\ V \sin \omega t\ d(\omega t) = \frac{2\sqrt{2}\ V}{\pi} \cos \alpha \quad \text{V}$$

The average value of i_a is then

$$i_{aav} = \frac{v_{tav} - e_a}{R_a} \quad \text{A} \tag{4.110}$$

Once again, the analysis of Section 4.7.1 may be applied for preliminary system design calculations. It must be emphasized, however, that for normal motors rated at less than 1 kW, the armature current is discontinuous even on full load.

The choice of motor for this type of service presents a problem, however, since the average torque developed is proportional to the average value of i_a, while the heating due to copper losses will depend upon the substantially higher rms value of i_a. Furthermore, the fluctuating armature current will produce additional core losses, while the commutator must have sufficient segments to accommodate the peak values of v_t shown in Fig. 4.62 without danger of flashover between segments.

All this means that a larger motor is required to deliver a given output power from a controlled-rectifier source than from a constant-potential source. Typically, a motor rated at 1 kW on a constant-potential source would be derated to about 700 W for use with a controlled rectifier. An alternative to such drastic derating is the inclusion of an inductor in the motor armature circuit to reduce the fluctuation in i_a.

The *form factor* of the motor current is the ratio of the rms to the average value. Figure 4.63 illustrates how form factor may vary at constant speed with variation of torque, and also how it may vary at constant torque with variation of speed for a particular system. The improved form factor obtained when the motor is operating at low speed and high torque is due to the fact that the current is large and therefore tends to be continuous. With totally enclosed nonventilated motors, the cooling is as good at low speed as it is at high speed. A consequence of the form-factor variation, therefore, is that the output torque permissible at low speed is greater than that permissible at high speed. This is shown in the speed–torque diagram for the system in Fig. 4.64.

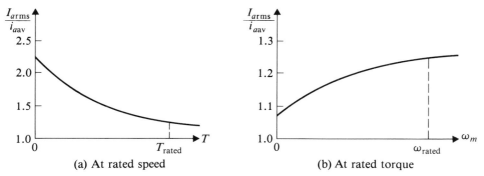

(a) At rated speed   (b) At rated torque

Fig. 4.63 Variation of form factor with torque and speed.

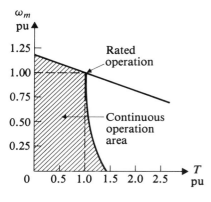

Fig. 4.64 Speed–torque diagram for motor supplied by controlled rectifier.

Discontinuous armature current is much more prevalent in single-phase rectifier drives than in chopper drives, since the frequency of the current pulses in a single-phase 60 Hz drive is 120 per second, while the chopper frequency can be made much higher, typically in the range 500–5000 Hz. In a three-phase rectifier drive, the pulse frequency is six per cycle or 360 per second. This approaches normal chopper frequencies. The waveforms for a large motor driven from a three-phase rectifier are illustrated in Fig. 4.65. In these waveforms, which apply to the circuit of Fig. 4.61,

$$v_{ab} = \sqrt{2}\, V \sin \omega t \quad \text{V}$$
$$v_{bc} = \sqrt{2}\, V \sin (\omega t + 2\pi/3) \quad \text{V} \qquad (4.111)$$
$$v_{ca} = \sqrt{2}\, V \sin (\omega t - 2\pi/3) \quad \text{V}$$

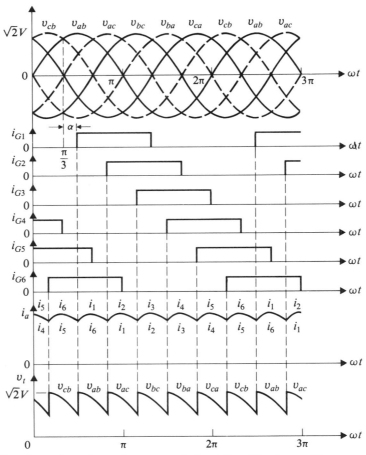

Fig. 4.65 Waveforms for a three-phase controlled rectifier drive ($\alpha = \pi/6$).

It will be noted that the thyristors are gated in sequence and that the delay angle α at which the reference thyristor Q_1 is turned on is measured from the datum point $\omega t = \pi/3$. From the waveform of v_t, it may therefore be seen that the average value of v_t is

$$v_{tav} = \frac{3}{\pi} \int_{\alpha+\pi/3}^{\alpha+2\pi/3} \sqrt{2} \, V \sin \omega t \, d(\omega t) = \frac{3\sqrt{2} \, V}{\pi} \cos \alpha \quad V \qquad (4.112)$$

The average value of i_a is again

$$i_{aav} = \frac{v_{tav} - e_a}{R_a} \quad A \qquad (4.113)$$

The analysis of Section 4.7.1 may be applied with much more assurance to a three-phase drive than to a single-phase drive, since i_a is inherently much less variable. Derating of the motor for such a drive need only be slight.

A controlled rectifier can transfer energy from the dc machine back to the ac system—that is, it can *invert*. Equations (4.109) and (4.112) for the full-wave and three-phase rectifier, respectively, show that for continuous-current operation, the average terminal potential difference of the motor becomes negative if $\alpha > \pi/2$. The rectifiers of Figs. 4.60 and 4.61 can provide only positive i_a. Thus the machine can provide electrical output for $\alpha > \pi/2$. If the field current is maintained in one direction only, the drive can operate in the first and fourth quadrants of Fig. 4.42.

The rectifier drives so far discussed do not provide directly for regenerative braking, for which the direction of rotation and v_t usually remain positive while i_a reverses. In some drives, this can be provided by reversing the field current, thus reversing the polarity of v_t. The reversal, however, takes place too slowly for adequate performance of many drives.

By employing a combination of two controlled rectifiers, as shown in Fig. 4.66, a drive capable of operation in all four quadrants can be achieved. Such an arrangement of rectifiers is called a *dual converter*.

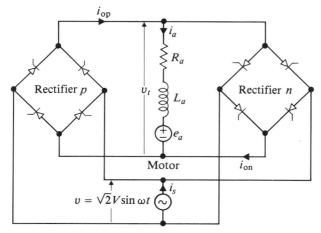

Fig. 4.66 Single-phase, dual converter drive.

Rectifier p provides only positive current i_{op} and, over the control range $0 < \alpha_p < \pi/2$, provides positive v_t, thus supplying power to the motor for first-quadrant operation. With $\alpha_p > \pi/2$, v_t is negative, while i_{op} is positive, providing operation in the fourth quadrant—that is, for reversed direction of rotation with regenerative braking. For operation in the second and third quadrants, rectifier p is shut off, and the supply is obtained from the reverse-connected rectifier n with i_{on} positive.

*4.9 RATING AND EFFICIENCY

The *rating* of a machine—that is, the power output that it is designed to provide, either as motor or generator—depends upon the amount of heat due to losses it can dissipate without any part of the machine reaching such a temperature that insulation deteriorates rapidly. The most common machine rating is the *continuous rating,* which specifies the output the machine can deliver for an indefinite period of time without overheating.

Short-time ratings are also employed. Such a rating specifies the period of time for which the rated output can be delivered, after which the machine runs on no load or is switched off for an equal period before load may again be applied. The advantage of this type of rating is that higher current and flux densities can be employed than is the case with continuously rated machines. A smaller machine may therefore be designed for a given output power. A short-time-rated machine never achieves a steady operating temperature, since the temperature is still rising when the time limit is reached and the load must be removed or the machine switched off. A typical example is the shop-crane hoist motor, which is usually ½-hour rated. This motor never runs continuously for so long, but experience has shown that a ½-hour rating is adequate, and so a small motor may be fitted into the confined space available on a crane trolley.

A common practical problem is that of specifying the machine rating required for a particular application. If the load is constant, specification is a simple matter. If the load varies considerably, then a typical work cycle for the machine must be determined. Usually an approximation to the work cycle of the form illustrated in Fig. 4.67 is acceptable. The vertical axis of the diagram is calibrated in units of instantaneous power output, p, while the horizontal axis is calibrated in time. One feature of the work cycle that may have a decisive influence on the rating of the machine chosen is the peak value of instantaneous power output.

Once a diagram of the work cycle such as is shown in Fig. 4.67 has been determined, the root-mean-square power for the work cycle may be obtained from the relationship

$$P_{rms} = \left[\frac{\Sigma(p^2 \times \text{Time})}{\text{Cycle time}} \right]^{1/2} \quad \text{W} \qquad (4.114)$$

where in the term ($p^2 \times$ Time) the time involved is that for which power p must be delivered. P_{rms} is then the continuous rating of the machine required to operate on

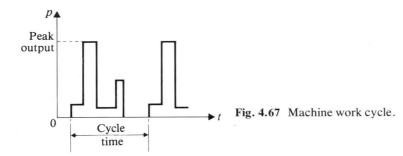

Fig. 4.67 Machine work cycle.

the given work cycle, assuming that the machine is capable of delivering the peak output demanded during the work cycle. Equation (4.114) is based on the assumption that losses in the motor are proportional to the square of armature current and that power is proportional to armature current.

Generally the peak output that a machine is capable of delivering is about twice its rated output, this limit being set by the ability of the machine to commutate the high currents occuring. If the peak output exceeds twice P_{rms}, then the machine rating must be about half the anticipated peak output required.

If the machine is stopped during part of the work cycle, the relationship given in Eq. (4.114) must be modified to allow for the poorer cooling that may take place when the machine is at standstill. The modified relationship is

$$P_{req} = \left[\frac{\Sigma(p^2 \times \text{Time})}{\text{Running time} + \dfrac{\text{Stopped time}}{q}} \right]^{1/2} \quad \text{W} \qquad (4.115)$$

where q is a factor that allows for the less effective cooling at standstill. For a totally enclosed machine, in which cooling is not significantly poorer at standstill, $q = 1$, and Eq. (4.115) becomes the same as Eq. (4.114). For an open type of machine, in which $q \simeq 4$, the effect of the factor is to increase the rating of the machine required to handle the work cycle.

Machines that are repeatedly started and stopped, or that are required to run at low speed for long periods, may be fitted with separate motor-driven fans that drive cooling air through them.

As has been seen for the transformer, per-unit efficiency may be expressed as

$$\eta = \frac{\text{Output}}{\text{Input}} = \frac{\text{Input} - \text{Losses}}{\text{Input}} = \frac{\text{Output}}{\text{Output} + \text{Losses}} \qquad (4.116)$$

Rotating machines are relatively efficient except at low loads. For integral horsepower or kilowatt ratings, at full load $0.75 < \eta < 0.95$.

The principal losses can be separated into (a) resistance losses and (b) rotational losses. Resistance losses occur in the field and armature windings and are readily determined when currents and winding resistances are known.

Rotational losses, made up of friction, windage, and core losses, are frequently assumed to be equal to the no-load input to the armature at the operating speed. Strictly, armature copper loss should be subtracted from this input, but it is not significant. In the case of a generator, the rotational losses are supplied by the prime mover through the shaft. However, it is convenient to determine the rotational losses for a generator by operating it as a motor. The largest part of the friction loss is due to brush friction on the commutator. Core losses occur for the most part in the rotor magnetic circuit, and are due to the flux reversals that take place as the rotor turns within the resultant stationary field produced by the stator and rotor mmf's. Core losses are therefore a function of both the pole flux and the speed of rotation. Rotor cores are always laminated to reduce eddy current loss; at least the faces of the stator pole pieces are also laminated. The stator yoke is rarely laminated except in machines which have frequent and rapid changes of field excitation.

Example 4.9 An open-type dc motor is required to operate a materials hoist. At each end of travel an electromagnetic brake sets, and the motor is switched off. When the conveyance is descending, it drives the motor as a generator, thus providing regenerative braking and returning energy to the supply system. The operating cycle is continuously repeated and may be approximately represented by the following table:

Motion	Shaft Horsepower	Time (s)
Ascending	50	30
Unloading	0	90
Descending	−20	25
Loading	0	180

The rate of heat dissipation for motors of this type at standstill is one quarter that at rated speed. Maximum permissible power output of the motor is twice the rated value. Determine the continuous rating of the smallest motor suitable for this service. Some standard motor sizes are 10, 15, 20, 25, 30, 40, 50 hp.

Solution In Eq. (4.115)

$$q = 4$$
$$\Sigma(p^2 \times \text{time}) = 50^2 \times 30 + (-20)^2 \times 25 = 85,000$$
$$P_{req} = \left[\frac{85,000}{55 + (270/4)} \right]^{1/2} = 26.4 \simeq 30 \text{ hp}$$

Since this is more than half the peak power requirement, a 30-hp machine is acceptable.

Example 4.10 A 4.5-kW, 125-volt, 1150 r/min shunt generator has an armature-circuit resistance of 0.37 Ω. When it is operated as a motor on no load at rated speed and terminal potential difference, the armature current is 3.2 A. The no-load saturation curve is shown below in Fig. 4.68.

Determine the full-load efficiency of the generator at rated speed with the field rheostat adjusted to give rated terminal potential difference.

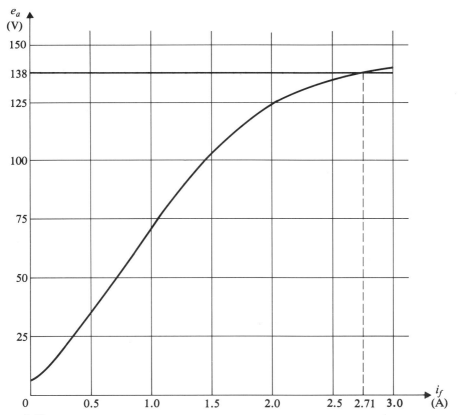

Figure 4.68

Solution When the machine is run as a motor,

$$e_a = v_t - R_a i_a = 125 - 0.37 \times 3.2 = 124 \quad \text{V}$$

Rotational losses are therefore

$$P_{\text{loss}} = e_a i_a = 124 \times 3.2 = 397 \quad \text{W}$$

Note that this differs by less than 1% from the armature input power, $v_t i_a$. When the machine is operated as a generator, the full-load current is

$$i_L = \frac{4500}{125} = 36 \quad \text{A}$$

It is now necessary to determine the field current required on full load. At that condition,

$$v_t = 125 = e_a - 0.37(36 + i_f)$$

from which

$$e_a = 138 + 0.37 \, i_f$$

This line may be drawn on the saturation curve of Fig. 4.68 to yield an intersection at (2.71, 139). Thus

$$i_a = i_f + i_L = 36.0 + 2.7 = 38.7 \quad \text{A}$$

$$\text{Copper losses} = v_t i_f + R_a i_a^2 = 125 \times 2.71 + 0.37 \times 38.7^2 = 893 \quad \text{W}$$

$$\eta = \frac{\text{Output}}{\text{Output + Losses}} = \frac{4.5 \times 10^3}{4.5 \times 10^3 + 397 + 893} = 0.777$$

Thus the percentage efficiency is approximately 78%.

*4.10 ARMATURE REACTION

In Section 4.2.1 it was stated that the effect of armature mmf on the induced emf in a machine is usually negligible. While this is true for values of armature current below rated value, the armature-reaction effects may be substantial during periods of high armature current. It is therefore desirable to examine these effects in more detail. In order to understand the effect of armature mmf, it is convenient to start with an examination of the field distribution that it produces in a magnetically linear machine.

A developed diagram illustrating two pole pitches of a stator and rotor of a machine is shown in Fig. 4.69(a). The direction of conductor current shown there applies for an operation as a generator. The distribution of stator and rotor mmf is shown in Fig. 4.69(b), where $\mathscr{F}_{rg}$ is rotor or armature mmf. The resultant mmf acting across the air gap is shown in Fig. 4.69(c).

The distribution of flux density around the rotor periphery may be deduced from the resultant mmf distribution. The reluctance of the magnetic path through the field

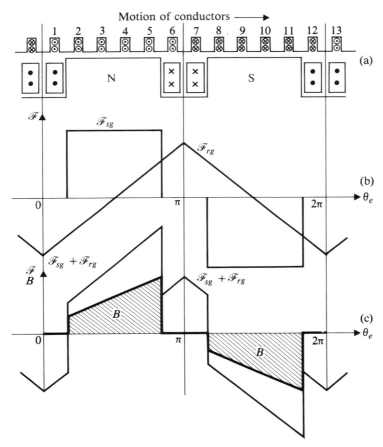

Fig. 4.69 Magnetomotive-force and flux-density distribution.

poles and across the air gap is quite low, and any mmf will produce an appreciable flux density between the stator pole faces and the rotor surface. The reluctance of the magnetic path through the interpolar space is extremely high; moreover, the mmf acting in this path is relatively low. For the present, therefore, it will be assumed that negligible flux density is produced in the interpolar space. If in addition the effect of magnetic fringing is neglected, then the flux-density distribution around the rotor periphery will be that indicated by the shaded areas marked "B" in Fig. 4.69(c). Since it is at present being assumed that the permeability of the machine steel is infinite, and that therefore magnetic saturation does not enter into the model of the machine under discussion, flux density will be directly proportional to the resultant mmf in the areas beneath the poles and will be zero in the interpolar space.

It is important to note that the total flux per field pole represented by one of the shaded areas in Fig. 4.69(c) is not changed by the armature mmf. Only the distribution of flux density under the field pole is changed. Since this illustrates generator operation, it may be seen that the armature mmf in a generator increases the flux density under the trailing pole tip and decreases it equally under the leading pole tip. In a motor, where the armature current is reversed, armature mmf increases the flux density under the leading pole tip and decreases that under the trailing pole tip.

When saturation in the magnetic system is taken into account, the armature mmf tends to reduce the flux per field pole. Figure 4.70(a) shows the mmf distribution for two poles of a loaded dc generator. The resultant mmf wave of $\mathscr{F}_{sg} + \mathscr{F}_{rg}$ is identical with that shown in Fig. 4.69(c). In Fig. 4.70(b) are shown the flux–density distribution on no load, marked B_{NL}, and that marked B_{prop}, which would occur when armature mmf was established, if the flux density were proportional to the mmf. Also shown is a curve marked B_{actual}, which represents the flux–density distribution under the poles that actually arises when armature mmf $\mathscr{F}_{rg}$ is established. The change in flux density at the leading and trailing pole tips is, in each case, due to a change in mmf of magnitude $\Delta\mathscr{F}$, which is shown in Fig. 4.70(a). The magnetic system of the machine must normally be operated in the saturated range of flux density. This is done to reduce the size of the machine by obtaining a

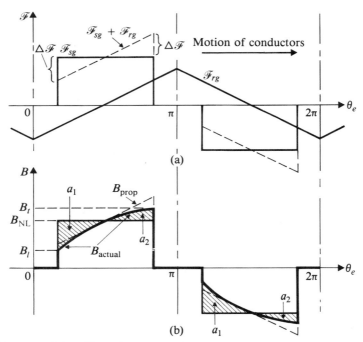

Fig. 4.70 Demagnetizing effect of armature reaction I.

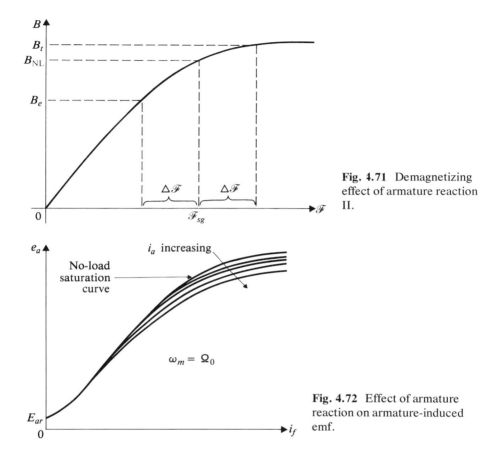

Fig. 4.71 Demagnetizing effect of armature reaction II.

Fig. 4.72 Effect of armature reaction on armature-induced emf.

given flux per pole from the smallest practicable area of pole face. The no-load flux density, B_{NL}, therefore lies on the curved part of the magnetization characteristic, as shown in Fig. 4.71. Equal changes of mmf of magnitude $\Delta \mathscr{F}$ on either side of that required to produce B_{NL} will produce unequal changes of flux density, due to the curvature of the magnetization characteristic. The resulting flux densities, B_l under the leading pole tip and B_t under the trailing pole tip, are shown in Figs. 4.71 and 4.70(b). In the latter figure, shaded area a_1 is greater than area a_2, and consequently the area under the curve B_{actual} is less than that under the curve B_{NL}, showing a reduction in total flux per pole. Thus, with varying load, ϕ is a function of i_a, and $e_a = k\Omega_0\phi$ is not constant if i_a is varied.

The demagnetizing effect of armature reaction is difficult to describe analytically If allowance is to be made for this effect, it must usually be done on the basis of results of tests on the machine in question or on a similar machine. One way of showing the results of such tests is to draw a family of curves of e_a versus i_f for various values of i_a, as illustrated in Fig. 4.72.

Example 4.11 A 7.5-kW, 230-V, 1250 r/min dc motor has an armature-circuit resistance of 0.65 Ω. The field current is adjusted until, on no load with a supply of 230 V, the motor runs at 1300 r/min and draws an armature current of 1.5 A. A load torque is then applied to the motor shaft, which causes the armature current to rise to 38 A and the speed to fall to 1200 r/min. Determine the percentage reduction in flux per pole due to armature reaction.

Solution The steady-state equivalent circuit of the motor is shown in Fig. 4.73, from which

$$v_t = e_a + R_a i_a = 230 \quad V$$

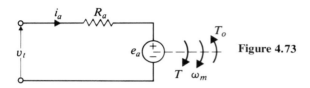

Figure 4.73

On no load,

$$e_a = 230 - 0.65 \times 1.50 = 229 \quad V$$

$$\omega_m = \frac{2\pi}{60} \times 1300 = \frac{130\pi}{3} \quad \text{rad/s}$$

$$k\phi_{\text{NL}} = \frac{e_a}{\omega_m} = \frac{229 \times 3}{130\pi}$$

On load,

$$e_a = 230 - 0.65 \times 38 = 205 \quad V$$

$$\omega_m = \frac{2\pi}{60} \times 1250 = \frac{125\pi}{3} \quad \text{rad/s}$$

$$k\phi_{\text{L}} = \frac{205 \times 3}{125\pi}$$

Thus

$$\frac{\phi_{\text{NL}}}{\phi_{\text{L}}} = \frac{229}{1300} \times \frac{1200}{205} = 1.031$$

The percent reduction is thus:

$$\frac{\phi_{\text{NL}} - \phi_{\text{L}}}{\phi_{\text{NL}}} \times 100\% = \frac{1.031 - 1.00}{1.031} \times 100\% = 3\%$$

*4.11 COMMUTATION

Discussions of the distribution of flux density in the air gap have, so far, been based on the assumption that all the flux passes normally across the gap between the stator pole faces and the rotor surface. The flux density in the interpolar space has been assumed to be zero.

While it is true that by far the greater part of the magnetic flux follows the lowest-reluctance path across the air gap, there is also a certain amount of fringing at the leading and trailing edges of the pole face and leakage between poles, as indicated in Fig. 4.74. This means that the flux-density distribution on no load cannot be truly represented by a rectangular wave, such as that representing the stator mmf $\mathcal{F}_{sg}$ in Fig. 4.75(a), but corresponds to a waveshape such as that shown in Fig. 4.75(b).

When the machine is on load and armature current is flowing, the rotor mmf illustrated for a loaded generator by the wave $\mathcal{F}_{rg}$ in Fig. 4.75(a) further modifies the flux-density distribution, and produces a small but significant flux density in the

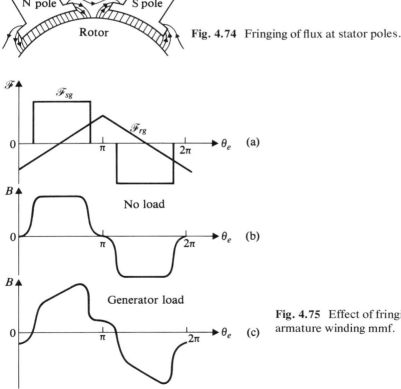

Fig. 4.74 Fringing of flux at stator poles.

Fig. 4.75 Effect of fringing and armature winding mmf.

interpolar space. Thus the flux-density distribution for a loaded generator is represented by a waveshape such as that shown in Fig. 4.75(c).

The flux density in the interpolar space, combined with the effect produced by the short-circuiting of the armature coils by the brushes as the sides of the coils pass through the neutral plane of the field structure, results in operating difficulties that may be discussed under the general heading of *commutation.*

Whereas the power that can be delivered continuously by a dc machine is limited by the heating in the machine due to the various losses, the substantially greater power that can be delivered for a short time is limited by the magnitude of the current that can pass between commutator and brushes without producing serious sparking due to excessive current density in the contact area. Sparking rapidly destroys the surface of the commutator and puts the machine in the repair shop.

Unacceptably high current densities over parts of the brush contact surface may be produced when a coil is commutated, unless appropriate precautions are taken.

*4.11.1 Interpoles or Commutating Poles

When a coil is commutated, the current flowing in it is reversed and this current reversal takes place while the coil is moving through the neutral plane and is short-circuited by the brush. Ideally, the current reversal should take place at a uniform rate, as indicated in Fig. 4.76(a). Two factors prevent this *linear commutation:*

a) The flux in the interpolar space, which induces an emf in the coil tending to maintain the current in its original direction.

b) The inductance of the coil, which also tends to maintain the current existing at the beginning of commutation.

The joint result of these two factors is that, as a commutator segment leaves a brush, the current density in the contact area between that segment and the trailing edge of the brush becomes extremely high and sparking is produced. Finally, as the

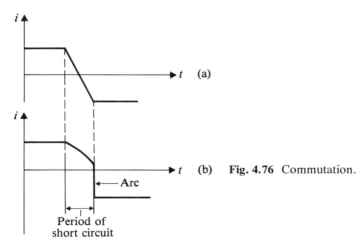

Fig. 4.76 Commutation.

segment leaves the brush, the remaining current is interrupted and an arc is formed momentarily. Such imperfect commutation is illustrated in Fig. 4.76(b).

In order to ensure that current is reversed by the end of the short-circuit period, an emf that opposes the current existing at the beginning of that period is intention-

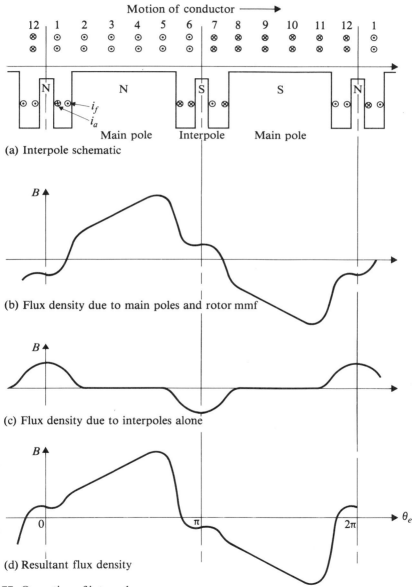

Fig. 4.77 Operation of interpoles.

ally induced in the short-circuited coil. This is done by fitting small narrow poles in the spaces between the main poles, as illustrated in Fig. 4.77(a).

Figure 4.77 illustrates the case of a loaded generator. The interpoles, which have the same polarity as the following main poles, reverse the flux density in the inter- polar space, and thus induce emf's that tend to reverse the current in the coils undergoing commutation. Since the armature mmf in the interpolar space is propor- tional to armature current, then the mmf of the interpole, which is to oppose the armature mmf, must also be made proportional to armature current. The interpole windings are therefore made of a small number of turns of conductor of large cross- section area and are connected in series with the armature of the machine. In order, moreover, that the flux density due to the interpoles may vary linearly with mmf, the interpole flux density is kept low, and the interpole air gap is made large.

If the machine illustrated in Fig. 4.77 were operating as a motor, the directions of currents in both the armature winding and the interpole windings would reverse without any change of connection, so that the interpole would continue to ensure satisfactory commutation for motoring as well as generating.

It is not essential to have as many interpoles as there are main poles; and, to reduce manufacturing costs, interpoles are often fitted in alternate interpolar spaces only. In such an arrangement, all interpoles have the same polarity, and the flux that they induce follows a magnetic path that includes the main poles of opposite polarity to the interpoles.

*4.11.2 Compensating Windings

The potential difference that may exist between adjacent commutator segments is limited. About 50 V/cm of commutator periphery is a common design criterion. Any greater potential gradient is liable to cause flashover between the segments due to carbon dust from the brushes and the existence of ionized air in their neighbor- hood. This factor therefore sets a lower limit to the number of segments that must be included in the commutator of a machine of given terminal potential difference.

In machines subjected to heavy overloads or rapid load changes, particularly when operating with a weak main field, flashover may occur for other reasons.

Figure 4.78 illustrates the kind of flux density distribution that may occur in a

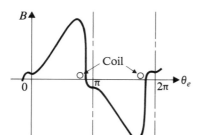

Fig. 4.78 Effect of heavy overload.

machine in which the armature currents are very high due to a momentary overload. The rate of change of flux linkage for a coil moving through the position indicated in the diagram will be very high, and consequently a large emf will be induced in the coil. If the armature is lap-wound, the ends of this coil will be connected to adjacent segments. If the armature is wave-wound, a number of coils in series, all with abnormally high emf's induced in them, will be connected to adjacent segments. The potential difference thus appearing between the segments may be sufficient to cause flashover. Once started, this may rapidly spread around the whole commutator from the positive to the negative brush.

Figure 4.79 illustrates the flux density distribution in the air gap that would be produced by armature mmf acting alone. If saturation is neglected, this flux may be considered as the component produced by the armature mmf of the total flux in the air gap. If a coil is situated symmetrically about the neutral plane of the stator as shown in Fig. 4.79, an emf will be induced in the coil by a change of flux due to a change in armature current and hence in armature mmf. If the change of armature current is rapid, and if the emf so induced aids the emf already induced in the coil by the rotation of the machine, an abnormally high resultant emf may be induced in the coil and a high potential difference will appear between adjacent commutator segments.

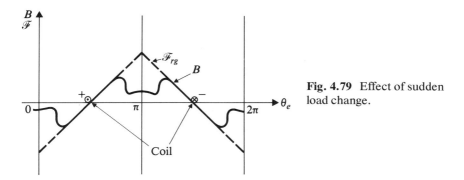

Fig. 4.79 Effect of sudden load change.

If the machine is acting as a generator, the direction of emf induced in the coil (shown in Fig. 4.79) by its rotation will be indicated by the $\pm$ at the coil conductors. Thus if the generator were suddenly open-circuited and consequently the flux linking the coil due to the rotor mmf decayed, an additional emf would be induced in the coil and would reinforce that due to rotation. Thus sudden removal of load from a generator may cause flashover between commutator segments. Conversely, sudden application of load to a motor, particularly a momentary heavy overload, may also cause flashover.

The flashover tendency, due either to distortion of the flux density distribution or to sudden load changes, may be largely eliminated by means of a compensating winding that neutralizes much of the rotor mmf. Such a winding also eliminates the demagnetizing effect of armature mmf. The compensating winding is fitted in slots

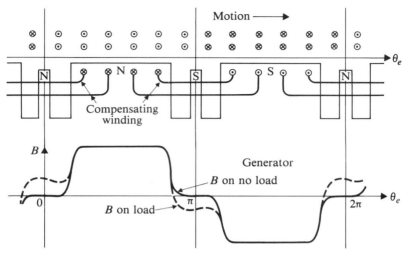

Fig. 4.80 Compensating winding.

Fig. 4.81 Large dc machine stator fitted with compensating windings. (*The Harland Engineering Company of Canada Ltd.*)

provided in the main pole faces, and its arrangement and the required current directions relative to those in the armature winding are shown in Fig. 4.80. Since its mmf should be proportional to the rotor mmf, it is excited with armature current. Figure 4.81 shows the stator of a large dc machine fitted with compensating windings.

Poleface windings are expensive. They are, therefore, fitted only to large machines or to machines subject to severe duty cycles. A typical example is the motor driving a steel rolling mill, which is both large and subjected to abrupt changes of load and speed.

A possible connection of the various windings on a machine is illustrated in Fig. 4.82.

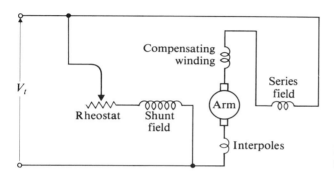

Fig. 4.82 Machine winding connections.

*4.12 PERMANENT-MAGNET MOTORS

Permanent-magnet (PM) motors are built with power ratings ranging from a few watts to 100 kW or more. For the purpose of analysis, they may be compared to the separately excited motor described in Section 4.7.1 operating with constant field current. The limitations to that analysis imposed by the demagnetizing effect of armature reaction do not affect a correctly applied PM motor. The only nonlinearities worthy of note are those due to brush-contact resistance and loss torque. The first of these has been seen to be negligible, while the latter may frequently be assumed to be a linear function of speed and included in the analysis. The PM motor may therefore usually be considered as a linear device over its entire operating range of torque and speed.

For the PM motor,

$$k\phi = \text{constant} \tag{4.117}$$

From Eqs. (4.60–4.65), the behavior of the machine may be described by

$$e_a = k\phi\omega_m \quad \text{V} \tag{4.118}$$

$$T = k\phi i_a \quad \text{N·m} \tag{4.119}$$

$$v_t = e_a + R_a i_a \quad \text{V} \tag{4.120}$$

$$T = T_{\text{loss}} + T_L \quad \text{N} \cdot \text{m} \tag{4.121}$$

$$\omega_m = \frac{v_t}{k\phi} - \frac{R_a T}{(k\phi)^2} \quad \text{rad/s} \tag{4.122}$$

Motor efficiencies are high, largely due to the elimination of the power loss in the field windings.

The speed–torque curve of a typical small permanent-magnet motor is shown in Fig. 4.83. Such a motor may be operated over the entire speed and torque ranges, but may only be subjected to the stalled torque for a very brief period; otherwise it will overheat. Any PM motor designed for this usage—for example, by switching on directly across the source without additonal armature-circuit resistance—may also be brought to a rapid standstill by dynamic braking obtained by simply shorting the armature terminals together. Neither this operation nor across-the-line starting may be carried out repeatedly, however, without overheating the machine. The brushes and commutator must be designed to carry six or seven times full-load current without excessive sparking.

Speed control of a PM motor can be obtained only by varying the armature's terminal potential difference v_t. This may be achieved by employing a variable dc source or by including an external resistor in the armature circuit. The disadvantages of this latter method have been explained in the discussions of wound-field motors in Sections 4.71 and 4.72. Figure 4.42 may also be used to illustrate the speed control of a PM motor.

Demagnetization, when it does occur, is due to the effect of armature reaction and may be illustrated by Fig. 4.69. When armature current flows, the flux density increases at one pole tip and decreases at the other. When the armature current again returns to zero, the flux density distribution should return to the rectangular no-load waveshape. This will occur only if the minimum value of flux density under the pole tip has not been less than the minimum value of flux density on the recoil line of the magnets.

For an Alnico material, the recoil line would be similar to that shown in Fig. 1.63, in which the minimum pole-tip flux density would not be less than that corre-

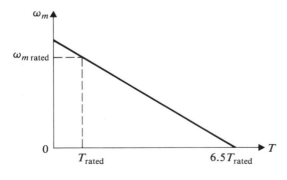

Fig. 4.83 Speed–torque curve of a PM motor.

sponding to magnetic field intensity H_s at the limit of the recoil line. If indeed the armature current were so great that this limit was exceeded, then a new recoil line lower and parallel to the original would be established for the tip of the pole piece. The no-load speed of the motor would then be increased, due to the permanent reduction in flux per pole. To restore the original operating characteristic, the magnets would have to be remagnetized, which might entail dismantling the machine.

For ferrite pole pieces, the recoil line is effectively coincident with the demagnetization curve, two of which are illustrated in Fig. 1.65. For these materials, the pole-tip flux density must not be reduced to a value below the limit of the straight-line part of the demagnetization curve. However, in the case of Ferrite D, even negative values of pole-tip flux density could be tolerated without permanent demagnetization.

*4.12.1 Construction of PM Motors

The construction of small PM motors up to ratings of a few kW differs markedly from that of wound-field shunt motors. In such small machines, ferrite magnets, which are grain-oriented, are normally employed and magnetized during manufacture—before they are fitted into a stator. Such a magnetic system has been discussed in Example 1.12.

For a given rating, it is usually necessary to make the armature of a PM motor somewhat larger than that of a shunt motor, because the air-gap flux density achieved with ferrite magnets is substantially less than that achieved with wound poles. However, owing to the saving of space due to elimination of field coils, the PM motor is nevertheless a good deal smaller than the equivalent wound machine. Typically, a 30% reduction in machine weight may be achieved by using permanent magnets instead of wound field poles.

In large PM motors, it is not possible to assemble the large magnets, ferrite or alloy, into the stator after they have been magnetized. It is therefore necessary to equip the pole pieces with magnetizing windings after the stator has been assembled with unmagnetized pole pieces. These windings need not be of large cross-section area, since they are not to be continuously excited. Nevertheless, their presence results in a machine structure somewhat similar to that of a shunt motor. Figure 4.84 shows the stator of a 4-pole, 30-hp PM motor with magnetizing coils fitted on the main field poles. To improve commutation (which is not a problem in small machines) interpoles are fitted, as also are compensating windings. These are wound in the normal way with coils that carry the armature current of the motor.

Note that the poles of the machine in Fig. 4.84 are fitted with pole shoes. These are laminated and made of high-permeability steel. Since the recoil permeability of the magnets is very low, and may indeed approach that of air, these pole shoes provide a low-reluctance shunt path for the component of flux produced by armature mmf, as illustrated in Fig. 4.85. In this way, the distortion of flux distribution in the permanent magnet due to armature reaction is much reduced, and large armature currents can be tolerated without demagnetization.

Fig. 4.84 Stator for 30-hp PM motor. *(Canadian Westinghouse Company Limited)*

When Alnico magnets are employed, their low coercive force requires that they be long in the direction of magnetization. The machine therefore approaches the proportions of an equivalent shunt motor, but is still appreciably smaller. Ferrite magnets, which have high coercive force, may be much shorter in the direction of magnetization; but owing to their lower residual flux density, they must be greater in cross-section area than the equivalent Alnico magnet. This increase in cross-section area is usually achieved by building a longer machine. The ratio of length to diameter for PM motors may vary over the range $4 > (l/2r) > 2$. For wound-field motors, the corresponding range is $1.5 > (l/2r) > 0.5$.

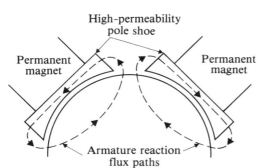

Fig. 4.85 Paths for flux produced by armature mmf.

A motor configuration radically different from that of the conventional machine is the *disk-armature* or *printed-circuit* motor, which has been widely used with permanent-magnet materials. Figure 4.86 illustrates the construction of such a machine. The rotor is formed of a disk of nonconducting, nonmagnetic material. On both sides of the disk, printed in copper, is the entire armature winding and commutator. One side of such a rotor is illustrated in Fig. 4.86(a); the other side of the

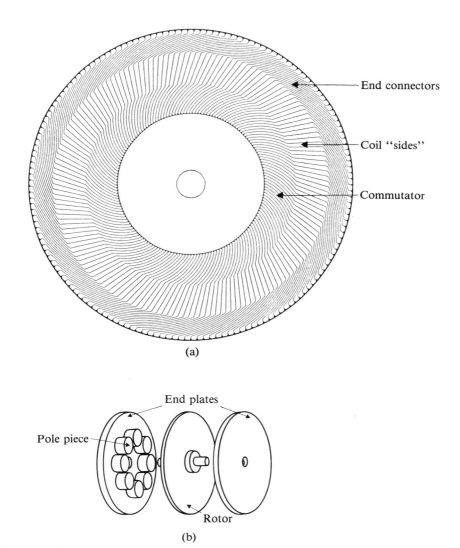

(a)

End connectors

Coil "sides"

Commutator

End plates

Pole piece

Rotor

(b)

Fig. 4.86 Printed-circuit motor.

rotor is identical with that shown. Connections from side to side are made by plated-through holes located at the ends of each of the separate conductors. The conductors form three distinct bands. The central band, in which the direction of the conductors is most nearly radial, form the "sides" of the single-turn armature coils. The inner and outer bands form the end connectors of the coils. The innermost band has the additional function of forming the commutator, against which axially oriented brushes are held by springs. The stator consists essentially of two ferromagnetic end plates each carrying the required number of permanent-magnet pole pieces. A great advantage of this configuration when the motor is to be employed in control systems is the low rotor inertia and consequent fast response of the motor.

*4.12.2 Motors with Rare-Earth/Cobalt Magnets

The rare-earth permanent-magnet materials discussed briefly in Section 1.24 possess both high residual flux and high coercivity, resulting in very high energy product. This is illustrated by the demagnetization curve for samarium–cobalt in Fig. 1.67. Thus this material appears ideally suited for the construction of PM motors. Samarium–cobalt magnets are still relatively expensive, but with increased application, the cost per unit of stored energy in the magnetic field for these magnets is expected to become comparable to that for ferrites. It therefore appears probable that this technique will permit the production of PM motors with improved performance and reduced bulk at a reasonable cost. In the meantime, motors up to about 1 kW rating are available with rare-earth/cobalt magnets for systems in which weight saving is of great importance.

Since the magnetic material is so costly, designers have sought a motor configuration that would employ the minimum bulk of magnet material for a given power output. Two configurations satisfy this requirement. One is the printed-circuit motor described in Section 4.12.1. The second configuration so far developed for samarium–cobalt PM motors is the "inside-out," or stationary-armature machine, in which the armature winding and commutator are placed on the stator and the field poles are on the rotor. In this case, the essentially cylindrical structure of the conventional machine is retained. Section 3.31 established that the function of the commutator of a dc machine is that of producing alternating current in the armature of the frequency required to satisfy the relationship

$$f = \frac{p}{2} n_r \quad \text{Hz} \qquad (4.123)$$

where n_r is the speed of rotation in revolutions per second. This relationship was satisfied at all speeds by mounting the commutator on the rotor shaft, so that its speed relative to the fixed brushes was always n_r r/s.

When the commutator is mounted on the stator, the essential relative speed may be obtained by mounting the brushes on the rotor shaft. The only minor complication is that the brushes must be connected to the stationary terminals of the dc source, and this must be done through additional brushes bearing on sliprings.

There are thus four brush contacts in series in the armature circuit of the inside-out motor, as compared with two in the conventional or printed-circuit configurations. This factor could introduce a significant nonlinearity into a low-potential system such as might be powered by a battery.

Two possible arrangements of the field poles on the rotor of the inside-out machine are illustrated in cross section in Fig. 4.87. Figure 4.87(a) shows the samarium-cobalt pole pieces cemented to the central rotor core. The space between the pole pieces may be filled with nonmagnetic material and the whole pole assembly enclosed in a nonmagnetic sleeve for protection and mechanical strength. This arrangement is suitable for machines with long narrow rotors and a small number of poles. In an alternative arrangement, illustrated in Fig. 4.87(b), the magnets are in the form of rectangular blocks assembled between ferromagnetic pole pieces. The flux through the magnets runs circumferentially between the pole pieces. In order that the flux may not be shunted through a path other than the air gaps and the machine stator, the core of the rotor and any end plates securing the magnets and pole pieces must be nonmagnetic.

The fixed-armature construction offers other advantages. Because the field poles require no coils, no resistive losses occur in the rotor, and no heat-dissipation problem arises. The armature, on the other hand, being mounted on the stator, can dissipate its resistive and core losses more easily than in the normal dc machine configuration. Furthermore, the stationary winding has no tendency to leave the slots due to centrifugal force, and therefore does not need the mechanical anchoring essential in a high-speed, rotating armature.

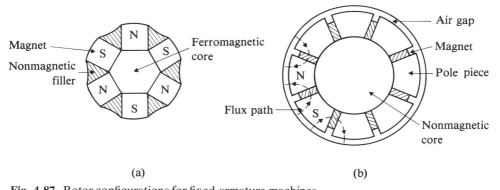

(a)   (b)

Fig. 4.87 Rotor configurations for fixed-armature machines.

PROBLEMS

4.1 A dc machine has four poles of 0.025 m² cross-section area and an air gap of effective length 6 mm. Each field coil has 1500 turns, and the coils are connected in series.

Assuming that the ferromagnetic materials have infinite permeability, determine the air-gap flux per pole per unit of field current.   *(Section 4.1)*

4.2 In the 6-pole machine of Fig. 4.3, each air gap has a reluctance of 0.4 MA/Wb. The leakage path between two adjacent pole tips has a reluctance of 2.5 MA/Wb.

a) Assuming the iron to have infinite permeability, draw a magnetic equivalent circuit for the machine.
b) Assume that each coil has 2000 turns and the coils are connected in series. Determine the inductance of the field circuit.
c) Determine the air-gap flux per pole per unit of field current.
d) Repeat parts (b) and (c) with the field coils connected in parallel.   *(Section 4.1.1)*

4.3 Figure 4.88 shows the dimensions of the teeth, slots, and air gap of a slotted-rotor machine. The stator and rotor each may be regarded as surfaces of constant magnetic potential.

a) Using the curvilinear squares technique, sketch the magnetic field around a tooth and slot.
b) Use the sketch in (a) to determine the approximate value of the effective air-gap length.
c) Compare this value with that obtained using Eq. (4.6).   *(Section 4.1.1)*

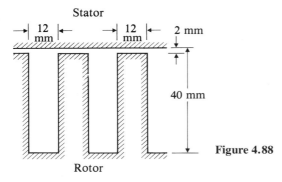

Figure 4.88

4.4 In Fig. 4.1(a), suppose each pole shoe has an arc length of 0.15 m and an axial length of 0.12 m. The air-gap length is 5 mm.

a) Assuming that the rotor is unslotted, determine the reluctance of one air gap.
b) Suppose the rotor diameter is 0.2 m and that it has 32 slots. Assuming the tooth width equals the slot width, determine the effective air-gap length and the reluctance of each air gap.
c) If each of the two series-connected field coils has 2500 turns and carries a current of 1.2 A, what is the flux in each pole?
d) If the magnetic field intensity in the iron can be considered zero up to a flux density of 1.4 T, at what value of flux per pole would the relationship between flux and field current be expected to depart from linearity?   *(Section 4.1.1)*

4.5 The armature of a motor has a radius of 0.12 m and an effective (stacked) length of 0.25 m. Its winding consists of 37 coils, each having six turns accommodated in 37 slots. The field structure has four poles which cover 70% of the armature periphery. The average flux density under each pole shoe is 0.8 T. The armature is lap wound.

a) Determine the constant k of Eq. (4.11) for this machine.
b) Determine the induced emf when the armature is rotated at a speed of 20 r/s.

c) Determine the torque when the armature current is 800 A.

d) Repeat parts (a), (b), and (c) for a wave-wound motor.   *(Section 4.2.1)*

4.6 A 1500-kW, 10-pole, dc generator has an open-circuit armature terminal potential difference of 250 V when driven at 300 r/min. There are 440 conductors on the armature, which is lap-wound. Determine the flux per pole under these conditions of operation.

(Section 4.2.1)

4.7 A 20-kW, 4-pole, dc generator has 558 conductors on the armature, which is wave-wound. If the flux per pole is 0.030 Wb, what is the speed at which the machine must be driven to give 440 V at the open-circuited armature terminals?   *(Section 4.2.1)*

4.8 Figure 4.89 gives the dimensions in millimeters of a dc machine. The rotor has 116 slots, each containing two conductors. The cross-section area of each conductor is 51.5 mm². A lap-winding is used. Each field coil has 1280 turns, and all field coils are connected in series.

a) Determine the reluctance of each air gap. The empirical expression of Eq. (4.6) may be used to find the effective air-gap length.

b) Assume that all reluctances except those of the air gaps are negligible, and determine the flux per pole per unit of field current.

c) Determine the armature induced emf per unit of the product of field current and speed on the basis of the assumption in (b).

d) Suppose the magnetization curve for the magnetic material in all parts of the machine can be represented for B > 1.9 T by the relation

$$B = 1.9 + 2\mu_0 H \quad \text{T}$$

Determine the value of flux per pole for which a flux density of 1.9 T is just reached at the base of the tooth. The objective in this part of the analysis is to predict the curve relating the flux per pole to the field current. It is evident from the dimensions given in Fig. 4.89 that the teeth will saturate well before other parts of the magnetic system. Assume that all the flux is confined to the teeth.

e) The magnetomotive force required for the tooth can be evaluated by integrating the magnetic field intensity H with respect to distance over the length of each tooth. With

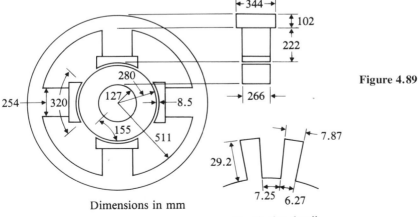

Dimensions in mm

Figure 4.89

Tooth-slot detail

the simplified B–H relation given in (d), only that part of the tooth having $B > 1.9$ T need be considered. Evaluate the tooth magnetomotive force for about three values of tooth-base flux density between 1.9 and 2.5 T. Determine the flux per pole and the field current for each of these conditions, and plot an approximate graph of ϕ versus i_f.

f) Suppose the machine is operated at a flux for which the field current is about 40% greater than that required for the air gap alone. Assuming that the armature conductors can be operated continuously at a current density of 4 A/mm², estimate the rated torque for the machine.

g) Assuming the speed of the machine is 1200 r/min, estimate the rated terminal potential difference and the rated power for the machine. *(Section 4.3)*

4.9 Figure 4.90 shows a cross section of a simple electrical brake. The rotor winding consists of two closed loops, each having a resistance of 0.005 Ω. Each of the stator poles covers a 90° arc of the rotor. The rotor length is 50 mm. The magnetic material in the stator and rotor may be considered ideal; fringing flux may be neglected; and the magnetic field produced by the closed loops may be ignored.

a) Derive expressions for the emf and for the current in each of the shorted loops during the interval that the loop sides are under the poles.

b) Show that the total power dissipated in the two loops is essentially constant throughout each revolution, provided the speed ω_m and the field current i_f are constant.

c) Derive an expression for the electromagnetic torque acting on the rotor.

d) Determine the field current required to provide a braking power of 200 W when the speed is 400 rad/s. *(Section 4.3)*

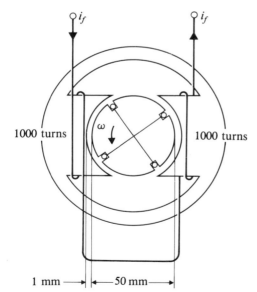

1000 turns   1000 turns

1 mm ——→ |←—— 50 mm ——→|

Figure 4.90

4.10 A dc machine has no nameplate to indicate its rating. Examination of the machine yields the following information: It has four poles, each with a pole-face area of 250 ×

200 mm; 150 bars on the commutator; armature coils wave-wound with two turns per coil. It is estimated that a reasonable value of flux density at the pole face would be 1 T. The cross-section area of each armature conductor appears to be adequate to carry about 50 A. Assuming that the machine is to be operated at 500 r/min, estimate

a) The rated terminal potential difference.
b) The rated torque.
c) The rated power. *(Section 4.3)*

4.11 A 4-pole, wave-wound dc machine is rated for 230 V and 5 kW at 1200 r/min. Its normal field current is 0.5 A.

a) If the speed could be increased to 1500 r/min, what would be the rated terminal potential difference and power of the machine?
b) If the armature coils were reconnected to form a lap winding, what would be the rated potential difference and power at 1200 r/min?
c) Suppose that a machine is built, identical in all respects to the one described, except that the axial length of the poles and the armature is doubled. Determine the rated potential difference, the rated power, and the required field current for this revised machine. *(Section 4.3)*

4.12 A 500-kW, 550-V, 6-pole, dc generator has 99 slots on the rotor with two conductors per slot. The armature is lap wound. When driven at 1000 r/min with the field current adjusted to give 0.150 Wb per pole, the machine delivers an output current of 850 A. Assuming the losses in the generator are negligible, determine the power developed by its prime mover under these conditions. *(Section 4.3)*

4.13 A dc machine has the following physical properties: 6 poles, 2000 turns per field coil, 10 Ω resistance per field coil, all field coils connected in series, armature radius 50 mm and axial length 50 mm, pole-face area 2000 mm², effective air-gap length 2 mm, armature wave-wound with 300 turns, each turn having a resistance of 8 mΩ. The rotor has a mass of 15 kg, and its effective radius of gyration is estimated to be 30 mm. The armature inductance, the friction and windage, and the saturation in the magnetic material may all be ignored. It is estimated that the leakage flux between adjacent poles is about 15% of the air-gap flux.

a) Determine the resistance and inductance of the field circuit, the constant relating armature emf, field current and speed, and the polar moment of inertia of the rotor.
b) Write three differential equations that relate the instantaneous values of the six terminal variables of this machine: terminal potential difference v_t, field potential difference v_f, armature current i_a, field current i_f, shaft torque T, and speed ω_m. Include the numerical values of all parameters in these equations. *(Section 4.5)*

4.14 A dc machine is driven at a constant speed of 1170 r/min. The open-circuit armature potential difference v_t is measured for a set of values of the field current i_f, giving the following results:

i_f (A)	0.1	0.2	0.3	0.4	0.5	0.6	0.7	0.8
v_t (V)	55	110	161	218	265	287	303	312

What internal torque will this machine develop with a field current of 0.7 A and an armature current of 25 A? *(Section 4.6.1)*

4.15 The open-circuit potential difference of a dc generator driven at 200 rad/s with rated field current is 85 V. The resistance of the armature circuit is 4.3 Ω. If a resistive load of 50 Ω is connected to the armature terminals, what will be the power delivered to this load when the machine is driven at 260 rad/s with rated field current?   *(Section 4.6.1)*

4.16 A 4½-kW, 125-V, 1150 r/min, separately excited dc generator has an armature-circuit resistance of 0.37 Ω. When the machine is driven at rated speed, the no-load saturation curve obtained is that shown in Fig. 4.26. A field-current regulator is required that will vary the field current so that rated terminal potential difference may be maintained from no load up to rated current. The prime mover drives the machine at rated speed on no load, but its speed falls 5% on full load.

Determine the range of field current that the regulator must provide.   *(Section 4.6.1)*

*4.17 The following data describe the no-load saturation curve of a dc generator driven at 1500 r/min:

i_f (A)	0	0.5	1.0	2.0	3.0	4.0	5.0
e_a (V)	10	40	80	135	172	199	220

The resistance of the field circuit is 44 Ω and that of the armature circuit is 0.035 Ω.

a) Suppose the field is supplied from a 200-V source in series with a field rheostat having a range from 0 to 25 Ω. What will be the maximum and minimum values of no-load terminal potential difference that can be obtained?

b) Suppose the machine is separately excited to provide a no-load terminal potential difference of 200 V. Determine the terminal potential difference when a load current of 200 A is being supplied.

c) Suppose the generator is self-excited, using the field rheostat of (a) in series with the field circuit. What will be the maximum and minimum values of no-load terminal potential difference that can be obtained?

d) Suppose the generator is self-excited and the rheostat adjusted to give a no-load terminal potential difference of 200 V. Determine the terminal potential difference when the load current is 200 A.

e) With the field rheostat adjusted as in (d), estimate the maximum load current that can be obtained from the machine terminals. What value of load-circuit resistance would produce this maximum-current condition?

f) Plot a few points on a curve relating terminal potential difference and terminal current for the machine operating as in (d). Suppose the machine is used to charge a battery that has an emf of 175 V and an internal resistance of 0.1 Ω. Estimate the magnitude of the battery current.   *(Section 4.6.2)*

*4.18 The machine of Problem 4.16 is connected as a shunt generator and driven at rated speed. The field rheostat is adjusted to give a no-load terminal potential difference of 140 V.

a) Plot curves of terminal potential difference v_t and field current i_f on the same sheet for $0 < i_L < 60$ A.

b) Determine the load-circuit resistance, the terminal potential difference, and the field current when the generator is delivering rated current. (The demagnetizing effect of armature reaction may be neglected.) *(Section 4.6.2)*

*4.19 The machine of Problem 4.16 is connected as a shunt generator and driven at 900 r/min. The field rheostat is adjusted to give a no-load terminal potential difference of 110 V.

Determine the possible terminal potential differences, line currents, and field currents when the generator armature current is 40.5 A; and determine the load-circuit resistance for the higher of these values of potential difference. *(Section 4.6.2)*

*4.20 A 5-kW, 125-V, dc series machine is to be employed as a series generator. In a no-load test in which the machine is driven at rated speed and the field winding is separately excited, the following data are obtained:

i_f (A)	0	1.4	2.8	5.7	11.4	17.1	22.8	28.6	34.3	40.0	45.7	51.5
e_a (V)	4	8	15	29	57	86	110	126	135	140	143	146

The armature-circuit resistance $R_a = 0.23\ \Omega$, and the field-circuit resistance $R_s = 0.12\ \Omega$.

a) Determine the external characteristic of the generator at rated speed as far as the data permit.
b) Determine the critical value of the external load-circuit resistance.
c) Determine the range of current over which the generator will not operate stably with a resistive load.
d) Determine the current delivered to a load circuit of 2.5-Ω resistance. *(Section 4.6.3)*

4.21 An industrial drive consists of a dc separately excited motor with the armature supplied from a source of variable potential difference. As this potential difference is varied from 0 to 600 V, the drive speed is to vary from 0 to 1600 r/min, the field flux being maintained constant. (All losses may be ignored.)

a) Determine the rated armature current of the motor if the load torque is to be held constant at 420 N·m.
b) The load is driven in the speed range from 1600 to 4000 r/min by weakening the field flux while the armature potential difference is held constant at 600 V. Determine the torque available from the drive at maximum speed. *(Section 4.7.1)*

4.22 A trolley bus is driven by a dc separately excited motor connected through a gear box to the 0.9-m diameter drive wheels. The drive is to produce the speed–thrust relation shown in Fig. 4.91 on page 364. The motor armature is supplied by a source that has a potential difference that may be varied from 0 to 600 V. (Electrical and mechanical losses may be neglected.)

a) If the maximum speed of the motor is not to exceed 4500 r/min, what is the required gear ratio?
b) Determine the required value of the quantity $k\phi$ if the speed is to be 12.5 m/s at an armature potential difference of 600 V.

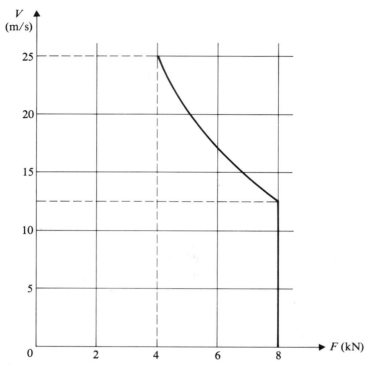

Fig. 4.91 Diagram for Problems 4.22 and 4.33.

c) Determine the required armature current to provide 8 kN thrust with the value of $k\phi$ obtained in (b).

d) The speed range of 12.5 to 25 m/s is provided by field weakening at a constant armature potential difference of 600 V. Determine the required value of $k\phi$ at a speed of 20 m/s.
(Section 4.7.1)

4.23 A radar antenna is supported on a turntable that is to be revolved at a variable speed in either direction of rotation. The following drive system is suggested. A diesel motor will drive a separately excited dc generator that, in turn, will supply power to the armature of a separately excited dc motor. The motor will be connected to the turntable through a 40:1 reduction gear. The effective load on the motor shaft is estimated to be 0.04 N·m per r/min of the motor shaft. The generator and motor are identical, each having an armature-circuit resistance of 0.47 Ω and negligible friction and windage losses. The no-load saturation curve at 1200 r/min of each of the machines is given by

i_f (A)	0	0.2	0.4	0.6	0.8	1.0	1.2
e_a (V)	0	108	183	230	254	267	276

The diesel motor has a no-load speed of 1000 r/min. Its speed drops by 2 r/min for each newton meter of its shaft torque.

  a) The radar system is to have a control by which the generator field current can be adjusted to any value up to 1.2 A in either direction. The motor field current will be set at 0.8 A. Provide a graph showing the relationship between antenna speed and generator field current.

  b) Assume that the dc machines are rated at 250 V, 5 kW, 1200 r/min, and discuss their ability to rotate the antenna continuously at maximum speed without overheating.   *(Section 4.7.1)*

4.24 A 10-kW, 230-V, 1800 r/min, dc shunt motor has an armature-circuit resistance of 0.25 Ω, and the rotational losses may be assumed constant at 600 W. The field rheostat is adjusted until the motor runs at 2000 r/min on no load.

By calculating a small number of points on the speed-torque characteristic, determine the speed of the motor when it is driving a load requiring 0.85 rated output torque.

(Section 4.7.2)

4.25 A dc shunt motor has an armature-circuit resistance of 0.46 Ω. It takes an armature current of 75 A from its 550 V supply when driving a load at 1350 r/min.

  a) Ignoring mechanical losses, calculate the torque applied to the load.

  b) Suppose the flux in the motor is reduced by 5%. Find the speed of the motor if the torque remains at the same value as in (a).   *(Section 4.7.2)*

4.26 An automatic starter is required for a 50-hp, 230-V, 1750 r/min, dc shunt motor. The armature-circuit resistance is 0.2 Ω. When delivering rated output, the armature takes a current of 205 A. The starter is to consist of resistors connected in series with the armature, with contactors to switch out the resistors in sequence. The armature current may not exceed 400 A, and a resistor can be switched out when the armature current drops to 250 A.

Determine the resistance of each of the resistors in the starter.   *(Section 4.7.2)*

*4.27 A 50-hp, 440-V, 4-pole, dc series motor has 31 slots on the rotor with 18 conductors per slot. The armature is wave-wound. When the motor current is 57 A, the flux per pole is 0.035 Wb. Permeability of the steel part of the magnetic system may be assumed to be infinite.

Determine the internal torque of the motor when the input current is 45 A.

(Section 4.7.3)

*4.28 A certain series motor is to be used to drive a load requiring a torque of 4 N·m at 25 r/s. A 200-V dc supply is available. A test on the machine at standstill shows that a shaft torque of 4 N·m can be produced by a terminal current of 5 A and a terminal potential difference of 10 V. The rotational losses of the machine are considered negligible.

What resistance is required in series with the 200-V supply to meet the required load condition?   *(Section 4.7.3)*

*4.29 A 4-pole, series-wound, fan motor rotates at 100 rad/s and takes 20 A from its 200-V supply when its field coils are connected in series. In order to increase the speed of the system, it is decided to reconnect the field coils into two parallel groups, each group having two coils in series. The fan load requires a mechanical power that varies as the cube of the speed. The motor may be considered magnetically linear, and all motor losses may be ignored.

Determine the new values of motor speed and motor current.   *(Section 4.7.3)*

*4.30 A load requiring a constant torque of 30 N·m is driven by a series motor from a 200-V source. The combined resistance of the armature and series field of the motor is 0.3 Ω. When tested at standstill, the motor produces a shaft torque of 10 N·m at a current of 25 A. Further tests show that the motor has negligible magnetic saturation up to a current of about 50 A. Its rotational losses are negligible.

Determine the speed at which the load is driven.   *(Section 4.7.3)*

*4.31 A dc series motor takes a current of 50 A from a 250-V supply to drive a mechanical load at 600 r/min. The resistance of the motor between its terminals is 0.35 Ω.

If the field flux at 15 A is half of its value at 50 A, what is the speed of the motor when it takes a current of 15 A from a 250 V supply?   *(Section 4.7.3)*

*4.32 A small universal series motor produces a standstill torque of 2 N·m when carrying a direct current of 3 A. The total resistance between its terminals is 2.5 Ω, and the total inductance is 0.04 H. Magnetic linearity may be assumed and rotational losses ignored. Suppose the machine is connected to a 115-V rms, 60-Hz ac source.

a) Determine the starting torque.
b) Determine the mechanical power produced when the motor current is 3 A rms.
c) Determine the input power factor for the condition of (b).   *(Section 4.7.4)*

*4.33 The speed–thrust relationship for a trolley-bus drive consisting of a separately excited motor with a chopper-supplied armature circuit is shown in Fig. 4.91 on page 364. The supply system source potential difference is 600 V. The motor base speed is 2250 r/min at an armature-circuit potential difference of 580 V. The armature-circuit resistance R_a is 0.125 Ω. The efficiency of the mechanical system from motor coupling to tire periphery may be assumed constant at 80%. By comparison, the mechanical losses of the motor may be considered negligible. Assuming the chopping frequency is 500 Hz and the armature current is maintained constant at above base speed, determine the required values of t_{ON} for speeds of

a) 7.5 m/s.
b) 20 m/s.   *(Section 4.8.1)*

*4.34 A 30-kW, 230-V dc motor has an armature circuit resistance $R_a = 0.055$ Ω. At the base speed of 1800 r/min, the full-load armature current is 147 A. The motor is to be driven as a separately excited machine with the armature supplied from a dual converter formed of two 3-phase, bridge-controlled rectifiers. The maximum average terminal potential difference available at the armature supply terminals is to be 250 V. A three-phase 550-V, 60-Hz source is available.

a) Determine the required turns ratio of a wye-delta–connected, 3-phase transformer to supply the converter.
b) Determine the required value of α_p at base speed and full load.
c) Determine the required values of α_p or α_n when
  i) The motor is drawing 100 A from the converter at 1200 r/min.
  ii) The motor is delivering 75 A to the converter at 1500 r/min.   *(Section 4.8.2)*

*4.35 A 120-V, 20-kW shunt generator requires a shaft torque of 12 N·m to drive it at rated speed of 1120 r/min when producing rated terminal potential difference on no load. The armature-circuit resistance is 0.12 Ω, and the field-circuit resistance is 75 Ω. Estimate the efficiency of the generator at rated load. *(Section 4.9)*

*4.36 A 200-V, 25-hp shunt motor requires an armature current of 8.2 A when operating at no load and rated speed. The armature-circuit resistance is 0.2 Ω, and the shunt field-circuit resistance is 80 Ω. Estimate the efficiency of the motor at rated load. *(Section 4.9)*

*4.37 Data for the no-load saturation curve at rated speed of a 10-kW, 230-V 1800 r/min, dc machine are as follows:

i_f (A)	0	0.25	0.50	0.75	1.00	1.25	1.50	1.75	2.00	2.25	2.50	2.75	3.00
e_a (V)	12	40	75	114	155	188	213	234	248	258	266	272	278

The armature-circuit resistance of the machine is 0.253 Ω.

The machine was driven at rated speed as a separately excited generator with an adjustable resistive load connected to its armature terminals. Armature potential difference was then recorded for a series of values of field current, the load resistance being adjusted in each case until the machine was delivering rated current. The readings obtained were as follows:

i_f (A)	0.5	1.0	1.5	2.0	2.5	3.0
v_t (A)	63	138	191	222	242	258

a) Draw the no-load saturation curve for this machine and, on the same axes, curves of terminal potential difference v_t and armature induced emf e_a for rated armature current as functions of i_f.

b) Determine the percent reduction in flux per pole caused by armature reaction at full-load current when the field current is that required to give rated terminal potential difference. *(Section 4.10)*

*4.38 A dc motor has a permanent-magnet field which maintains constant air-gap flux under all operating conditions. With no shaft load, the motor operates at a speed of 6000 r/min when its armature is connected to a 120-V dc supply. The armature-circuit resistance is 2.5 Ω. Rotational losses may be neglected.

Determine the speed of the motor when it is connected to a 60-V dc supply and delivering a mechanical torque of 0.5 N·m. *(Section 4.12)*

*4.39 A permanent-magnet tachometer consists of a conducting disk rotating in the axially directed magnetic field of an annular ferrite magnet, as shown in Fig. 4.92. The ferrite material is described by the curve for Ferrite D in Fig. 1.65. The emf induced in the disk is brought out to terminals by the use of one brush on the outer edge of the disk and one on the shaft.

Assuming the yoke and shaft to have very high permeability, and ignoring fringing flux, determine the terminal potential difference per rad/s of rotational speed. *(Section 4.12)*

*4.40 The printed-circuit motor of Fig. 4.86 has eight poles made of the samarium–cobalt material described in Fig. 1.67. Each pole is 20 mm in diameter.

a) Assuming that the armature disk has a thickness of 4 mm and that the clearance on each side of the disk is 1 mm, determine the required pole length to operate the magnetic material at its point of maximum energy product. Fringing and end-plate reluctance may be ignored.

b) Determine the flux per pole.

c) Assuming a wave winding, determine the armature emf at 5000 r/min.

d) Assuming that each printed conductor can carry a continuous current of 8 A, estimate the power rating of the motor at 5000 r/min.   *(Section 4.12)*

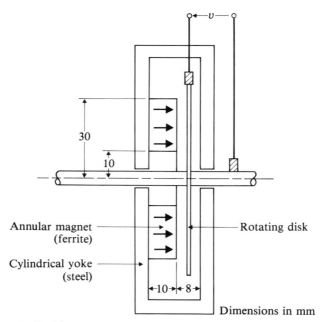

Fig. 4.92 Diagram for Problem 4.39.

5 / Induction Machines

In this chapter, the principal types of induction machine, introduced in Chapter 3, are analyzed in sufficient detail to develop equivalent-circuit models for prediction of their steady-state performance.

The section on windings, which immediately follows, applies not only to induction machines but also to the stator windings of synchronous machines that are analyzed in Chapter 6.

5.1 ALTERNATING-CURRENT MACHINE WINDINGS

Section 3.2.1 explained how a winding placed in a small number of uniformly distributed slots could produce an essentially sinusoidal distribution of flux density around the air gap of a machine. The concentric or spiral winding described there is by no means suitable for all types and sizes of machines; indeed, it is restricted, for reasons that will become apparent, to the stator windings of small motors and the rotor windings of very large, high-speed ac generators.

It was also explained in Section 3.2.1 that, since the arrangement of the winding makes all harmonic components of the mmf space wave small in amplitude in comparison with the fundamental component, the effect of the harmonic components can be neglected in any analysis of the behavior of the machine. This assumption will also be made in discussing the effect of the winding arrangements described in this chapter. However, in Chapter 3, the ferromagnetic system of the machine was assumed to be ideal, and magnetic saturation was ignored. To make efficient use of the magnetic material in the machine, it is generally necessary to operate it at reasonably high flux density, the highest values of which occur in the stator and rotor teeth. It is therefore necessary to make allowance in the winding design for the reluctance of at least that part of the ferromagnetic system.

As a result of saturation, the sinusoidal fundamental component of the mmf space wave does not, in fact, produce a corresponding sinusoidal fundamental component of the flux-density wave around the air gap. Because of saturation in the teeth, the air-gap flux-density wave will tend to be flat topped, so that harmonic components will exist in the flux-density wave that do not exist in the mmf wave. However, it can be shown that the harmonic components of the flux-density wave make no contribution to the flux linkage of a sinusoidally distributed winding. The winding can only be linked by a flux density that has the same spatial period of sinusoidal distribution as the winding. Therefore, in a sinusoidally wound machine, a balanced set of sinusoidal currents in its windings produces sinusoidally varying flux linkages, even though the magnetic system is nonlinear.

In the following discussion of ac machines, the effect of magnetic saturation will usually be included by the choice of a value of the magnetic reluctance of the machine appropriate for the peak air-gap flux density, a value that is nearly constant for machines operated at constant ac potential. In special cases, where variation in saturation has a significant effect on the machine performance, this factor will also be included.

With the exception of the concentric winding, the principles underlying the winding of ac machines are the same as those for dc machines, in that the pitch of each coil must be approximately the same as the pole pitch, and the coils must be connected together so that the emf's induced in them act in the same direction. Windings on ac machines are not as complicated as dc windings, in which good commutation requires a large number of coils. All dc windings, moreover, are closed windings, having a continuous path around the entire armature. In contrast, ac windings are usually open-circuit windings: there is a continuous path through the conductors of each phase, with both the beginning and ending of the phase winding free. In a single-phase machine, these free endings constitute the terminals of the machine, whereas in a 3-phase machine the free ends of the phases are connected in wye or delta.

Only a brief survey of the principal types of winding can be undertaken here.

*5.1.1 Half-Coil and Whole-Coil Windings

Figure 5.1 shows a developed diagram of a half-coil, single-phase, concentrated winding for a 6-pole machine. Positive directions of current in the coil sides are

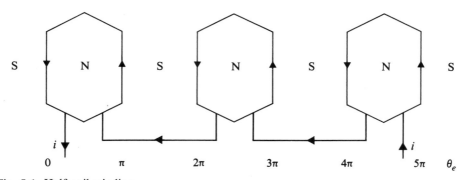

Fig. 5.1 Half-coil winding.

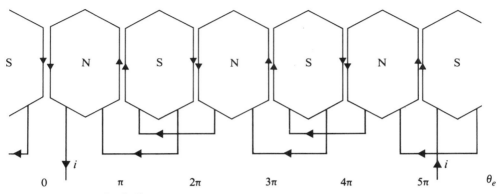

Fig. 5.2 Whole-coil winding.

shown, as also are the magnetic polarities resulting from these current directions. The winding is said to be *concentrated* because all the conductors on one side of each pole are grouped together in one slot. There are half as many coils as there are poles, so the winding is said to be *half-coil*.

Figure 5.2 shows a developed diagram of a whole-coil, single-phase, concentrated winding. In this winding there are as many coils as there are poles. If the coils are full pitch—that is, the distance between the two sides of one coil is exactly one pole pitch—this must be a double-layer winding, since each slot must contain two coil sides. The whole-coil winding has the advantage that the end connections are more uniformly distributed than are those of the half-coil winding.

Each of these concentrated windings would produce a rectangular space wave of mmf in the air gap. Since the ideal is a sinusoidal space wave, windings are usually distributed in a number of adjacent slots in order to give a better approximation to a sinusoidal distribution of mmf.

*5.1.2 Use of Identical Coils

The manufacture of coils and their installation in integral horsepower machines is greatly simplified if all the coils are identical in shape and number of turns. Such a coil is shown in Fig. 5.3. While this might appear to impose severe limitations, an approximately sinusoidal mmf distribution can, nevertheless, be obtained from identical-coil windings whose conductor distribution is far from sinusoidal. Part of a single-phase, whole-coil, distributed winding made up of identical coils is shown in Fig. 5.4.

Figure 5.5(a) shows a single concentrated coil fitted on the stator of a machine. With current in the coil in the direction indicated in the diagram, the rectangular mmf wave shown in Fig. 5.5(b) will be produced in the air gap. If the coil has N_s turns and is carrying current i_s, then the amplitude of the rectangular wave will be

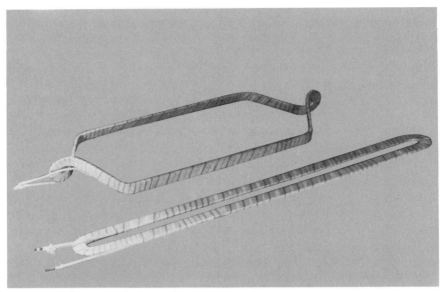

Fig. 5.3 Multiturn coil for an ac winding before and after pulling. *(The Harland Engineering Company of Canada Ltd.)*

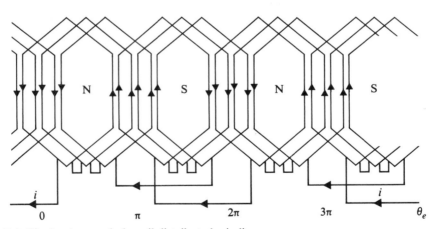

Fig. 5.4 Single-phase, whole-coil distributed winding.

$N_s i_s / 2$. If a p-pole concentrated winding were being considered and there were N_s turns per phase, then the peak amplitude of the rectangular wave would be

$$\hat{\mathscr{F}}_{sg} = \frac{N_s}{p} i_s \quad \text{A} \tag{5.1}$$

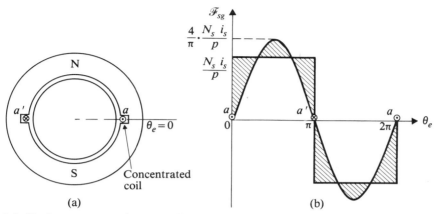

Fig. 5.5 Single concentrated stator coil.

where *peak* amplitude is the greatest amplitude of the space distribution of mmf when current i_s is flowing. This current may be anything between zero, when the peak amplitude would be zero, and $\hat{i}_s$, when the peak amplitude would also be equal to the *maximum* amplitude occurring during the current cycle.

The rectangular wave of Fig. 5.5(b) may be described by a Fourier series of space harmonics:

$$\mathscr{F}_{sg} = \frac{4}{\pi} \frac{N_s}{p} i_s \left(\sin \theta_e + \frac{1}{3} \sin 3\theta_e + \frac{1}{5} \sin 5\theta_e + \cdots \right) \quad \text{A} \tag{5.2}$$

Thus the peak amplitude of the fundamental component of the rectangular wave is

$$\hat{\mathscr{F}}_{sg1} = \frac{4}{\pi} \frac{N_s}{p} i_s \quad \text{A} \tag{5.3}$$

The fundamental component is shown in Fig. 5.5(b) for comparison with the rectangular wave. The shaded areas in the diagram indicate the contributions of the harmonic terms to the total mmf wave. If the coil were excited with alternating current, then the amplitude of the fundamental component would vary sinusoidally with time. If the rms magnitude of the coil current were I_s, then the maximum amplitude of the fundamental component would be

$$\mathscr{F}_{sg1 \text{ max}} = \frac{4}{\pi} \frac{N_s}{p} \sqrt{2} I_s \quad \text{A} \tag{5.4}$$

*5.1.3 Three-Phase Distributed Windings

In a 3-phase winding, the total available slot space is divided equally between the phases, and all slots are filled. Stator and rotor laminations for 3-phase distributed induction motor windings are shown in Fig. 5.6. Figure 5.7 shows sections of the

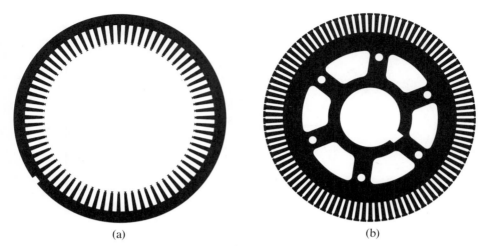

(a)   (b)

Fig. 5.6 Induction motor laminations. (a) Stator. (b) Rotor. *(Canadian Westinghouse Company Limited)*

Fig. 5.7 Stator sections for a 170,000 kVA, 13,300 V, 150 rpm waterwheel generator. *(Canadian Westinghouse Company Limited)*

stator winding of a large machine. Since the winding is made up of a number of similar concentrated coils, it is possible to determine the shape of the mmf wave produced by the winding by superposing all the rectangular waves produced by the individual concentrated coils.

A simple example of a 2-pole, 3-phase, double-layer, full-pitch winding is shown in Fig. 5.8(a), and a developed diagram of the winding is shown in Fig. 5.8(b). Coil sides that are in adjacent slots and are connected in the same phase constitute a *phase belt*. The end connections of the coils are carried around the periphery of the machine and, for the 2-pole machine illustrated in Fig. 5.8, must traverse half the circumference of the stator. In multipole machines, the end connections are shorter. For example, in a 4-pole machine the end connections would need to traverse only

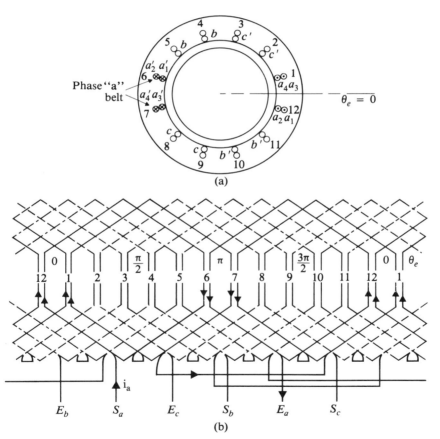

Fig. 5.8 Two-pole, three-phase, double-layer full-pitch winding (S = start of phase winding; E = end).

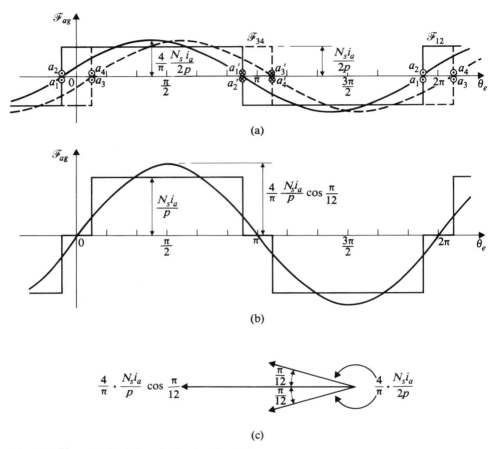

Fig. 5.9 Phase "a" of the winding in Fig. 5.8.

one quarter of the stator's circumference. Thus an increase in the number of poles introduces a saving in idle copper, reducing both the cost and the resistance of the winding.

Figure 5.9(a) shows the mmf waves resulting from current i_a flowing in the phase "a" coils in the directions indicated in Fig. 5.8(a). Each of the rectangular mmf waves shown—one in full line and one in broken line—is produced by two coils. Thus, if N_s is the number of stator turns per phase, the amplitude of each rectangular wave will be $N_s i_a/2p$, since each rectangular wave is produced by only half of the turns per pole per phase. The fundamental component of each of the rectangular waves is also shown; the amplitude of the fundamental component is $(4/\pi)(N_s i_a/2p)$.

The resultant mmf wave due to the four coils is shown in Fig. 5.9(b), as is the fundamental component of that wave. The fundamental component of the resultant wave is necessarily the resultant of the fundamental components of the two separate rectangular waves shown in Fig. 5.8(a). Its peak amplitude may therefore be determined by means of the vector diagram in Fig. 5.8(c), from which

$$\hat{\mathscr{F}}_{ag1} = 2 \cdot \frac{4}{\pi} \cdot \frac{N_s i_a}{2p} \cos \frac{\pi}{12} \quad \text{A} \tag{5.5}$$

Thus if an rms current I_a flows in phase "a," the maximum amplitude of the fundamental component is

$$\mathscr{F}_{ag1 \text{ max}} = \frac{4}{\pi} \frac{N_s \sqrt{2} I_a}{p} \cos \frac{\pi}{12} \quad \text{A} \tag{5.6}$$

This may be expressed as

$$\mathscr{F}_{ag1 \text{ max}} = K_{1W} \cdot \frac{4}{\pi} \frac{N_s}{p} \sqrt{2} I_a \quad \text{A} \tag{5.7}$$

where K_{1W}, the winding factor for the fundamental component is defined as

$$K_{1W} = \frac{\begin{array}{c}\text{Fundamental mmf amplitude for}\\ \text{distributed winding of } N_s \text{ turns}\end{array}}{\begin{array}{c}\text{Fundamental mmf amplitude for}\\ \text{concentrated winding of } N_s \text{ turns}\end{array}} \tag{5.8}$$

In this particular case $K_{1W} = \cos(\pi/12) = 0.966$. Necessarily, $K_{1W} \lesssim 1$ for any possible arrangement of coils.

To represent a rectangular mmf wave by its fundamental component would be a considerable approximation. To represent the wave in Fig. 5.9(b) by its fundamental component is a less drastic approximation, since certain of the harmonics of the two rectangular waves of Fig. 5.9(a) have partially cancelled one another. The approximation may be improved if windings with more coils distributed in more slots are employed. The result of employing more coils is to change K_{1W}, which may readily be determined for any coil arrangement.

*5.1.4 Short-Pitch or Chorded Windings

If all the coil sides in the tops of the slots in Fig. 5.8(a) are shifted one slot clockwise, then each coil spans only $\frac{5}{6}$ of one pole pitch, or 150 electrical degrees. Such coils are said to be *short-pitch,* or *chorded,* coils, and one consequence of employing them is that the phase belts overlap. Two useful purposes may be served by employing short-pitch coils. First, the harmonic content of the mmf wave may be reduced further than it would be with a full-pitch distributed winding. Second, the length of the end connections of the coils is reduced.

The resultant mmf wave may again be built up from the rectangular waves due to the concentrated coils, and the result of so doing is shown as a stepped wave in Fig. 5.10(a). In this figure the datum line for $\theta_e = 0$ has been moved half a slot pitch

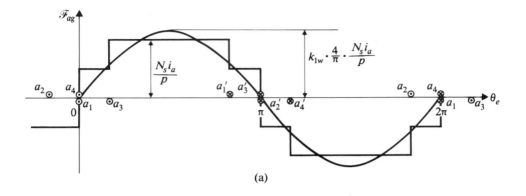

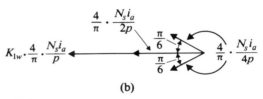

Fig. 5.10 Short-pitch winding.

in a clockwise direction so that the origin of the diagrams again coincides with the midpoint of the phase "a" belt. The fundamental component of the resultant wave is again the resultant of the fundamental components of the separate rectangular waves and is shown in Fig. 5.10(a). Note that the stepped wave in Fig. 5.10(a) provides a better approximation to a sinusoid than the stepped wave in Fig. 5.9(b).

The amplitude of the fundamental component of the mmf wave produced by the chorded winding may be determined by means of the vector diagram in Fig. 5.10(b), from which

$$\mathscr{F}_{ag1 \ peak} = \frac{4}{\pi} \left(\frac{1}{2} + \frac{1}{2} \cos \frac{\pi}{6}\right) \frac{N_s i_a}{p} \quad \text{A} \tag{5.9}$$

Thus if an rms current I_a flows in phase "a," the maximum amplitude of the fundamental component is

$$\mathscr{F}_{ag1 \ max} = K_{1W} \frac{4}{\pi} \frac{N_s}{p} \sqrt{2} I_a \quad \text{A} \tag{5.10}$$

where

$$K_{1W} = \left(\frac{1}{2} + \frac{1}{2} \cos \frac{\pi}{6}\right) = 0.933 \tag{5.11}$$

This winding factor is less than that obtained for the full-pitch winding employing the same number of coils. Thus a penalty associated with the improved mmf waveform is a reduction in the amplitude of its fundamental component for the same total phase mmf.

It is convenient to represent the winding factor as

$$K_{1W} = K_{1d} \cdot K_{1p} \tag{5.12}$$

where K_{1d} is a factor due to the distribution of the winding—that is, the number of slots per pole per phase; and K_{1p} is a factor due to the pitch of the coils. To some extent, K_{1d} and K_{1p} may be varied in relation to one another, but they are coupled by their dependence on slot pitch. With a small number of slots per phase belt the range of choice of K_{1d} and K_{1p} is limited. It is, however, possible to reduce the most troublesome low-order harmonics to very low amplitudes by choosing a suitable combination of K_{1d} and K_{1p}.

The resultant mmf wave produced by a three-phase winding is a better approximation to a sinusoid than the wave that is produced by one phase alone; and this approximation improves as the number of phases is increased. For example, consider the three-phase distributed short-pitch winding discussed in the foregoing. For the convenient instant illustrated in Fig. 5.11, phase "a" is carrying peak positive current, so that $i_a = \sqrt{2} I_s$, where I_s is the rms value of the stator phase current. At this instant,

$$i_b = i_c = -\frac{i_a}{2} = -\frac{I_s}{\sqrt{2}}.$$

The mmf waves caused by the three separate phase windings at this instant are shown in Fig. 5.12(a), (b), and (c). They are combined to give the resultant mmf wave in Fig. 5.12(d). The amplitude of the fundamental component of the resulting rotating wave is

$$\mathscr{F}_{sg1 \text{ max}} = \frac{3}{2} \cdot \mathscr{F}_{ag1 \text{ max}} \quad A \tag{5.13}$$

The improvement in approximation to a sinusoid may be seen by comparing this diagram with that in Fig. 5.10(a).

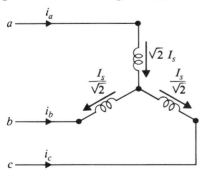

Fig. 5.11 Winding currents when $i_a = \hat{i}_a$.

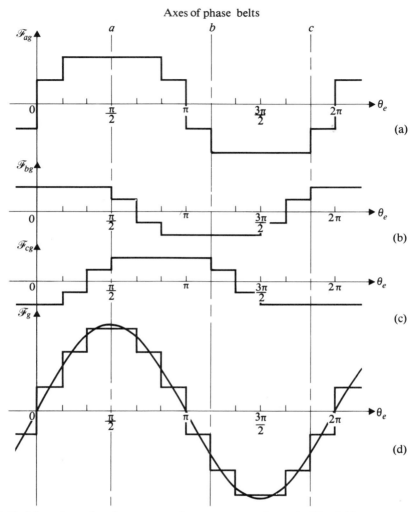

Fig. 5.12 Magnetomotive-force waves due to currents shown in Fig. 5.11.

*5.1.5 Effective Turns of a Winding

In the analysis of machine performance that follows, only the fundamental components of mmf will be considered, on the assumption that the design of the winding is good enough to permit any harmonic components to be neglected. The effective peak mmf of a phase winding is thus considered to be that of the fundamental component of the winding mmf. This will not be equal to the product of the number of turns in the winding and the winding current. Thus, the effective maximum value of the rotating mmf wave produced by a three-phase winding is

$$\mathscr{F}_{sg1 \text{ max}} = \frac{3}{2} \frac{N_{se}}{p} \sqrt{2} \, I_s \quad \text{A} \tag{5.14}$$

From Eq. (5.10), (5.13), and (5.14) it may be seen that

$$N_{se} = K_{1W} \frac{4}{\pi} N_s \tag{5.15}$$

where N_{se} is the effective turns per phase of the stator winding; in general this will not be an integer, as explained in Section 3.2.1.

In the case of a machine with three-phase windings on both stator and rotor, the effective turns ratio of the windings would then be N_{se}/N_{re}, where N_{re} is the effective turns per phase of the rotor winding. This effective turns ratio would not necessarily be equal to N_s/N_r, the ratio of the actual turns of the two windings.

Example 5.1 A 3-phase, 4-pole machine stator has 36 slots. The winding is double-layered and shortpitched, being made up of 6-turn coils whose pitch is 140° electrical. Determine the winding factor K_{1W} and the effective turns per phase N_{se}.

Solution Since there are 36 slots with two coil-sides per slot, there will be 36 coils in the winding, giving 12 coils per phase. Thus

$$N_s = 6 \times 12 = 72 \text{ turns}$$

The winding is 4-pole; there will, therefore, be three coils per pole per phase and three slots in a phase belt. The angle subtended by one slot pitch at the rotor axis is

$$\frac{2\pi}{36} \text{ radians} = \frac{4\pi}{36} \text{ electrical radians} = 20° \text{ electrical}$$

The coil pitch is 140° electrical, so that each coil spans seven slot pitches. The arrangement of the coil sides of phase "a" in the slots will therefore be as shown in Fig. 5.13(a).

The peak amplitude of the stepped wave produced by this winding is $N_s i_a/p$. Since 6 coils produce this amplitude, the amplitude of the rectangular wave produced by one coil is $N_s i_a/6p$. The amplitude of the fundamental component produced by one coil is thus $(4/\pi)(N_s i_a/6p)$ A. The vector diagram from which the resultant fundamental component may be obtained is shown in Fig. 5.13(b), where it is seen that the fundamental component of the stepped wave produced by the whole phase "a" winding is

$$\frac{4}{\pi} \frac{N_s i_a}{6p} (2 + 2 \cos 20° + 2 \cos 40°) \quad \text{A}$$

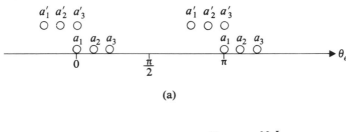

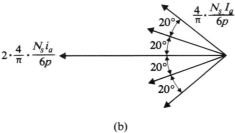

(b)

Fig. 5.13 Diagram for Example 5.1.

From Eq. (5.3), it is seen that the fundamental mmf amplitude for a concentrated winding of N_s turns would be $(4/\pi)(N_s i_a/p)$. Thus from Eq. (5.8),

$$K_{1W} = \frac{1}{6}(2 + 2\cos 20° + 2\cos 40°) = 0.902$$

From Eq. (5.15),

$$N_{se} = K_{1W}\frac{4}{\pi}N_s = 0.902 \times \frac{4}{\pi} \times 72 = 83.5 \text{ turns}$$

This paradoxical result, namely that $N_{se} > N_s$, is due to the fact that N_{se} is calculated from the amplitude of the fundamental component. As may be seen from Fig. 5.10(a), the amplitude of the fundamental component is greater than that of the stepped wave.

5.2 THREE-PHASE INDUCTION MOTORS

The three-phase induction motor in its most elaborate form consists of a cylindrical ferromagnetic structure with slotted stator and rotor. The stator carries three identical, symmetrically placed, phase windings so distributed in the slots that an effectively sinusoidal distribution of mmf is produced in the air gap. The rotor also carries three identical, symmetrically placed, phase windings, similar to those on the stator.

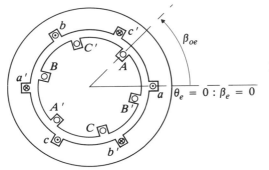

Fig. 5.14 Three-phase induction motor.

$\theta_e = 0 : \beta_e = 0$

A diagram representing a two-pole motor is shown in Fig. 5.14. In this diagram only a single pair of conductors for each phase is shown, and these represent the center of a belt of conductors that produces a sinusoidal distribution of mmf around the air gap due to the current in that phase. Now a model of the machine in the form of an equivalent circuit may be developed that will permit performance prediction by means of simple calculations.

5.2.1 Stationary Motor

If the stator windings of the motor in Fig. 5.14 are excited from a balanced three-phase source, then a rotating flux wave is produced in the air gap as described in Section 3.3.5. This rotating flux wave links the phase windings of the rotor in sequence and induces in them emf's that are sinusoidal functions of time. Thus, if the rotor is stationary, the motor simply acts as a three-phase transformer. It may be assumed that the primary (stator) and secondary (rotor) windings of this "transformer" are connected in wye, regardless of their true connection. Indeed, at least one of them is usually connected in delta; however, this makes no difference to the analysis of the operation of the machine, as will be seen when the method of determining the equivalent-circuit parameters is discussed.

As has been seen in Section 2.12.1, the operation of a three-phase transformer may be discussed in terms of a per-phase equivalent circuit obtained by detaching one phase from the three-phase diagram. Since the stationary motor acts as a three-phase transformer, it also can be represented by a per-phase equivalent circuit similar to that of a single-phase transformer.

Figure 5.15 shows an equivalent circuit that represents the relationship between phase "a" on the stator and phase "A" on the rotor of the machine in Fig. 5.14. A similar circuit would show the relationship between any two phases, but these two are taken as typical. As in the case of the transformer, core loss is initially neglected in the early stages of analysis and is not represented by a circuit element in Fig. 5.15. The same equivalent circuit applies to a p-pole motor, since the phase "a"

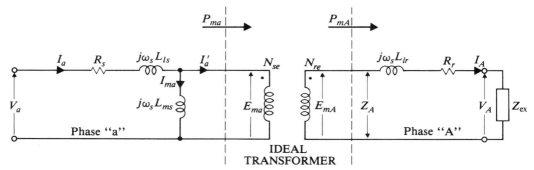

Fig. 5.15 Per-phase equivalent circuit of the stationary motor.

windings for all poles are simply connected in series. The physical diagrams in Fig. 5.14 and in later figures are drawn for two-pole machines simply for the sake of avoiding complication.

Assume the rotor windings are open-circuited and a balanced three-phase supply of angular frequency ω_s is applied to the stator terminals. Potential V_a is applied to phase "a." The resulting current I_a, in conjunction with the currents in the other two stator phases, produces a rotating flux wave in the air gap that induces emf E_{ma} in the winding of phase "a." A further emf is induced in the phase "a" winding due to the rate of change of its leakage flux linkage, and this emf is represented by the potential difference appearing across the leakage inductance L_{ls} in the equivalent circuit. In addition, a potential drop occurs in the winding due to the winding resistance R_s.

With the rotor windings open-circuited, there is no current in the ideal transformer, and I_a is simply the magnetizing current I_{ma}. The rotating flux wave that induces E_{ma} in phase "a" of the stator will induce emf E_{mA} in phase "A" of the rotor. Necessarily,

$$\frac{E_{mA}}{E_{ma}} = \frac{N_{re}}{N_{se}} \tag{5.16}$$

However, due to the angle β_{0e} between the phase "a" winding on the stator and the phase "A" winding on the rotor as shown in Fig. 5.14, the emf induced in the rotor phase winding will not reach its peak value at the same instant as that induced in the stator phase winding, but will be delayed by a time β_{0e}/ω_s so that the ratio of potential differences of the ideal transformer in the equivalent circuit of Fig. 5.15 is

$$\frac{\overline{E}_{mA}}{\overline{E}_{ma}} = \frac{N_{re}}{N_{se}} \underline{\big\lfloor -\beta_{0e}} \tag{5.17}$$

If a balanced external circuit of Z_{ex} per phase is connected to the three-phase rotor terminals and the stator is excited from a balanced three-phase source, then balanced currents of angular frequency $\omega_r = \omega_s$ will flow in the rotor windings. The rotor currents may be described by the equations

$$i_A = \hat{i}_r \sin(\omega_r t + \alpha_r) \quad \text{A}$$

$$i_B = \hat{i}_r \sin\left(\omega_r t + \alpha_r - \frac{2\pi}{3}\right) \quad \text{A} \qquad (5.18)$$

$$i_C = \hat{i}_r \sin\left(\omega_r t + \alpha_r - \frac{4\pi}{3}\right) \quad \text{A}$$

These currents will, by analogy with Eqs. (3.146) and (3.147), produce a rotating mmf wave

$$\mathscr{F}_{rg} = \frac{3N_{re}\hat{i}_r}{2p} \cos(\omega_r t + \alpha_r + \beta_{0e} - \theta_e) \quad \text{A} \qquad (5.19)$$

Since $\omega_r = \omega_s$, Eq. (5.19) may be written as

$$\mathscr{F}_{rg} = \hat{\mathscr{F}}_{rg} \cos(\omega_s t + \alpha_r + \beta_{0e} - \theta_e) \quad \text{A} \qquad (5.20)$$

where

$$\hat{\mathscr{F}}_{rg} = \frac{3N_{re}\hat{i}_r}{2p} \quad \text{A} \qquad (5.21)$$

In a single-phase transformer, the mmf produced by the secondary current I_2 may be considered to be equal and opposite to that produced by a "load component" I_2' of the primary current, such that

$$N_1\bar{I}_2' = N_2\bar{I}_2 \quad \text{A} \qquad (5.22)$$

In the stationary motor acting as a three-phase transformer, the mmf produced by the rotor phase currents I_A, I_B, I_C may be considered equal and opposite to that produced by load components I_a', I_b', I_c', of the stator phase currents, where

$$\bar{I}_a = \bar{I}_a' + \bar{I}_{ma} \quad \text{A}$$

$$\bar{I}_b = \bar{I}_b' + \bar{I}_{mb} \quad \text{A} \qquad (5.23)$$

$$\bar{I}_c = \bar{I}_c' + \bar{I}_{mc} \quad \text{A}$$

and

$$i_a' = \hat{i}_s' \sin(\omega_s t + \alpha_s') \quad \text{A}$$

$$i_b' = \hat{i}_s' \sin\left(\omega_s + \alpha_s' - \frac{2\pi}{3}\right) \quad \text{A} \qquad (5.24)$$

$$i_c' = \hat{i}_s' \sin\left(\omega_s t + \alpha_s' - \frac{4\pi}{3}\right) \quad \text{A}$$

The mmf wave produced by these load components of stator currents will be

$$\mathcal{F}'_{sg} = \hat{\mathcal{F}}'_{sg} \cos(\omega_s t + \alpha'_s - \theta_e) \quad \text{A} \tag{5.25}$$

where

$$\hat{\mathcal{F}}'_{sg} = \frac{3N_{se}\hat{i}'_s}{2p} \quad \text{A} \tag{5.26}$$

If the mmf wave described by Eq. (5.25) is to be equal to that described by Eq. (5.20), then it is necessary that

$$\frac{3N_{se}\hat{i}'_s}{2p} = \frac{3N_{re}\hat{i}_r}{2p} \quad \text{A} \tag{5.27}$$

and

$$\alpha'_s = \alpha_r + \beta_{0e} \quad \text{rad} \tag{5.28}$$

This condition is illustrated in Fig. 5.16, where the directions of winding currents at a particular instant of time are shown.

The induced emf's in the stator conductors of Fig. 5.16 are in the same directions as the induced emf's in the adjacent rotor conductors. This is to be expected, since both stator and rotor emf's are induced by the same rotating flux wave. However, as may be seen from the positive directions of emf's and current in the equivalent circuit of Fig. 5.15, if the induced emf's are in the same directions, the currents are necessarily in opposite directions, as shown in Fig. 5.16.

For convenience in later analysis, the phasors of current i_A and i'_a in Eqs. (5.18) and (5.24) are defined, with reference to phasor $\overline{V}_a = V_a \underline{|0}$, as

$$\overline{I}_A = \frac{\hat{i}_r}{\sqrt{2}} \underline{|\alpha_r} \quad \text{A} \tag{5.29}$$

and

$$\overline{I}_a = \frac{\hat{i}'_s}{\sqrt{2}} \underline{|\alpha'_s} \quad \text{A} \tag{5.30}$$

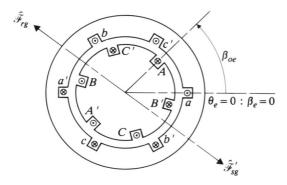

Fig. 5.16 Opposition of rotor mmf to that due to load component of stator current.

Thus

$$\frac{\overline{I}_A}{\overline{I}_a} = \frac{\hat{i}_r}{\hat{i}_s} \underline{\left| \alpha_r - \alpha_s' \right.}$$ (5.31)

From Eqs. (5.27), (5.28), and (5.31),

$$\frac{\overline{I}_A}{\overline{I}_a'} = \frac{N_{se}}{N_{re}} \underline{\left| -\beta_{0'e} \right.}$$ (5.32)

Equation 5.32 gives the current ratio of the ideal transformer of Fig. 5.15.

The impedance ratio of the ideal transformer is given by the ratio of potential differences divided by the current ratio. Thus, from Eqs. (5.17) and (5.32),

$$\frac{\overline{Z}_A}{\overline{Z}_A'} = \left[\frac{N_{re}}{N_{se}} \right]^2$$ (5.33)

where $\overline{Z}_A'$ is total rotor-circuit impedance, $\overline{Z}_A$, referred to the stator of the machine.

Finally, the power entering one phase of the ideal transformer primary is equal to the power leaving one phase of the secondary. That is,

$$\frac{P_{mA}}{P_{ma}} = 1$$ (5.34)

From the equivalent circuit of Fig. 5.15 and the ratios of Eqs. (5.17) and (5.32), it is evident that a stationary three-phase machine can be used as a phase shifter. It may also be used as a variable three-phase source when connected as an induction regulator, as shown in Fig. 5.17(a). In this application, the primary windings, usually on the rotor, are connected to the supply through flexible leads; the stator windings are connected in series with the supply. As the rotor is shifted through 180°, the phasor of the output potential difference V_2 follows a circular locus, as shown in Fig. 5.17(b). With identical stator and rotor windings, V_2 may be adjusted approximately over the range $0 < V_2 < 2V_1$. The advantage of induction regulators over variable autotransformers (Section 2.13.1) is that continuous stepless variation of the output potential difference can be obtained and no sliding electrical contacts are required. The disadvantages of induction regulators arise from their higher leakage inductance and higher magnetizing current.

Example 5.2 A three-phase induction motor has the following per-phase equivalent-circuit parameters:

Stator winding resistance R_s	$= 0.3\Omega$
Rotor winding resistance R_r	$= 0.4\Omega$
Stator leakage inductance L_{ls}	$= 3.2 \times 10^{-3}$ H
Rotor leakage inductance L_{lr}	$= 2.1 \times 10^{-3}$ H
Stator magnetizing inductance L_{ms}	$= 40 \times 10^{-3}$ H
Effective turns ratio N_{se}/N_{re}	$= 1.2$

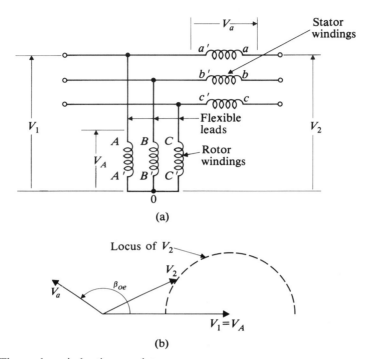

(a)

(b)

Fig. 5.17 Three-phase induction regulator.

A balanced resistive circuit consisting of three two-ohm resistors arranged in wye is connected to the rotor terminals; and with the rotor stationary, a balanced three-phase 60-Hz supply of 110 volts line-to-line is connected to the stator terminals.

Determine the ratio of the power dissipated as heat due to copper losses in the motor to the power dissipated as heat in the externally-connected resistors.

Solution   The equivalent circuit of the motor is that in Fig. 5.15, in which for this case $Z_{ex} = R_{ex} = 2\ \Omega$. As $\omega_s = 120\pi$, the various equivalent-circuit reactances are

$$X_{ls} = 120\pi \times 3.2 \times 10^{-3} = 1.21\quad \Omega$$
$$X_{lr} = 0.791\ \Omega;\ X_{ms} = 15.1\quad \Omega$$

Power dissipated internally $= 3(R_s I_a^2 + R_r I_A^2)$ W

Power dissipated externally $= 3R_{ex}\, I_A^2$ W

Thus

$$\frac{\text{Internal Power}}{\text{External Power}} = \frac{R_s\,(I_a/I_A)^2 + R_r}{R_{ex}}$$

To determine the ratio I_a/I_A, assume $\bar{I}_A = 1\,\lfloor 0\quad$ A
Then

$$\bar{E}_A = R_{ex}\bar{I}_A = 2\,\lfloor 0\quad \text{V}$$

$$\bar{E}_{mA} = \bar{E}_A + (R_r + jX_{lr})\bar{I}_A = 2\,\lfloor 0 + (0.4 + j0.791)1\,\lfloor 0 = 2.4 + j0.791 \quad \text{V}$$

$$\bar{E}_{ma} = \frac{N_{se}}{N_{re}}\bar{E}_{mA}\,\lfloor \beta_{oe} = 1.2(2.4 + j0.791)\,\lfloor \beta_{oe}$$

$$\quad = (2.88 + j0.950)\,\lfloor \beta_{oe} = 3.04\,\lfloor 18.3° + \beta_{oe}\quad \text{V}$$

$$\bar{I}'_a = \frac{N_{re}}{N_{se}}\,\lfloor \beta_{oe} \times \bar{I}_A = \frac{1}{1.2}\,\lfloor \beta_{oe} = 0.833\,\lfloor \beta_{oe}\quad \text{A}$$

$$\bar{I}_{ma} = \frac{\bar{E}_{ma}}{jX_{ms}} = \frac{3.04\,\lfloor 18.3° + \beta_{oe}}{15.1\,\lfloor 90°} = 0.201\,\lfloor -71.7° + \beta_{oe}\quad \text{A}$$

$$\bar{I}_a = \bar{I}'_a + \bar{I}_{ma} = (0.833\,\lfloor 0 + 0.201\,\lfloor -71.7°)\,\lfloor \beta_{oe} = 0.917\,\lfloor -12.0° + \beta_{oe}\quad \text{A}$$

Thus

$$\frac{I_a}{I_A} = \frac{0.917}{1.0}$$

$$\frac{\text{Internal Power}}{\text{External Power}} = \frac{0.3 \times 0.917^2 + 0.4}{2} = 0.326$$

5.2.2 Rotating Motor

Rotation of the motor does not change its inductance and resistance parameters, but it does change the relationships between stator and rotor emf's and frequencies. In the discussion of the two-pole induction machine in Chapter 3, it was seen that the frequency of the emf's induced in the rotor windings was

$$\omega_r = \omega_s - \omega_m \quad \text{rad/s} \tag{5.35}$$

For the p-pole motor, this relationship also applies if rotor speed is expressed in electrical radians per second. However, if it is to be expressed in mechanical radians per second, which is usually convenient, then Eq. (5.35) becomes

$$\omega_r = \omega_s - \frac{p}{2}\,\omega_m \quad \text{rad/s} \tag{5.36}$$

The rotating three-phase motor may therefore be represented by the equivalent circuit of Fig. 5.18. Instead of having a balanced impedance connected to them, the rotor terminals may simply be shorted together. In such a case, the rotor-circuit impedance Z_A consists only of the rotor winding resistance and its leakage reactance at the frequency of the rotor currents.

The two-pole induction machine operates as a motor when $\omega_s - \omega_m > 0$. The corresponding condition for a p-pole machine is

$$\frac{2}{p}\,\omega_s - \omega_m > 0 \quad \text{rad/s} \tag{5.37}$$

Thus the rotor must rotate more slowly than the rotating magnetic field produced by the combined stator and rotor mmf's. Under these conditions, the rotor is said to *slip* backwards with respect to the rotating field. The term "slip" is therefore employed to denote the speed at which the rotating field gains on the rotor in per unit of the synchronous or rotating-field speed. Slip may be signified by the symbol s and defined as

$$s = \frac{(2/p)\omega_s - \omega_m}{(2/p)\omega_s} = \frac{\omega_s - (p/2)\omega_m}{\omega_s} = \frac{\omega_r}{\omega_s} \tag{5.38}$$

This factor appears in the emf, impedance, and power ratios of the three-phase motor.

As was shown in Section 3.4.2, the amplitude of the induced rotor emf's is directly proportional to the rotor frequency. Thus the ratio of Eq. (5.17), which applies to the stationary motor, and for which $\omega_r = \omega_s$, must be modified in the case of the rotating motor to the form

$$\frac{\overline{E}_{mA}}{\overline{E}_{ma}} = \frac{\omega_s - (p/2)\omega_m}{\omega_s} \cdot \frac{N_{re}}{N_{se}} \bigg\lfloor -\beta_{0e} = s\,\frac{N_{re}}{N_{se}} \bigg\lfloor -\beta_{0e} \tag{5.39}$$

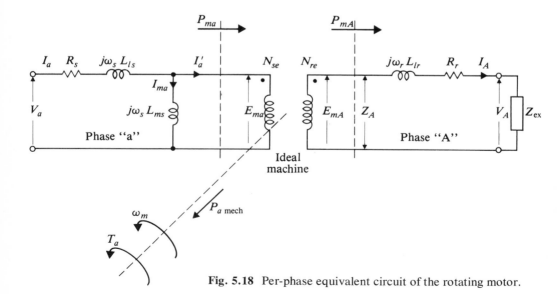

Fig. 5.18 Per-phase equivalent circuit of the rotating motor.

The current ratio is unaffected, and remains

$$\frac{\bar{I}_A}{\bar{I}_a} = \frac{N_{se}}{N_{re}} \underline{\left| -\beta_{0e} \right.} \tag{5.40}$$

The impedance ratio is again obtained by dividing the emf ratio by the current ratio, so that

$$\frac{Z_A}{Z'_A} = s \left[\frac{N_{re}}{N_{se}} \right]^2 \tag{5.41}$$

It is evident from the potential ratio of Eq. (5.39) and from the frequency relationship existing between the two parts of the equivalent circuit that the ideal transformer in the equivalent circuit for the stationary motor has now been replaced by a device with properties other than those of simple emf and current transformation. In particular, a frequency transformation now takes place such that, from Eq. (5.38),

$$\omega_r = s\,\omega_s \quad \text{rad/s} \tag{5.42}$$

This is often called the *slip frequency*.

In order to determine the power ratio existing between the stator and rotor phases, it is convenient to employ the concept of *complex power* (see Appendix C). A certain amount of the active power absorbed by one stator phase is dissipated in the winding resistance. The small amount dissipated in stator core loss is being neglected, since no corresponding branch has been included in the equivalent circuit. Thus the useful active power absorbed in the phase "a" winding is

$$P_{ma} = \mathscr{R}e\, (\bar{I}'_a\, \bar{E}^*_{ma}) \quad \text{W} \tag{5.43}$$

and the useful three-phase active power is

$$3P_{ma} = \mathscr{R}e\, (3\, \bar{I}'_a\, \bar{E}^*_{ma}) \quad \text{W} \tag{5.44}$$

This three-phase power may be considered to be transferred across the air gap and be absorbed in the rotor. It is therefore commonly called the *air-gap power*. The power dissipated as heat in the rotor circuits is

$$3P_{mA} = \mathscr{R}e\, (3\, \bar{I}_A\, \bar{E}^*_{mA}) \quad \text{W} \tag{5.45}$$

Substitution from Eqs. (5.39) and (5.40) yields

$$3P_{mA} = \mathscr{R}e\, \left(3\, \frac{N_{se}}{N_{re}}\, \bar{I}'_a \underline{\left| -\beta_{0e} \right.} \times s\, \frac{N_{re}}{N_{se}}\, \bar{E}^*_{ma} \underline{\left| +\beta_{0e} \right.} \right)$$
$$= \mathscr{R}e\, (3\, s\bar{I}'_a\, \bar{E}^*_{ma}) = 3sP_{ma} \quad \text{W} \tag{5.46}$$

Thus

$$\frac{P_{mA}}{P_{ma}} = s \tag{5.47}$$

Of the total power developed at the air gap of the motor, only a fraction, s, is dissipated as heat in the rotor circuits. By the conservation of energy, the remaining part of the air-gap power must be converted to mechanical power. This is

$$P_{\text{mech}} = 3(P_{ma} - P_{mA}) = 3(1 - s)P_{ma} \quad \text{W} \tag{5.48}$$

From Eq. (5.38),

$$s = 1 - \frac{p}{2} \frac{\omega_m}{\omega_s} \tag{5.49}$$

and substitution in Eq. (5.48) yields

$$P_{\text{mech}} = 3 \frac{p}{2} \frac{\omega_m}{\omega_s} P_{ma} \quad \text{W} \tag{5.50}$$

The ideal three-phase machine in the equivalent circuit of Fig. 5.18 therefore develops mechanical power in the form of a torque acting at rotor speed such that

$$P_{\text{mech}} = T \omega_m \quad \text{W} \tag{5.51}$$

and

$$T = \frac{P_{\text{mech}}}{\omega_m} = 3 \left(\frac{p}{2}\right) \left(\frac{P_{ma}}{\omega_s}\right) \quad \text{N} \cdot \text{m} \tag{5.52}$$

Thus the torque is equal to the air-gap power divided by the speed of rotation of the rotating field, which is $(2/p)\omega_s$ rad/s.

For steady-state operation, the power absorbed by the stator is constant. Similarly, the power dissipated in the rotor circuits is also constant. It follows, therefore, that the mechanical power developed by the machine is constant and takes place at constant speed and torque.

The mechanism of the induction motor is analogous to a slipping clutch, where the torque is the same on both sides of the clutch. Power input is proportional to the input speed, but power output is proportional to the output speed. Correspondingly, the power dissipated as heat within the clutch is proportional to the relative speed of the slipping plates.

With no-load torque applied to the shaft coupling, an induction motor normally runs with negligible slip at essentially synchronous speed. When a mechanical load torque is imposed on the motor, the rotor is decelerated, so that slip, rotor frequency, induced rotor emf, rotor current, and developed torque increase, until the internal torque developed by the machine matches the external applied torque. If, on the other hand, the rotor is driven at a speed greater than synchronous speed, the slip is negative, the induced rotor emf is reversed, and the rotor current is also reversed. The internal torque then opposes the rotation of the machine. Thus, above synchronous speed the machine operates as an asynchronous generator if it is connected to an ac system.

In an induction motor with shorted rotor terminals, only a small induced emf is required to produce rated rotor current and rated torque. Thus only a small slip is necessary (0.005 to 0.05), and speed is essentially constant, falling only slightly with increased load torque. This feature, combined with the ability to start from standstill, makes the induction motor particularly suitable for applications in which speed variation is not required. The speed of an induction motor may be controlled, however, and methods of doing so are discussed later.

The 3-phase induction motor is simple, cheap, and rugged and is therefore very widely used. The magnetic system of the rotor and stator is laminated to reduce eddy-current loss. To reduce magnetizing current the air gap is made as small as possible, and the tops of the slots are often partially closed. Different numbers of slots are used in the stator and rotor to avoid reluctance torques, and the rotor slots are frequently skewed one slot pitch (made helical instead of straight) to reduce vibration and noise.

There are two major types of rotor winding—the wound-rotor and the squirrel-cage. Stator windings—and rotor windings on wound-rotor machines—may be connected in wye or delta, and usually at least one on a wound-rotor motor is in delta. For nearly all practical purposes, however, the motor may be treated as if both sets of windings were connected in wye, and this assumption has been made in the foregoing analysis.

If the rotor terminals of the equivalent circuit of Fig. 5.18 are short-circuited, and the rotor parameters and variables are referred to the stator of the ideal machine, then a per-phase equivalent circuit of the form shown in Fig. 5.19 is obtained. The impedance of the rotor branch in Fig. 5.19 may be determined by applying the impedance ratio of Eq. (5.41) to the rotor branch impedance of Fig. 5.18. Thus

$$\frac{Z_A}{\overline{Z}'_A} = \frac{R_r + j\omega_r L_{lr}}{\overline{Z}'_A} = s \left[\frac{N_{re}}{N_{se}} \right]^2 \tag{5.53}$$

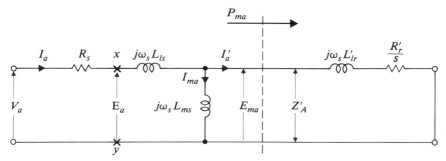

Fig. 5.19 Per-phase equivalent circuit referred to the stator.

or

$$\overline{Z}'_A = \frac{1}{s}\left[\frac{N_{se}}{N_{re}}\right]^2 R_r + j\frac{\omega_r}{s}\left[\frac{N_{se}}{N_{re}}\right]^2 L_{lr} = \frac{R'_r}{s} + j\omega_s L'_{lr} \quad \Omega \qquad (5.54)$$

where

$$R'_r = \left[\frac{N_{se}}{N_{re}}\right]^2 R_r \quad \Omega \qquad (5.55)$$

and

$$L'_{lr} = \left[\frac{N_{se}}{N_{re}}\right]^2 L_{lr} \quad \Omega \qquad (5.56)$$

The potential difference applied to the rotor branch of Fig. 5.19 is necessarily equal to E_{ma}, and the current I'_a may be obtained in terms of the actual rotor current I_A from the current ratio of Eq. (5.40).

In the process of referring the rotor phase impedance to the stator of the ideal machine, the ideal machine itself has been eliminated from the equivalent circuit, and this has two important consequences. The first is that the entire equivalent circuit of Fig. 5.19 must operate at supply frequency ω_s. This is confirmed by the expression $j\omega_s L'_{lr}$ for the referred rotor leakage impedance. The second consequence is that the per-phase mechanical output P_{amech} has been eliminated, yet the per-phase air-gap power P_{ma} has not been changed. The power, which in Fig. 5.18 is divided into two parts, P_{mA} (dissipated as heat in the rotor resistance R_r) and P_{amech}, is now entirely dissipated in the fictitious resistance R'_r/s. Thus

$$P_{ma} = \frac{R'_r}{s}(I'_a)^2 = R'_r(I'_a)^2 + \frac{(1-s)}{s}R'_r(I'_a)^2 \quad W \qquad (5.57)$$

From Eqs. (5.40) and (5.55), it may be seen that

$$R'_r(I'_a)^2 = R_r I_A^2 = P_{mA} \quad W \qquad (5.58)$$

Thus the first term on the right-hand side of Eq. (5.57) represents the power per phase dissipated in rotor resistance loss. By the principle of the conservation of energy, therefore, the second term on the right-hand side of Eq. (5.57) must be the mechanical power developed per phase. Thus

$$P_{amech} = \frac{(1-s)}{s}R'_r(I'_a)^2 \quad W \qquad (5.59)$$

The fictitious resistance R'_r/s in Fig. 5.19, therefore, could be represented as two resistances in series: the first, R'_r, representing the rotor-circuit resistance referred to the stator; and the second, $R'_r(1-s)/s$, representing the energy sink that absorbs the mechanical power developed per phase.

Example 5.3 The motor of Example 5.2 is 4-pole. Determine the frequency of the rotor currents and the torque that is developed at a speed of 1000 r/min with the same balanced resistive circuit connected to the rotor terminals.

Solution

$$\omega_m = \frac{2\pi}{60} \times 1000 = 105 \text{ rad/s}$$

$$s = \frac{\omega_s - (p/2)\omega_m}{\omega_s} = \frac{120\pi - 2 \times 105}{120\pi} = 0.443$$

Rotor frequency $\omega_r = s\omega_s = 0.443 \times 120\pi = 167$ rad/s
$$= 26.6 \text{ Hz}$$

The equivalent circuit of the motor is that in Fig. 5.19 with the referred external resistance R_{ex}/s shown between the rotor terminals. From Eq. (5.54),

$$Z_A' = \frac{R_r' + R_{\text{ex}}'}{s} + j\omega_s L_{lr}' \quad \Omega$$

where, from Eqs. (5.55) and (5.56),

$$R_r' = 1.2^2 \times 0.4 = 0.576 \quad \Omega$$
$$R_{\text{ex}}' = 1.2^2 \times 2 = 2.88 \quad \Omega$$
$$L_{lr}' = 1.2^2 \times 2.1 \times 10^{-3} = 3.02 \times 10^{-3} \quad \text{H}$$

Thus

$$Z_A' = \frac{0.576 + 2.88}{0.443} + j120\pi \times 3.02 \times 10^{-3} = 7.88 \,\underline{|8.31°} \quad \Omega$$

Assume

$$\bar{I}_a' = 1 \,\underline{|0} \quad \text{A}$$

Then

$$\bar{E}_{ma} = Z_A' \bar{I}_a' = 7.88 \,\underline{|8.31°}$$

$$\bar{I}_{ma} = \frac{\bar{E}_{ma}}{j\omega_s L_{ms}} = \frac{7.88 \,\underline{|8.31°}}{120\pi \times 40 \times 10^{-3} \,\underline{|90°}} = 0.523 \,\underline{|-81.7°} = 1.0756 - j0.517 \quad \text{A}$$

Thus

$$\bar{I}_a = \bar{I}_a' + \bar{I}_{ma} = 1.0756 - j0.517 = 1.19 \,\underline{|-25.7°}$$
$$\bar{V}_a = \bar{E}_{ma} + (R_s + j\omega_s L_{ls}) \bar{I}_a$$

$$R_s + j\omega_s L_{ls} = 0.3 + j120\pi \times 3.2 \times 10^{-3} = 1.24 \,\underline{|76.0°} \quad \Omega$$

and
$$\overline{V}_a = 7.88 \,\underline{|8.31°} + 1.24 \,\underline{|76.0°} \times 1.19 \,\underline{|-25.7°} = 9.04 \,\underline{|14.6°} \quad V$$

The per-phase applied potential difference is, in fact, $110/\sqrt{3}$ V, and if its phasor is taken as the reference,
$$\overline{V}_a = 63.5 \,\underline{|0} \quad V$$

Thus, to obtain the true values, all equivalent-circuit variables calculated on the assumption that $\overline{I}_a' = 1\,\underline{|0}$ A must be multiplied by the factor

$$\overline{K} = \frac{63.5 \,\underline{|0}}{9.04 \,\underline{|14.6°}} = 7.02 \,\underline{|-14.6°}$$

In particular,
$$\overline{I}_a' = \overline{K} \times 1 \,\underline{|0} = 7.02 \,\underline{|-14.6°} \quad A$$

From Eq. (5.59), the mechanical power developed per phase is

$$P_{amech} = \left(\frac{1-s}{s}\right)(R_r' + R_{ex}')(I_a')^2 = \frac{(1-0.443)}{0.443} \times 3.46 \times (7.22)^2 = 214 \quad W$$

and
$$T = \frac{3\,P_{amech}}{\omega_m} = \frac{3 \times 214}{105} = 6.11 \quad N \cdot m$$

Alternatively,
$$\overline{E}_{ma} = 7.02 \,\underline{|-14.6°} \times 7.88 \,\underline{|8.31°} = 55.3 \,\underline{|-6.3°} \quad V$$

and from Eq. (5.43),
$$P_{ma} = \mathcal{R}e\,(\overline{I}_a'\,\overline{E}_{ma}^*) = \mathcal{R}e\,(7.02 \,\underline{|-14.6°} \times 55.3 \,\underline{|6.3°}) = \mathcal{R}e\,(3.88 \,\underline{|-8.3°}) = 384 \quad W$$

So from Eqs. (5.50) and (5.52),

$$T = 3\frac{p}{2}\frac{P_{ma}}{\omega_s} = 3 \times \frac{4}{2} \times \frac{384}{120\pi} = 6.11 \quad N \cdot m$$

5.2.3 Approximate Equivalent Circuit

The two-mesh equivalent circuit of Fig. 5.19 is somewhat inconvenient for expressing a machine's internal torque in terms of the terminal variables. However, for many practical purposes, a sufficiently accurate equivalent circuit may be obtained by means of an approximation already employed in the case of single-phase transformers. That is to say, the magnetizing reactance $\omega_s L_{ms}$ is simply moved to the stator terminals. The remaining circuit parameters of Fig. 5.19 and the source potential difference remain unchanged. The resulting approximate equivalent circuit is shown in Fig. 5.20. The correspondence between the parameters of Fig. 5.19 and those of Fig. 5.20 may be expressed as follows:

Fig. 5.20		Fig. 5.19	
R_s	=	R_s	Ω
X_M	=	$\omega_s L_{ms}$	Ω
X_L	=	$\omega_s(L_{ls} + L'_{lr})$	Ω
R'_R	=	R'_r	Ω

$$(5.60)$$

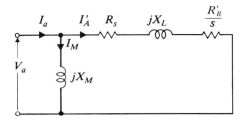

Fig. 5.20 Equivalent circuit employed for analysis.

Since the per-unit magnetizing reactance of an induction motor is considerably less than that of a transformer, due to the air gap in the magnetic system, the approximation described in the preceding paragraph is greater than in the case of the transformer. Nevertheless, for motors of some 10 kW rating and above, this approximation is usually acceptable, at least for preliminary calculations.

5.2.4 Improved Equivalent Circuit

The predictions of performance based on the approximate equivalent circuit described in the preceding section may differ by more than 5% from those based on much more lengthy calculations employing the equivalent circuit of Fig. 5.19. This is particularly the case for small motors. The labor of employing the circuit of Fig. 5.19 may be avoided, however, and accurate results obtained by employing the circuit of Fig. 5.20 with the following correspondence of circuit parameters:

Fig. 5.20		Fig. 5.19		
R_s	=	R_s	Ω	(5.61)
X_M	=	$\dfrac{\omega_s L_{ms}}{k}$	Ω	(5.62)
X_L	=	$\omega_s \left[\dfrac{L_{ls}}{k} + \dfrac{L'_{lr}}{k^2} \right]$	Ω	(5.63)
R'_R	=	$\dfrac{R'_r}{k^2}$	Ω	(5.64)

where

$$k = \frac{X_{ms}}{X_{ms} + X_{ls}} = \left[\frac{E_{mA}}{V_a} \cdot \frac{E_{ma}}{V_A} \right]^{1/2} \qquad (5.65)$$

The justification for the use of Eqs. (5.61) to (5.65) is given in Appendix H, where it is also shown that to a close approximation,

$$
\begin{array}{|c c c|}
\hline
\text{Fig. 5.20} & & \text{Fig. 5.19} \\
\hline
R'_R(I'_A)^2 & = & R'_r(I'_a)^2 \quad \text{W} \\
\hline
\end{array}
\tag{5.66}
$$

From Eqs. (5.57) and (5.66), therefore, the power transferred to the rotor across the air gap is

$$
P_{ma} = R'_R (I'_A)^2 + \frac{(1-s)}{s} R'_R (I'_A)^2 \quad \text{W}
\tag{5.67}
$$

The first term on the right-hand side of Eq. (5.67) represents the per-phase resistive loss in the rotor winding, and the second term represents the per-phase mechanical power developed in the machine—that is,

$$
P_{amech} = \frac{(1-s)}{s} R'_R (I'_A)^2 \quad \text{W}
\tag{5.68}
$$

5.2.5 Steady-State Operation

The relationship between the internal torque T and the rotor speed ω_m may be derived from the equivalent circuit of Fig. 5.20. From Eq. (5.67)

$$
P_{ma} = \frac{R'_R}{s} (I'_A)^2 \quad \text{W}
\tag{5.69}
$$

and from Eq. (5.52),

$$
T = 3 \cdot \frac{p}{2} \cdot \frac{P_{ma}}{\omega_s} \quad \text{N} \cdot \text{m}
\tag{5.70}
$$

From Fig. 5.20,

$$
\bar{I}_A = \frac{\bar{V}_a}{(R_s + (R'_R/s)) + jX_L} \quad \text{A}
\tag{5.71}
$$

Substitution from Eqs. (5.69) and (5.71) in Eq. (5.70) then yields

$$
T = \frac{3}{\omega_s} \cdot \frac{p}{2} \cdot \frac{R'_R}{s} \frac{V_a^2}{[R_s + (R'_R/s)]^2 + X_L^2} \quad \text{N} \cdot \text{m}
\tag{5.72}
$$

A typical relationship between speed and torque, or slip and torque, may be determined from Eq. (5.72). At speeds approaching synchronous speed, the slip s approaches zero, so that $R'_R/s >> R_s$, and $R'_R/s >> X_L$. Thus, from Eq. (5.72),

$$T \simeq \frac{3}{\omega_s} \cdot \frac{p}{2} \frac{s}{R_R'} V_a^2 \quad \text{N} \cdot \text{m} \tag{5.73}$$

and from Eq. (5.71),

$$I_A' \simeq \frac{s}{R_R'} V_a \quad \text{A} \tag{5.74}$$

Thus, near synchronous speed, torque and the load component of current are essentially proportional to slip. The proportionality between torque and slip is indicated in Fig. 5.21 by the broken line passing through the point $(0, (2/p)\omega_s)$.

Since Eq. (5.72) shows that for constant applied terminal potential difference, torque is a function of slip only, then maximum and minimum values of torque and the slip s at which they occur may be determined by setting $dT/ds = 0$ and solving for s. If this is done, the solution is

$$s = \pm \frac{R_R'}{(R_s^2 + X_L^2)^{1/2}} \tag{5.75}$$

Substitution of the positive value of s from Eq. (5.75) in Eq. (5.72) gives the maximum positive or motoring torque:

$$T_{m\text{max}} = \frac{3}{\omega_s} \cdot \frac{p}{4} \cdot \frac{V_a^2}{(R_s^2 + X_L^2)^{1/2} + R_s} \quad \text{N} \cdot \text{m} \tag{5.76}$$

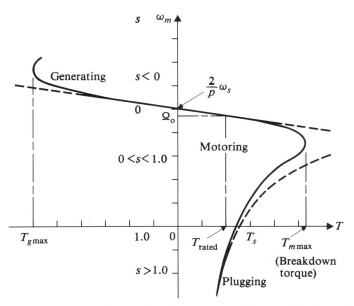

Fig. 5.21 Speed–torque curve of a wound-rotor machine with rotor slip-rings shorted together.

This is called the *breakdown torque* of the motor. Substitution of the negative value of s gives the maximum negative or generating torque as

$$T_{g\max} = -\frac{3}{\omega_s} \cdot \frac{p}{4} \cdot \frac{V_a^2}{(R_s^2 + X_L^2)^{1/2} - R_s} \quad \text{N} \cdot \text{m} \tag{5.77}$$

so that $|T_{g\max}| > |T_{m\max}|$. In large machines, $R_s \ll X_L$, and

$$|T_{m\max}| \simeq |T_{g\max}| \simeq \frac{3}{\omega_s} \cdot \frac{p}{4} \cdot \frac{V_a^2}{X_L} \quad \text{N} \cdot \text{m} \tag{5.78}$$

Equation (5.78) shows that the maximum torque which an induction motor can develop is limited principally by the total leakage reactance and is independent of rotor-circuit resistance.

It may also be noted that, at large values of slip s, $X_L \gg R_s + R_R'/s$, and Eq. (5.72) can be approximated by

$$T = \frac{3}{\omega_s} \frac{p}{2} \frac{R_R'}{s} \left(\frac{V_a}{X_L}\right)^2 \quad \text{N} \cdot \text{m} \tag{5.79}$$

That is, torque is inversely proportional to slip, as shown by the curve in broken line in Fig. 5.21.

Substitution of $s = 1$ in Eq. (5.72) gives the locked-rotor or starting torque of the motor as

$$T_s = \frac{3}{\omega_s} \cdot \frac{p}{2} \cdot \frac{R_R' V_a^2}{(R_s + R_R')^2 + X_L^2} \quad \text{N} \cdot \text{m} \tag{5.80}$$

For standard wound-rotor induction motors with no external resistance in the rotor circuit,

$$T_s \simeq 1.25 \, T_{\text{FL}} \qquad T_{\max} \simeq 2.70 \, T_{\text{FL}} \quad \text{N} \cdot \text{m} \tag{5.81}$$

where T_{FL} is the full-load or rated shaft torque of the motor.

Example 5.4 A 10-hp, 220-V, 60-Hz, 3-phase, wound-rotor induction motor has a rated speed of 1725 r/min. The parameters of the equivalent circuit of Fig. 5.20 for this machine are:

$$R_s = 0.20 \, \Omega \qquad X_L = 0.839 \, \Omega$$
$$R_R' = 0.256 \, \Omega \qquad X_M = 12.9 \, \Omega$$

Determine the internal torque and line current at rated speed.

Solution For the rated speed,

$$s = \frac{1800 - 1725}{1800} = 0.0417$$

Substitution of the given quantities in Eq. (5.72) yields

$$T = \frac{3}{2\pi \times 60} \times \frac{4}{2} \times \frac{0.256}{0.0417} \times \frac{127^2}{[0.20 + (0.256/0.0417)]^2 + 0.839^2} = 38.6 \quad \text{N} \cdot \text{m}$$

Substitution in Eq. (5.71) yields

$$\overline{I}_A' = \frac{127 \underline{|0}}{[0.20 + (0.256/0.0417)] + j0.839} = 19.7 - j2.61 \quad \text{A}$$

From Fig. 5.20,

$$\overline{I}_M = \frac{\overline{V}_a}{jX_M} = \frac{127 \underline{|0}}{j12.9} = -j9.84$$

so that

$$\overline{I}_a = \overline{I}_M + \overline{I}_A' = 19.7 - j12.5$$

and

$$I_a = 23.3 \quad \text{A}$$

5.2.6 Mechanism of Torque Production

The analysis of Section 5.2.5 permits accurate prediction of the performance of an induction motor. If to this is added a physical picture of the machine's operation, then understanding should be increased, as well as an appreciation of the similarities to and differences from other machines. A simple model is required for this purpose and is obtained by ignoring the stator resistance and leakage inductance in the equivalent circuit of Fig. 5.18. If in addition, the rotor terminals are short circuited, so that $Z_{ex} = 0$, the resulting model is shown in Fig. 5.22(a).

The circuit model of Fig. 5.22(a) represents the conditions in one phase only of the machine. There are two other similar phases for which similar circuits could be drawn and in which all of the variables would be shifted in phase $\pm 120°$ relative to those in phase "a" of the stator and phase "A" of the rotor. The three rotor phase currents, I_A, I_B, I_C, produce a rotating wave of mmf at the air gap of the machine. The three power components of the stator phase currents, I_a', I_b', I_c', must produce an exactly equal and opposite rotating wave of mmf so that the resultant mmf on the ideal machine is zero. From this it follows that in the ideal machine

$$N_{se} I_a' = N_{re} I_A \quad \text{A} \tag{5.82}$$

In the real machine the two mmf waves do not balance out, and their resultant in the circuit of Fig. 5.22(a) is considered to be a wave produced by the magnetizing components of the stator currents I_{ma}, I_{mb}, I_{mc}. The total stator current in phase "a" is expressed by

$$\overline{I}_a = \overline{I}_{ma} + \overline{I}_a' \quad \text{A} \tag{5.83}$$

The magnetizing components of the stator currents produce the rotating flux wave in the air gap; this flux wave induces E_{ma}, E_{mb}, and E_{mc} in the 3-phase stator winding.

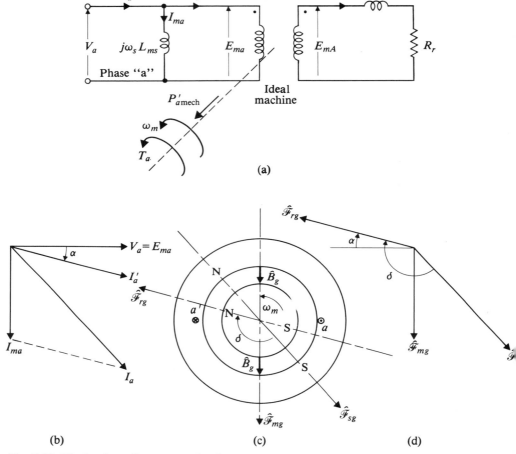

Fig. 5.22 Mechanism of torque production.

In the circuit model of Fig. 5.22(a), the emf induced in phase "a" of the stator winding is

$$E_{ma} = V_a = \text{constant} \quad \text{V} \tag{5.84}$$

If the rotor is rotating at the same speed and in the same direction as the rotating flux wave, then slip $s = 0$, and the emf induced in rotor phase "A" is

$$E_{mA} = s\frac{N_{re}}{N_{se}} E_{ma} = 0 \quad \text{V} \tag{5.85}$$

Under these circumstances, I_A and I'_a are both zero, and the entire stator current in phase "a" is magnetizing current I_{ma}. This current lags potential difference V_a by 90°, as shown in Fig. 5.22(b).

Now consider that the rotor is rotating at a speed lower than the synchronous speed ω_s of the flux wave: that is, $\omega_m < \omega_s$, so that $s > 0$. The rotor emf E_{mA} is therefore no longer zero, and produces a rotor current at frequency $\omega_r = s\omega_s$. This current is

$$\bar{I}_A = \frac{\bar{E}_{mA}}{R_r + j\omega_r L_{lr}} \quad \text{A} \tag{5.86}$$

The corresponding power component of the stator current is

$$\bar{I}'_a = \frac{N_{se}}{N_{re}} \bar{I}_A \quad \text{A} \tag{5.87}$$

The current I'_a lags the induced emf E_{ma} by the same angle as current I_A lags emf E_{mA}—that is, by angle

$$\alpha = \tan^{-1} \frac{\omega_r L_{lr}}{R_r} \quad \text{rad} \tag{5.88}$$

Equation (5.83) applies and is illustrated in the phasor diagram of Fig. 5.22(b). An mmf space vector diagram may be drawn corresponding to this phasor diagram. For this purpose it is helpful to employ the physical diagram of Fig. 5.22(c) as an intermediate step.

As shown in Section 3.3.5, the peak value of the mmf produced by a three-phase stator winding lies on the axis of a particular phase winding when that phase is carrying its peak current. At the instant at which the magnetizing current I_{ma} has its peak positive value, then, for the positive current direction shown in Fig. 5.22(c), the mmf wave $\mathscr{F}_{mg}$ due to I_{ma}, I_{mb}, and I_{mc} has its peak value acting vertically downward. This peak value $\hat{\mathscr{F}}_{mg}$ will be proportional to the magnitude of I_{ma} in the phasor diagram of Fig. 5.22(b). The corresponding space vector is shown in Fig. 5.22(d). Necessarily the peak flux density $\hat{B}_g$ coincides with $\mathscr{F}_{mg}$ and is shown in the physical diagram.

The complete stator mmf wave produced by currents I_a, I_b, and I_c will have a peak value $\hat{\mathscr{F}}_{sg}$ proportional to I_a in Fig. 5.22(b), and will be shifted forward in space relative to $\mathscr{F}_{mg}$ by the angle between phasors $\bar{I}_{ma}$ and $\bar{I}_a$ in the same diagram. The stator peak mmf $\hat{\mathscr{F}}_{sg}$ is therefore in the direction shown in the physical diagram, producing the N and S poles on the stator. The corresponding space vector of mmf is shown in Fig. 5.22(d).

The position of the space vector due to the load components of stator currents I'_a, I'_b, I'_c could also be obtained from the phasor diagram, and the space vector of mmf could be added to Fig. 5.22(d). However, what is of more interest is the equal and opposite space vector due to the rotor currents I_A, I_B, and I_C. The corresponding vector $\mathscr{F}_{rg}$ is shown in the vector diagram, and the N and S poles that this mmf produces on the rotor are shown in the physical diagram. From Fig. 5.22(d),

$$\vec{\mathcal{F}}_{mg} = \vec{\mathcal{F}}_{sg} + \vec{\mathcal{F}}_{rg} \quad \text{A} \tag{5.89}$$

The torque produced by the interaction of two mmf's is given in Eq. (3.130) as

$$T = -K \; \hat{\mathcal{F}}_{sg} \; \hat{\mathcal{F}}_{rg} \sin \delta \quad \text{N·m} \tag{5.90}$$

where δ is the angle measured from the stator to the rotor mmf axis. In Fig. 5.22(b)–(d), which represent motor action, δ is negative; and a positive torque acting in the direction of rotation is developed.

As the slip s of the machine increases, E_{mA} increases proportionately. This increases the rotor current I_A and mmf $\hat{\mathcal{F}}_{rg}$. A stator current I_a is then established to produce a stator mmf $\hat{\mathcal{F}}_{sg}$ sufficient to balance the rotor mmf $\hat{\mathcal{F}}_{rg}$ and, in addition, to supply the constant magnetizing mmf $\hat{\mathcal{F}}_{mg}$ in Fig. 5.22(d). The result of an increased slip is therefore to increase $\hat{\mathcal{F}}_{rg}$, $\hat{\mathcal{F}}_{sg}$, and also the angle δ. This produces an increase in torque, provided that the increase in the product $\hat{\mathcal{F}}_{sg} \hat{\mathcal{F}}_{rg}$ in Eq. (5.90) is sufficient to outweigh the effect of the decrease in $\sin \delta$.

The variation of torque with speed can be appreciated better with the aid of the vector diagram of Fig. 5.23. From this diagram it may be seen that

$$\hat{\mathcal{F}}_{sg} \cos(\delta - 90°) = -\hat{\mathcal{F}}_{sg} \sin \delta = \hat{\mathcal{F}}_{mg} \cos \alpha \quad \text{A} \tag{5.91}$$

Substitution in Eq. (5.90) then yields

$$T = K \; \hat{\mathcal{F}}_{rg} \; \hat{\mathcal{F}}_{mg} \cos \alpha \quad \text{N·m} \tag{5.92}$$

Equation (5.84) shows that for constant terminal potential difference, E_{ma} is constant. This means that the amplitude of the rotating flux wave, and therefore of $\hat{\mathcal{F}}_{mg}$, must also be constant. At low values of slip (near synchronous speed), $R_r \gg \omega_r L_{lr}$ and, from Eq. (5.88), $\alpha \approx 0$. In Fig. 5.22(d), $\hat{\mathcal{F}}_{rg}$ is approximately perpendicular to $\hat{\mathcal{F}}_{mg}$—that is, to the axis of the flux wave. This is the optimum condition for torque production for a given flux and rotor current magnitude. As slip is increased, $\hat{\mathcal{F}}_{rg}$ increases in smaller proportion because of the effect of inductive reactance $\omega_r L_{lr}$. The angle α also increases. Thus the increase in torque, expressed in Eq. (5.92), is less than proportional to the increase in slip. At high values of slip, $\omega_r L_{lr} \gg R_r$. From Eqs. (5.85) and (5.86),

$$I_A = \frac{s \, (N_{se}/N_{re}) \, \overline{E}_{ma}}{R_r + j s \omega_s L_{lr}} \simeq \frac{N_{se}}{N_{re}} \frac{\overline{V}_a}{j \omega_s L_{lr}} = \text{constant} \quad \text{A} \tag{5.93}$$

Thus $\hat{\mathcal{F}}_{rg}$ tends toward a constant magnitude, but α tends toward zero. The net effect of these variations can be seen from Eq. (5.92) to be a decrease in torque as slip increases and speed decreases. At very high slip (in the plugging region of Fig. 5.21), $\hat{\mathcal{F}}_{rg}$ and $\hat{\mathcal{F}}_{mg}$ in Fig. 5.23 are nearly 180° out of phase. Thus as slip increases, torque initially increases, reaches a maximum (breakdown) value, and then decreases as shown in Fig. 5.21.

From Fig. 5.22(a) and (b) it is seen that the power per phase entering the air gap of the machine is

$$P_{ma} = E_{ma} I'_a \cos \alpha \quad \text{W} \tag{5.94}$$

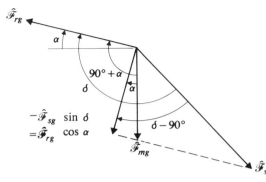

Fig. 5.23 Variation of torque with slip.

Thus substitution in Eq. (5.70) gives the torque developed in a p-pole machine as:

$$T = 3 \frac{p}{2} \frac{P_{ma}}{\omega_s} = \frac{3}{2} \frac{p}{\omega_s} E_{ma} I'_a \cos \alpha \quad \text{N·m} \tag{5.95}$$

This expression may now be compared with that in Eq. (5.92). Note that E_{ma} is proportional to the product $\omega_s \hat{\mathscr{F}}_{mg}$, and $\hat{\mathscr{F}}_{rg}$ is proportional to I'_a.

As usual, the illustrations in Fig. 5.22(c) and (d) are for a 2-pole machine. For a p-pole machine, the space-vector diagram would be unaltered if the angles in it were expressed in electrical degrees. The physical diagram in (c) could be replaced by one showing p poles on both stator and rotor. A similar set of diagrams and a discussion similar to the foregoing may be developed for the machine acting as a generator in the second quadrant of Fig. 5.21.

*5.2.7 Circle Diagram

From the equivalent circuit of Fig. 5.20,

$$\bar{I}_a = \frac{\overline{V}_a}{jX_M} + \frac{\overline{V}_a}{R_s + (R'_R/s) + jX_L} \quad \text{A} \tag{5.96}$$

The first component of I_a is constant in magnitude and lags the applied terminal potential difference V_a by 90°. In practice, if the stator were excited and the rotor driven at synchronous speed $(2/p)\omega_s$ rad/s, so that $s = 0$, and the second component of current in Eq. (5.96) became zero, the measured line current would contain a small power component in phase with V_a. This power component would be due to core loss, which has been neglected in deriving the equivalent circuit, and to a small stator resistance loss due to the exciting current. (This resistance loss is ignored at $s = 0$ in the approximation made in employing the circuit of Fig. 5.20 in place of the equivalent circuit of Fig. 5.19.)

The second component of the current in Eq. (5.96) is I'_A, and this may be expressed as

$$\bar{I}'_A = \frac{\overline{V}_a}{R + jX_L} \quad \text{A} \tag{5.97}$$

where

$$R = R_s + \frac{R'_R}{s} \quad \Omega \tag{5.98}$$

The resistance R is a function of rotor speed or slip, while X_L is constant. Now let

$$R + jX_L = Z \underline{|\theta} \quad \Omega \tag{5.99}$$

where

$$\theta = \tan^{-1} \frac{X_L}{R} \quad \text{rad} \tag{5.100}$$

Then

$$I'_A = \frac{V_a}{(R^2 + X_L^2)^{1/2}} = \frac{V_a}{X_L [(R^2/X_L^2) + 1]^{1/2}} = \frac{V_a}{X_L (\cot^2\theta + 1)^{1/2}} \quad \text{A} \tag{5.101}$$

or

$$I'_A = \frac{V_a}{X_L} \sin \theta \quad \text{A} \tag{5.102}$$

Equation (5.102) is the polar equation of a circle of diameter V_a/X_L, and since

$$\bar{I}_a = \bar{I}_M + \bar{I}'_A \quad \text{A} \tag{5.103}$$

the locus of phasor $\bar{I}_a$ is a circle that may be calibrated in slip s. A diagram illustrating the relationships in Eqs. (5.102) and (5.103) is shown in Fig. 5.24. It is very useful in visualizing the relationship between V_a, I_a, and s for all conditions of operation of the machine. For example, point a represents the condition of maximum power input to the machine, since the in-phase component of I_a is a maximum. Point b represents the condition of maximum stator power factor, since angle ϕ is a minimum.

*5.2.8 Speed Control of a Wound-Rotor Motor

The classical method of speed control of an induction motor is by variation of the rotor-circuit resistance.

Equation (5.75) shows that the slip at which maximum torque occurs is a function of R'_R, which is directly proportional to the rotor resistance per phase. Equation 5.76 shows that the maximum torque is independent of rotor resistance, since it is a function of X_L, which is dependent upon the stator and rotor leakage reactances. The terminals of the rotor windings of a wound-rotor motor are connected to slip rings so that external impedances may be introduced into the rotor circuits. Thus, if per-phase external resistance R_{ex} is introduced into the rotor circuit, then from Eq. (5.75) the slip for $T = T_{\max}$ is given by

$$s = \frac{R'_R + R'_{EX}}{(R_s^2 + X_L^2)^{1/2}} \tag{5.104}$$

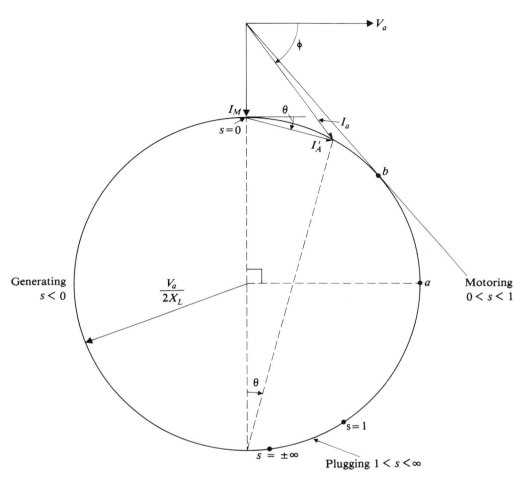

Fig. 5.24 Induction motor circle diagram.

where

$$R'_{EX} = \frac{1}{k^2} \left[\frac{N_{se}}{N_{re}} \right]^2 R_{ex} \quad \Omega \qquad (5.105)$$

The introduction of external resistance into the rotor circuit will therefore not change the breakdown torque, but will simply increase the slip at which it occurs.

The family of speed–torque curves shown in Fig. 5.25 illustrates the effect of the variation of rotor-circuit resistance. Maximum torque may be produced at standstill if necessary. Alternatively, the standstill or starting torque may be reduced below the rated full-load value by addition of sufficient external rotor-circuit resistance.

The magnitude of the stator current corresponding to any operating condition may be visualized by reference to the circle diagram of Fig. 5.24. The effect of introducing external rotor-circuit resistance is that of moving a point on the circle corresponding to a particular slip around the circle toward the $s = 0$ point. Thus, by the addition of sufficient external rotor-circuit resistance, the stator current at standstill may be reduced to the full-load value or less. High starting torque at low current can therefore be obtained, undesirable heating of the machine during starting is avoided, and rapid acceleration of the machine and driven load is ensured.

By variation of rotor-circuit resistance, speed control of the loaded motor is achieved, as may be seen from the intersections of the motor speed–torque curves with that of a typical load marked $T_{\text{loss}} + T_L$ in Fig. 5.25. However, the intersections with the curve of T_{loss} show that the machine on no load will run up to nearly synchronous speed, regardless of how much rotor-circuit resistance is introduced.

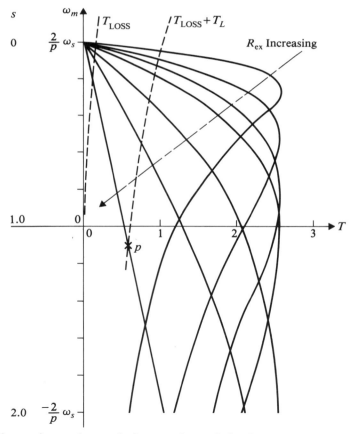

Fig. 5.25 Rheostatic speed control of a wound-rotor induction motor.

It may also be seen that with high external resistance in the circuit, large speed variations may result from relatively small variations in the load torque. If the load characteristic extends into the fourth quadrant and if the rotor-circuit resistance is made too great, then the motor may be driven backwards by the load, as illustrated by point p in Fig. 5.25.

Speed control by means of variable rotor-circuit resistance is often employed, but the system is inefficient, since much energy is dissipated as heat in the external resistors. Low efficiency is acceptable, however, when only a small proportion of the work cycle calls for low-speed running. A typical application of this type of control is the hoist drive of a shop crane.

The so-called *ideal efficiency* of an induction motor, in which all losses except rotor-circuit resistance losses are neglected, is expressed by

$$\eta_{\text{ideal}} = \frac{P_{amech}}{P_{ma}} \tag{5.106}$$

Substitution from Eqs. (5.67) and (5.68) in Eq. (5.106) yields

$$\eta_{\text{ideal}} = 1 - s \tag{5.107}$$

From Eq. (5.107), it is clear that an induction motor must operate near synchronous speed if the efficiency is to be high.

The effect of adding inductive reactance to the external rotor circuit would be that of increasing the reactance X_L in Eq. (5.102), and thus reducing the diameter V_a/X_L of the circle in Fig. 5.24. This would reduce torques and currents over the whole speed range and cause the machine to operate at a lower power factor.

Example 5.5 A 50-hp, 2200-V, 13.5-A, 1160 r/min., 60-Hz, wound-rotor induction motor has the following equivalent-circuit parameters:

$$R_s = 2.22 \ \Omega \qquad X_L = 14.2 \ \Omega$$
$$R'_R = 2.97 \ \Omega \qquad X_M = 324.0 \ \Omega$$

The effective turns ratio is $N_{se}/N_{re} = 1.22$ and $k = 0.979$.

a) Determine the external rotor-circuit resistance per phase required to hold down the starting current to three times the rated value.

b) Determine the speed at which the motor would develop an internal torque equal to 125% of the rated torque with this starting resistance still in circuit.

c) Determine the line current that would result if the starting resistance were shorted out of circuit at the speed determined in (b).

Solution

a) Permissible starting current is

$$I_a = 3 \times 13.5 = 40.5 \ \text{A}$$

The corresponding value of I'_A at standstill may be obtained from a circle diagram. From the equivalent circuit of Fig. 5.20,

$$I_M = \frac{2200}{324.0\sqrt{3}} = 3.920 \quad A$$

The radius of the circle diagram is:

$$\frac{V_a}{2X_L} = \frac{2200}{2 \times 14.2\sqrt{3}} = 44.72 \quad A$$

The circle diagram for starting conditions is shown in Fig. 5.26, from which

$$\cos \alpha = \frac{40.5^2 + 48.6^2 - 44.7^2}{2 \times 40.5 \times 48.6} = \frac{40.5^2 + 3.92^2 - (I'_A)^2}{2 \times 40.5 \times 3.92}$$

From which

$$I'_A = 38.7 \text{ A}$$

The impedance at standstill of the right-hand branch of the circuit of Fig. 5.20 is

$$\overline{Z} = 2.22 + j\,14.2 + 2.97 + R'_{EX} = (5.19 + R'_{EX}) + j\,14.2$$

Thus

$$Z^2 = (5.19 + R'_{EX})^2 + 14.2^2 = \left(\frac{2200}{38.7\sqrt{3}}\right)^2$$

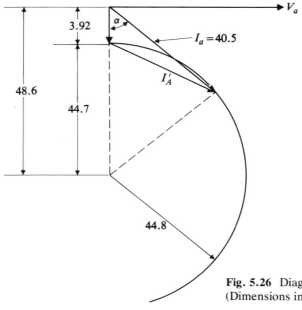

Fig. 5.26 Diagram for example 5.5
(Dimensions in amperes; not drawn to scale.)

From which

$$R'_{EX} = 24.4 \ \Omega$$

Thus

$$R_{ex} = k^2 \left(\frac{N_{re}}{N_{se}}\right)^2 R'_{EX} = \frac{0.979^2}{1.22^2} \times 24.5 = 15.8 \quad \Omega$$

b)

$$\text{Rated speed} = \frac{2\pi}{60} \times 1160 = 121.5 \ \text{rad/s}$$

$$\text{Rated torque} = \frac{50 \times 746}{121.5} = 307.1 \quad \text{N} \cdot \text{m}$$

From Eq. (5.72), $(R'_R + \mathbf{R}'_{EX} = 27.4$ must be substituted for R'_R:

$$1.25 \times 307.1 = \frac{3}{120\pi} \times \frac{6}{2} \times \frac{27.4}{s} \times \frac{(2200/\sqrt{3})^2}{(2.22 + 27.4/s)^2 + 14.2^2}$$

From which $s = 0.293$.

$$\omega_m = (1 - s)\frac{2}{p}\ \omega_s = (1 - 0.293)\frac{2}{6} \times 120\pi = 88.8 \ \text{rad/s}$$
$$= 848 \ \text{r/min}$$

c) Impedance of the right-hand branch of the equivalent circuit at $s = 0.293$, and with R_{ex} shorted out is

$$\overline{Z} = 2.22 + \frac{2.97}{0.293} + j\ 14.2 = 12.36 + j\ 14.2 = 18.82\ \underline{|48.96°}$$

$$\overline{I}_A = \frac{2200\ \underline{|0}}{\sqrt{3} \times 18.82\ \underline{|48.96°}} = 67.49\ \underline{|-48.96°} = 44.3 - j50.9$$

$$\overline{I}_a = \overline{I}_A + \overline{I}_M = 44.3 - j50.9 - j3.92 = 44.2 - j54.8$$

from which

$$I_a = 70.4 \ \text{A}$$

Since this is more than three times the rated current, the whole external resistance may not be shorted out. It must be reduced in steps.

*5.2.9 Squirrel-Cage Motors

When a simple, essentially constant-speed drive is required with only short periods of low-speed operation, when starting is particularly frequent, or when the starting torque required is particularly high, a wound-rotor motor with a variable external rotor-circuit resistor is often employed, as described in Section 5.2.8. When a constant-speed drive with no continuous low-speed running is required, the much simpler and therefore cheaper squirrel-cage motor is normally used.

Squirrel-cage motors are divided into four classes: A, B, C, D. The characteristics of each class are laid down in the specifications of the electrical manufacturers associations NEMA and CEMA and result in machines of markedly different characteristics. These different characteristics are achieved by a variety of different rotor constructions, and machines of one class may have several different rotor types, although there is some correspondence between class and construction.

The manner of specifying the speed-torque characteristics of squirrel-cage machines is illustrated in Fig. 5.27, and Table 5.1 lists the specifications for 230-volt, 50-hp, 4-pole machines.

Note that the specified breakdown torque for a wound-rotor machine of similar rating is 2.25. Also note that the quantities specified in Table 5.1 vary with the machine rating. Manufacturers' specifications must therefore be consulted before a squirrel-cage motor is chosen for a particular application.

Frequently squirrel-cage motors are started by switching them directly across the line. For this purpose, however, the supply system must be such that the relatively large starting current does not cause an appreciable drop in motor terminal potential, which reduces the starting torque and also interferes with other drives connected to the same line. When the supply potential drop would be excessive, some kind of across-the-line starter that reduces the terminal potential difference and, hence, the starting current, is required. For this purpose a three-phase step-

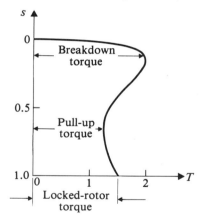

Fig. 5.27 Specification of speed–torque characteristics.

Table 5.1 Squirrel-Cage Motor Torques in Per Unit of Full-Load Torque

Class	Locked Rotor	Breakdown	Pull Up
A	· 1.40	Over 2.00	1.00
B	1.40	2.00	1.00
C	2.00	1.90	Over 1.40
D	2.75	Not specified	Not specified

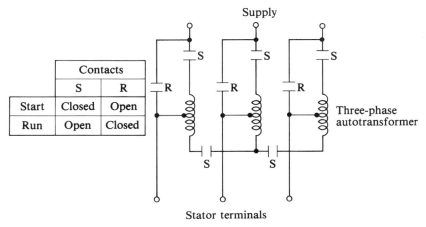

Fig. 5.28 Autotransformer starter for squirrel-cage motors.

down autotransformer may be employed, as illustrated in Fig. 5.28. The autotransformer is switched out of circuit, and its star point opened as the motor approaches full speed. The transformer may be small, since it is only required for very short-time operation, and its losses may therefore be large.

Another method of starting is by a *star-delta switch*. Both ends of each phase of the stator winding must be brought out to terminals, so that at standstill the stator windings may be connected in star or wye. As the motor approaches full speed, the switch is operated and reconnects the stator windings in delta, and the motor runs at full speed. A star-delta starter is equivalent to an autotransformer of ratio $n = 1/\sqrt{3}$.

One of the chief problems of the designer of squirrel-cage induction motors is, thus, that of providing high starting and pull-up torque with low starting current. An additional desirable feature is low slip at full load, since this results in efficient operation. The attempt to reconcile these conflicting requirements by compromises in design has resulted in the motor classes listed in Table 5.1. The most significant design variable in the motor is the effective resistance of the closed rotor cage circuits, which may be controlled by choice of the rotor conductor material, dimensions, and—to some extent—shape.

a) *Class A Motors.* These motors have low rotor-circuit resistance and therefore operate efficiently with small slip ($0.005 < s < 0.01$) at full load. Their disadvantage lies in their relatively low starting torque and high starting current. Their speed–torque characteristic resembles that of a wound-rotor motor with the rotor terminals shorted. Such a characteristic is shown in Fig. 5.25. These machines are suitable for applications in which they are required to start with very low load torque, since they then achieve full speed rapidly and do not overheat during starting.

b) *Class D Motors.* The starting current of a squirrel-cage motor may be kept low and the starting torque made high by forming the rotor bars of high-resistance material such as brass instead of copper. This results in a motor with a speed–torque characteristic similar to that of a wound-rotor motor with some external resistance in the rotor circuit, as illustrated in Fig. 5.25. The breakdown torque occurs at a slip of 0.5 or even higher. The disadvantages of this type of motor are as follows: first, that it runs at high slip (about 0.1) on full load, and is therefore inefficient; second, that the losses in the rotor circuit necessitate building a large and therefore expensive machine for a given power. There are applications, however, in which these disadvantages are outweighed by the desirability of the "soft" speed–torque characteristic. Such applications include intermittent drives requiring high acceleration, and high-impact loads, such as punch presses. In this latter application, a flywheel is fitted that delivers some of its kinetic energy during the punching of the work piece, since the motor decelerates appreciably under the load.

c) *Class B and Class C Motors.* While equivalent circuits and circle diagrams similar to those for the wound-rotor motor may be constructed for Class A and Class D machines, the more elaborate rotors required to provide a combination of high starting torque, low starting current, and low full-load slip are not so simply represented, as will be apparent from a description of their construction. There is no exact correlation between construction and class of Class B and Class C motors.

Class C motors have higher starting torque per ampere of starting current than do Class B motors. Both run at slips of less than 0.05 at full load, with slip as low as 0.005 for large machines, which are therefore very efficient. Class B motors are suitable for driving fans, blowers, centrifugal pumps, motor-generator sets, and—in general—systems requiring low or medium starting torque. Class C motors, which are designed to start at full-load torque, are suitable for driving conveyors, crushers, reciprocating pumps, compressors, and so forth.

d) *Deep-Bar Rotors.* If the rotor conductors are formed of deep, narrow bars, then leakage flux that crosses the slots in the rotor will be distributed as indicated in Fig. 5.29, which shows a cross section of a bar in a rotor slot. To appreciate the effect of this arrangement, consider the bar to be made up of a number of thin horizontal layers. Two such layers—one at the bottom and one at the top of the bar—are shaded in Fig. 5.29.

The leakage inductance of the bottom layer of the conductor is greater than that of the top layer, because the bottom layer is linked by more leakage flux. The current in the low-reactance upper layers will therefore be greater than that in the high-reactance lower layers, and the current density will be greater at the top than at the bottom of the conductor bar. This nonuniform current distribution results in an increase in the effective resistance of the bar. Since this unequal current distribution depends upon reactance, it is more pronounced at high frequency (that is, when the rotor is stationary) than at low frequency (that is, when the rotor is running at full speed). A rotor with deep bars may be designed to have an effective resistance at standstill that is several times its effective resistance at rated speed.

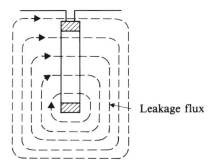

Leakage flux

Fig. 5.29 Deep-bar rotor construction.

Rotor bars may also be employed that are broader at the base than at the top of the slot. This further accentuates the variation in effective resistance with rotor speed.

Double-Cage Rotors. High starting torque with low starting current may also be obtained by building machines whose rotors carry two cage windings, one buried more deeply in the rotor than the other. A possible arrangement of the inner- and outer-cage conductors is illustrated in Fig. 5.30. The effect is similar to that of the deep-bar winding, but the leakage inductance of the inner cage may be much increased by narrowing the slots above it. The outer cage is made of relatively high-resistivity material and the inner cage of low-resistivity material.

Both the deep-bar and the double-cage machine have a "stiff" characteristic with low slip and consequent high efficiency at full load. Equivalent circuits for these machines in which the rotor branch is elaborated may be developed by the designer, but are not obtainable from simple tests. It follows that no corresponding circle diagram can be developed.

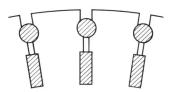

Fig. 5.30 Double-cage rotor construction.

*5.2.10 Speed Control of Squirrel-Cage Motors

A squirrel-cage motor operating from a constant-potential, constant-frequency source is essentially a constant-speed machine. However, this type of motor is so simple, cheap, and robust that in some applications quite elaborate methods of control are justifiable if the result is a satisfactory variable-speed drive employing a squirrel-cage motor.

a) *Pole-Changing.* The stator winding of an induction motor can be so designed that changes in coil connections can change the number of poles. Since a squirrel-cage rotor does not have a specific number of poles, it will operate with any number of poles on the stator. In the simplest case, either of two synchronous speeds may be selected. The basic principle of the pole-changing winding is illustrated in Fig. 5.31, in which conductors at the center of the phase belt of only one phase winding are shown. In Fig. 5.31(a), the coils are connected to produce a 4-pole field. In Fig. 5.31(b), the current in one coil has been reversed by switching, so that the coils produce a two-pole field. With two independent polyphase stator windings, each arranged for pole-changing, four synchronous speeds are available in a single motor. Unlike the wound-rotor motor with variable rotor–circuit resistance, the pole-changing motor gives speed control on no load. Its elaborate stator winding makes it expensive, however.

Pole-changing motors can be designed to have speed–torque characteristics similar to those of any of the four classes of squirrel-cage motors. They can therefore be employed with the types of load specified for those classes when speed change is an additional requirement.

b) *Variation of Terminal Potential Difference.* Equation (5.72) shows that the torque of an induction motor at any speed is proportional to the square of the terminal potential difference. Thus, if the load torque is constant and V_a is varied, the slip s also must vary in order to keep the developed torque of the motor constant. One simple means of varying the terminal potential difference is to introduce impedances into the stator supply lines, but this has disadvantages. If the impedance is resistive, energy is wasted; if it is inductive, the motor and impedance combined operate at a low power factor. Even so, primary impedance is sometimes acceptable for short-time running.

If a variable source such as a variable-ratio autotransformer, is available, then the speed–torque curve may be varied over a wide range. A Class D motor provides

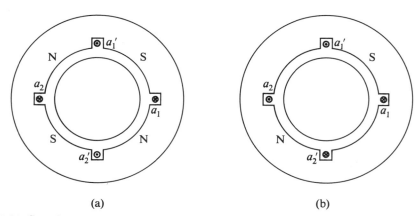

(a)   (b)

Fig. 5.31 Speed control by pole-changing.

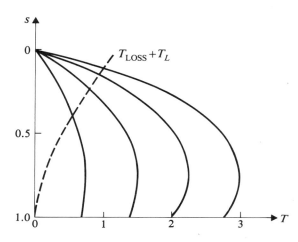

Fig. 5.32 Speed control by variation of source potential difference.

a good speed control for a load such as a fan or a centrifugal pump, for which the starting torque is virtually zero and torque varies approximately as the square of the speed. The operating characteristics of such an arrangement are illustrated in Fig. 5.32. In this system, operation at high slip brings with it the penalty of low efficiency, but since this occurs at low power operation, it may be acceptable.

*5.2.11 Determination of Equivalent-Circuit Parameters

The parameters required to predict performance are shown, referred to the stator, in Fig. 5.20. For Class A and Class D squirrel-cage induction motors and wound-rotor motors, measurements are required from tests, the chief of which are the *No-Load Test* (comparable to the transformer open-circuit test) and the *Locked-Rotor Test* (comparable to the transformer short-circuit test). These two tests are carried out at rated frequency.

a) *No-Load Test.* Run the machine on no load at rated terminal potential difference (with the rotor terminals short-circuited for a wound-rotor machine). Measure

$$V_{NL} = \text{line-to-line stator potential difference}$$
$$I_{NL} = \text{line current}$$
$$P_{NL} = \text{3-phase input power}$$

b) *Locked-Rotor Test.* Clamp the rotor shaft so the rotor remains stationary (with the rotor terminals short-circuited for a wound-rotor machine). Apply a reduced potential difference to the stator terminals so that full-load current flows in the stator windings. Measure

$$V_{lk} = \text{line-to-line stator potential difference}$$
$$I_{lk} = \text{line current}$$
$$P_{lk} = \text{3-phase input power}$$

c) *Stator Winding Resistance.* While the windings are hot after the locked-rotor test, measure the resistance between each pair of stator terminals by means of a bridge. Half of the average of the three resistance measurements gives the per-phase resistance of the stator windings, on the assumption that they are wye-connected.

d) *Squirrel-Cage Motor Calculations.* The necessary per-phase equivalent-circuit parameters may be calculated from the results of tests (a) and (b) as follows:

On no load, s approaches zero, so that in Fig. 5.20,

$$\frac{R'_R}{s} >> |R_s + jX_L| \quad \Omega \tag{5.108}$$

Consequently the equivalent circuit under these conditions may be represented to a close approximation by that shown in Fig. 5.33(a).

In the analysis yielding the circuit of Fig. 5.20, core losses in the machine have been neglected. However, as in the case of the approximate transformer equivalent circuit of Fig. 2.21(a), core losses may be accounted for by a resistance connected across the stator terminals. This core-loss resistance may be considered to be combined with the parallel resistance R'_R/s to give a resistance R_{rot}, which accounts for the rotational losses of the motor due to friction, windage, and core loss. The resulting circuit is shown in Fig. 5.33(b).

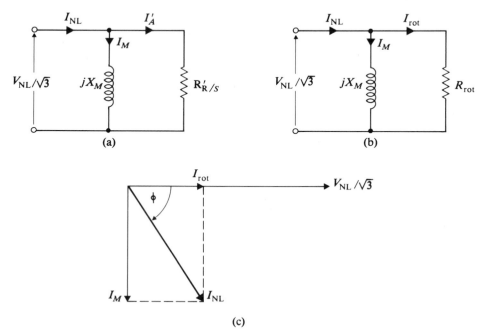

Fig. 5.33 Induction motor no-load test.

The combined resistance $R_{\rm rot}$ is substantially greater than the magnetizing reactance of the motor, which operates on no load at a low power factor. The phasor diagram for this condition of operation is illustrated in Fig. 5.33(c), where

$$\cos \phi = \frac{P_{\rm NL}}{\sqrt{3}\,V_{\rm NL}\,I_{\rm NL}} \tag{5.109}$$

$$I_{\rm rot} = I_{\rm NL} \cos \phi \quad {\rm A} \tag{5.110}$$

$$I_M = I_{\rm NL} \sin \phi \quad {\rm A} \tag{5.111}$$

Then

$$X_M = \frac{V_{\rm NL}}{\sqrt{3}\,I_M} \quad \Omega \tag{5.112}$$

In most machines, $I_M \gg I_{\rm rot}$, so that the magnetizing reactance is very little greater than the no-load impedance as seen from the stator terminals. That is,

$$X_M \simeq Z_{\rm NL} = \frac{V_{\rm NL}}{\sqrt{3}\,I_{\rm NL}} \quad \Omega \tag{5.113}$$

Under locked-rotor conditions, $s = 1$. For most induction machines,

$$X_M \gg |(R_s + R'_R) + jX_L| \quad \Omega \tag{5.114}$$

Thus the branch X_M may be considered an open circuit, and the effective equivalent circuit is that shown in Fig. 5.34. Locked rotor resistance is

$$R_{lk} = \frac{P_{lk}}{3I_{lk}^2} \quad \Omega \tag{5.115}$$

Locked-rotor impedance is

$$Z_{lk} = \frac{V_{lk}}{\sqrt{3}\,I_{lk}} \quad \Omega \tag{5.116}$$

Locked-rotor reactance is

$$X_{lk} = (Z_{lk}^2 - R_{lk}^2)^{1/2} \quad \Omega \tag{5.117}$$

and to a close approximation,

$$R_{lk} + jX_{lk} \simeq R_s + R'_R + jX_L \quad \Omega \tag{5.118}$$

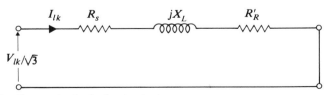

Fig. 5.34 Induction motor locked-rotor test.

Since R_s has been measured directly, R'_R and X_L may be obtained from Eq. (5.118). However, if inequality (5.114) is not satisfied, a correction for X_M may be made to the quadrature component of I_{lk}.

If a wound-rotor motor is to be employed, it will be intended to introduce external rotor circuit resistance for the purpose of starting and/or speed control. For this purpose it is necessary to know the effective turns ratio, the factor k, and the actual rotor-circuit resistance. This necessitates further tests.

e) *Ratios of Terminal Potential Difference to Induced Electromotive Force.* Clamp the rotor. Open circuit the rotor terminals and apply rated potential difference to the stator terminals. Measure the potential difference appearing at the rotor terminals. As may be seen from Fig. 5.15, this will give the ratio V_a/E_{mA}.

Open circuit the stator terminals and apply rated rotor potential difference to the rotor terminals (If this is not specified, the value of E_{mA} in the preceding test will do.) From Fig. 5.15, this will give the ratio E_{ma}/V_A. Normally,

$$X_{ls} = \omega_s L_{ls} >> R_s \quad \Omega \tag{5.119}$$

and

$$X_{lr} = \omega_s L_{lr} >> R_r \quad \Omega \tag{5.120}$$

Since

$$X_{ms} = \omega_s L_{ms} \quad \Omega \tag{5.121}$$

then

$$\frac{V_a}{E_{mA}} = \frac{N_{se}}{N_{re}} \frac{V_a}{E_{ma}} \approx \frac{N_{se}}{N_{re}} \left(\frac{X_{ls} + X_{ms}}{X_{ms}} \right) \tag{5.122}$$

Similarly

$$\frac{E_{ma}}{V_a} = \frac{N_{se}}{N_{re}} \frac{E_{mA}}{V_A} \approx \frac{N_{se}}{N_{re}} \left(\frac{X_{ms}}{X_{ms} + X'_{lr}} \right) \tag{5.123}$$

For a wound-rotor motor, on the grounds of symmetry of the leakage flux paths, it may be assumed that

$$X_{ls} = X'_{lr} \quad \Omega \tag{5.124}$$

Then, from Eqs. (5.122) and (5.123), the turns ratio is given by

$$\frac{N_{se}}{N_{re}} = \left[\frac{V_a}{E_{mA}} \frac{E_{ma}}{V_A} \right]^{1/2} \tag{5.125}$$

The factor k may also be obtained from Eqs. (5.122) and (5.123).

$$\left[\frac{X_{ms}}{X_{ms} + X_{ls}} \right]^2 = \frac{E_{ma}}{V_A} \frac{E_{mA}}{V_a} \tag{5.126}$$

and from the defining Eq. (5.65):

$$k = \frac{X_{ms}}{X_{ms} + X_{ls}} = \left[\frac{E_{ma}}{V_A} \frac{E_{mA}}{V_a} \right]^{1/2} \tag{5.127}$$

f) *Rotor Winding Resistance.* The actual winding resistance R_r is measured in the same manner as the stator winding resistance, but the measurements must be made between the slip rings (not between rotor terminals, since this would introduce brush contact resistance into the measurements). Then from Eq. (5.64),

$$R'_R = \frac{R'_r}{k^2} \quad \Omega \tag{5.128}$$

This value should correspond closely with that obtained from Eq. (5.118).

For Class B and Class C squirrel-cage induction machines, the locked-rotor test described in Section 5.2.11(b) does *not* give values of leakage inductance and rotor resistance appropriate for the prediction of normal load performance. The reason is that, with rotor currents at rated line frequency, the rotor resistance is substantially increased over its dc value and the rotor leakage inductance is somewhat reduced, as described in Section 5.2.9(c). Calculation of the performance of the motor on normal load conditions requires a knowledge of the leakage inductance and rotor resistance when the rotor current is at low values of slip frequency. These values can be obtained by performing the locked-rotor test of Section 5.2.11(b) using a low frequency source with a frequency approximately equal to the expected slip frequency under rated load conditions.

*5.2.12 Rating and Efficiency

The discussion of rating appearing in Section 4.9 and relating to dc machines is equally applicable to induction motors. Per-unit efficiency may again be expressed as

$$\eta = \frac{\text{Input} - \text{Losses}}{\text{Input}} = \frac{\text{Output}}{\text{Output} + \text{Losses}} \tag{5.129}$$

The power input to an induction motor on no load is absorbed in stator resistance loss, in stator and rotor core losses, and in the friction and windage loss incurred in driving the rotor at nearly synchronous speed. There is also a very small rotor resistance loss.

At full speed, the frequency of flux reversals in the rotor approaches zero, so that rotor core loss is negligible. At lower speeds, rotor core loss is no longer negligible; thus, since stator core loss is effectively constant at constant source potential difference, the total core loss increases as speed falls. At full speed, friction and windage losses have their maximum value; and as speed falls, these losses decrease. As an acceptable approximation, therefore, it may be assumed that, for a machine operating on constant source potential difference, the sum of the core losses and the friction and windage losses is constant at all operating speeds. These are then lumped together as the constant *rotational losses* of the machine.

Rotational losses may be determined from the results of the no-load test, since the energy dissipated in R_{rot} of Fig. 5.33(b) is

$$P_{rot} = P_{NL} - 3R_s I_{NL}^2 \quad W \tag{5.130}$$

The second term on the right-hand side of Eq. (5.130) is the stator resistance loss, and since I_{NL} is small,

$$P_{rot} \simeq P_{NL} \quad W \tag{5.131}$$

The resistive losses at any speed are, from Fig. 5.20,

$$P_{res} = 3(R_s + R_R')(I_A')^2 \quad W \tag{5.132}$$

Thus P_{res} may readily be determined for any value of slip. The three-phase mechanical power developed for this same value of slip is, from Eq. (5.68),

$$P_{mech} = \frac{3(1-s)}{s} R_R'(I_A')^2 \quad W \tag{5.133}$$

This mechanical power includes the rotational losses P_{rot}. Thus the efficiency of the machine at a particular slip is given by

$$\eta = \frac{\text{Power Output}}{\text{Power Input}} = \frac{P_{mech} - P_{rot}}{P_{mech} + P_{res}} \tag{5.134}$$

Example 5.6 The test results for a 5-kW, three-phase, 220-V, 60-Hz, 1750 r/min., wound-rotor induction motor are as follows:

No-Load Test	Locked-Rotor Test	Winding Resistance	Ratios
$V_{NL} = 220$ V	$V_{lk} = 56.5$ V	$R_s = 0.158\ \Omega$	Stator excited $V_a/E_{mA} = 1.20$
$I_{NL} = 7.75$ A	$I_{lk} = 20.3$ A	$R_r = 0.150\ \Omega$	Rotor excited $E_{ma}/V_A = 1.05$
$P_{NL} = 490$ W	$P_{lk} = 464$ W		

a) Determine the parameters of the equivalent circuit of Fig. 5.20.

b) Determine the power output and efficiency of the motor at $s = 0.03$.

Solution

a) If the procedure described in Section 5.2.11 is followed and Fig. 5.33(c) is used,

$$\phi = \cos^{-1} \frac{490}{\sqrt{3} \times 220 \times 7.75} = 80.4°$$

$$I_M = 7.75 \sin 80.4° = 7.64 \quad A$$

$$X_M = \frac{220}{\sqrt{3} \times 7.64} = 16.6 \quad \Omega$$

$$Z_{NL} = \frac{220}{\sqrt{3} \times 7.75} = 16.4 \quad \Omega$$

If the approximation of Eq. (5.113) had been employed, an error of little more than 1% in the value of this impedance would have been incurred.

$$R_{lk} = \frac{464}{3 \times 20.3^2} = 0.375 \quad \Omega$$

$$Z_{lk} = \frac{56.5}{\sqrt{3} \times 20.3} = 161 \quad \Omega$$

$$X_{lk} = (1.61^2 - 0.375^2)^{1/2} = 1.57 \quad \Omega$$

Thus, from Eq. (5.118),

$$X_L = 1.57 \quad \Omega$$

$$R'_R = 0.375 - 0.158 = 0.217 \quad \Omega$$

The effective turns ratio of the machine is

$$\frac{N_{se}}{N_{re}} = (1.20 \times 1.05)^{1/2} = 1.12$$

and the factor k is

$$k = \left[\frac{1.05}{1.20} \right]^{1/2} = 0.935$$

With the measured value of R_r and the turns ratio

$$R'_R = \frac{R'_r}{k^2} = \frac{[N_{se}/N_{re}]^2 \, R_r}{k^2} = \frac{1.12^2}{0.935^2} \times 0.150 = 0.215 \quad \Omega$$

which corresponds closely with the value obtained from the locked-rotor test.

b) $$P_{rot} = 490 - 3 \times 0.158 \times 7.75^2 = 462 \quad W$$

The approximation of Eq. (5.131) would thus have introduced an error of some 6% in this quantity.

At $s = 0.03$, the impedance of the right-hand branch of the equivalent circuit of Fig. 5.20 is

$$Z = 0.158 + j1.57 + \frac{0.215}{0.03} = 7.32 + j1.57 = 7.49 \,\underline{|12.1°}$$

$$I'_A = \frac{220}{\sqrt{3} \times 7.49} = 17.0 \quad A$$

$$P_{\text{res}} = 3(0.158 + 0.215) \times 17.0^2 = 323 \quad \text{W}$$

$$P_{\text{mech}} = \frac{3(1 - 0.03)}{0.03} \times 0.215 \times 17.0^2 = 6027 \quad \text{W}$$

Power output is therefore

$$P_{\text{out}} = 6027 - 462 = 5565 \quad \text{W}$$

$$\eta = \frac{5565}{6027 + 323} = 0.876$$

*5.3 SOLID-STATE DRIVES FOR INDUCTION MOTORS

As compared with dc motors, induction motors are robust, easily maintained and cheap. These characteristics make it desirable to employ them in variable-speed drives, even when doing so requires a somewhat complicated control system. Such systems have been developed in recent years for both squirrel-cage and wound-rotor machines, and some of the principal types are described in the following paragraphs.

*5.3.1 Squirrel-Cage Motor with Alternating-Current Power Controller

Section 5.2.10 described the speed of a loaded squirrel-cage motor controlled by varying the stator terminal potential difference. An alternative system employs a constant potential source and an ac *power controller*.

The operation of a power controller, while simple to understand in physical terms, is decidedly complicated to analyze. It is therefore best to begin by considering the operation of the single-phase controller shown in Fig. 5.35(a). Control of the power developed in the load circuit is achieved by variation of the point in the cycle at which the thyristors are turned on. For the circuit of Fig. 5.35(a), with a delay angle α of 60° and a purely resistive load, the terminal potential difference applied to the load circuit and also the load current would have the waveform shown in Fig. 5.35(b). If, however, the load circuit possessed inductance as well as resistance, there would no longer be an instantaneous rise in current at $\omega t = \alpha$. Instead, the current would grow approximately exponentially from zero. Moreover, the inductance would also prolong the current beyond the point $\omega t = \pi$.

If, in Fig. 5.35(a), α were made small or the inductance of the circuit were large, then the current of one half-cycle would be prolonged until the beginning of the succeeding half-cycle. This means that each of the two thyristors would be conducting for a complete half-cycle, and the circuit would behave as if no thyristors were present. This situation occurs when

$$\alpha = \phi = \tan^{-1} \frac{\omega L}{R} \quad \text{rad} \tag{5.135}$$

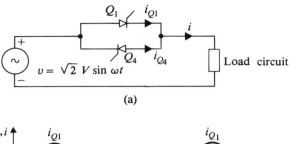

(a)

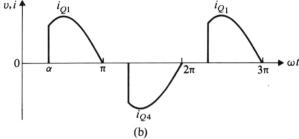

(b)

Fig. 5.35 Output waveform of an ac power controller.

Under these circumstances, or when $\alpha < \phi$, the potential and current waveforms are purely sinusoidal. The effective control range is therefore $\phi < \alpha < \pi$. Even when α is considerably greater than ϕ, a load circuit possessing appreciable inductance draws a current whose waveform has a fundamental component of much greater amplitude than that of any of the higher (necessarily odd) current harmonics.

A three-phase power controller circuit suitable for use with a squirrel-cage induction motor is shown in Fig. 5.36. The thyristors are turned on in sequence

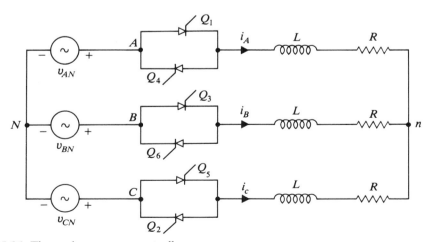

Fig. 5.36 Three-phase power controller.

from Q_1 to Q_6 at intervals of one-sixth of a cycle; as a consequence, 3-phase currents flow. The waveforms are by no means as simple as those shown in Fig. 5.35(b), but a controllable source of control range $\phi < \alpha < \pi$ is nevertheless provided. The fundamental component of the potential difference applied to the motor is controllable from zero to the value of the supply, and speed–torque curves such as those shown in Fig. 5.32 can be produced. A Class D induction motor is normally used, so that the torque will increase with increasing slip over a wide speed range. A simple closed-loop speed control system is illustrated in Fig. 5.37.

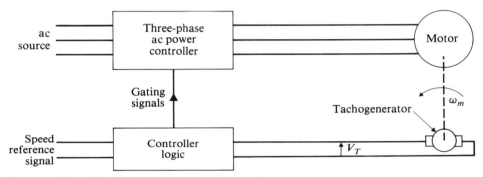

Fig. 5.37 Class D induction motor speed control system.

*5.3.2 Squirrel-Cage Motor with Inverter

The synchronous speed of a squirrel-cage motor may also be varied by varying the source frequency. If currents are to be maintained at normal values and the magnetic system is not to become heavily saturated, then the terminal potential difference must also be varied approximately in proportion to frequency, as may be seen from the equivalent circuit of Fig. 5.19. Only at low frequencies, where the resistance as compared with reactance of the motor windings becomes large, is it necessary to increase the ratio V_a/ω_s to maintain I_{ma} at a constant value, thus maintaining the air-gap flux density.

One possible combination of power semiconductor converters that would give the required variable-frequency, variable-potential excitation from a fixed-frequency, fixed-potential ac source is illustrated in the open-loop system in Fig. 5.38. An alternative combination of controllers would consist of an uncontrolled diode rectifier and an inverter with internal control of its output potential difference as well as frequency. While a 3-phase source is shown, if the power required is not too great, a single-phase source would provide the necessary power input to the rectifier supplying the 3-phase inverter. The potential control of the rectifier ensures that the potential difference V_o varies with frequency ω_s as described previously.

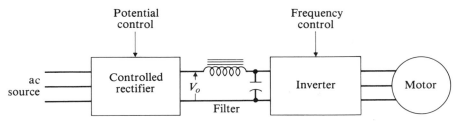

Fig. 5.38 Induction motor speed control by frequency variation.

Figure 5.39 shows a family of motor speed–torque characteristics for various values of ω_s in the continuously variable range that would be available from the system of Fig. 5.38. The shapes of these characteristics can be understood in the light of the analysis of Section 5.2.5.

Equation (5.78), in which the stator resistance R_s has been neglected, may be written in the form

$$|T_{mmax}| = |T_{gmax}| \simeq \frac{3p}{4} \frac{V_a^2}{\omega_s^2 L_L} \quad \text{N} \cdot \text{m} \tag{5.136}$$

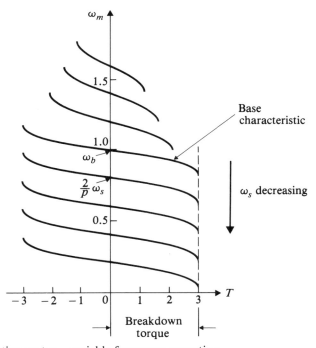

Fig. 5.39 Induction motor—variable frequency operation.

From Eq. (5.63),

$$\frac{X_L}{\omega_s} = L_L = \frac{L_{ls}}{k} + \frac{L'_{lr}}{k^2} \quad \text{H} \tag{5.137}$$

Equation (5.136) shows that, as long as the ratio V_a/ω_s is held constant, the breakdown torque developed by the motor does not change with speed. If once again stator resistance is neglected, then the slip at which the breakdown torque occurs is seen from Eq. (5.75) to be

$$s = \pm \frac{R'_R}{\omega_s L_L} \tag{5.138}$$

Thus $s\omega_s$, the rotor frequency at breakdown torque, is a constant quantity. From the definition of slip in Eq. (5.38), therefore, this constant value means that the quantity $((2/p)\omega_s - \omega_m)$ is constant, where ω_m is the rotor speed at which breakdown torque occurs. As ω_s is varied and the intercept at $(2/p)\omega_s$ made by the speed-torque curve on the axis of ω_m or s varies, the vertical distance between that intercept and the point of breakdown torque does not change. In other words, within the speed range for which R_s may be neglected, the shape of the speed-torque curve does not change with variation of synchronous speed.

The speed ω_b is the synchronous speed of the motor when the maximum terminal potential difference from the inverter of Fig. 5.38 is applied to the motor terminals. This potential difference might well be the rated terminal potential difference of the motor and would be applied at rated frequency, thus establishing the ratio V_a/ω_s. The resulting characteristic may be designated the "base characteristic." Characteristics with synchronous speed less than ω_b represent operating conditions in which V_a/ω_s is held constant. Characteristics with synchronous speed greater than ω_b represent operating conditions in which V_a is fixed while ω_s increases. Thus, from Eq. (5.136),

$$|T_{m\max}| \simeq |T_{g\max}| \propto \frac{1}{\omega_s^2} \quad \text{N·m} \tag{5.139}$$

The curves in Fig. 5.39 show that, at speeds exceeding ω_b, the load torque must be reduced in order to avoid exceeding the breakdown torque of the motor at that frequency.

The induction motor in a variable-frequency, speed-control system may be required to operate in any one of the four quadrants of the speed-torque characteristic shown in Fig. 5.21. Operation in the first two quadrants may be visualized from Fig. 5.40. The speed-torque characteristic shown in full line is that corresponding to source frequency ω_{s1}. On this characteristic, the motor is operating at point p_1 and developing motoring torque T_1. A small reduction of the source frequency to ω_{s2}, for which the corresponding characteristic is shown in broken line, causes the motor to operate at point p_2, developing braking or regenerating torque T_2. The motor would therefore decelerate along the ω_{s2} characteristic until an equilibrium condition was reached in the first quadrant.

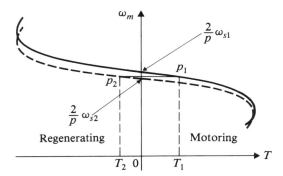

Fig. 5.40 Induction motor— regenerative braking.

Operation in the second quadrant of Fig. 5.39 would not be possible with the source illustrated in Fig. 5.38, since regeneration demands that power be fed back to the supply system. This regeneration requires a negative current at the dc terminals of the controlled rectifier, which it is not able to accept. For a regenerative system, the single controlled rectifier must be replaced by a combination of a rectifier and an inverter that is able to accept the negative current and thus return power to the supply system. Such a combination is known as a *dual converter* and has been described in Section 4.8.2 and Fig. 4.66. The necessary system is illustrated by the block diagram of Fig. 5.41. When the motor regenerates, negative current flows from the inverter to the dual converter, which then returns energy to the ac power source. Operation can be achieved in the third and fourth quadrants of Fig. 5.39 by reversing the phase sequence of the inverter output. This reversal would be made by means of the inverter logic.

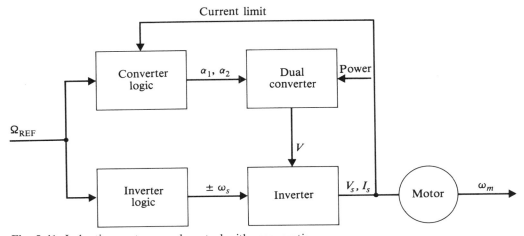

Fig. 5.41 Induction motor speed control with regeneration.

*5.3.3 Rotor Slip Frequency Control

The availability of power semiconductor converters permits a method of induction-motor speed control in which efficiency need not be sacrificed at low-speed operation, and a speed unaffected by load variation may be obtained using squirrel-cage motors.

From Eqs. (5.36) and (5.42),

$$\omega_r = \frac{\omega_r}{s} - \frac{p}{2} \omega_m \quad \text{rad/s} \tag{5.140}$$

Rearrangement yields

$$\frac{s}{1-s} \cdot \omega_m = \frac{2}{p} \omega_r \quad \text{rad/s} \tag{5.141}$$

From Eq. (5.59), the three-phase mechanical power of the machine is

$$3P_{a \text{ mech}} = \frac{3(1-s)}{s} R'_r(I'_a)^2 = T\omega_m \quad \text{W} \tag{5.142}$$

Substitution from Eq. (5.141) in Eq. (5.142) then gives

$$T = \frac{3p}{2} \frac{R'_r(I'_a)}{\omega_r} (I'_a)^2 \quad \text{N} \cdot \text{m} \tag{5.143}$$

This expression for torque shows that if rotor frequency ω_r is held constant, the internal torque per rotor ampere will also be constant. Moreover, if ω_r is held at a lower value, then the torque per rotor ampere will be large. Also, from Eq. 5.36,

$$\omega_s = \omega_r + \frac{p}{2} \omega_m \quad \text{rad/s} \tag{5.144}$$

Thus, if ω_r in a control system is determined by a constant reference signal Ω_r, and if the required value of stator frequency ω_s can be computed from this reference signal and a motor-speed command signal Ω_m, then an exact value of motor speed ω_m, equal to the command signal, can be obtained.

A control system of the type described in the preceding paragraph is illustrated in the block diagram of Fig. 5.42. This is a zero-error system, with integration taking place in both the speed-control unit and the converter logic unit. Consider the system at rest, but with power switched on. Since both Ω_m and ω_m are zero, error signal $k_T(\Omega_m - \omega_m)$ is also zero, and the output of the polarity sensor is +1. The inverter logic is therefore supplying gating signals to the inverter at frequency $\omega_s = \Omega_r$, and Eq. (5.144) is satisfied. Since $k_T(\Omega_m - \omega_m)$ is also the input to the speed control, I_{REF} is also zero, and the converter logic holds the dual-converter output potential difference to zero. The system in this condition is awaiting a finite signal, Ω_m, commanding acceleration of the motor from standstill.

When Ω_m is applied, $k_T(\Omega_m - \omega_m)$ has a positive value. The output of the polarity sensor remains at +1, so that Eq. (5.144) is satisfied for motor action in the first quadrant of Fig. 5.21. Signal I_{REF} increases as a ramp function from zero, error

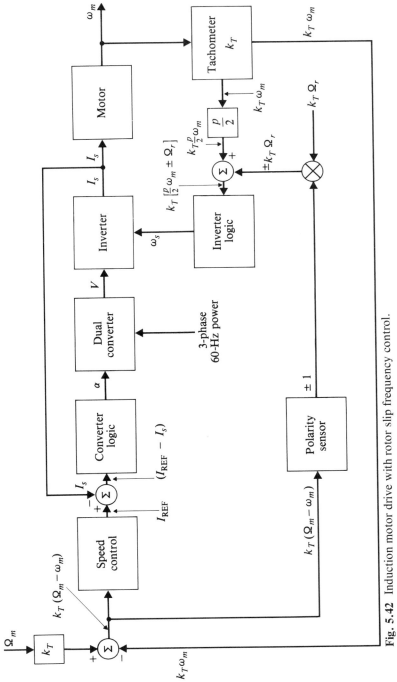

Fig. 5.42 Induction motor drive with rotor slip frequency control.

$(I_{REF} - I_s)$ is positive, and dual-converter output V has a positive value. Energy is supplied to the inverter, and a current at or below the current limit I_{REF} is supplied to the motor, which accelerates. As ω_m approaches Ω_m, $k_T(\Omega_m - \omega_m)$ becomes zero. Equation (5.144) continues to be satisfied. I_{REF} falls until $I_{REF} - I_s$ is zero, and the motor runs in a steady state.

If Ω_m is now reduced so that $k_T(\Omega_m - \omega_m)$ is negative, the polarity sensor output become -1, and Eq. (5.144) is satisfied for negative slip s—that is, for the regenerative condition of operation illustrated in the second quadrant of Fig. 5.21. The motor feeds energy back through the inverter and dual converter to the power source, and the motor speed ω_m decreases until a steady state is again reached.

*5.3.4 Rotor Slip Energy Recovery

The function of the external resistors introduced into the rotor circuit in the rheostatic speed control of Section 5.2.8 is to produce potential differences of rotor frequency that oppose the emf's induced in the rotor windings. If the energy dissipated in the external resistors could instead be returned to the ac source, the overall efficiency of the speed-control system would be very much increased. A method by which this may be done is illustrated in Fig. 5.43(a).

The diode rectifier connected to the 3-phase rotor terminals accepts the power that in the rheostatic control is dissipated in the resistors. The smoothed output of this rectifier is then applied to the dc terminals of an inverter operating at the frequency of the power supply system. The three-phase power from the ac terminals of the inverter is then fed to the induction-motor stator source.

A further advantage of the system of Fig. 5.43(a) is that it permits closed-loop speed control of the motor and the imposition of a current limit. A block diagram of such a system is shown in Fig. 5.43(b). The current feedback signal I_T is obtained from a current transformer in one of the ac lines from the inverter. The speed signal V_T is obtained from a tachogenerator driven by the motor.

If an increase of speed is called for, that is, $V_R - V_T > 0$, then the speed controller in Fig. 5.43(b) generates a current reference signal I_R that is greater than the signal I_T, representing the output current of the controlled rectifier. The gating circuits then call for a lower inverter input potential difference V_{DR} for a given output current I_o. This permits the rotor current to increase, and the increase of rotor current is reflected in an increase of stator current. The consequence is an increase in the developed torque of the motor, which accelerates.

5.4 SINGLE-PHASE INDUCTION MOTORS

In domestic applications, where only a single-phase ac source is normally available, and in a very large number of low-power drives in industry, it is either necessary or convenient to employ a single-phase induction motor. This machine resembles a small, three-phase, squirrel-cage motor except for the fact that when it is running at full speed, only a single winding on the stator is usually excited.

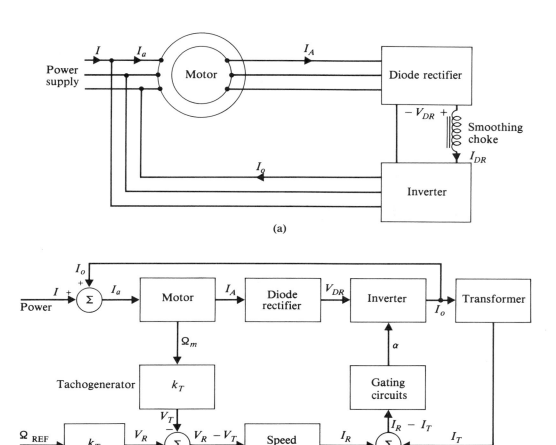

Fig. 5.43 Slip energy recovery system.

Single-phase induction motors are usually 2- or 4-pole and rated at two horse-power or less. Slower and larger motors may be manufactured, however, but are not generally stock items. The large volume of single-phase motors produced results in their being cheaper than 3-phase motors of the same rating. A typical single-phase motor is shown in Fig. 5.44.

*5.4.1 Steady-State Operation

The windings of a 2-pole, single-phase induction motor are represented by the diagram in Fig. 5.45, where only the center turn of the single stator winding is shown. The diagram also shows the chosen positive direction of current in that winding.

Fig. 5.44 Single-phase induction motor. *(Canron Limited)*

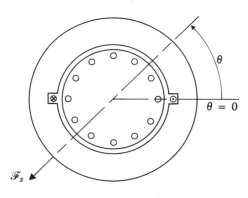

Fig. 5.45 Single-phase induction motor.

The stator mmf axis necessarily coincides with the axis of the stator winding and is therefore stationary. Since the winding is excited with alternating current, the resulting stator mmf wave is stationary but pulsates in amplitide, so that the magnitude of the stator mmf acting across the air gap at any angle θ is a sinusoidal function of time.

When the rotor is stationary, currents are induced in its squirrel-cage winding by transformer action in such a direction as to produce a rotor mmf opposing that caused by the stator winding. The rotor mmf axis coincides with that of the stator; consequently the relationship of Eq. (3.130),

$$T = -K \; \hat{\mathscr{F}}_{sg} \, \hat{\mathscr{F}}_{rg} \sin \delta \quad \text{N·m} \tag{5.145}$$

shows that no torque is produced, since $\delta = 0$. Thus the stationary motor simply behaves as a single-phase transformer with an air gap in the magnetic system and a short-circuited secondary winding.

The air-gap mmf due to the stator winding at any angle θ shown in Fig. 5.45 is

$$\mathscr{F}_{sg} = \frac{N_s i_s}{2} \sin \theta \quad \text{A} \tag{5.146}$$

where $N_s i_s$ is the peak magnitude of the stator mmf wave at any instant. If

$$i_s = \hat{i}_s \cos \omega_s t \quad \text{A} \tag{5.147}$$

then

$$\mathscr{F}_{sg} = \frac{N_s \hat{i}_s}{2} \cos \omega_s t \sin \theta = \frac{N_s \hat{i}_s}{4} \sin (\theta + \omega_s t) + \frac{N_s \hat{i}_s}{4} \sin (\theta - \omega_s t) \quad \text{A} \tag{5.148}$$

As was shown in Section 3.3.5, each of the terms on the right-hand side of Eq. (5.148) describes a constant-amplitude, sinusoidally distributed, rotating mmf wave similar to that produced by a 3-phase stator winding. The two waves are rotating in opposite directions, however. The directions and magnitudes of the peak amplitudes of these two rotating stator mmf waves and the resulting peak amplitude of the actual stationary stator mmf wave are illustrated for a particular instant by the space-vector diagram in Fig. 5.46. Thus the pulsating flux wave produced in the air gap of the stationary machine may be considered to be the resultant of two rotating flux waves of equal amplitude. Each of these flux waves may be considered to

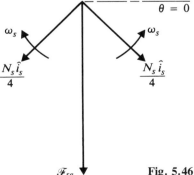

Fig. 5.46 Space-vector diagram of stator mmf waves.

produce induction-motor action on the stationary squirrel-cage rotor, but the corresponding torques are in opposite directions and of equal magnitude. Thus the net torque on the stationary rotor is zero.

The emf induced in the stator winding is that produced by the stationary, pulsating, air-gap flux wave. The flux wave is produced by the resultant of the stator and rotor mmf's. Since the pulsating flux wave may be resolved into two rotating flux waves, the equivalent circuit that represents the effect of the air-gap flux may be split into two parts representing the effects of the forward and backward flux waves respectively. The resulting equivalent circuit is shown in Fig. 5.47. Both parts of the equivalent circuit carry the stator current I_s, and the impedances have been halved because the total induced emf must be equal to the applied potential difference. Thus, for the stationary machine,

$$E_{mf} = E_{mb} = \frac{V_s}{2} \quad \text{V} \tag{5.149}$$

Now consider what happens when the rotor is given some rotation in the forward direction and is running at slip s_f with respect to the forward rotating flux wave. By the definition of slip,

$$s_f = \frac{(2/p)\omega_s - \omega_m}{(2/p)\omega_s} \tag{5.150}$$

where $(2/p)\omega_s$ is the speed of rotation of the forward flux wave. From Eq. (5.150),

$$\omega_m = (1 - s_f)\frac{2}{p}\omega_s \quad \text{rad/s} \tag{5.151}$$

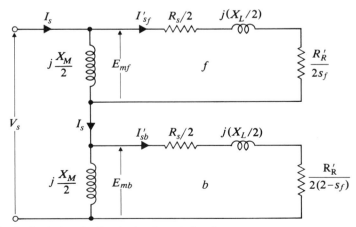

Fig. 5.47 Equivalent circuit of a single-phase induction motor.

The rotor currents induced by the forward flux wave are of slip frequency $s_f \omega_s$. These rotor currents produce a rotating field traveling forward with respect to the rotor at forward slip speed $s_f(2/p)\omega_s$ rad/s and hence at speed $(2/p)\omega_s$ with respect to the stator, just as in a 3-phase machine. The circuit branch carrying the rotor current I'_{sf} has an impedance to currents induced by the forward wave of

$$\overline{Z}_{Rf} = \frac{R_s}{2} + \frac{1}{2}\frac{R'_R}{s_f} + j\frac{X_L}{2} \quad \Omega \tag{5.152}$$

This is shown in parallel with $jX_M/2$ in Fig. 5.47.

The speed of the backward flux wave with respect to the stator is $-(2/p)\omega_s$, so that the rotor slip with respect to the backward flux wave is

$$s_b = \frac{-(2/p)\omega_s - \omega_m}{-(2/p)\omega_s} = 1 + \frac{\omega_m}{(2/p)\omega_s} \tag{5.153}$$

But, from Eq. (5.151)

$$\frac{\omega_m}{(2/p)\omega_s} = 1 - s_f \tag{5.154}$$

Thus,

$$s_b = 2 - s_f \tag{5.155}$$

Rotor currents induced by the backward flux wave are of slip frequency $(2 - s_f)\omega_s$. These rotor currents produce a rotating field traveling backward with respect to the rotor at backward slip speed $(2 - s_f)(- 2/p)\omega_s$, and hence at speed $-(2/p)\omega_s$ with respect to the stator. The circuit branch carrying the rotor current I'_{sb} has an impedance to currents induced by the backward wave of

$$\overline{Z}_{Rb} = \frac{R_s}{2} + \frac{1}{2}\frac{R'_R}{(2 - s_f)} + j\frac{X_L}{2} \quad \Omega \tag{5.156}$$

and this is shown in parallel with $jX_M/2$ in Fig. 5.47. At normal operating speed, the impedance offered to I_s by the "f" parallel circuit in Fig. 5.47, in which $R'_R/2s_f$ is large, is greater than that offered by the "b" parallel circuit, in which $R'_R/2(2 - s_f)$ is small. Thus $E_{mf} > E_{mb}$. This means that as the motor speed increases, the forward air-gap flux wave increases in amplitude, while the backward wave decreases. At all speeds, however,

$$\overline{E}_{mf} + \overline{E}_{mb} = \overline{V}_s \quad V \tag{5.157}$$

Thus with the rotor in motion, the torque due to the forward rotating flux wave increases over that at standstill, while the torque due to the backward flux wave decreases. In terms of a 3-phase motor characteristic, the driving torque is increased, and the plugging torque is decreased. The two speed–torque curves, which should be combined to give the resultant speed–torque curve of the single-phase motor, are therefore as illustrated in Fig. 5.48.

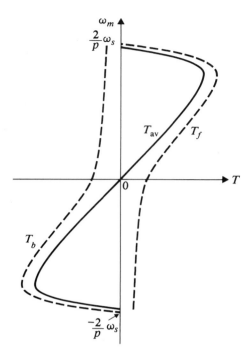

Fig. 5.48 Speed–torque curve of a single-phase induction motor.

It must be realized that the torque shown in the curve for the single-phase motor is the average value of a pulsating torque, because the instantaneous power in a single-phase circuit is pulsating. If the average value of power is P, then

$$P = T_{\mathrm{av}}\omega_m \quad \mathrm{W} \tag{5.158}$$

Discussions of the operation of the 3-phase induction motor have shown that the internal torque developed by the motor is given by the power dissipated in the fictitious rotor-circuit resistance R_R'/s divided by the synchronous speed of the motor. This principle may be applied to both sections of the equivalent circuit of Fig. 5.47 to determine the net torque developed by the single-phase motor. If the circuit of Fig. 5.47 is solved for a particular pair of values of V_s and ω_m, then the forward and backward torques are given by

$$T_f = \frac{p}{2}\cdot\frac{1}{\omega_s}\,(I_{sf}')^2\,\frac{R_R'}{2s_f} \quad \mathrm{N\cdot m} \tag{5.159}$$

$$T_b = -\frac{p}{2}\,\frac{1}{\omega_s}\,(I_{sb}')^2\,\frac{R_R'}{2(2-s_f)} \quad \mathrm{N\cdot m} \tag{5.160}$$

and the average internal torque of the single-phase motor is

$$T_{\mathrm{av}} = T_f + T_b \quad \mathrm{N\cdot m} \tag{5.161}$$

When the rotor is stationary and $s_f = 1$, then $I'_{sf} = I'_{sb}$, and $T_{av} = 0$, so that a single winding on the stator develops no starting torque. This starting torque may be provided by means of a second stator winding. Such starting methods are discussed in Section 5.4.3.

Due to the backward torque, a single-phase motor develops a lower net torque and runs at a greater slip for a given load than would a 3-phase motor of the same physical size. Thus for a given power rating, a single-phase motor is slightly larger and less efficient than a 3-phase motor.

*5.4.2 Determination of Equivalent-Circuit Parameters

In order to determine the parameters of the circuit of Fig. 5.47, tests may be carried out on the main winding of a single-phase motor that are similar to those carried out on a three-phase squirrel-cage motor. If the approximations suggested in Section 5.2.11 are accepted, then the no-load test determines $X_M/2$ and the rotational losses, while the locked-rotor test gives jX_M in parallel with $R_s + jX_L + R'_R$. The assumption in the locked-rotor test for large three-phase motors that

$$X_M >> | R_s + jX_L + R'_R | \quad \Omega \qquad (5.162)$$

is less valid in the case of small single-phase machines.

Example 5.7 Tests on the main winding of a one-hp, 4-pole, 220-V, 60-Hz, single-phase induction motor gave the following results:

$$
\begin{aligned}
V_{NL} &= 220 \quad V & V_{lk} &= 80 \quad V \\
I_{NL} &= 4.2 \quad A & I_{lk} &= 10.4 \quad A \\
P_{NL} &= 192 \quad W & P_{lk} &= 404 \quad W \\
& & R_s &= 1.54 \quad \Omega
\end{aligned}
$$

Determine the line current, power factor, shaft torque, and efficiency of the motor at a speed of 1725 r/min.

Solution The equivalent circuit of the motor must first be determined.

$$\frac{X_M}{2} \simeq \frac{V_{NL}}{I_{NL}} = \frac{220}{4.2} = 52.4 \quad \Omega$$

$$P_{rot} \simeq P_{NL} = 192 \quad W$$

The equivalent circuit and phasor diagram for the locked-rotor test are shown in Fig. 5.49.

$$\phi_{lk} = \cos^{-1} \frac{P_{lk}}{V_{lk}I_{lk}} = \cos^{-1} \frac{404}{80 \times 10.4} = 60.9°$$

Let

$$\overline{V}_{lk} = 80 \underline{|0} \; V$$

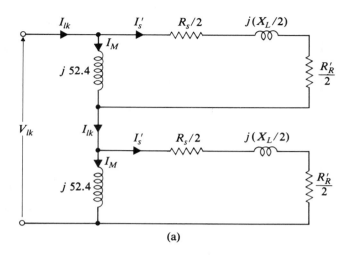

(a)

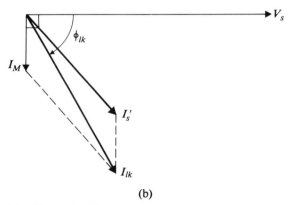

(b)

Fig. 5.49 Diagram I for Example 5.7.

Then

$$\bar{I}_{lk} = 10.4 \underline{|-60.95°} = 5.05 - j9.09 \quad \text{A}$$

$$\bar{I}_M = \frac{\bar{V}_{lk}}{jX_M} = \frac{80 \underline{|0}}{j104.8} = -j0.763 \quad \text{A}$$

$$\bar{I}'_s = \bar{I}_{lk} - \bar{I}_M = 5.05 - j8.33 = 9.74 \underline{|-58.8°} \quad \text{A}$$

$$Z_R = \frac{V_{lk}}{I'_s} = \frac{80}{9.74} = 8.21 \quad \Omega$$

$$R_s + R'_R = \frac{P_{lk}}{(I'_s)^2} = \frac{404}{9.74^2} = 4.26 \quad \Omega$$

$$R'_R = 4.26 - R_s = 4.26 - 1.54 = 2.72 \quad \Omega$$

$$X_L = [(Z'_R)^2 - (R_s + R'_R)^2]^{1/2} = [8.21^2 - 4.26^2]^{1/2} = 7.02 \quad \Omega$$

The equivalent circuit of the motor is shown in Fig. 5.50.
At 1725 r/min,

$$s_f = \frac{1800 - 1725}{1800} = 0.0417$$

$$\overline{Z}_{Rf} = 0.770 + \frac{1.36}{0.0417} + j3.51 = 33.4 + j3.51 = 33.6 \underline{|6.0°} \quad \Omega$$

$$\overline{Z}_f = \frac{j(X_M/2)\overline{Z}_{Rf}}{j(X_M/2) + \overline{Z}_{Rf}} = \frac{j52.4 \times 33.6 \underline{|6.0°}}{j52.4 + 33.4 + j3.51} = \frac{52.4 \underline{|90°} \times 33.6 \underline{|6.0°}}{65.1 \underline{|59.1°}}$$

$$= 27.0 \underline{|36.9°} = 21.6 + j16.2 \quad \Omega$$

$$\overline{Z}_{Rb} = 0.770 + j3.51 + \frac{1.36}{2 - .0417} = 1.47 + j3.51 = 3.80 \underline{|67.3°} \quad \Omega$$

$$\overline{Z}_b = \frac{j52.4 \times 3.80 \underline{|67.3°}}{j52.4 + 1.47 + j3.51} = \frac{52.4 \underline{|90°} \times 3.80 \underline{|67.3°}}{55.9 \underline{|88.5°}}$$

$$= 3.56 \underline{|68.8°} = 1.29 + j3.32 \quad \Omega$$

$$\overline{Z} = \overline{Z}_f + \overline{Z}_b = 21.6 + j16.2 + 1.29 + j3.32 = 22.9 + j19.5 = 30.1 \underline{|40.4°} \quad \Omega$$

Let

$$\overline{V}_s = 220 \underline{|0} \quad V$$

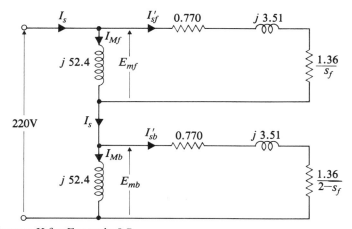

Fig. 5.50 Diagram II for Example 5.7.

Then

$$\overline{I}_s = \frac{220\,\underline{|0}}{30.1\,\underline{|40.4°}} = 7.31\,\underline{|-40.4°}\quad A$$

$$PF = \cos 40.4° = 0.762$$

$$\overline{E}_{mf} = \frac{\overline{Z}_f}{\overline{Z}} \times \overline{V}_s = \frac{27.0\,\underline{-|36.9°}}{30.1\,\underline{|40.0°}} \times 220\,\underline{|0} = 197\,\underline{|-3.5°}\quad V$$

$$\overline{I}_{sf} = \frac{\overline{E}_{mf}}{\overline{Z}_{Rf}} = \frac{197\,\underline{|-3.5°}}{33.6\,\underline{|6.0°}} = 5.86\,\underline{|-9.5°}\quad A$$

$$\overline{E}_{mb} = \frac{\overline{Z}_b}{\overline{Z}} \times \overline{V}_s = \frac{3.56\,\underline{|68.8°}}{30.1\,\underline{|40.4°}} \times 220\,\underline{|0} = 26.0\,\underline{|28.4°}\quad V$$

$$I'_{sb} = \frac{\overline{E}_{mb}}{\overline{Z}_{Rb}} = \frac{26.0\,\underline{|28.4°}}{3.80\,\underline{|67.3°}} = 6.84\,\underline{|-38.9°}\quad A$$

From Eqs. (5.159) to (5.161), the internal torque is

$$T_{av} = \frac{4}{2}\frac{1}{120\pi}\left[5.86^2 \times \frac{1.36}{0.0417} - 6.84^2 \times \frac{1.36}{2 - 0.0417}\right] = 5.77\quad N\cdot m$$

Torque absorbed in rotational losses is

$$T_{rot} = \frac{P_{NL}}{\omega_m} = \frac{192}{1725} \times \frac{60}{2\pi} = 1.06\quad N\cdot m$$

Shaft torque is

$$T_{av} - T_{rot} = 5.77 - 1.06 = 4.71\quad N\cdot m$$

Power output is

$$P_o = 4.71 \times 1725 \times \frac{2\pi}{60} = 851\quad W$$

Power input is

$$V_s I_s(PF) = 220 \times 7.31 \times 0.762 = 1225\quad W$$

Efficiency is

$$\eta = \frac{851}{1225} = 0.695$$

The motor is, thus, overloaded. If the load were reduced to the rated output of 746 W, the speed would increase, and probably the efficiency would increase also.

5.4.3 Starting of Single-Phase Induction Motors

As may be seen from the curve of T_{av} versus ω_m for a single-phase induction motor in Fig. 5.48, if the stator winding is excited and the rotor is given some speed of rotation in either direction, then the motor will develop an internal torque that tends to maintain that rotation. Thus, if the rotor is rotated simply by twisting the shaft quickly enough by hand when the stator winding is excited, it will run up to full speed on no load.

It has been seen in Section 3.3.4 that with two windings displaced $\pi/2$ radians from one another on the stator and excited with equal currents $\pi/2$ radians out of phase with one another, a rotating mmf wave of constant amplitude is produced. The locus of the end of the space vector representing the maximum resultant mmf due to these two windings may be compared to the circular pattern obtained on an oscilloscope screen from two sinusoidal input signals of equal amplitude displaced in phase by $\pi/2$ radians from one another. If the frequency of the input signals is so low that the direction of rotation of the spot on the screen may be observed, then it may also be observed that if the phase sequence of the signals is reversed, so also is the direction of rotation of the spot.

If the currents in the two stator windings are of unequal amplitude or if the windings have unequal numbers of turns, then a rotating mmf wave is still produced, but its peak amplitude varies as it rotates, having a maximum value on the axis of one winding and a minimum value on the other. The locus of the end of the space vector may be compared to the elliptical pattern with major axis either vertical or horizontal obtained on an oscilloscope screen from two sinusoidal input signals of unequal amplitude displaced in phase by $\pi/2$ radians from one another.

If, in addition, the currents in the two stator windings have a phase displacement of some angle other than $\pi/2$ radians, then the variation in peak amplitude of the rotating mmf wave is further increased, and the maximum and minimum values no longer occur on the axes of the windings. Now the locus may be compared to the diagonally oriented ellipse obtained on an oscilloscope from two sinusoidal signals of the same frequency but of unequal amplitude with a phase displacement other than $\pi/2$ radians.

The last combination of stator excitations corresponds to the commonest method of starting a single-phase induction motor—by the addition of an auxiliary stator winding displaced $\pi/2$ electrical radians from the main p-pole winding. The two windings are excited by currents displaced in phase from one another. The resulting "elliptical" rotating field can be made to produce an average torque at standstill well in excess of the full-load torque of the motor.

a) *Split-Phase Motor.*   In the split-phase motor, both stator windings are excited from the same source, and the necessary phase displacement of their currents is obtained by designing the auxiliary winding with a higher ratio of resistance to reactance than is the case for the main winding. A circuit diagram for the motor is shown in Fig. 5.51. A phasor diagram corresponding to typical starting conditions is shown in Fig. 5.52, in which I_a is the auxiliary-winding current, I_m is the main winding current, and the phasor sum of these two is I_s, the line current to the stator. The auxiliary winding is excited during acceleration from standstill up to about 0.8 of synchronous speed, when a centrifugally operated switch opens. The shape of a typical speed–torque curve is shown in Fig. 5.53. The requisite high ratio of resis-

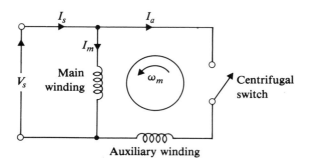

Fig. 5.51 Circuit of a split-phase motor.

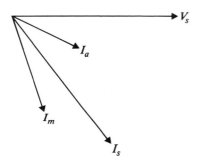

Fig. 5.52 Starting conditions for a split-phase motor.

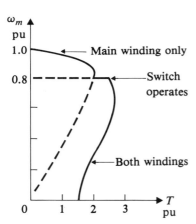

Fig. 5.53 Speed–torque curve for a split-phase motor.

tance to reactance in the auxiliary winding is obtained by making it of smaller wire than the main winding. The smaller wire is acceptable, since the winding is excited for a short period only, and high current density may be employed without risk of overheating.

b) *Capacitor Motors.*   In the commonest type of capacitor motor, the auxiliary winding is connected to the source in series with a capacitor (approx. 20 μF for a

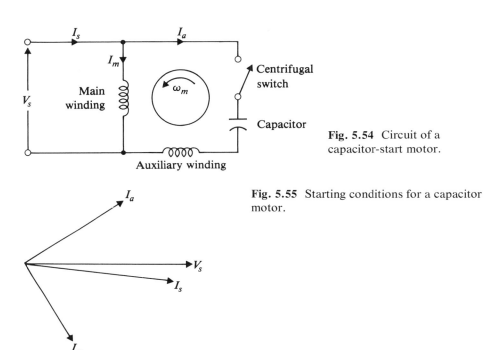

Fig. 5.54 Circuit of a capacitor-start motor.

Fig. 5.55 Starting conditions for a capacitor motor.

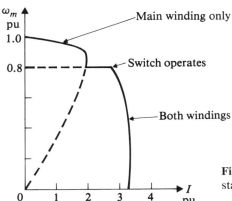

Fig. 5.56 Speed–torque curve for a capacitor-start motor.

1-hp motor) as indicated in the circuit diagram of Fig. 5.54. A phasor diagram corresponding to typical starting conditions is shown in Fig. 5.55. Again the auxiliary winding is disconnected after starting, and consequently it and the capacitor may be designed for intermittent service. The shape of a typical speed–torque characteristic is shown in Fig. 5.56. Motors are also built in which the capacitor and auxiliary winding are not disconnected after starting. This simplifies construction of the motor by eliminating the centrifugally operated switch, but necessitates a more expensive auxiliary winding and capacitor for continuous service. Quiet running, due to reduced torque pulsations, and a high operating power factor can be obtained by this means. Capacitor-start motors may also be fitted with two capacitors, one of which is switched out and one left permanently in the auxiliary-winding circuit. This is expensive, but gives the optimum combination of starting and running performance.

c) *Shaded-Pole Motor.* The shaded-pole motor has salient poles carrying the stator winding, as illustrated in Fig. 5.57. The main windings are coils resembling those of the field windings of a small dc machine. One portion of each pole is surrounded by a short-circuited turn of copper called a *shading coil.* Induced currents in the shading coil cause the flux in the shaded portion of the pole to lag in phase behind the flux in the unshaded portion. Thus the stator has in effect two axes of magnetization with resultant mmf's, which are displaced in phase, acting along these axes. An "elliptical" rotating field is thus produced.

The shaded-pole principle is employed only in very small motors needing low starting torque, such as those used for driving fans. It may also be employed as a starting method for synchronous reluctance motors such as those used in electric clocks. The efficiency of shaded-pole motors is low, but they are simple, reliable, and cheap. A typical motor speed–torque characteristic and a fan-load characteristic are shown in Fig. 5.58.

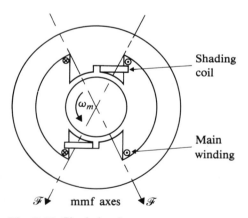

Fig. 5.57 Shaded-pole motor.

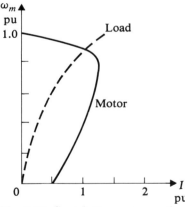

Fig. 5.58 Speed–torque curves for a shaded-pole motor and typical load.

*5.5 SYNCHROS

The class name "synchro" designates units manufactured under a variety of trade names (Selsyn, Autosyn, Magslip, etc.). These units superficially resemble small electric motors and vary in size from roughly that of a typical 1-hp capacitor motor down to small control-system elements of 10 mm in diameter or less. While a number of specialized units for particular purposes are available, these are for the most part variations of a common basic type consisting of a rotor carrying a single winding enclosed in a stator on which are three wye-connected windings arranged with their axes displaced 120 electrical degrees from one another.

Synchros may be employed as data-transmitting elements of servomechanisms and as torque-transmitting elements for remote-position indicating or the transmission of small amounts of power.

a) *Data Transmission.* The simplest data-transmission link consists of two synchros interconnected as shown in Fig. 5.59. When the input terminals, R1 R2, of the transmitter synchro are connected to an ac source, the resulting alternating flux linking the stator windings induces in them emf's whose magnitude and phase uniquely define the angular position of the rotor relative to the stator. This position may be measured from an arbitrarily-chosen reference position, giving the angle θ_i shown in Fig. 5.59.

When the stator windings of the control-transformer synchro are excited by the terminal potentials of the transmitter, an alternating flux similar to that of the transmitter is produced in the control transformer. This alternating flux induces emf's in the control transformer stator windings equal and opposite to the applied potentials.

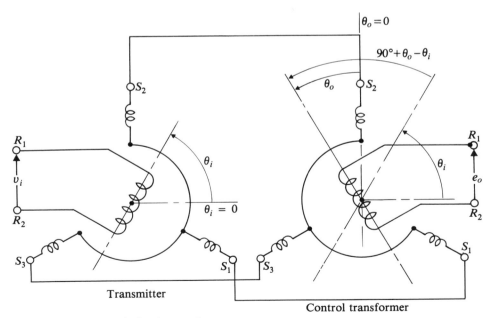

Fig. 5.59 Data transmission by synchros.

This last statement ignores the effects of exciting currents and leakage impedances of the synchros; but, since these are small, the alternating flux induced in the control transformer defines to a high degree of accuracy the position of the transmitter rotor. The flux at angle θ_i in the control transformer of Fig. 5.59 will therefore be the same as that at angle θ_i of the transmitter, provided the two synchros are identical.

The alternating flux in the control transformer induces an emf in the rotor winding that will be zero when the axis of that winding is at an angle of 90 electrical degrees to the axis of the induced flux. The reference position of the control transformer may be conveniently taken as $+90°$ from the transmitter reference position, and the actual position of the control-transformer rotor will be θ_o measured from this reference position. Thus, if

$$v_i = \hat{v}_i \sin \omega_c t \quad \text{V} \tag{5.163}$$

where ω_c is the angular frequency of the carrier, then, ignoring leakage impedance,

$$e_o = \hat{v}_i \sin \omega_c t \cos (90 + \theta_o - \theta_i) \quad \text{V} \tag{5.164}$$

If the *error* angle is defined as

$$\theta_e = \theta_i - \theta_o \quad \text{rad} \tag{5.165}$$

then

$$e_o = \hat{v}_i \sin \omega_c t \sin \theta_e \quad \text{V} \tag{5.166}$$

If θ_e is small,

$$e_o = \acute{V}_i \, \theta_e \sin \omega_c t \quad \text{V} \tag{5.167}$$

that is, the output potential difference at the control transformer rotor terminals is directly proportional to error angle θ_e. The data transmission link indicates the direction of the error, since a change of sign in the error angle results in a phase reversal of e_o. This is illustrated in Fig. 5.60.

Figure 5.61 shows a schematic for a position-control servomechanism in which the transmitter potential v_i is in phase quadrature with the constant reference potential v_d applied to the ac control motor. The output of the control transformer, e_o, when amplified, provides the source of control potential v_q applied to the second stator winding of the motor. For any transmitter angle θ_i, there will be two positions of the control-transformer rotor that yield zero values of e_o. One of these positions represents a condition of stable equilibrium of the system where $\theta_o = \theta_i$, since a small error displacement on either side of it will result in a potential v_q applied to the motor of such a phase relationship to v_d that the motor drives the control transformer back to the stable equilibrium position. The second position represents a condition of unstable equilibrium, where $\theta_o = \pi + \theta_i$, since a small angular displacement on either side of it will result in the motor's driving in such a direction as to increase the displacement until the stable equilibrium position is reached.

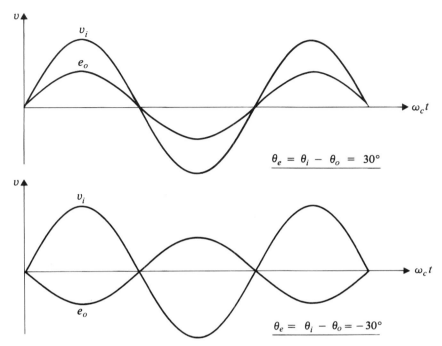

$$\theta_e = \theta_i - \theta_o = 30°$$

$$\theta_e = \theta_i - \theta_o = -30°$$

Fig. 5.60 Phase discrimination by synchros.

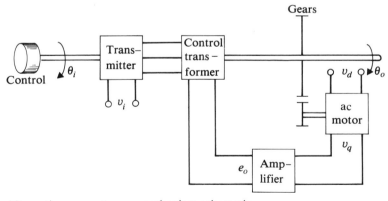

Fig. 5.61 Alternating-current servomechanism schematic.

A transmitter rotor usually has salient poles providing space for a winding giving a large mmf, and this reduces the effect on the transmitter of any current which may be drawn from it by the control transformer. A control transformer usually has a cylindrical rotor, so that its input impedance is independent of rotor position. It has a substantially higher impedance than the transmitter, so that if necessary several control transformers may be connected to one transmitter.

b) *Torque Transmission.* When torque transmission is required, a circuit similar to that in Fig. 5.59 is employed, but the control transformer is replaced by a receiver of essentially the same form as the transmitter. The rotor of both the transmitter and receiver are then excited from the same ac source. As a result of this excitation, currents flow in the stator windings of the two elements unless the two rotors are at the same angle relative to their stator windings (magnetizing currents are again neglected here). When stator currents flow, the elements develop internal torques tending to reduce the angle between the rotors. If the transmitter rotor is fixed, only the receiver rotor moves to reduce the difference angle to zero.

Again by employing high-impedance receivers, it is possible to connect several receivers to one transmitter. Such an arrangement is frequently employed for position indicators, in which case quite small elements may give sufficient torque to operate the indicators.

Large 3-phase machines may be used when transmission of substantial torque is required. For example, two remote shafts may be maintained in synchronism by coupling to them wound-rotor induction motors whose stators are corrected to a common source, while the rotor windings are connected in parallel.

*5.6 LINEAR INDUCTION MOTORS

The word "linear" in the title of this section is not intended to signify the absence of magnetic saturation in a machine—that is, an assumed condition that permits linear mathematical analysis of its performance. The word refers, instead, to the kind of motion produced by the motors to be discussed here. This motion is translational, or linear—that is, in a line—as opposed to the rotational motion of the motors previously discussed. The motion of the armature of the actuator in Fig. 3.1 is, of course, "linear" in this sense, but is restricted in extent. The motion produced by a linear motor can proceed indefinitely, limited only by the length of its track.

Many different configurations of the linear induction motor (LIM) have been devised. The configuration that will be most fully discussed here has been developed for the purpose of transportation drives. Its operation is most easily understood by deriving its configuration from that of a squirrel-cage rotary machine.

*5.6.1 LIM Configurations

Section 3.4.2 stated that the limiting case of the squirrel-cage rotor winding, illustrated in Fig. 3.48, is that of a cylinder of conductor enclosing the rotor's ferromagnetic core, as shown in Fig. 5.62(a).

The terms "stator" and "rotor" are inappropriate when discussing the operation of a LIM. They will therefore be replaced by the terms "primary" and "secondary," respectively, with the understanding that either the primary or the secondary may be the moving member.

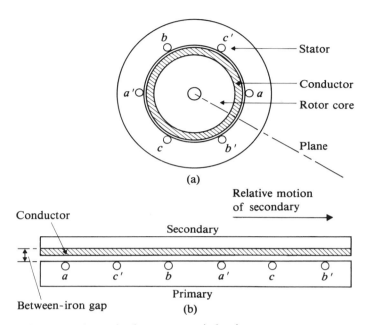

Fig. 5.62 Development of a squirrel-cage rotary induction motor.

The LIM configuration of most interest may be derived from the rotary motor by the development process, shown in detail in Fig. 4.7, to yield the arrangement shown in Fig. 5.62(b). The flux-density wave produced by the primary excitation in the machine of Fig. 5.62(a) and rotating in a positive (counterclockwise) direction, thus producing positive (counterclockwise) torque on the secondary, now becomes a traveling flux-density wave moving from left to right in the machine of Fig. 5.62(b). If the primary is fixed and the secondary is free, the force acting on the secondary will drive it to the right at some speed less than that of the traveling wave. If the secondary is fixed and the primary is free, the force of reaction on the primary will drive it to the left.

For a continuous force to be developed over a given distance, one or the other member of the LIM must be extended over the same distance. Which member is extended depends upon the particular application. For example, in transportation applications it is usual to have a "short" primary on the vehicle and a "long" secondary on the track. In this case, power must be conveyed to the vehicle through collectors running on rails or wires. The arrangement of the motor members in this case is shown in Fig. 5.63, and Fig. 5.64 shows a transportation test vehicle fitted with such a LIM. In some systems it is preferable to have a fixed continuous or discontinuous primary and a short moving secondary. In this case, the secondary may consist of conducting and ferromagnetic sheets, or it may be made in the form of a developed squirrel cage, since this is more effective than a sheet.

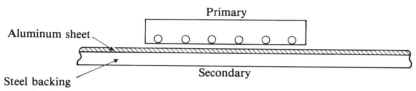

Fig. 5.63 Single-sided LIM configuration for transportation.

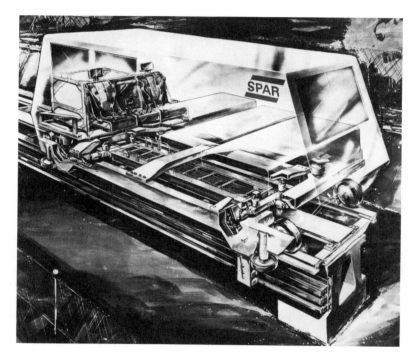

Fig. 5.64 Test vehicle for single-sided, 200 kW LIM. *(Spar Aerospace Limited)*

An alternative LIM form, in which there are two similar primary members, one on either side of a conducting sheet acting as secondary, is illustrated in Fig. 5.65. This configuration is said to be *double-sided,* whereas that of Fig. 5.-63 is *single-sided*. An experimental double-sided motor intended for transportation use is illustrated in Fig. 5.66. A short section of its fixed secondary sheet is seen projecting from the left-hand end of the primary. This machine was designed for 2500 hp and has operated at its designed speed of 250 miles per hour.

Large LIM's usually have 3-phase primary windings and have been employed for transportation, materials handling, extrusion presses, and the pumping of liquid

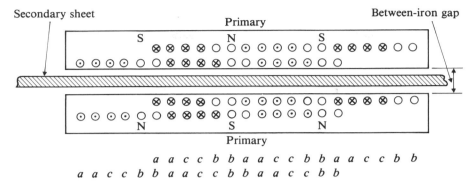

Conductor designations for both primaries

$$I_a > 0 : I_b = 0 : I_c < 0$$

Fig. 5.65 Double-sided, three-pole LIM configuration.

Fig. 5.66 Experimental double-sided LIM. *(Airesearch Manufacturing Company of California, a division of Garrett Corporation)*

metal. Small LIM's are employed as curtain pullers, sliding-door closers, etc. A small machine is illustrated in Fig. 5.67, both mounted on its secondary and also dismounted with the top cover removed. The inside of the top cover is smooth and slides along the upper flange of the secondary. A tachogenerator may be seen at the right-hand end of the mounted LIM, being driven by a pulley bearing on the secondary flange. The model illustrated in Fig. 5.67 happens to be excited by 3-phase power, but small LIM's may have two-winding primaries excited from a single-phase source in the same way as the single-phase induction motors discussed in Section 5.4.

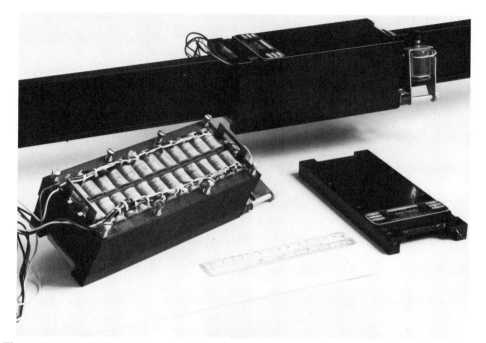

Fig. 5.67 A small, three-phase, double-sided LIM.

*5.6.2 LIM Performance

The characteristic curve of rotational speed versus torque for a rotating induction motor is, for a LIM, replaced by a curve of velocity versus thrust, the synchronous velocity of the linear motor being that of the traveling wave produced by the primary windings relative to those windings. As in the case of the rotating motor, the traveling wave moves through two pole pitches during one cycle of the ac source. The synchronous velocity of a LIM is expressed by

$$v_s = 2\tau f \quad \text{m/s} \tag{5.168}$$

where τ is the pole pitch, and f is the source frequency. If relative motion takes place between the two members of the LIM, then a motoring force is normally developed when the relative velocity v is less than the synchronous velocity v_s. The slip of the LIM is therefore expressed by

$$s = \frac{v_s - v}{v_s} \tag{5.169}$$

Since pole pitch τ can be made any desired value over a wide range, the synchronous velocity of a LIM for a given frequency can also have any desired value and

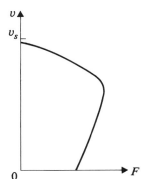

Fig. 5.68 Velocity-thrust curve of a LIM.

is not governed by the number of poles, as is the case with the rotating machine. Furthermore, the number of poles on the primary need not be even, or indeed integral, as can be seen in Fig. 5.65.

A typical velocity-thrust curve for a LIM, shown in Fig. 5.68, has the same general shape as the speed–torque curve for an induction motor with a reasonably high-resistance rotor. A reason for the high effective resistance of the secondary of a LIM can be inferred from Fig. 5.69, which illustrates the paths taken by the currents in the secondary sheet of a four-pole motor. While the general direction of the current beneath the primary is transverse, as it should be to maintain the desired orthogonal relationship between current, flux, and motion, there are substantial areas where this orthogonality does not exist. In these areas, little or no thrust in the direction of motion develops. Much energy is therefore dissipated as resistive losses without the development of corresponding mechanical power. The motor, thus, operates at high slip with correspondingly low efficiency.

A second factor that strongly influences the operating characteristics of a LIM is the large gap between the ferromagnetic surfaces of the primary and secondary members in a single-sided LIM or between the two primary members in a double-sided LIM. Typically, this *between-iron gap* may be about 25 mm for a single-sided transportation LIM as compared with about 1 mm for a typical rotary induction motor of the same power rating. This gap calls for a large magnetizing current and causes a relatively low primary power factor.

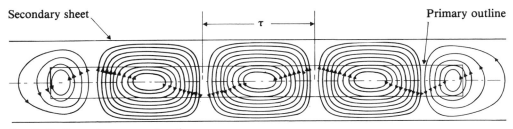

Fig. 5.69 Secondary current paths.

A phenomenon present in the LIM, which has no parallel in the rotational induction motor, is known as the *end effect*. The LIM primary has an entry edge at which new secondary conductor continuously enters the between-iron gap, and an exit edge at which secondary conductor continuously leaves. By Lenz's law, the secondary current in the region of entry acts to prevent the buildup of gap flux. As a result, when the motor is moving, the average gap-flux density for the first pair of poles near the entry edge may be significantly lower than that under later poles in the motor. Also, by Lenz's law, a current that tends to maintain the flux persists in the secondary conductor after it has left the exit edge. This current results in resistive loss without any corresponding development of thrust. The end effect, therefore, causes the motor to have a higher effective secondary resistance with consequent lower efficiency. It also reduces the maximum thrust that the motor can produce. The effect is naturally most pronounced at high speed.

*5.6.3 LIM Equivalent Circuit

A multipole LIM may be represented to a good approximation by an equivalent circuit of the form shown in Fig. 5.19. However, the physical configuration of a LIM results in equivalent-circuit parameters differing widely in per-unit value from those of a rotary motor. The magnetizing reactance is low, due to the large between-iron gap. The effective secondary resistance is relatively high because of the current distribution in the secondary and also because of end effect. An increase of the secondary conductor thickness to reduce this resistance has the adverse effect of increasing the between-iron gap and the magnetizing current. The secondary leakage inductance tends to be high because of the secondary current distribution, where much of the current does not link the whole of the air-gap flux.

Since the flux density in a LIM is low, owing to the large between-iron gap, core losses are usually negligible. Friction does not exist; and if windage is neglected, a LIM has no equivalent to rotational losses. Thus the entire power delivered to the secondary of the machine is converted to output power or secondary resistive losses. By analogy with the rotary motor, therefore, from Eq. (5.57),

$$P_{ma} = \frac{R_2'}{s}(I_A')^2 = R_2'(I_A')^2 + \left(\frac{1-s}{s}\right)R_2'(I_A')^2 \quad \text{W} \tag{5.170}$$

The second term on the right-hand side of Eq. (5.170) represents output power per phase. Thus for a 3-phase machine,

$$P_{\text{mech}} = Fv = \frac{3(1-s)}{s}R_2'(I_A')^2 \quad \text{W} \tag{5.171}$$

This power may be determined by analysis of the equivalent circuit.

PROBLEMS

*5.1 Determine the current required in a single-phase, concentrated, 4-pole winding of a machine to produce an air-gap flux density of 0.8 T across an air gap of effective length 1.2 mm. The winding has a total of 280 turns. *(Section 5.1.2)*

*5.2 A 3-phase, 6-pole machine stator has 36 slots. The coils each have 10 turns; each coil has a pitch of 5 slots. Determine the number of effective turns per phase.   *(Section 5.1.5)*

5.3 For the 3-phase, wound-rotor induction motor of Fig. 5.14, the ends a', b', c' of the stator windings are connected together, as are the ends A', B', C' of the rotor windings. The neutral points so formed may be designated as o and O. The effective turns ratio of the machine is $N_{re}/N_{se} = 1.5$, and the machine may be considered ideal.

   a) Draw a phasor diagram showing phasors $\overline{V}_{ao}$, $\overline{V}_{bo}$, $\overline{V}_{co}$, $\overline{V}_{AO}$, $\overline{V}_{BO}$, and $\overline{V}_{CO}$ when the rotor is in position $\beta = \beta_{oe}$.

   b) On one set of axes draw curves of v_{ao} and v_{AO} as functions of time for $\beta = \beta_{oe}$.   *(Section 5.2.1)*

5.4 A 3-phase, 4-pole, 60-Hz wound-rotor motor is to be used as a phase shifter. The stator leakage reactance is 0.1 Ω/phase, the rotor leakage reactance is 0.064 Ω/phase, and the magnetizing reactance is 2.1 Ω/phase. The winding resistances may be ignored. The machine has an effective rotor-to-stator turns ratio N_{re}/N_{se} of 0.8. The stator is connected to a 220-V line-to-line supply. The rotor is displaced in the direction of the rotating field through 15 mechanical degrees from alignment of corresponding phase windings.

   Determine the phasor emf and the impedance of a single-phase Thévenin equivalent circuit for the machine as seen from its rotor terminals.   *(Section 5.2.1)*

5.5 A four-pole wound-rotor induction motor has the following per-phase equivalent-circuit parameters:

Stator winding resistance R_s	$= 0.3$ Ω
Rotor winding resistance R_r	$= 0.4$ Ω
Stator leakage inductance L_{ls}	$= 3.2$ mH
Rotor leakage inductance L_{lr}	$= 2.1$ mH
Stator magnetizing inductance L_{ms}	$= 40$ mH
Effective turns ratio N_{se}/N_{re}	$= 1.2$

The machine is employed as a phase shifter. The stator windings are excited from a 60-Hz, 3-phase source of 110 V line-to-line, and the rotor is then turned into a position where the axis of one phase winding of the stator coincides with that of one phase winding of the rotor. Assuming that a 3-phase, wye-connected, balanced impedance of $1.5 + j\,1.0$ Ω/phase is then connected to the rotor terminals, and the rotor is turned 25 mechanical degrees in the direction of the stator rotating wave, determine the relative amplitudes and phase relationship of the stator and rotor terminal potential differences.   *(Section 5.2.1)*

5.6 The machine of Problem 5.5 is employed as an induction regulator. It is excited from a 100-V, line-to-line, 60-Hz, 3-phase source, and the rotor position for which V_{ao}, V_{bo}, and V_{co} have their maximum values on open circuit is determined. A 3-phase, wye-connected impedance of $1.5 + j\,1.0$ Ω/phase is then connected to stator terminals a, b, c; and the rotor is turned 45 degrees in the direction of the stator rotating wave.

   Draw a phasor diagram and determine the amplitude and phase of potential difference $\overline{V}_{ao}$ under these conditions, using $\overline{V}_{AO}$ as the reference phasor.   *(Section 5.2.1)*

5.7 For the induction regulator, load circuit, and rotor position of Problem 5.6, determine the ratio

$$\frac{V_{ao}(\text{no load}) - V_{ao}(\text{load})}{V_{ao}(\text{no load})}$$

   (Section 5.2.1)

5.8 A 2-pole, 60-Hz, wound-rotor motor has a stator leakage reactance of 0.1 Ω/phase, a rotor leakage reactance of 0.1 Ω/phase, and a magnetizing reactance of 2.0 Ω/phase. The resistances of the windings are negligible, and the effective turns ratio of the machine is unity. The machine is employed as a variable 3-phase inductive load by connecting it as shown in Fig. 5.70. The impedance per phase is a function of the angle β_e, the rotor displacement.

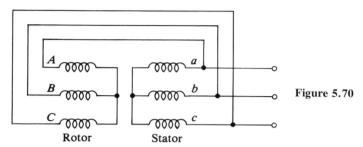

Figure 5.70

a) Determine the inductive reactance per phase at 60 Hz as a function of β_e.
b) Determine the angle β_e required if the load is to draw 25 kVA from a 60-Hz, 115-V line-to-line source.   *(Section 5.2.1)*

5.9 The machine of Problem 5.8 is employed as an induction regulator, being connected as shown in Fig. 5.17(a). The source line-to-neutral potential difference V_1 is 133 V.

a) Determine V_2 on no-load as a function of angle β_{0e}, the rotor displacement.
b) Determine the per-phase effective reactance of the machine viewed from terminals a, b, c.
c) Determine the per-phase Thévenin equivalent circuit of the regulator and source.
d) At what angle should the regulator be set to provide a line-to-neutral potential of 50 V to a balanced resistive load of 0.8 Ω/phase?   *(Section 5.2.1)*

5.10 A 3-phase, 6-pole, wound-rotor motor is to be employed as a variable-frequency source. For this purpose the stator is excited from a 60-Hz, 440-V line-to-line, 3-phase supply; and the machine is driven at a variable speed so that the rotor terminals constitute the variable-frequency source. When the rotor is stationary, the line-to-line open-circuit rotor potential difference is 220 V.

a) Determine the speed range required to give a frequency range $20 < f < 150$ Hz.
b) Determine the corresponding range of open-circuit rotor potential difference.
c) Regard the machine as ideal, and determine the relative magnitudes of the power supplied ($+ve$) or absorbed ($-ve$) at both limits of the frequency range by
   (i) the stator source,
   (ii) the driving machine,
   (iii) the rotor circuit.   *(Section 5.2.2)*

5.11 A 4-pole, 3-phase, wound-rotor induction motor has the following per-phase equivalent-circuit parameters:

$$\begin{aligned}
&\text{Stator winding resistance } R_s = 0.260 \ \Omega \\
&\text{Rotor winding resistance } R_r = 0.182 \ \Omega \\
&\text{Stator leakage inductance } L_{ls} = 1.06 \ \text{mH} \\
&\text{Rotor leakage inductance } L_{lr} = 0.803 \ \text{mH}
\end{aligned}$$

$$\text{Magnetizing inductance } L_{ms} = 19.9 \quad \text{mH}$$
$$\text{Effective turns ratio } N_{se}/N_{re} = 1.15$$

The machine is to be employed as a frequency converter and for this purpose is mechanically coupled to an 8-pole, 60-Hz, synchronous motor. The set is then driven by the synchronous motor, and the stator terminals of the induction machine are connected to a 3-phase, 60-Hz source of 220 V line-to-line.

A 3-phase, wye-connected load, with a resistance of 5 Ω and an inductance of 7 mH per phase, is then connected to the rotor terminals. Assume that the set is running in the direction that gives the higher of the two possible rotor frequencies, and determine the current in the load circuit.   *(Section 5.2.2)*

5.12 Given the frequency-conversion system of Problem 5.11, for the two possible frequencies, determine

a) The three-phase active power delivered to the load circuit.
b) The three-phase active power input to the induction motor terminals.
c) The power output of the synchronous motor.
Rotational losses of the induction machine may be neglected.   *(Section 5.2.2)*

5.13 A 3-phase, 6-pole, 220-V, 60-Hz induction motor has an effective rotor-to-stator turns ratio N_{re}/N_{se} of 0.8. Regard the machine as ideal. A balanced wye-connected load of 3 Ω resistance in parallel with 2200 μF capacitance in each phase is connected to the rotor terminals. The motor is rotating at 350 r/min. Determine

a) The effective impedance per phase as seen from the stator side.
b) The total power delivered by the supply.
c) The power delivered to the rotor load.
d) The mechanical power.
e) The shaft torque.   *(Section 5.2.2)*

5.14 A 5-hp, 3-phase, 110-V, 60-Hz, 1140 r/min, wound-rotor induction motor has the following per-phase equivalent-circuit parameters referred to the stator:

$$\text{Stator winding resistance } R_s = 0.074 \ \Omega$$
$$\text{Rotor winding resistance } R_r' = 0.144 \ \Omega$$
$$\text{Stator leakage reactance } X_{ls} = 0.232 \ \Omega$$
$$\text{Rotor leakage reactance } X_{lr}' = 0.232 \ \Omega$$
$$\text{Magnetizing reactance } X_{ms} = 6.0 \ \Omega$$

The rotor terminals are shorted together.

a) From the approximate equivalent circuit of Section 5.2.3, determine the internal breakdown torque and the internal starting torque of the motor in per unit of the rated output torque.
b) Assuming the motor is operating under rated conditions, determine the air-gap power, stator resistance loss, and rotor resistance loss.   *(Section 5.2.5)*

5.15 A three-phase, two-pole, 110-V, 60-Hz, wound-rotor induction motor has the following circuit parameters for the approximate equivalent circuit of Section 5.2.3:

$$R_s = 0.260 \ \Omega \qquad X_l = 1.58 \ \Omega$$
$$R_r' = 0.408 \ \Omega \qquad X_{ms} = 36.6 \ \Omega$$

The motor is running with the rotor terminals shorted together at a slip of 0.05 when the phase sequence of the stator supply is reversed by switching. Determine the internal plugging torque, which is developed by the motor immediately after switching, and the power factor at the terminals.   *(Section 5.2.5)*

5.16 The motor of Problem 5.15 is employed in a small hoist. When the load is being lowered, the motor is driven at more than synchronous speed. Assume the rotor terminals are shorted together, and determine the maximum internal braking torque that the motor can develop under these conditions, and the speed in r/min at which it occurs.   *(Section 5.2.5)*

5.17 A 3-phase, 6-pole, 440-V, 60-Hz induction motor has the following per-phase equivalent-circuit parameters referred to the stator:

$$\begin{aligned}
\text{Stator winding resistance } R_s &= \text{negligible} \\
\text{Rotor winding resistance } R_r' &= 0.3 \ \Omega \\
\text{Stator leakage inductance } L_{ls} &= 3 \ \text{mH} \\
\text{Rotor leakage inductance } L_{lr}' &= 3 \ \text{mH} \\
\text{Magnetizing inductance } L_{ms} &= 0.1 \ \text{H}
\end{aligned}$$

Rotational losses are negligible. Assuming the stator is connected to a 440-V, 60-Hz source, employ the improved equivalent circuit of Section 5.2.4 to determine

- **a)** The starting torque.
- **b)** The breakdown torque.
- **c)** The speed at which breakdown torque occurs.
- **d)** The torque per unit of slip for operation near synchronous speed.
- **e)** The speed at which the motor will drive a load demanding a torque of 2 N·m per rad/s of speed.
- **f)** The rotor frequency for the condition in (e).
- **g)** The motor efficiency for the condition in (e).   *(Section 5.2.5)*

5.18 A 3-phase induction motor has the following parameters:

$$\begin{aligned}
R_s &= 0.3 \ \Omega & L_{lr} &= 2.1 \ \text{mH} \\
R_r &= 0.4 \ \Omega & L_{ms} &= 40 \ \text{mH} \\
L_{ls} &= 3.2 \ \text{mH} & N_{se}/N_{re} &= 1.2
\end{aligned}$$

- **a)** Draw the approximate equivalent circuit of Section 5.2.3.
- **b)** Derive the parameters for the improved equivalent circuit of Section 5.2.4.
- **c)** Assume the stator supply is 110-V, line-to-line at 60 Hz, and determine the torque, the line current, and the power factor at a slip of 0.05 for the two equivalent circuits of (a) and (b) and compare the results.   *(Section 5.2.5)*

5.19 A small 2-pole, 110-V, 60-Hz, 3-phase induction motor has the following parameters for the equivalent circuit of Section 5.2.4:

$$\begin{aligned}
R_s &= 0.22 \ \Omega & X_L &= 1.6 \ \Omega \\
R_R' &= 0.42 \ \Omega & X_M &= 34 \ \Omega
\end{aligned}$$

- **a)** Determine the mechanical power developed at a speed of 3500 r/min.
- **b)** Determine the input power factor for the condition in part (a).
- **c)** Determine the breakdown torque for the motor and the speed at which it occurs.
- **d)** Determine the starting torque.   *(Section 5.2.5)*

5.20 Equations (5.76) and (5.77) show that the maximum motoring and generating torque are equal if the stator resistance R_s is negligible. Assume the rotor resistance is independent of rotor frequency, and show that for such a machine the torque T produced at any slip is given by

$$T = \frac{2T_{max}}{s'/s + s/s'} \quad \text{N·m}$$

where s' is the slip at which T_{max} occurs.   *(Section 5.2.5)*

5.21 A 3-phase, 4-pole, 550-V, 60-Hz induction motor has a breakdown torque of 100 N·m that occurs at a slip of 0.25 per unit. The stator resistance is negligible, and the rotor resistance does not vary with frequency. The machine is to be operated on a 400-V, 50-Hz supply. For these conditions, determine

a) The new breakdown torque. Compare it with that for 550-V, 60-Hz operation.
b) The speed at which breakdown torque occurs.
c) The new starting torque. Compare it with that for 550-V, 60-Hz operation.
d) The new rated torque, basing your determination on the fact that rotor resistive losses limit the maximum continuous load that can be accepted.   *(Section 5.2.5)*

5.22 A water-wheel turbine that develops 1 MW of mechanical power at a speed between 600 and 625 r/min drives an asynchronous generator in the form of a 3-phase, 12-pole, squirrel-cage induction machine. The generator supplies power to a 4000-V, line-to-line, 60-Hz distribution system. A no-load test on the machine gives a per-phase impedance of $0 + j\,33\Omega$. A locked-rotor test gives a per-phase impedance of $0.5 + j\,3.0\Omega$. The resistance of the stator windings is negligibly small.

a) Derive an equivalent circuit for a generator like that in Fig. 5.20.
b) If you use the equivalent circuit of (a), what is the generator speed when the turbine is delivering 1 MW? (Assume that generator rotational losses are negligible.)
c) The terminal power factor of the generating station is brought to unity under the conditions of (b) in order to reduce the current in the distribution line. If this is done by means of wye-connected capacitors at the generator terminals, what is the per-phase capacitance required?   *(Section 5.2.5)*

*5.23 For the motor of Problem 5.19, draw a circle diagram and determine:

a) The stator current at starting.
b) The power factor at starting.
c) The maximum power factor.
d) The input power at maximum power factor.   *(Section 5.2.7)*

*5.24 A 15-kW, 3-phase, 220-V, 60-Hz, four-pole, wound-rotor induction motor has the following circuit parameters for the equivalent circuit of Fig. 5.20:

$$R_s = 0.118\ \Omega \qquad X_L = 0.416\ \Omega$$
$$R'_R = 0.102\ \Omega \qquad X_M = 10.7\ \Omega$$
$$N_{se}/N_{re} = 1.10 \qquad k = 0.98$$

Employ the circle diagram to determine the external rotor-circuit resistance per phase that is required to make the motor run at a slip of 0.10 when it is drawing a stator current of 50 A.   *(Section 5.2.7)*

*5.25 For the motor of Problem 5.17, determine

a) The external rotor-circuit resistance per phase required to produce breakdown torque at standstill.

b) The speed at which the motor with the resistance obtained in (a) will drive a load demanding a torque of 10 N · m/rad/s.

c) The efficiency of the machine operating as in (b).   *(Section 5.2.8)*

*5.26 A 6-pole, 3-phase, 220-V, 60-Hz, squirrel-cage induction motor has the following parameters:

$$R_s = 0.06 \ \Omega \qquad X_L = 0.2 \ \Omega$$
$$R'_R = 0.1 \ \Omega \qquad X_M = 5 \ \Omega$$

The motor drives a fan that requires a torque of 250 N · m at a speed of 1200 r/min. The torque for the fan is proportional to the square of the speed.

a) Estimate the speed and power of the fan with rated potential supplied to the motor.

b) Repeat (a) with 75% of rated potential applied to the motor.   *(Section 5.2.10)*

*5.27 A ship is propelled by a variable-speed steam turbine driving a synchronous generator. This generator provides power for two induction motors, each driving a propeller. The synchronous speed of the induction motors, when the turbogenerator set is running at full speed, is 300 r/min. Under these conditions each induction motor runs at 240 r/min and produces its maximum torque of 100×10^3 N · m. When the turbogenerator set runs at half speed, the terminal potential difference of the generator is half the full-speed value.

a) When the turbogenerator set runs at half speed, what are the synchronous speed, maximum torque, and operating speed for maximum torque of the induction motors? (The magnetizing current and stator resistance of the motors may be neglected.)

b) Draw motor speed-torque curves for full- and half-speed operation of the turbine.

c) The power delivered to each propeller varies approximately as the cube of the propeller speed. Estimate the power delivered to each propeller at full- and half-speed of the turbine, if each propeller demands a torque of 30,000 N · m at turbine full-speed.   *(Section 5.2.10)*

*5.28 A 3-phase, 4-pole, 220-V, 60-Hz, squirrel-cage induction motor has the following equivalent-circuit parameters in Fig. 5.20:

$$R_s = \text{negligible} \qquad X_L = 0.6 \ \Omega$$
$$R'_R = 0.3 \ \Omega \qquad X_M = 15 \ \Omega$$

The motor is supplied from a 3-phase, variable-frequency inverter that provides rated potential at rated frequency and a constant ratio of potential to frequency for all frequency values.

a) Sketch speed–torque curves for the motor at 60, 20, and 3 Hz.

b) Determine the inverter frequency to drive at 100 r/min a load requiring a constant internal motor torque of 150 N · m.

c) Determine the motor terminal potential difference for the operating conditions of (b).   *(Section 5.2.10)*

*5.29 The following test results were obtained from a 3-phase, 10-hp, 4-pole, 440-V, 60-Hz, squirrel-cage induction motor:

No-Load Test	Locked-Rotor Test
$V_{NL} = 440$ V	$V_{lk} = 110$ V (line-to-line)
$I_{NL} = 3.1$ A	$I_{lk} = 10.5$ A
$P_{NL} = 480$ W	$P_{lk} = 1260$ W

The average resistance measured by dc bridge between a pair of stator terminals was found to be 0.61 Ω.

a) Determine the per-phase stator resistance.
b) Determine the effective rotor resistance and total leakage reactance per phase.
c) Determine the per-phase magnetizing reactance under rated conditions.
d) Estimate the rotational losses near synchronous speed.
e) Obtain the equivalent circuit of Section 5.2.3.
f) Determine the stator current, the torque at the coupling, and the efficiency of the motor when it is operating under rated conditions at a slip of 0.05 per unit.
g) In view of the fact that the locked-rotor test was made at rated frequency, is the predicted starting torque from the equivalent circuit likely to be more accurate than the torque obtained in (f)? Explain.   *(Section 5.2.11)*

*5.30 The following test results were obtained from a 3-phase, 4-pole, 220-V, 60-Hz, Class-D squirrel-cage induction motor:

No-Load Test	Locked-Rotor Test
$V_{NL} = 220$ V	$V_{lk} = 46.8$ V (line-to-line)
$I_{NL} = 7.05$ A	$I_{lk} = 26.0$ A
$P_{NL} = 553$ W	$P_{lk} = 1350$ W

The average resistance measured by dc bridge between a pair of stator terminals was found to be 0.502 Ω.

a) Obtain the equivalent-circuit parameters of Section 5.2.3.
b) By calculating output power for a few values of slip in the neighborhood of $s = 0.1$, determine by interpolation the speed at which the motor provides its rated output. Rotational losses may be considered constant and equal to the no-load power input.   *(Section 5.2.11)*

*5.31 The following test results were obtained from a 3-phase, 6-pole, 440-V, 60-Hz, wound-rotor induction motor:

No-Load Test with Stator Energized
$V_{NL} = 440$ V
$I_{NL} = 3.2$ A
$P_{NL} = 440$ W

Locked-Rotor Test with Stator Energized
$V_{lk} = 110$ V
$I_{lk} = 10.5$ A
$P_{lk} = 1250$ W

Open-Circuit Zero Speed Test	
Stator Energized	Rotor Energized
$V_{stator} = 440$ V	$V_{rotor} = 220$ V
$E_{rotor} = 201$ V	$E_{stator} = 396$ V

Stator resistance = 0.6 Ω between a pair of terminals.

Derive the parameters for the equivalent circuit of Fig. 5.18, assuming equal stator and referred rotor-leakage inductance. *(Section 5.2.11)*

*5.32 For comparative purposes, the parameters of a machine are usually best expressed in per unit of the ratings of the machine. A 60-Hz, 6-pole, 440-V, 12.1-A, 7.5-kW induction machine has a stator resistance R_s of 0.15 Ω, a leakage reactance X_L of 4.3 Ω, a magnetizing reactance X_M of 48 Ω, and an effective rotor resistance R'_R of 0.4 Ω.

a) Determine the base value of line-to-neutral potential, impedance per phase, synchronous speed, power per phase, and total apparent power.

b) Determine the per-unit value of the parameters of an equivalent circuit of the form of Fig. 5.20. *(Section 5.2.11)*

*5.33 The following test results were obtained from a three-phase, 8-pole, 2300-V, 60-Hz, Class-A, squirrel-cage induction motor:

No-Load Test	Locked-Rotor Test
$V_{NL} = 2300$ V	$V_{lk} = 550$ V (line-to-line)
$I_{NL} = 5.75$ A	$I_{lk} = 33.6$ A
$P_{NL} = 3400$ W	$P_{lk} = 12\,800$ W

The average resistance measured by dc bridge between a pair of stator terminals was found to be 3.58 Ω.

Determine the actual efficiency and the ideal efficiency of this motor at a slip of 0.03 per unit. *(Section 5.2.12)*

*5.34 A 220-V, 60-Hz, 6-pole, single-phase induction motor has the following parameters for the equivalent circuit of Fig. 5.20:

$$X_M = 90\ \Omega \qquad X_L = 14.5\ \Omega$$
$$R_s = 3.1\ \Omega \qquad R'_R = 6.6\ \Omega$$

The no-load losses in the motor are 85 W. Assuming the speed of the motor is 1125 r/min, determine the line current, the power factor, the shaft power, and the efficiency.

(Section 5.4.1)

*5.35 Two identical 440-V, 60-Hz, 2-pole, wound-rotor induction machines are connected with their stator windings in parallel, and with their rotor windings in parallel, to form a synchro tie. Each machine has the following parameters:

$$R_s \simeq 0 \qquad X_M = 20 \ \Omega$$
$$X_L = 2.2 \ \Omega \qquad R_R = 0.4 \ \Omega$$

The rated potential is applied to the stator windings.

a) Determine the torque produced in each machine if the angular positions β_1 and β_2 of the two rotors differ by 5° and the rotors are at standstill.

b) Determine the maximum torque that will be produced as the angle between the two rotors is increased at standstill. *(Section 5.5)*

*5.36 A linear induction motor has a pole pitch of 240 mm and has 5 poles. The parameters for an equivalent circuit of the form in Fig. 5.20 are

$$R_s = 0.11 \ \Omega \qquad L_M = 6 \ \text{mH}$$
$$R'_R = 0.24 \ \Omega \qquad L_L = 1.2 \ \text{mH}$$

a) Determine the synchronous velocity of this motor when operated on a 20-Hz supply.

b) Determine the thrust, mechanical power, input current, and power factor when the motor is operated on 300-V line-to-line, 40-Hz supply at a speed of 45 km/H. *(Section 5.6)*

6 / Synchronous Machines

The most important machine discussed in this chapter is the three-phase synchronous machine with wound rotor and stator. The stator windings are similar to those of an induction machine, and therefore Section 5.1 may also be applied to synchronous machines. Once again, the objective is to analyze synchronous machines in sufficient detail to develop equivalent-circuit models that may then be used to make sufficiently accurate predictions of performance without unduly complicated calculations.

6.1 THREE-PHASE SYNCHRONOUS MACHINES

A necessary condition for the production of torque and consequent energy conversion in a 3-phase, p-pole, synchronous machine was expressed in Eq. (3.168) as

$$\omega_m = \frac{2}{p} \omega_s \quad \text{rad/s} \tag{6.1}$$

where ω_m is the mechanical speed and ω_s is the angular frequency of the supply. This relationship arises from the fact that in the synchronous machine the rotor winding is excited with direct current, and $\omega_r = 0$. In Eq. (6.1), ω_m is the *synchronous speed* for the p-pole machine.

It is, of course, possible to build a synchronous machine in which the stator winding is excited with direct current and the rotor winding with alternating current, but this would be less convenient than the normal configuration in which the rotor winding is excited with direct current. Only a single rotor winding is required to produce the desired sinusoidal distribution of rotor mmf around the air gap. This winding usually operates at relatively low potential and low current, and has a large number of turns per pole. It can be excited without difficulty through slip rings, and accommodated in the limited space available on the rotor. By analogy with the dc machine, the rotor winding is usually called the *field* winding. In large machines, the field current is frequently supplied by a dc generator built on the rotor shaft.

The stator winding of a synchronous machine is similar to that of an induction motor. It may be a high-potential winding requiring much space for insulation, or it may carry high currents that are best conveyed to it through fixed terminals rather than sliding contacts. It may also be subjected to large forces during transient operating conditions and can be strongly mounted on the stator, where adequate space is available. Once again, by analogy to the dc machine, the stator winding is sometimes called the *armature* winding.

Synchronous machines may be divided into two broad classes; these are (a) high-speed machines with cylindrical rotors, and (b) low-speed machines with salient-pole rotors—that is, with rotors carrying a number of poles similar to the main field poles of a dc machine that project radially from a central hub.

Class (a) includes only very large generators with two, or sometimes four, poles. These are usually driven by steam turbines. To an engineer not engaged in the electric power industry, these machines are of secondary interest. They are the source of the energy he or she utilizes.

Class (b) machines may be further divided into two subclasses. The first are very large generators with a number of poles of the order of fifty. These machines are driven by water turbines and are employed exclusively for power generation. The second subclass consists of smaller independent generators (such as may be used in emergency power supplies) and synchronous motors, usually with ratings in the range of 50 to 5000 kW, that operate at various submultiples of 3600 r/min.

The typical synchronous machine is thus a salient-pole machine, in which all of the coils on the salient poles are connected in series in such a way as to alternate the polarities around the rotor. When interest lies in utilization of relatively small machines, rather than in bulk power generation by very large machines, sufficiently accurate predictions of performance for many purposes can be made by applying the simpler analysis of cylindrical-rotor machines to salient-pole machines. This will be done in the first part of this section.

6.1.1 Equivalent Circuit

Figure 6.1 represents the arrangement of the winding of a cylindrical rotor, 2-pole, 3-phase synchronous machine. As usual, each pair of conductors shown in the diagram represents the center of a belt of conductors producing an essentially sinusoidal distribution of mmf around the air gap. Each of the stator phase windings has N_{se} effective turns. The single rotor winding has N_{re} effective turns.

When the machine operates at synchronous speed, the position of the rotor winding at any instant, expressed in electrical radians for a p-pole machine, is

$$\beta_e = \omega_s t + \beta_{0e} \quad \text{rad} \tag{6.2}$$

Thus at instant $t = 0$, $\beta_e = \beta_{0e}$. The peak value of air-gap mmf due to the rotor winding is

$$\hat{\mathscr{F}}_{rg} = \frac{N_{re}}{p} i_A \quad \text{A} \tag{6.3}$$

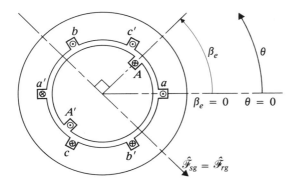

Fig. 6.1 Cylindrical-rotor, two-pole, three-phase synchronous machine.

For the direction of rotor current shown in Fig. 6.1, the peak value of mmf occurs along an axis at $\pi/2$ electrical radians from the plane of the rotor coil. Thus, since mmf acting from stator to rotor is positive, at $t = 0$, $\hat{\mathscr{F}}_{rg}$ occurs at angle

$$\theta_e = \beta_{0e} + \frac{\pi}{2} \quad \text{rad} \tag{6.4}$$

Assume that an mmf equal to the rotor mmf is produced by currents I'_{ma}, I'_{mb}, I'_{mc} added to the three-phase stator currents, where

$$i'_{ma} = \hat{i}'_s \sin(\omega_s t + \alpha'_s) \quad \text{A}$$

$$i'_{mb} = \hat{i}'_s \sin\left(\omega_s t + \alpha'_s - \frac{2\pi}{3}\right) \quad \text{A} \tag{6.5}$$

$$i'_{mc} = \hat{i}'_s \sin\left(\omega_s t + \alpha'_s - \frac{4\pi}{3}\right) \quad \text{A}$$

The directions of these additional currents are indicated in Fig. 6.1 for the instant at which the rotor position is that shown in the diagram. These components produce an mmf wave

$$\mathscr{F}'_{sg} = \hat{\mathscr{F}}'_{sg} \cos(\omega_s t + \alpha'_s - \theta_e) \quad \text{A} \tag{6.6}$$

where

$$\hat{\mathscr{F}}'_{sg} = \frac{3 N_{se}\, \hat{i}'_s}{2p} \quad \text{A} \tag{6.7}$$

If the mmf of Eq. (6.6) is to be equal to, and is incident with, that due to I_A, then it is necessary that

$$\frac{3 N_{se}\, \hat{i}'_s}{2p} = \frac{N_{re}}{p} i_A \quad \text{A} \tag{6.8}$$

and

$$\alpha'_s = \beta_{0e} + \frac{\pi}{2} \quad \text{rad} \tag{6.9}$$

From Eq. (6.8),

$$\hat{i}'_s = \frac{2}{3}\frac{N_{re}}{N_{se}}i_a \quad \text{A} \tag{6.10}$$

The phasor of the stator current I'_{ma} will be defined, with reference to phasor $\overline{V}_a = V_a\lfloor 0$, as

$$I'_{ma} = \frac{\hat{i}'_s}{\sqrt{2}}\lfloor\alpha'_s \quad \text{A} \tag{6.11}$$

Substitution from Eqs. (6.9) and (6.10) yields

$$\overline{I}'_{ma} = \frac{\sqrt{2}}{3}\frac{N_{re}}{N_{se}}i_A\left\lfloor\beta_{0e} + \frac{\pi}{2}\right. \quad \text{A} \tag{6.12}$$

Since the rotor current is supplied from an external source, it is convenient to designate it as a field current and write

$$i_f = -i_A \quad \text{A} \tag{6.13}$$

so that

$$\overline{I}'_{ma} = n'i_f\left\lfloor\beta_{oe} - \frac{\pi}{2}\right. \quad \text{A} \tag{6.14}$$

where

$$n' = \frac{\sqrt{2}}{3}\frac{N_{re}}{N_{se}} \tag{6.15}$$

Thus n' is the ratio relating the rms magnitude of the current I'_{ma} to the magnitude of the direct field current i_f.

An equivalent circuit for the synchronous machine may now be drawn and is shown in Fig. 6.2. The form of the stator section of the equivalent circuit is unchanged from that for the induction motor in Fig. 5.18. Inductance L_{ms} is shown as nonlinear because it is often necessary to take into consideration the saturation

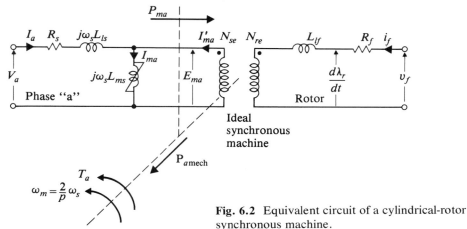

Fig. 6.2 Equivalent circuit of a cylindrical-rotor synchronous machine.

of the magnetic system when predicting the performance of a synchronous machine.

In the rotor section of the equivalent circuit, the emf $d\lambda_r / dt$ is induced by any change of flux linking the rotor winding. Under stead-state conditions, when the rotor and stator currents are constant and the rotor is moving in synchronism with the rotating flux wave, the rotor winding flux linkage is constant, this emf is zero, and

$$i_f = \frac{v_f}{R_f} \quad \text{A} \tag{6.16}$$

The rotor section of the equivalent circuit and the ideal synchronous machine in Fig. 6.2 may thus be replaced by a current source of frequency ω_s, as shown in Fig. 6.3(a). The magnitude of this current source is directly proportional to the field current as shown in Eq. (6.14).

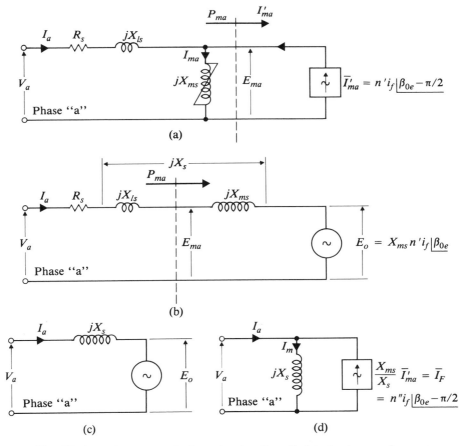

Fig. 6.3 Simplified steady-state equivalent circuits of a cylindrical-rotor synchronous machine.

During steady-state operation $d\lambda_r/dE$ is zero, and no energy is transferred across the air gap to the rotor winding; and, hence, all of the air-gap power is converted to mechanical form.

In general, the magnetizing reactance $\omega_s L_{ms}$ must be regarded as a nonlinear function of the magnetizing current I_{ma} or of the induced emf E_{ma}. If, however, the machine is operated over a reasonably narrow range of values of E_{ma}, a suitable average value of the magnetizing reactance may be chosen for this range. If the equivalent circuit of Fig. 6.3(a) is linearized in this way, an alternative form of equivalent circuit may be derived by the use of Thévenin's theorem. This circuit, shown in Fig. 6.3(b), consists of an impedance in series with a source of alternating emf E_o. This is called the *excitation emf*, and its phasor is

$$\bar{E}_o = jX_{ms}\bar{I}_a = X_{ms}n'i_f\underline{|\beta_{0e}} \quad \text{V} \tag{6.17}$$

Note that E_o is directly proportional to the field current i_f, and that its angle is equal to the rotor position at $t = 0$.

The sum of the stator leakage reactance and the magnetizing reactance is commonly known as the *synchronous reactance, X_s,* of the machine.

$$X_s = X_{ls} + X_{ms} \quad \Omega \tag{6.18}$$

In many analyses, it is simpler not to separate the two components of the synchronous reactance. But the magnetizing reactance X_{ms} in Eqs. (6.17) and (6.18) must be assigned a value appropriate to the operating value of emf E_{ma}.

Figures 6.3(c) and 6.3(d) show other equivalent circuits that are useful in analyses for which the magnetizing reactance is considered constant and the stator resistance is neglected. In Fig. 6.3(d),

$$I_F = \frac{n'X_{ms}}{X_s}i_f \quad \text{A} \tag{6.19}$$

6.1.2 Steady-State Operation

A synchronous machine normally operates at constant speed and constant ac terminal potential difference. Thus, in the analysis of its steady-state behavior, the variables to be considered are torque, stator current, and field current. The necessary relationships may be derived from the equivalent circuit of Fig. 6.3(d), in which

$$\bar{I}_F = \frac{\bar{E}_o}{jX_s} = \frac{n'X_{ms}}{X_s}i_f\underline{|\beta_{0e} - \pi/2} = n''i_f\underline{|\beta_{0e} - \pi/2} \quad \text{A} \tag{6.20}$$

In this equation, i_f is the actual dc rotor current, and n' is defined in Eq. (6.15).

In Fig. 6.3(d),

$$\bar{I}_a = \bar{I}_m - \bar{I}_F = \frac{V_a\underline{|0}}{jX_s} - I_F\underline{|\beta_{0e} - \pi/2}\quad\text{A} \tag{6.21}$$

The complex power per phase at the stator terminals is

$$\bar{S} = P + jQ = \bar{I}_a\bar{V}_a^*\quad\text{VA} \tag{6.22}$$

Where $\bar{V}_a^*$ is the conjugate of $\bar{V}_a$. From Eqs. (6.21) and (6.22),

$$S = \frac{V_a^2}{X_s}\underline{|-\pi/2} - I_F V_a\underline{|\beta_{0e} - \pi/2}\quad\text{VA} \tag{6.23}$$

and

$$P_a = -I_F V_a \cos(\beta_{0e} - \pi/2) = -I_F V_a \sin\beta_{0e}\quad\text{W} \tag{6.24}$$

while

$$Q_a = \frac{-V_a^2}{X_s} + I_F V_a \cos\beta_{0e}\quad\text{VAR} \tag{6.25}$$

Since stator losses are being neglected at this stage of analysis, the three-phase power developed at the terminals of the stator is also the air-gap power, which in turn is equal to the mechanical power. Thus,

$$P = T\omega_m = \frac{2}{p}\omega_s T = 3P_a\quad\text{W} \tag{6.26}$$

from which

$$T = -\frac{p}{2}\frac{3}{\omega_s}I_F V_a \sin\beta_{0e}\quad\text{N·m} \tag{6.27}$$

The relationship expressed in Eq. (6.27) is shown graphically in Fig. 6.4. The maximum torque for a given excitation, called the *pull-out torque*, is

$$T_{\max} = \frac{p}{2}\frac{3}{\omega_s}I_F V_a\quad\text{N·m} \tag{6.28}$$

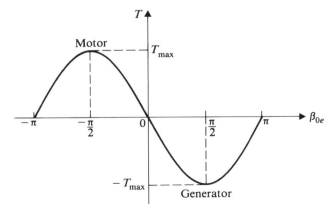

Fig. 6.4 Synchronous machine torque as a function of β_{0e}.

which is a function of I_F and, hence, from Eq. (6.20), of i_f. High field current will enable the machine to develop a high torque before it is pulled out of synchronism by excessive load torque or driven to super-synchronous speed by excessive driving torque. For a given load torque, a high field current will also result in a low value of β_{0e}.

The speed–torque characteristic of a synchronous machine consists simply of a straight line parallel to the torque axis, as shown in Fig. 6.5.

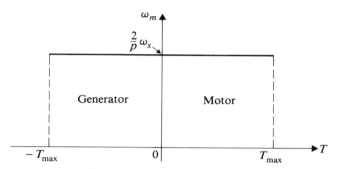

Fig. 6.5 Speed–torque relationship for a synchronous machine.

6.1.3 Mechanism of Torque Production

In order to draw a physical diagram illustrating the operation of the machine, a simple model is required. This is obtained by ignoring the stator resistance and leakage inductance in Fig. 6.3(a), so that an approximate equivalent circuit of the form of Fig. 6.3(d) is obtained. Since it has been assumed that $X_{ls} = 0$, it follows that $X_s = X_{ms}$, $I_F = I'_a$, and $I_M = I_{ma}$. Also,

$$V_a = X_s I_M = X_{ms} I_{ma} = E_{ma} \quad \text{V} \tag{6.29}$$

On the basis of these approximations, it is convenient to use the circuit and symbols of Fig. 6.3(d).

If the synchronous motor, with the stator terminals open-circuited and with current i_f in its field winding, is driven by a prime mover at speed $\omega_m = (2/p)\omega_s$, an emf will be generated in the stator windings. This emf will appear as the per-phase terminal potential difference V_a, since $I_a = 0$. The angular frequency of this potential difference will be ω_s. The field winding with its current i_f rotating at speed $\omega_m = (2/p)\omega_s$ produces the same rotating field as a sinusoidal current $\overline{I}_F$ of frequency ω_s in a set of stationary rotor windings, as indicated in the simplified equivalent circuit of Fig. 6.3(d). With the stator current I_a equal to zero, this equivalent source current $\overline{I}_F$ is equal to the magnetizing current $\overline{I}_M$. By adjustment of i_f, the magnitude of I_F, and therefore of V_a, may be adjusted. If the prime mover speed is changed, the frequency of V_a is changed.

If the power system has constant potential difference and constant frequency, the synchronous machine may be connected to the ac system in the following way. First, i_f is adjusted to make the magnitude of V_a equal to the system potential difference. Second, the difference in the phase of the two sets of 3-phase potentials is observed by the use of appropriate indicating equipment. Third, the machine potentials are brought into phase with the system potentials by a momentary alteration of the prime-mover speed. Fourth and finally, the two sets of terminals are connected to each other.

This process, called *synchronizing the machine onto the bus,* is illustrated in the phasor diagram of Fig. 6.6, which shows a condition where the machine and system potential differences are equal in magnitude, but where the machine must be decelerated momentarily to permit connection.

Immediately after synchronization, I_a is still zero, and the prime-mover torque is just sufficient to overcome the friction, windage, and core losses of the machine. The machine is said to be *floating on the bus.* This condition may be interpreted in terms of the approximate equivalent circuit of Fig. 6.3(d), in which, for $I_a = 0$,

$$\overline{V}_a = V_a \underline{|0} = jX_s\overline{I}_F \quad \text{V} \tag{6.30}$$

The phasor diagram for this condition is shown in Fig. 6.7(a). Figure 6.7(b) shows a physical diagram of the machine. All of the field mmf $\mathscr{F}_{rg}$ is used to magnetize the machine, and the axis of the air-gap flux density wave B_g, indicated by $\hat{B}_g$, is therefore coincident with the axis of $\mathscr{F}_{rg}$, indicated by $\hat{\mathscr{F}}_{rg}$. Figure 6.7(c) shows the corresponding mmf space vector diagram for the instant $t = 0$. Since $I_a = 0$, the stator mmf $\mathscr{F}_{sg}$ is zero.

If the prime mover is now uncoupled, the machine will run as a motor supplying its own rotational losses. If an external load torque is imposed, the rotor will be momentarily decelerated, then will resume rotation at speed $\omega_m = (2/p)\omega_s$ rad/s. Observation of the rotor during this momentary deceleration by means of a strobo-

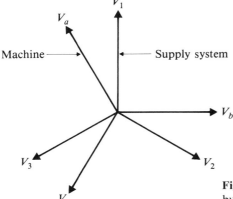

Fig. 6.6 Synchronizing the machine onto the bus.

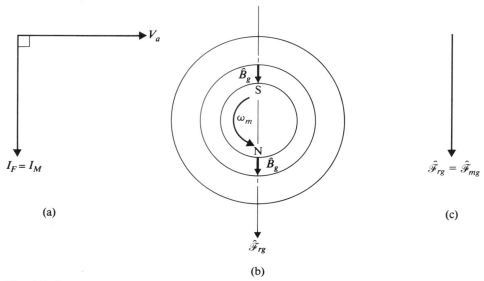

(a)

(c)

(b)

Fig. 6.7 Synchronous machine floating on the bus.

scope excited from potential difference V_a would show that the rotor had fallen back in space through a mechanical angle β_0 from the position it held when floating on the bus. From Eq. (6.1), the corresponding electrical angle is

$$\beta_{0e} = \frac{p}{2}\beta_0 \quad \text{rad} \tag{6.31}$$

A phasor diagram illustrating this condition of operation for the equivalent circuit of Fig. 6.3(d) is shown in Fig. 6.8 with corresponding physical and mmf space vector diagrams. Field current i_f has been increased to raise the input power factor.

From Eq. (3.151), the torque is

$$T = -K\hat{\mathscr{F}}_{sg}\hat{\mathscr{F}}_{rg}\sin\delta \quad \text{N·m} \tag{6.32}$$

where δ is the angle between the two mmf axes. Figure 6.8(a) shows that power is now entering the stator, and its per-phase value is

$$P_{ma} = V_a I_a \cos\phi \quad \text{W} \tag{6.33}$$

From energy conservation, the torque may also be expressed as

$$T = \frac{P_{\text{mech}}}{\omega_m} = \frac{3P_{ma}}{(2/p)\omega_s} \quad \text{N·m} \tag{6.34}$$

As load torque is increased, the angle β_{0e} increases. The net magnetizing mmf $\mathscr{F}_{mg}$, corresponding to the magnetizing current I_m in Fig. 6.3(d), remains constant, since V_a and ω_s are constant. A stator current therefore flows, producing a stator mmf $\mathcal{F}_{sg}$ such that

$$\hat{\mathcal{F}}_{mg} = \hat{\mathcal{F}}_{sg} + \hat{\mathcal{F}}_{rg} \quad \text{A} \tag{6.35}$$

The stator mmf $\hat{\mathcal{F}}_{sg}$ at angle δ from $\hat{\mathcal{F}}_{rg}$ then produces the torque of Eq. (6.32).

Although Eq. (6.32) expresses the torque, it is not convenient for visualizing how the torque varies and what its limiting value is, since both $\mathcal{F}_{sg}$ and δ vary when the load torque imposed at the coupling is varied. A better appreciation of the variation of torque with rotor position β_{0e} can be obtained with the aid of the space vector diagram of Fig. 6.8(c). The geometry of the diagram shows that since β_{0e} is a negative angle, then

$$\hat{\mathcal{F}}_{sg} \cos (\delta + 90°) = - \hat{\mathcal{F}}_{sg} \sin \delta = - \hat{\mathcal{F}}_{mg} \sin \beta_{0e} \quad \text{A} \tag{6.36}$$

Substitution in Eq. (6.32) then gives

$$T = -K \hat{\mathcal{F}}_{rg} \hat{\mathcal{F}}_{mg} \sin \beta_{0e} \quad \text{N·m} \tag{6.37}$$

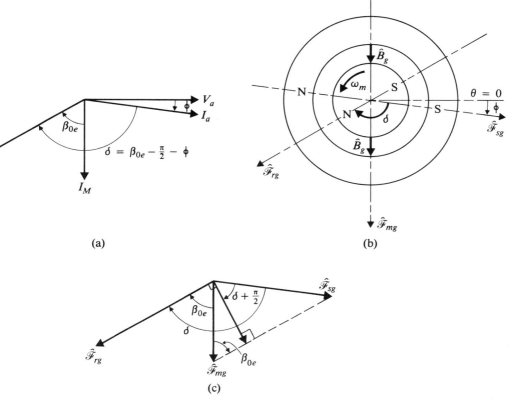

(a)

(b)

(c)

Fig. 6.8 Synchronous machine—motor operation.

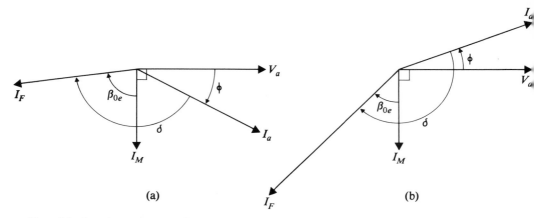

Fig. 6.9 Synchronous machine—effect of varying rotor excitation. (a) Decrease. (b) Increase.

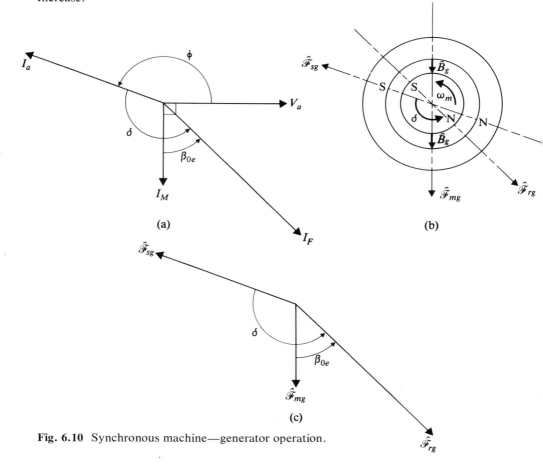

Fig. 6.10 Synchronous machine—generator operation.

The rotor mmf $\hat{\mathscr{F}}_{rg}$ is proportional to the field current i_f or to its ac equivalent I_F shown in the equivalent circuit of Fig. 6.3(d). The magnetizing mmf $\hat{\mathscr{F}}_{mg}$ is proportional to the magnetizing current I_m of Fig. 6.3(d), which in turn is proportional to V_a / ω_s. Thus, the torque expression of Eq. (6.37) is seen to be consistent with that of Eq. (6.27).

If the load torque is maintained constant and i_f is now reduced, the phasor diagram becomes that of Fig. 6.9(a), which shows the machine is operating at a lower lagging power factor. If on the other hand, i_f is increased, the phasor diagram becomes that of Fig. 6.9(b), which shows the motor can operate at a leading power factor.

If the load torque were now removed, the prime mover recoupled, and a driving torque again applied, the rotor could be observed by stroboscope to accelerate momentarily and take up a positive angle β_{0e} relative to its position immediately after synchronization. Figure 6.10 illustrates this condition of operation. $\overline{I}_a$ is now more than $\pi/2$ out of phase with $\overline{V}_a$, indicating that power is leaving the stator terminals, and the machine is generating. In Eq. (6.31), δ is now positive, and the air-gap torque opposes rotation.

Example 6.1 A 3-phase, 2300-V, 100-kVA, 60-Hz, 6-pole synchronous generator has a stator leakage reactance X_{ls} of 7.9 Ω and a magnetizing reactance X_{ms} of 56.5 Ω. The winding resistances are negligible. Immediately after the machine was synchronized onto the bus, the field current was 23 A and the power input at the prime-mover coupling was 3.75 kW.

a) Determine the field current required when the generator is supplying rated output at 0.9 lagging power factor (that is, the load on the generator is inductive).

b) Determine the internal torque developed by the machine when it is delivering an output current of 15 A, and the field current is 20 A.

c) Determine the power input at the shaft required to drive the machine out of synchronism at the 20-A field current.

Solution The synchronous reactance of the machine is

$$X_s = 7.9 + 56.5 = 64.4 \quad \Omega$$

On no load,

$$V_a = \frac{2300}{\sqrt{3}} = 1328 \quad V$$

and

$$I_F = I_M = \frac{V_a}{X_s} = \frac{1328}{64.4} = 20.62 \quad A$$

Thus

$$n'' = \frac{I_F}{i_f} = \frac{20.62}{23} = 0.897$$

a) For the condition of operation described, the phasor diagram will have the form shown in Fig. 6.10, where

$$\overline{V}_a = 1328 \underline{|0} \quad V$$

$$\phi = 180° - \cos^{-1} 0.9 = 154.2°$$

$$I_a = \frac{100 \times 10^3}{3 \times 1328} = 25.10 \quad A$$

$$\overline{I}_M = \frac{\overline{V}_a}{jX_s} = \frac{1328 \underline{|0}}{j64.4} = 20.62 \underline{|-90°} \quad A$$

$$\overline{I}_F = \overline{I}_M - \overline{I}_a = 20.62 \underline{|-90°} - 25.10 \underline{|154.2°} = 41.91 \underline{|-49.8°} \quad A$$

$$i_f = \frac{41.91}{0.897} = 46.7 \quad A$$

b) Since the field current is less than that on synchronization, the phasor diagram corresponding to this operating condition has the form shown in Fig. 6.11. As in (a),

$$\overline{V}_a = 1328 \underline{|0} \quad V$$

$$\overline{I}_M = 20.62 \underline{|-90°} \quad A$$

$$I_F = 0.897 \times 20 = 17.94 \quad A$$

$$I_a = 15 \quad A$$

From Fig. 6.11,

$$\cos \beta_{0e} = \frac{I_F^2 + I_M^2 - I_a^2}{2 I_F I_M} = \frac{17.94^2 + 20.62^2 - 15^2}{2 \times 17.94 \times 20.62} = 0.706$$

$$\beta_{0e} = 45.12°$$

$$T = -\frac{p}{2} \times \frac{3}{\omega_s} \times I_F V_a \sin \beta_{0e}$$

$$= -\frac{6}{2} \times \frac{3}{120\pi} \times 17.94 \times 1328 \sin 45.12° = 403 \quad N \cdot m$$

c)

$$T_{max} = -\frac{6}{2} \times \frac{3}{120\pi} \times 17.94 \times 1328 = 569 \quad N \cdot m$$

Air-gap power is

$$P = \frac{2}{p} \omega_s T = \frac{2}{6} \times 120\pi \times 569 = 71.50 \quad kW$$

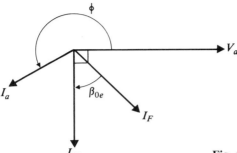

Fig. 6.11 Diagram for Example 6.1.

Rotational loss

$$P_{\text{rot}} = 3.75 \quad \text{kW}$$

The required input at the shaft is therefore

$$P_{\text{shaft}} = 71.50 + 3.75 = 75.25 \quad \text{kW}$$

6.1.4 Synchronous Motor Power Factor

If a synchronous motor is driving against a constant load torque, then the power converted by the machine is constant, no matter what the value of i_f. Thus, from Eq. (6.27), let

$$T = -\frac{p}{2}\frac{3}{\omega_s} I_F V_a \sin \beta_{0e} = \text{constant} \quad \text{N·m} \tag{6.38}$$

Then

$$I_F \sin \beta_{0e} = \text{constant} \quad \text{A} \tag{6.39}$$

As i_f is varied, therefore, the phasor $\bar{I}_F$ follows the vertical locus shown in Fig. 6.12. Also, since stator losses are being neglected, from Eq. (6.33),

$$P = 3V_a I_a \cos \phi = \text{constant} \quad \text{W} \tag{6.40}$$

and

$$I_a \cos \phi = \text{constant} \quad \text{A} \tag{6.41}$$

Thus, as i_f is varied, the phasor of $\bar{I}_a$ follows the other vertical locus shown in Fig. 6.12.

A low value of field current i_f produces the condition $I_F \cos \beta_{0e} < I_M$, and some of the mmf required to produce flux density $\hat{B}_g$ in the air gap of the machine must be contributed by the stator currents. Under these conditions, the motor is said to be *under-excited* and operates at a lagging power factor, as illustrated in Fig. 6.9(a).

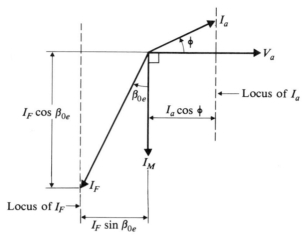

Fig. 6.12 Synchronous motor power factor.

A high value of i_f, such that $I_F \cos \beta_{0e} > I_M$ produces the condition shown in Fig. 6.9(b), where the machine is *over-excited* and operates at a leading power factor. For *normal excitation* the field current is adjusted until the motor operates at unity power factor for a particular load. (Note that this is not the field current required for synchronization, but is somewhat higher.)

Synchronous machines may be specially designed to operate on no load as over-excited motors and are then called *synchronous capacitors*. Such machines are connected at some point in a supply system at which a load with a low lagging power factor (PF) exists. The leading current drawn by the synchronous capacitor brings the current in the supply line more nearly into phase with the supply potential difference, thus giving *power factor correction*. Since the only "load" torque on such a machine is due to its own rotational losses, the component of stator current $\overline{I}_a$ in phase with stator terminal potential $\overline{V}_a$ is very small. A curve of I_a versus i_f for a synchronous capacitor is shown marked "No load" in Fig. 6.13.

A synchronous motor employed to drive a mechanical load can also provide some power factor correction if its windings have been designed to carry the necessary currents. Such machines are normally designed to operate at a leading power factor of 0.8 on full mechanical load. For such a motor the "V" curves illustrated in Fig. 6.13 may be deduced directly from the phasor diagram of Fig. 6.12; however, only the parts of these curves giving leading power factor are used. For each value of mechanical load, I_a is a minimum when the motor is operating at unity power factor. On no load, the motor operates as a synchronous capacitor. Since the greater part of a normal plant load consists of induction motors that draw a lagging current, a few synchronous motors may be installed to improve the power factor of the overall plant load when the mechanical drive characteristics are suitable. Since

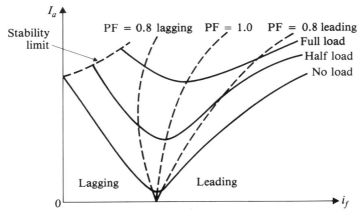

Fig. 6.13 Synchronous motor "V" curves.

this installation reduces the kVA rating of feeders and transformers required to supply the plant, power companies sell energy more cheaply if it is taken at high, rather than low, power factor. The advantages of employing a synchronous motor for any large constant-speed drive would therefore appear to be obvious. The machines themselves, however, are much more expensive than squirrel-cage induction motors of equal speed and rating, so that the decision as to which type of machine to install involves economic problems of some complexity.

Example 6.2 The synchronous machine of Example 6.1 is to be used as a 75-kW synchronous motor with a leading power factor of 0.8 on full load. Determine

a) The field current required for full-load operation.

b) The pull-out torque at the field current determined in (a).

c) The full-load efficiency of the motor if $R_s = 0.81$ Ω and the field-winding resistance $R_F = 1.2$ Ω.

Solution

a) The rotational losses of the motor are 3.75 kW. If the stator resistive losses are neglected, then the input power on full load is

$$P_{in} = 75 + 3.75 = 78.75 \quad kW$$

Thus

$$\sqrt{3} \times 2300 \, I_a \times 0.8 = 78,750$$

from which the full load current is

$$I_a = 24.71 \quad \text{A}$$

The phasor diagram for full load will have the form of that shown in Fig. 6.12, in which

$$\phi = \cos^{-1} 0.8 = 36.87°$$
$$\overline{V}_a = 1328 \underline{|0} \quad \text{V}$$
$$\overline{I}_a = 24.71 \underline{|36.87°} \quad \text{A}$$

Once again neglecting stator resistance, from Example 6.1,

$$\overline{I}_M = 20.62 \underline{|-90°} \text{ A}$$
$$\overline{I}_F = \overline{I}_M - \overline{I}_a = 40.59 \underline{|-119.2°} \quad \text{A}$$

Thus

$$i_f = \frac{I_F}{n''} = \frac{40.59}{0.897} = 45.3 \quad \text{A}$$

b) From Eq. (6.28),

$$T_{\max} = \frac{6}{2} \times \frac{3}{120\pi} \times 40.59 \times 1328 = 1287 \quad \text{N·m}$$

c) P_{in} may now be calculated including the stator resistive losses.

$$P_{in} = 75,000 + 3750 + 3 \times 0.81 \times 24.71^2 + 1.2 \times 45.3^2 = 82,700 \quad \text{W}$$

$$\eta = \frac{75,000}{82,700} = 0.907$$

6.1.5 Determination of Equivalent-Circuit Parameters

This section describes the tests that must be carried out in order to determine the circuit parameters required in applying the analysis of the preceding sections. This analysis referred to the simplified equivalent circuits of Fig. 6.3(c) and (d), in which the synchronous reactance X_s was assumed to be constant over the normal working range of excitations of the machine. Two tests that require the machine be driven at rated speed are carried out as follows:

Test 1—Open-Circuit Test. Drive the machine at synchronous speed on open circuit, and determine the field current required to give rated terminal potential difference.

Test 2—Short-Circuit Test. Reduce the field current to a minimum, using the field rheostat, and then open the field-supply circuit breaker. Short the stator terminals of the machine together through three ammeters; close the field-circuit breaker; and raise the field current to the value noted in Test 1 while maintaining synchronous speed. Record the three stator currents. (This test should be carried out quickly, since the stator currents may be greater than the rated value.)

The synchronous reactance may be calculated in terms of the equivalent circuits of Fig. 6.3(c) and (d). On the assumption that the synchronous reactance X_s and the excitation emf E_o have the same magnitudes in both tests, then

$$X_s = \frac{\text{Open-circuit line-to-neutral potential difference in Test 1}}{\text{Average short-circuit current in Test 2}} \quad \Omega \quad (6.42)$$

From Eq. (6.20),

$$n'' = \frac{I_F}{i_f} \tag{6.43}$$

This ratio may be obtained directly from the results of Test 2 since, from the equivalent circuit of Fig. 6.3(d), I_F is seen to be equal to the per-phase short-circuit current. Thus

$$n'' = \frac{\text{Average short-circuit current in Test 2}}{\text{Field current}} \tag{6.44}$$

Example 6.3 The following test results were obtained from a 3-phase, 60-Hz, 6-pole, 15-kVA, 220-V synchronous motor:

Open-Circuit Test	Short-Circuit Test
Field current = 6.7 A	Field current = 6.7 A
Terminal potential difference = 220 V	Average line current = 57 A

The motor is designed to operate at a full-load leading power factor of 0.8. Determine

a) The required full-load field current.

b) the reactive power absorbed by the motor on full load.

Solution

a) Per-phase synchronous reactance is

$$X_s = \frac{220}{\sqrt{3} \times 57} = 2.23 \quad \Omega$$

$$n'' = \frac{I_F}{i_f} = \frac{57}{6.7} = 8.51$$

Full-load current is

$$I_a = \frac{15 \times 10^3}{220\sqrt{3}} = 39.4 \quad \text{A}$$

On full load, the phasor diagram for the motor will have the form illustrated in Fig. 6.9(b), where

$$\phi = \cos^{-1} 0.8 = 36.9°$$

If

$$V_a = \frac{220}{\sqrt{3}} \underline{|0}$$

then from Fig. 6.3(d),

$$\bar{I}_M = \frac{220\underline{|0}}{\sqrt{3} \times 2.23\underline{|90°}} = 57.0\underline{|-90°} \quad \text{A}$$

In the phasor diagram,

$$\bar{I}_F = \bar{I}_M - \bar{I}_a = 57.0\underline{|-90°} - 39.4\underline{|36.9°} = 86.6\underline{|-111°} \quad \text{A}$$

Thus

$$i_f = \frac{86.6}{8.51} = 10.0 \quad \text{A}$$

b) Reactive power absorbed by the motor is

$$Q = 3 \times \frac{220}{\sqrt{3}} \times 39.4 \sin 36.9° = 9.00 \quad \text{kVAR}$$

*6.1.6 Independent Generators

So far, synchronous machines have been considered only when connected to large ac systems, but small synchronous generators may be required to supply independent electric loads. The chief requirement in such systems is that the terminal potential difference of the generator shall remain constant with varying load, and this is generally accomplished by including an automatic regulator that varies the generator field current. The prime-mover speed must be held constant to maintain constant frequency.

In order to determine what range of field current is required for operation at constant terminal potential difference, it is necessary to consider how the machine would operate if no regulator were present. The effect of load impedance variation may be determined from analysis of the equivalent circuit of Fig. 6.14, in which the reference direction of I_a is the reverse of that previously employed, so that

$$\bar{I}_F = \bar{I}_M + \bar{I}_a \quad \text{A} \tag{6.45}$$

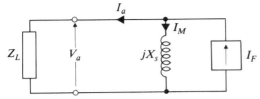

Fig. 6.14 Independent ac generator.

The power factor at which an independent generator operates is determined by the nature of the load circuit, which also determines the relationship between the magnitudes of V_a and I_a. The two curves of Fig. 6.15 illustrate the relationship between V_a and I_a for a purely resistive and purely inductive load circuit with fixed field current. These represent the limits between which practical systems usually lie.

If the load circuit is a pure resistance R_L, then from Fig. 6.14,

$$V_a = R_L I_a \quad \text{V} \tag{6.46}$$

and

$$I_a = \frac{X_s I_F}{(R_L^2 + X_s^2)^{1/2}} \quad \text{A} \tag{6.47}$$

Substitution for R_L in Eq. (6.47) and rearrangement yields

$$\frac{V_a^2}{(X_s I_F)^2} + \frac{I_a^2}{I_F^2} = 1 \tag{6.48}$$

Equation (6.48) describes the unity power factor curve in Fig. 6.15. This is a quarter ellipse with rectangular coordinates V_a and I_a. For $I_a = 0$—that is, for open circuit operation—$V_a = X_s I_F$. For $V_a = 0$—that is for short-circuit operation—$I_a = I_F$. This is the short-circuit current of the generator for the particular field current yielding I_F. The phasor diagram for operation with a resistive load will have the form shown in Fig. 6.16(a).

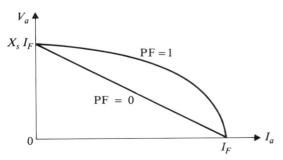

Fig. 6.15 External characteristics of an independent generator.

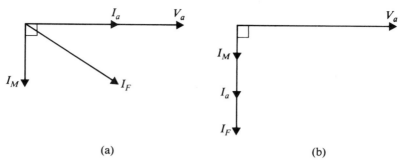

(a)   (b)

Fig. 6.16 Operation of an independent generator. (a) Resistive load. (b) Inductive load.

If the load circuit is pure inductive reactance X_L, then

$$V_a = X_L I_a \quad \text{V} \tag{6.49}$$

and

$$I_a = \frac{X_s I_F}{X_s + X_L} \quad \text{A} \tag{6.50}$$

Substitution for X_L in Eq. (6.50) yields

$$V_a = X_s(I_F - I_a) \quad \text{V} \tag{6.51}$$

Equation (6.51) describes the zero-power-factor curve in Fig. 6.15. The phasor diagram for operation with an inductive load would be that shown in Fig. 6.16(b).

The curves of Fig. 6.15 show that as load current increases, the terminal potential difference at constant field current decreases; that is, the generator has regulation that becomes more marked as the load circuit becomes more inductive and the operating power factor falls. This generator regulation is defined in the same way as for a transformer in Eq. (2.86):

$$\text{Regulation} = \frac{V_{a(\text{no load})} - V_{a(\text{rated load})}}{V_{a(\text{rated load})}} \times 100\% \tag{6.52}$$

This value may readily be determined from the phasor diagram for full-load operation. If regulation is excessive, automatic control of field current may be employed to maintain a nearly constant terminal potential difference as load is varied.

Example 6.4   The machine of Example 6.3 is employed as an independent generator, being driven at constant speed by a gasoline engine and excited by a small dc shunt generator coupled to its own shaft.

Assuming the field current is adjusted to give rated terminal potential difference on no load, determine

a) The regulation when the generator is supplying rated current at 0.9 power factor lagging.

b) The increase of field current required to produce rated terminal potential difference at rated current.

Solution The equivalent circuit of the machine is that shown in Fig. 6.14. The phasor diagram at rated current will have the form shown in Fig. 6.17.

a) Employing quantities calculated in Example 6.3, on no load $V_a = X_s I_F$ so that

$$I_F = \frac{220}{\sqrt{3} \times 2.23} = 57.0 \quad \text{A}$$

and

$$i_f = \frac{57.0}{8.51} = 6.7 \quad \text{A}$$

Rated current is

$$I_a = \frac{15 \times 10^3}{220 \sqrt{3}} = 39.4 \quad \text{A}$$

$$\cos^{-1} 0.9 = 25.8°.$$

Thus if

$$\overline{V}_a = \frac{220}{\sqrt{3}} \underline{|0} = 127 \underline{|0} \quad \text{V}$$

then on full load

$$\overline{I}_a = 39.4 \underline{|-25.8°} \quad \text{A}$$

On full load

$$\overline{I}_F = 57.0 \underline{|\psi} \quad \text{A}$$

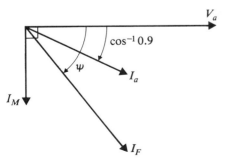

Fig. 6.17 Diagram for Example 6.4.

and

$$I_F \cos \psi = 0.9 \, I_a$$

so that

$$\psi = \cos^{-1} \frac{0.9 \times 39.4}{57.0} = 51.5°$$

$$I_M = 57.0 \sin 51.5° - 39.4 \sin 25.8° = 27.5 \quad A$$

and

$$V_a = X_s I_M = 2.23 \times 27.5 = 61.3 \quad V$$

$$\text{Regulation} = \frac{127 - 61.3}{61.3} \times 100\% = 107\%$$

Such a value of regulation would normally be unacceptable.

b) The required value of I_M at rated load is

$$I_M = \frac{127}{2.23} = 57.0 \quad A$$

and

$$\bar{I}_F = \bar{I}_M + \bar{I}_a = 57.0 \, \underline{|-90°} + 39.4 \, \underline{|-25.8°} = 82.2 \, \underline{|-64.4°} \quad A$$

The required field current is thus

$$i_f = \frac{82.2}{8.51} = 9.66 \quad A$$

*6.1.7 Effect of Magnetic Saturation

The analysis of Sections 6.1.2 to 6.1.4 was developed employing the equivalent circuits of Fig. 6.3(c) and (d) that embody the assumption that the synchronous reactance of the machine is sensibly constant over the anticipated range of its operating conditions. If the operating conditions are expected to vary widely, more accurate predictions of performance need to be made. For this purpose it is necessary to use the equivalent circuit of Fig. 6.3(a), in which the value of the magnetizing reactance X_{ms} is a function of emf E_{ma} or current I_{ma}, and is therefore dependent not only on the field current i_f but also upon the line current I_a. Some method of determining the nature of this dependence and also of determining the magnitude of the stator leakage reactance X_{ls} and the ratio n' is therefore necessary. The tests required to obtain these parameters will now be described.

Test 1—Open-Circuit Test. In this test, the machine is driven at synchronous speed with field current i_f in the rotor winding. The average value of the three line-to-line terminal potential differences divided by $\sqrt{3}$ then gives emf E_{ma} in Fig. 6.3(a). Since $I_a = 0$,

$$I_{ma} = I'_a = n'i_f \quad \text{A} \tag{6.53}$$

The phasor diagram for this condition of operation is shown in Fig. 6.18. A series of readings of E_{ma} and i_f are taken as i_f is increased from zero to the maximum value that the field winding is designed to carry. This will yield a value of terminal potential difference well in excess of the rated value. Figure 6.19 shows a typical curve of E_{ma} versus i_f. From Eq. (6.53), the abscissa of the curve is a measure of I_{ma}/n'. For a pair of values of E_{ma} and i_f, the value of the magnetizing reactance is

$$X_{ms} = \frac{E_{ma}}{n'i_f} \tag{6.54}$$

When the current ratio n' is determined, it can be inserted into Eq. (6.54) to give X_{ms} for a known value of current I_{ma}.

The curve of Fig. 6.19 shows a small effect of residual magnetism near $i_f = 0$. A significant part of the curve is approximately linear, and the constant value of X_{ms}, which applies over this linear region, is known as the *unsaturated magnetizing reactance*. To make efficient use of the ferromagnetic material, most synchronous machines are operated at values of E_{ma} that are in the nonlinear or saturated region of Fig. 6.19.

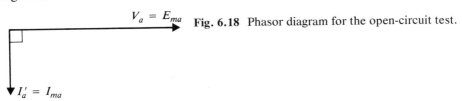

Fig. 6.18 Phasor diagram for the open-circuit test.

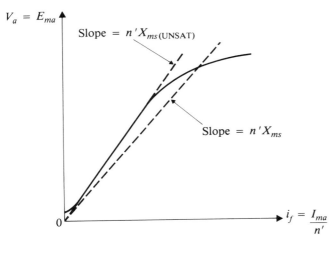

Fig. 6.19 Open-circuit magnetization curve of a synchronous machine.

Test 2—Zero-Power-Factor Test.   To determine the three parameters X_{ms}, X_{ls}, and n', two further relationships are needed in addition to that of Eq. (6.54).

For this test, a balanced, variable, 3-phase, "purely inductive" load (in practice PF < 0.2 is acceptable) is connected to the stator terminals. If stator resistance R_s and whatever resistance is possessed by the inductive load are neglected, then from Fig. 6.3(a), the per-phase equivalent circuit for this test will be that shown in Fig. 6.20, where the reference direction of current I_a appropriate to the operation of an independent generator has been employed. From Fig. 6.20,

$$\overline{I}_{ma} = \overline{I}_a - \overline{I}_a \quad \text{A} \tag{6.55}$$

and

$$\overline{V}_a = \overline{E}_{ma} - jX_{ls}\overline{I}_a \quad \text{V} \tag{6.56}$$

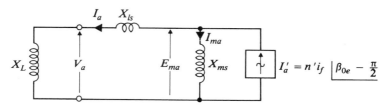

Fig. 6.20 Equivalent circuit for the zero power-factor test.

The test consists of driving the machine at synchronous speed and taking a series of readings of terminal potential difference (from which V_a is obtained) and of i_f as the field current is increased from zero to the maximum. The initial reading is taken with the stator terminals short-circuited so that $V_a = 0$; the field current is adjusted to give rated line current at the shorted terminals. Subsequent readings are then taken at increasing values of load-circuit inductance, and the field current at each load setting is again adjusted to give rated line current. A phasor diagram for one such operating condition is shown in Fig. 6.21. For the readings taken with the stator terminals short-circuited, from Fig. 6.20,

$$I_a = \frac{X_{ms}}{X_{ls} + X_{ms}} n'i_f \quad \text{A} \tag{6.57}$$

Note that the prime-mover power required for this test is small, since the inductive load circuit absorbs no power.

The curve of V_a versus i_f obtained from the zero-power-factor test is shown in Fig. 6.22. On the same axes is the curve obtained from the no-load test, for which $V_a = E_{ma}$. The intercept of the inductive-load curve on the i_f axis represents short-circuit operation.

If the inductive-load curve is compared with the open-circuit curve in Fig. 6.22, it will be seen that they are of similar shape, but that the inductive-load curve is

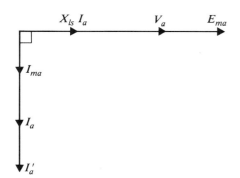

Fig. 6.21 Phasor diagram for the zero-power-factor test.

shifted downward and to the right. This shift can be explained by reference to Eqs. (6.55) and (6.56) and the phasor diagrams for the two tests. The emf E_{ma} is constant for a given value of I_{ma}, regardless of the loading of the machine or the magnitude of the field current.

Consider point p of coordinates (i_{fp}, V_{ap}) on the open-circuit characteristic of Fig. 6.22. From the phasor diagram for this test in Fig. 6.18,

$$V_{ap} = E_{ma} \quad \text{V} \tag{6.58}$$

and

$$i_{fp} = \frac{I'_{ap}}{n'} = \frac{I_{ma}}{n'} \quad \text{A} \tag{6.59}$$

Let point q on the inductive load curve represent a condition of operation for which I_{ma} and hence E_{ma} are the same as for point p. From phasor diagram Fig. 6.21,

$$\overline{V}_{aq} = \overline{E}_{ma} - jX_{ls}\overline{I}_a \quad \text{V} \tag{6.60}$$

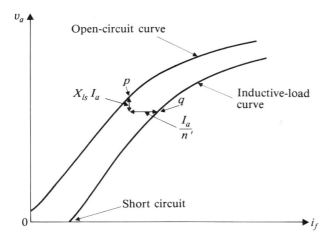

Fig. 6.22 Determination of X_{ls} and n'.

That is,

$$V_{aq} = E_{ma} - X_{ls}I_a \quad \text{V} \tag{6.61}$$

and from Eq. (6.58),

$$V_{aq} = V_{ap} - X_{ls}I_a \quad \text{V} \tag{6.62}$$

Also from Fig. 6.21,

$$\bar{I}_{ma} = \bar{I}'_{aq} - \bar{I}_a \quad \text{A} \tag{6.63}$$

that is,

$$I_{ma} = I'_{aq} - I_a \quad \text{A} \tag{6.64}$$

from which

$$\frac{I'_{aq}}{n'} = \frac{I_{ma}}{n'} + \frac{I_a}{n'}$$

Substitution from Eq. (6.59) gives

$$i_{fq} = i_{fp} + \frac{I_a}{n'} \quad \text{A} \tag{6.65}$$

The relative positions of points p and q are therefore as shown in Fig. 6.22. The practical problem remains of identifying two such points as p and q on the curves so that the quantities $X_{ls} I_a$ and I_a/n' may be measured. The procedure is as follows: Trace the inductive-load curve and its axes on transparent paper. Position the traced curve on the open-circuit curve, obtaining the best fit and ensuring that the axes of both curves remain parallel. Note the vertical interval in volts and the horizontal interval in amperes by which the origin of the tracing is displaced from the origin of the open-circuit curve. Since the magnitude of I_a is known, X_{ls} may be calculated from the vertical interval and n' from the horizontal interval. Finally, X_{ms} may be calculated from Eq. (6.57). A curve of X_{ms} as a function of E_{ma} may now be determined from Fig. 6.19.

Test 3—Stator-Resistance Measurement.   Although stator resistance has been neglected in the foregoing tests, measurements of it will be required if the efficiency of the machine is to be predicted for a given condition of operation. The resistance should be measured by means of a dc bridge immediately after the zero-power-factor test, while the windings are hot. If it is assumed that the stator windings are wye-connected, then half of the average line-to-line resistance gives R_s.

Alternating current resistance may be somewhat greater than the dc value, particularly if the stator conductors are large, since eddy currents may be induced in them, and the current density will not be uniform.

Test 4—Measurement of Rotational Losses.   For this measurement, the machine should be run as a motor at rated terminal potential difference, with the field current adjusted to give minimum stator current. The 3-phase input power for this condition provides a good approximation to the friction, windage, and core losses, all of which may be regarded as constant. The small stator resistance loss involved in this test may be neglected.

Test 5—Field-Resistance Losses. The energy dissipated in resistive losses in the field winding and rheostat is obtained from the product of the field current and the field source potential difference.

Example 6.5 Figure 6.23 shows the results of an Open-Circuit Test and a Zero-Power-Factor Test carried out on a 110-V, 11-kVA, 60-Hz, 1200 r/min synchronous machine. The Zero-Power-Factor Test was carried out at a stator current of 58 A, the last reading being taken with the stator terminals short-circuited. Stator resistance is 0.02 Ω/phase. The input power required to drive the machine as a motor at rated terminal potential difference was 980 W.

Determine the equivalent circuit of the machine. Also determine field current and regulation when the machine is operated as an independent generator with the field current adjusted to give rated terminal potential difference on full load with a lagging power factor of 0.9.

Solution From Fig. 6.23,

$$\frac{I_a}{n'} = \frac{58}{n'} = 1.7 \quad \text{A}$$

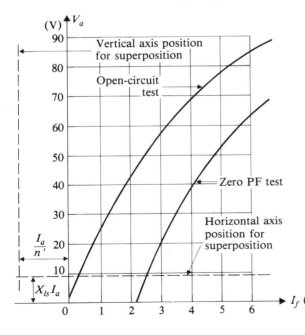

Fig. 6.23 Diagram I for Example 6.5.

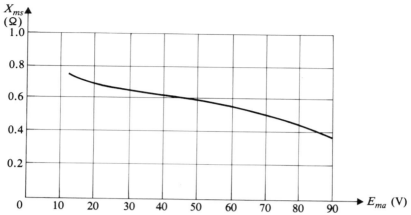

Fig. 6.24 Diagram II for Example 6.5.

from which

$$n' = \frac{58}{1.7} = 34.1$$

Also from Fig. 6.23,

$$X_{ls}I_a = 58 \ X_{ls} = 9.1 \quad \text{V}$$

from which

$$X_{ls} = \frac{9.1}{58} = 0.157 \quad \Omega$$

Equation (6.54) and the open-circuit curve in Fig. 6.23 may now be employed to determine the relationship between E_{ma} and X_{ms} shown in Fig. 6.24.

The phasor diagram for the equivalent circuit of Fig. 6.3(a) with I_a reversed is shown in Fig. 6.25. Rated current is

$$I_a = \frac{11 \times 10^3}{110\sqrt{3}} = 57.7 \quad \text{A}$$

Let

$$\overline{V}_a = \frac{110}{\sqrt{3}} \lfloor 0 = 63.5 \lfloor 0 \quad \text{V}$$

$$\cos^{-1}0.9 = 25.8°$$

Then

$$\overline{I}_a = 57.7 \lfloor -25.8° \quad \text{A}$$

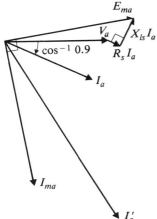

Fig. 6.25 Diagram III for Example 6.5.

$$\overline{Z}_s = R_s + jX_{ls} = 0.026 + j0.157 = 0.159 \,\lfloor 80.6° \quad \Omega$$

$$\overline{E}_{ma} = \overline{V}_a + \overline{Z}_s \overline{I}_a = 63.5 \lfloor 0 + 0.159 \,\lfloor 80.6° \times 57.7 \,\lfloor -25.8° = 69.2 \,\lfloor 6.21° \quad V$$

From Fig. 6.24,

$$X_{ms} = 0.456 \quad \Omega$$

Thus

$$\overline{I}_{ma} = \frac{\overline{E}_{ma}}{jX_{ms}} = \frac{69.2 \lfloor 6.21°}{0.456 \,\lfloor 90°} = 152 \,\underline{-84.0°} \quad A$$

$$\overline{I}_a' = \overline{I}_{ma} + \overline{I}_a = 152 \,\lfloor -84.0° + 57.7 \,\lfloor -25.8° = 189 \,\lfloor -68.9° \quad A$$

$$i_f = \frac{189}{34.1} = 5.54 \quad A$$

At this field current, $E_{ma} = 82.5$ V; that is, on no load the line-to-line potential difference is 143 V.

$$\text{Regulation} = \frac{82.5 - 63.5}{63.5} \times 100\% = 29.9\%$$

Example 6.6 The machine of Example 6.5 is operating as a motor on a 110-V, line-to-line, 60-Hz system. Determine

a) The excitation required to cause the machine to operate at unity power factor when it is drawing rated current from the supply.

b) The efficiency of the motor under the conditions of (a).

Solution

a) The phasor diagram for the equivalent circuit of Fig. 6.3(a), corresponding to full-load operation at unity power factor, is shown in Fig. 6.26. From that diagram, if $\overline{V}_a = 63.5 \lfloor \underline{0}$ V then, with the quantities calculated in Example 6.5,

$$\overline{E}_{ma} = 63.5 \lfloor \underline{0} - 57.7 \lfloor \underline{0} \times 0.159 \lfloor \underline{80.6°} = 62.7 \lfloor \underline{-8.3°} \quad \text{V}$$

From Fig. 6.24,

$$X_{ms} = 0.545 \quad \Omega$$

$$\overline{I}_{ma} = \frac{62.7 \lfloor \underline{-8.3°}}{0.545 \lfloor \underline{90°}} = 115 \lfloor \underline{-98.3°} \quad \text{A}$$

From the phasor diagram,

$$\overline{I}_a = 115 \lfloor \underline{-98.3°} - 57.7 \lfloor \underline{0} = 136 \lfloor \underline{123°} \quad \text{A}$$

Thus

$$i_f = \frac{136}{34.1} = 3.99 \quad \text{A}$$

b) Copper losses $= 3 \times 0.026 \times 57.7^2 = 260 \quad \text{W}$

Rotational losses $= 980 \quad \text{W}$

Power input $= 11 \times 10^3 \quad \text{W}$

Efficiency is $\eta = \dfrac{11 \times 10^3 - 980 - 260}{11 \times 10^3} = 0.887$

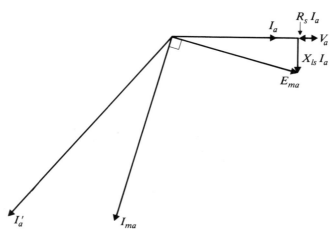

Fig. 6.26 Diagram for Example 6.6.

*6.1.8 Permanent-Magnet Synchronous Machines

As in the case of dc machines, the field excitation of a synchronous machine may be provided by means of permanent magnets, which eliminate the need for a dc source for excitation.

Figure 6.27 shows a cross section of a permanent-magnet (PM) synchronous machine with a cylindrical rotor. The stator is similar to that of a normal synchronous machine, while the uniform magnetization of the rotor produces a distribution of flux density in the air gap that, to a close approximation, is sinusoidal. While a solid permanent-magnet rotor, as in Fig. 6.27, is possible, the rotor is normally made up of a combination of permanent-magnet material, such as Alnico or ferrite, with soft iron parts to complete the magnetic path.

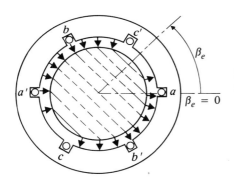

Fig. 6.27 Cylindrical-rotor, permanent-magnet machine.

The equivalent circuit shown in Fig. 6.3(a) may be employed as a starting point for the analysis of PM machines. Since the stators of the two machines are similar, the parameters R_s, X_{ls}, and X_{ms} may be as shown in the equivalent circuit of Fig. 6.28(a). In this case, X_{ms} is a magnetizing reactance that includes the effect of air-gap and stator reluctance, but not the effects of the magnetic flux path in the permanent magnet, which must be represented separately.

The rotor of a PM machine must be designed so that the rotor material is on the recoil line forming the linear part of the curve in Fig. 1.68(b). Section 1.25 showed that the magnet could be represented by an mmf source $\mathscr{F}_0$ in series with a reluctance $\mathscr{R}_0$, as illustrated in Fig. 1.68(c), or by a flux source in parallel with a reluctance $\mathscr{R}_0$, as illustrated in Fig. 1.68(d). The flux-source representation is more appropriate for the rotor of a synchronous machine, since a very large change in mmf in the region of the recoil line produces only a very small change in flux.

Section 2.10.2 showed how an electric circuit, equivalent to a given electromagnetic system, could be developed from the magnetic equivalent circuit. If these principles are applied to the flux source and parallel reluctance $\mathscr{R}_0$ in Fig. 1.68(d), the electric equivalent circuit obtained is an emf source E_o in series with a linear reactance X_o, and these parameters may therefore be included in the equivalent circuit of Fig. 6.28(a). As for the normal equivalent circuit of Fig. 6.3(b), if the rotor position is

$$\beta_e = \omega_s t + \beta_{0e} \quad \text{rad} \tag{6.66}$$

then

$$\overline{E}_o = E_o \underline{|\beta_{0e}} \quad \text{V} \tag{6.67}$$

If the reactance X_{ms} is assumed to be constant, the Thévenin equivalent circuit of Fig. 6.28(b) may be employed, where

$$E_e = \frac{X_{ms}}{X_{ms} + X_o} E_o \quad \text{V} \tag{6.68}$$

and

$$X_e = X_{ls} + \frac{X_{ms} X_o}{X_{ms} + X_o} \quad \Omega \tag{6.69}$$

The theory developed for the normal synchronous machine and derived from the equivalent circuit of Fig. 6.3(b) or (c) may then be applied.

The field excitation of a PM synchronous machine cannot be varied. If such a machine is to be used as a generator, it is therefore important that the equivalent reactance X_l should be as small as possible. As may be seen from Eq. (6.69), this calls for low values of X_o and X_{ls}, while the value of X_{ms} should be high. Low X_o may be obtained by employing permanent-magnet material with a low value of

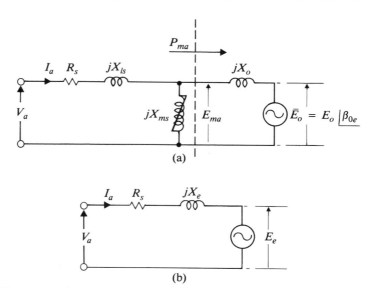

Fig. 6.28 Equivalent circuits for a permanent-magnet machine.

recoil permeability; therefore, modern ferrite magnets are appropriate. A low value of X_{ls} may be obtained by ensuring that the path for stator leakage flux has high reluctance; and, to this end, the stator is constructed with relatively open slots. The magnetizing reactance X_{ms} may be made large by making the air gap small.

The linear model of Fig. 6.28 is applicable only within the limits of the straight recoil line of the magnetic material—that is, the flux must not fall below the value ϕ_a shown in Fig. 1.68(b). If the machine is short-circuited at any time, there is a danger that the consequent high stator currents will drive the flux below that value. This indicates that materials with characteristics similar to that of Ferrite D (Fig. 1.65) or that of samarium-cobalt (Fig. 1.67) would be most suitable for PM synchronous generators. When a material such as Alnico is used, it may be necessary to have reasonably large values of X_{ls} and X_o in order to limit the short-circuit current.

Permanent-magnet rotors frequently have salient poles, as illustrated in Fig. 6.29. No large reluctance torque is produced, because the low recoil permeability of the magnets presents a high-reluctance path to flux produced in them by the stator mmf. For this reason, the analysis of the cylindrical-rotor machine gives reasonably accurate results when applied to a salient-pole PM machine.

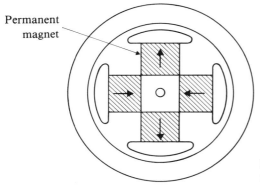

Permanent magnet

Fig. 6.29 Salient-pole, permanent-magnet machine.

*6.1.9 Hysteresis Motors

A hysteresis motor is a synchronous machine with a 3-phase stator and a rotor of special construction that carries no winding. The rotor consists of a central core of nonmagnetic material upon which are mounted a number of rings of special steel alloy. These rings form a cylindrical outer covering that is laminated in a plane normal to the rotor shaft on which the core is mounted.

The rotor acts as a permanent magnet for continuous synchronous-speed running. However, during starting, the stator current is large, as in an induction machine, and the ring material is cycled around its hysteresis loop. The flux produced in the rotor is delayed in phase behind the rotating magnetomotive force, and the angle between the stator and rotor mmf axes results in the development of torque. Since the angle by which the rotor flux lags the mmf is dependent only on the loop shape and not on the frequency at which it is cycled, this motor has an essentially constant torque during starting, from standstill up to synchronous speed. When synchronous speed is reached, the stator current falls, and the motor runs as a permanent-magnet machine.

Hysteresis motors find wide application in timing devices and in drives requiring a very smooth start, free from torque oscillation.

Example 6.7 Figure 6.30 shows the per-phase equivalent circuit of a PM ac generator in which R_s is neglected. The permanent-magnet material is Ferrite D, as described in the $B-H$ characteristic of Fig. 1.65.

a) Determine the open-circuit terminal potential difference of the generator and find the operating point on the $B-H$ curve of the magnetic material.

b) Determine the short-circuit current for the generator and find the $B-H$ operating point during short circuit.

c) Three capacitors, each having a reactance of 3.75 Ω, are connected in series with the generator to improve its regulation. If the load connected to the generator and series capacitors is short-circuited, will the rotor magnets be demagnetized?

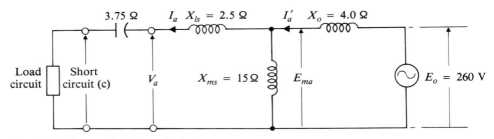

Fig. 6.30 Diagram for Example 6.7.

Solution

a) On open circuit,

$$V_a = E_{ma} = \frac{15}{15 + 4} \times 260 = 205.3 \quad V$$

Thus the line-to-line terminal potential difference is 355.6 V. From Fig. 1.65, $B_0 = 0.34$ T, which corresponds to $E_o = 260$ V. Thus $E_{ma} = 205.3$ corresponds to

$$B = \frac{205.3}{260} \times 0.34 = 0.27 \quad T$$

From Fig. 1.65, $H = -40$ kA/m.

b) On short circuit, $V_a = 0$,

$$I'_a = \frac{260}{[4 + (2.5 \times 15)/(2.5 + 15)]} = 42.33 \quad A$$

and

$$I_a = \frac{15}{2.5 + 15} \times 42.33 = 36.28 \quad A$$

Thus

$$E_{ma} = 36.28 \times 2.5 = 90.7 \quad V$$

and this corresponds to

$$B = \frac{90.7}{260} \times 0.34 = 0.119 \quad T$$

From Fig. 1.65, $H = -170$ kA/m.

c) With the capacitors in series with the generator

$$X_{ls} + X_c = 2.5 - 3.75 = -1.25 \quad \Omega$$

and

$$I'_a = \frac{260}{[4 + (-1.25 \times 15)/(-1.25 + 15)]} = 98.6 \quad A$$

$$I_a = \frac{15}{15 - 1.25} \times 98.6 = 107.6$$

$$E_{ma} = 107.6 (2.5 - 3.75) = -134.5 \quad V$$

Thus

$$B = \frac{-134.5}{260} \times 0.34 = -0.176$$

From Fig. 1.65, $H \simeq -325$ kA/m. Since this operating point is beyond the straight-line part of the characteristic, the rotor magnets will be demagnetized.

*6.1.10 Three-Phase Synchronous Reluctance Motors

Section 3.1.3 showed that a primitive machine of the form illustrated in Fig. 3.11 would operate as an energy converter, provided that its speed of rotation, ω_m, was the same as the angular frequency, ω_s, of the stator source—in other words, that it would operate as a synchronous machine, despite the fact that the alternating field produced by its stator winding acted along a stationary axis. This was a synchronous reluctance machine.

It is not difficult to visualize the addition of further coils spaced around the stator that, if they were excited with alternating currents having suitable phase relationships to that of the original ac excitation, would also contribute to the energy conversion process. If there were three such coils, the rotor would, in effect, be enclosed in a three-phase stator. The result would be a machine like that illustrated in Fig. 6.31(a), from which it is easy to visualize the rotating stator poles "catching up" the ferromagnetic rotor and rotating it at synchronous speed.

Three-phase synchronous reluctance motors have become increasingly important in recent years, since it is possible to drive a large number of them in exact synchronism from a three-phase source of controllable frequency, such as a static inverter. Such systems find ready application in the textile industry in particular. It

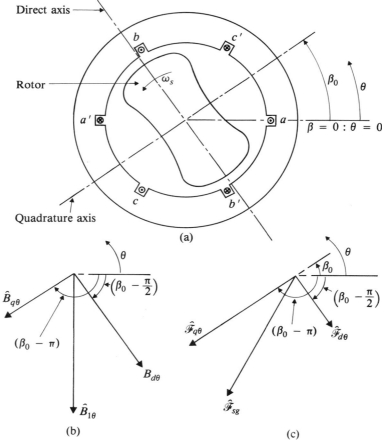

Fig. 6.31 Three-phase reluctance machine. (a) Rotor position at $t = 0$. (b) Flux-density space-vector diagram. (c) Magnetomotive-force space-vector diagram.

is therefore desirable to derive an equivalent circuit that may be employed to predict the performance of these motors.

a) *Analysis and Equivalent Circuit.* It is possible to approach the analysis of a reluctance motor by means of the procedure employed with the induction and wound-rotor synchronous machines, but the result is unsatisfactory because the magnitudes of some of the equivalent-circuit parameters depend upon the phase angle of the current. The model is therefore not readily applied. A substantial step of idealization will therefore be taken initially in assuming that not only the stator resistance but also the stator leakage reactance may be neglected. This means that the per-phase applied terminal potential difference V_a is identical to the per-phase induced stator emf E_{ma}.

Assume that the currents in the stator windings of the machine in Fig. 6.31(a) are

$$i_a = \hat{i}_s \sin(\omega_s t + \alpha_s) \quad \text{A}$$

$$i_b = \hat{i}_s \sin\left(\omega_s t + \alpha_s - \frac{2\pi}{3}\right) \quad \text{A} \tag{6.70}$$

$$i_c = \hat{i}_s \sin\left(\omega_{st} + \alpha_s - \frac{4\pi}{3}\right) \quad \text{A}$$

Then, from Eqs. (3.146) and (3.147), the net mmf acting across the air gap at any angle θ is given by

$$\mathcal{F}_{sg} = \hat{\mathcal{F}}_g \cos(\omega_s t + \alpha_s - \theta) \quad \text{A} \tag{6.71}$$

where, for a 2-pole machine,

$$\hat{\mathcal{F}}_g = \frac{3N_{se}\hat{i}_s}{4} \quad \text{A} \tag{6.72}$$

In Eq. (6.72), N_{se} is as before the effective turns per phase of the stator winding. On the assumption that the permeability of the magnetic material of stator and rotor is infinite, then the flux density B_θ in the air gap at angle θ is

$$B_\theta = \frac{\mu_0 \mathcal{F}_{sg}}{g_\theta} \quad \text{T} \tag{6.73}$$

where g_θ is the effective air-gap length at angle θ. If β_0 is the angular position of the rotor at $t = 0$, then Fig. 6.32 shows the mmf and flux density distribution in the air gap at instant $t = 0$, when the direct axis of the rotor is at position $\beta = \beta_0 + \pi/2$.

Since the air gap is not uniform, space harmonics of flux density exist, giving the distribution of B_θ shown in Fig. 6.32. However, as has already been pointed out in Section 5.1, the harmonic components of the flux-density wave make no contribution to the flux linkage of a sinusoidally distributed winding. This means that only the fundamental component $B_{1\theta}$ shown in Fig. 6.32 contributes to the stator flux linkage.

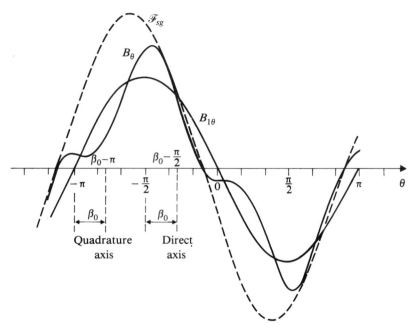

Fig. 6.32 Magnetomotive-force and flux-density distribution in the air gap ($t = 0$).

It is convenient to employ the phasor of the terminal potential difference as a reference, so that $\overline{V}_a = V_a\underline{|0}$. And since the effects of stator resistance and leakage reactance are being neglected, it follows that $\overline{E}_{ma} = E_{ma}\underline{|0}$, and

$$e_{ma} = \hat{e}_{ma}\cos\omega_s t \quad \text{V} \tag{6.74}$$

This emf is induced by the flux density $B_{1\theta}$, which must therefore lag e_{ma} by $\pi/2$ rad. Thus the distribution of the fundamental component of the flux density in the air gap is expressed by

$$B_{1\theta} = \hat{B}_1 \cos\left(\omega_s t - \frac{\pi}{2} - \theta\right) \quad \text{T} \tag{6.75}$$

This flux-density distribution may be considered to consist of two components: the first, a sinusoidal distribution $B_{d\theta}$ with its peak values lying on the direct axis of the rotor, so that

$$B_{d\theta} = \hat{B}_1 \cos\beta_0 \cos\left(\omega_s t + \beta_0 - \frac{\pi}{2} - \theta\right) \quad \text{T} \tag{6.76}$$

The other sinusoidal distribution is $B_{q\theta}$, with its peak values on the quadrature axis, so that

$$B_{q\theta} = \hat{B}_1 \sin\beta_0 \cos(\omega_s t + \beta_0 - \pi - \theta) \quad \text{T} \tag{6.77}$$

In Fig. 6.31(b), a space-vector diagram drawn for the instant $t = 0$ shows the directions of the peak values of these flux-density components and of the resultant peak flux density $\hat{B}_{1\theta}$.

The two sinusoidal components of flux density, $B_{d\theta}$ and $B_{q\theta}$, may be considered to be due to sinusoidally distributed mmf waves $\mathscr{F}_{d\theta}$ and $\mathscr{F}_{q\theta}$. Fig. 6.31(c) shows the directions and magnitudes of the peak values of these mmf waves and of their resultant, $\mathscr{F}_{\theta}$. Necessarily,

$$\frac{\mathscr{F}_{q\theta}}{B_{q\theta}} > \frac{\mathscr{F}_{d\theta}}{B_{d\theta}} \quad \text{A/T} \tag{6.78}$$

because the air gap along the quadrature axis is much greater than that along the direct axis. The resultant mmf $\mathscr{F}_{\theta}$ is displaced from the flux density $B_{1\theta}$ and thus contains not only a magnetizing component, but also a power component. This power component accounts for the energy conversion and develops the corresponding internal torque.

A phasor diagram of the idealized reluctance machine, shown in Fig. 6.33, is based on the assumption that the flux-density waves described in Eqs. (6.75) and

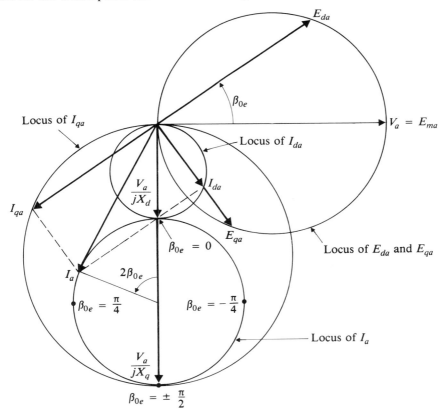

Fig. 6.33 Phasor diagram for three-phase reluctance machine (regenerating).

(6.76) induce emf's E_{da} and E_{qa}. These emf's then lead the flux-density waves by $\pi/2$ rad, so that their phase relationships to their resultant E_{ma} and, therefore, the terminal potential difference V_a are that shown in the diagram. Thus, if electrical angular position β_{0e} is used for the rotor of a p-pole machine,

$$\overline{E}_{da} = V_a \cos \beta_{0e} \underline{|\beta_{0e}} \quad \text{V} \tag{6.79}$$

$$\overline{E}_{qa} = V_a \sin \beta_{0e} \underline{|\beta_{0e} - \pi/2} \quad \text{V} \tag{6.80}$$

The mmf waves $\mathscr{F}_{d\theta}$ and $\mathscr{F}_{q\theta}$ may be considered to be produced by two components of the stator current that together produce the mmf wave $\mathscr{F}_\theta$. The phase relationships of the stator current I_a and its two components I_{da} and I_{qa} are necessarily those of the mmf vectors shown in Fig. 6.31(c).

Since $\overline{E}_{da}$ leads $\overline{I}_{da}$ by $\pi/2$ rad, and $\overline{E}_{qa}$ leads $\overline{I}_{qa}$ by the same angle, the *direct-axis reactance* X_d may be defined by

$$X_d = \frac{E_{da}}{I_{da}} \quad \Omega \tag{6.81}$$

Similarly, the *quadrature-axis reactance* may be defined by

$$X_q = \frac{E_{qa}}{I_{qa}} \quad \Omega \tag{6.82}$$

The total stator current in phase a is

$$\overline{I}_a = \overline{I}_{da} + \overline{I}_{qa} \quad \text{A} \tag{6.83}$$

As may be seen from Fig. 6.33, the phasors $\overline{E}_{da}$, $\overline{E}_{qa}$, $\overline{I}_{da}$, and $\overline{I}_{qa}$ all follow circular loci as β_{0e} varies. Consequently, current I_a also follows a circular locus.

If the relationship of $\overline{V}_a$ and $\overline{I}_a$, together with the circular locus of $\overline{I}_a$, are compared with the induction motor circle diagram of Fig. 5.24, a striking similarity is noticeable. From this it may be concluded that an appropriate equivalent circuit for a reluctance motor may have the same form as the induction motor equivalent circuit of Fig. 5.22, in which R_s is neglected. The equivalent circuit for the reluctance motor therefore consists of a branch X_d in parallel with an impedance of constant reactance and variable resistance. These two parameters must therefore be defined. From Eq. (6.79) to Eq. (6.83),

$$\begin{aligned}
\overline{I}_a &= \frac{V_a \cos \beta_{0e} \underline{|\beta_{0e}}}{jX_d} + \frac{V_a \sin \beta_{0e} \underline{|\beta_{0e} - \pi/2}}{jX_q} \\
&= V_a \left[\frac{\cos^2\beta_{0e} + j \cos \beta_{0e} \sin \beta_{0e}}{jX_d} + \frac{\sin^2\beta_{0e} - j \sin \beta_{0e} \cos \beta_{0e}}{jX_q} \right] \\
&= V_a \left[\frac{1}{jX_d} + \frac{X_d - X_q}{X_d X_q} (- \sin \beta_{0e} \cos \beta_{0e} - j \sin^2\beta_{0e}) \right] \\
&= V_a \left[\frac{1}{jX_d} + \frac{X_d - X_q}{X_d X_q (- \cot \beta_{0e} + j)} \right] \quad \text{A} \tag{6.84}
\end{aligned}$$

The impedance of the branch in parallel with X_d is thus

$$Z = - \left(\frac{X_d X_q}{X_d - X_q} \right) \cot \beta_{0e} + \frac{jX_d X_q}{X_d - X_q} \quad \Omega \tag{6.85}$$

The equivalent circuit is shown in Fig. 6.34. When the angle β_{0e} is zero, the resistive part of this impedance is infinite, so that no current flows in the branch, no energy is absorbed from or delivered to the source; and, as may be seen from the phasor diagram of Fig. 6.32, I_a lags V_a by $\pi/2$ rad and is equal to V_a/X_d. For a positive value of β_{0e}—that is, where the rotor direct axis is being driven ahead of the field flux—the resistive part of Z is negative, and the machine acts as a generator. For negative β_{0e}, it acts as a motor. Also from the phasor diagram it is seen that the component of I_a in phase with V_a reaches a positive or negative maximum when $\beta_{0e} = -\pi/4$ or $+\pi/4$. The first condition represents the power limit when the machine is motoring; the second, when it is regenerating. It may be observed that Figs. 6.31 and 6.33 are drawn for a condition of regeneration; that is, the connected load is driving the motor and is being braked by it.

b) *Terminal Characteristics.* At constant terminal potential difference, the important terminal characteristics are those relating the stator current I_a, the internal torque T, and the angle β_{0e} between the direct axis of the rotor and the axis of the 3-phase stator mmf.

The per-phase input power is

$$P_a = \mathcal{Re} \; [\bar{I}_a \bar{V}^*_a] = \mathcal{Re} \left[\frac{X_d - X_q}{X_d X_q} \cdot \frac{V_a^2}{(-\cot \beta_{0e} + j)} \right]$$

$$= - \frac{X_d - X_q}{2 X_d X_q} V_a^2 \sin 2\beta_{0e} \quad W \tag{6.86}$$

Since stator resistance is being neglected, P_a is also the air-gap power. The internal torque for a p-pole machine is thus

$$T = 3 \cdot \frac{p}{2} \cdot \frac{P_a}{\omega_s} = - \frac{3}{\omega_s} \cdot \frac{p}{2} \left(\frac{X_d - X_q}{X_d X_q} \right) V_a^2 \sin 2\beta_{0e} \quad N \cdot m \tag{6.87}$$

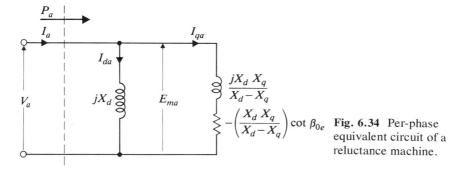

Fig. **6.34** Per-phase equivalent circuit of a reluctance machine.

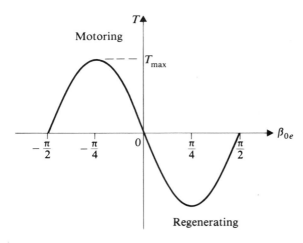

Fig. 6.35 Variation of internal torque with β_{0e}.

This relationship is illustrated in Fig. 6.35. It may be compared with that in Fig. 3.13 for the single-phase reluctance motor. From Eq. (6.87) it may be seen that the maximum torque developed by the 3-phase machine is

$$|T_{\max}| = \frac{3}{\omega_s} \frac{p}{2} \left(\frac{X_d - X_q}{X_d X_q} \right) V_a^2 \quad \text{N} \cdot \text{m} \tag{6.88}$$

The stable operating point for zero torque is at $\beta_{0e} = 0$, since application of a mechanical load slowing the motor momentarily produces a negative value of β_{0e}, resulting in a motoring torque that tends to restore β_{0e} to zero.

Reluctance motors, which have no controllable dc rotor excitation, necessarily operate at a lagging power factor. Detailed attention to the design of the magnetic system of these machines has, however, raised their full-load power factor to about 0.75.

Like other synchronous machines, the reluctance motor has no starting torque; it may, however, be started by means of a pole-face, squirrel-cage winding as described for the normal synchronous motor in Section 3.4.3.

c) *Determination of X_d and X_q.* Machines to be driven from a variable-frequency and, therefore, variable-potential source are subject to a varying degree of magnetic saturation. In general, the direct-axis reactance X_d is much more affected by saturation than is the quadrature-axis reactance X_q.

At any terminal potential difference, the values of X_d and X_q can be measured by applying that potential difference to the stator terminals and observing the variation of rms stator current while the machine is driven at slightly less than synchronous speed (using, for example, an induction motor of the same synchronous speed). The maximum value of the observed current variation then gives V_a/X_q, and the minimum value gives V_a/X_d. While X_d will fall appreciably as V_a is increased, it is probable that X_q will vary little and may be considered constant at some average value for the range of potential difference employed. Direct- and quadrature-axis inductances may then be calculated, and equivalent-circuit parameters for any desired frequency obtained. These measured inductances account for the effect of the stator leakage inductances as well as that of the air gap.

Example 6.8 A 6-pole, 60-Hz, 3-phase reluctance motor is connected to a 220-V line-to-line supply. Its direct-axis reactance is 15 Ω, and its quadrature-axis reactance is 3 Ω. Stator resistance and rotational losses may be ignored.

a) Develop an equivalent circuit for the motor.

b) Determine the maximum torque that the motor can supply before losing synchronism.

c) Determine the maximum power factor at which the motor will operate and the mechanical power output for this condition.

Solution

a)
$$X_d = 15 \ \ \Omega; \qquad X_q = 3 \ \ \Omega$$

Let
$$\overline{V}_a = \frac{220}{\sqrt{3}} \underline{|0} = 127 \underline{|0} \quad \text{V}$$

$$\frac{X_d X_q}{X_d - X_q} = \frac{15 \times 3}{15 - 3} = 3.75 \ \Omega$$

The equivalent circuit is that of Fig. 6.34 with this reactance value and that of X_d inserted.

b) From Eq. 6.88,
$$T_{\text{max}} = \frac{3}{2\pi \times 60} \times \frac{6}{2} \times \frac{127^2}{3.75} = 102.7 \quad \text{N} \cdot \text{m}$$

c) Maximum power factor occurs when the phasor of I_a in Fig. 6.33 is tangential to its locus in the fourth quadrant of the phasor diagram, as illustrated in Fig. 6.36. From Fig. 6.33, the dimensions of Fig. 6.36 are
$$0a = \frac{V_a}{2}\left(\frac{1}{X_q} + \frac{1}{X_d}\right) \qquad = \frac{V_a}{2} \times \frac{X_d + X_q}{X_d X_q}$$
$$ab = \frac{V_a}{2}\left(\frac{1}{X_q} - \frac{1}{X_d}\right) \qquad = \frac{V_a}{2} \times \frac{X_d - X_q}{X_d X_q}$$

From Fig. 6.36, the power factor is
$$\cos \phi = \frac{ab}{0a} = \frac{X_d - X_q}{X_d + X_q} = \frac{15 - 3}{15 + 3} = 0.667$$

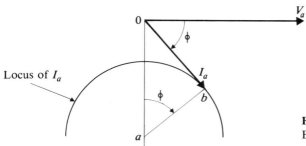

Fig. 6.36 Diagram for Example 6.8.

Also

$$I_a = 0a \sin \phi = \frac{127}{2} \times \frac{18}{15 \times 3} (1 - 0.667^2)^{1/2} = 18.93 \quad \text{A}$$

$$P_{ma} = V_a I_a \cos \phi = 127 \times 18.93 \times 0.667 = 1604 \quad \text{W}$$

$$P_{\text{mech}} = 3P_{ma} = 4.81 \quad \text{kW}$$

*6.1.11 Salient-Pole Synchronous Machines

A salient-pole synchronous machine may be regarded as a reluctance machine with field windings added to the rotor poles. The construction of such a rotor may be seen in Fig. 3.56. It has also been shown in Eqs. (6.15) and (6.20) that the effect of the rotor excitation may be represented in a per-phase equivalent circuit by means of a current source

$$\bar{I}_F = n' \frac{X_{ms}}{X_s} i_f \lfloor \beta_{0e} - \pi/2 = n'' i_f \lfloor \beta_{0e} - \pi/2 \quad \text{A} \tag{6.89}$$

Such a source is shown in Fig. 6.3(d). If a current source representing the effect of the rotor current is combined with the equivalent circuit for a 3-phase reluctance machine, shown in Fig. 6.34, a model may be obtained that permits investigation of the behavior of a salient-pole synchronous machine. An approximate equivalent circuit for a salient-pole synchronous machine is therefore that shown in Fig. 6.37.

From the equivalent circuit,

$$\bar{I}_a = \bar{V}_a \left[\frac{1}{jX_d} + \frac{X_d - X_q}{X_d X_q(-\cot \beta_{0e} + j)} \right] - I_F \lfloor \beta_{0e} - \pi/2$$

$$= \bar{V}_a \left[\frac{1}{jX_d} - \frac{(X_d - X_q)}{X_d X_q} \sin \beta_{0e} \lfloor \beta_{0e} \right] - I_F \lfloor \beta_{0e} - \pi/2 \quad \text{A} \tag{6.90}$$

The per-phase air-gap power, which is also the power entering the stator terminals, is

$$P_a = \mathscr{R}e \ [\bar{I}_a \bar{V}_a^*] = -\frac{(X_d - X_q)}{2X_d X_q} V_a^2 \sin 2\beta_{0e} - V_a I_F \sin \beta_{0e} \quad \text{W} \tag{6.91}$$

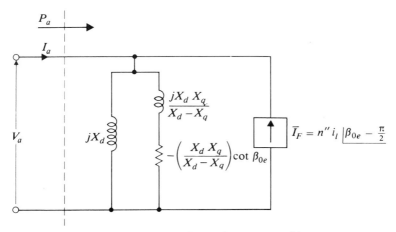

Fig. 6.37 Equivalent circuit for a salient-pole synchronous machine.

The air-gap torque is thus

$$
T = 3 \cdot \frac{p}{2} \cdot \frac{P_a}{\omega_s}
$$

$$
= \frac{3}{\omega_s} \frac{p}{2} \left[-\frac{X_d - X_q}{2 X_d X_q} V_a^2 \sin 2\beta_{0e} - V_a I_F \sin \beta_{0e} \right] \quad \text{N} \cdot \text{m} \qquad (6.92)
$$

It is clear from this expression that the torque is made up of two components that vary with β_{0e}. The first depends upon the difference between the direct- and quadrature-axis synchronous reactance, the applied terminal potential difference, and the angle β_{0e}. Its variation at a particular terminal potential difference is therefore similar to that shown in Fig. 6.35 for the reluctance machine. The second component depends upon the field current, the applied terminal potential difference, and the angle β_{0e}. Its variation at a particular terminal potential difference is therefore similar to that shown in Fig. 6.4 for a cylindrical-rotor synchronous machine. The resultant variation of the developed torque for three values of field current is illustrated in Fig. 6.38.

The magnitude of the field current i_f may have some influence on the reluctance-torque component, since it may affect the degree of saturation of the magnetic system and therefore influence the values of X_d and X_q. However, in practice the ratio X_d / X_q for a salient-pole synchronous machine is generally less than 2, which means that the reluctance-torque component is a relatively small proportion of the whole. In particular it does not greatly affect the value of the maximum torque $T_{\max}$. The most marked effect of the reluctance torque is the steepening of the curve of T versus β_{0e} in the neighborhood of $\beta_{0e} = 0$. This enables the machine to respond rapidly to changes in shaft torque.

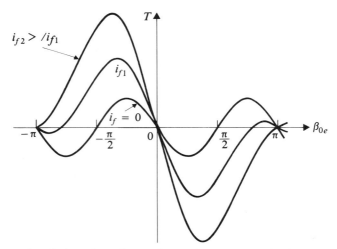

Fig. 6.38 Torque-angle relations for salient-pole synchronous machine in the steady state with several values of field current i_f.

The operation of a synchronous capacitor has already been discussed in Section 6.1.4, where it was seen that an unloaded and over-excited synchronous motor would draw a large leading current from the line. There are situations in some electric power systems where it is convenient at certain times during the 24 hours to have a machine that draws a large lagging current to assist in maintaining the system potential difference within the normal range. An under-excited synchronous machine will fulfill this function. However, if the field current of a cylindrical-rotor machine is reduced too far, the developed torque fails to equal the loss torque, and the machine falls out of synchronism. With a salient-pole machine employed for this purpose, the reluctance torque assists in keeping the machine in synchronism to such an extent that the field current may be reduced to zero or even reversed. This capability greatly increases the magnitude of the lagging current that the machine can draw—that is to say, it increases its ability to act as a "synchronous inductor."

The steady-state behavior of the machine, which may be predicted from the equivalent circuit or from the expressions in Eqs. (6.91) and (6.92) derived from that circuit, is illustrated by the phasor diagram of Fig. 6.39. This diagram is drawn for particular values of V_a and i_f. The values of X_d and X_q for zero field current may be determined as functions of V_a in the manner described in Section 6.1.9.

Example 6.9 A salient-pole synchronous generator is rated at 100 MVA, 13.8 kV, 60 Hz, and 180 r/min. Its direct-axis reactance is 2.38 Ω, and its quadrature-axis reactance is 1.33 Ω. A field current of 1200 A is required to generate rated terminal potential difference on no load.

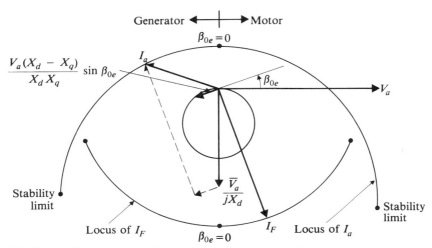

Fig. 6.39 Phasor diagram for a salient-pole synchronous machine.

a) Assuming that the field current is 2000 A and that the generator is connected to a 13.8 kV bus, determine the power delivered to the bus and the power factor when the angle $\beta_{0e} = 30°$. Machine losses may be ignored.

b) Assuming that the field current is 1200 A and that the terminal potential difference is 13.8 kV, determine both the maximum torque that can be applied to the generator shaft before it loses synchronism, and the shaft power at this maximum torque.

Solution

a)
$$\text{Let } V_a = \frac{13,800}{\sqrt{3}} \lfloor 0 = 7967 \lfloor 0 \quad V$$

On no load,

$$i_f = 1200 \text{ A} \quad \text{and} \quad I_F = \frac{V_a}{X_d} = \frac{7967}{2.38} = 3348 \quad A$$

At $i_f = 2000$ A,

$$I_F = \frac{2000}{1200} \times 3348 = 5579 \quad A$$

From Eq. (6.90),

$$\bar{I}_a = 7967 \lfloor 0 \left[\frac{1}{j2.38} - \frac{2.38 - 1.33}{2.38 \times 1.33} \sin 30° \lfloor 30° \right] - 5579 \lfloor -60° = 4018 \lfloor 168.2° \quad A$$

The generator output power is

$$P_{out} = -\mathcal{R}e\ [3\overline{I}_a\overline{V}^*_a] = -3 \times 4018 \times 7967 \cos 168.2 = 94 \quad MW$$

Power factor is

$$PF = |\cos 168.2| = 0.979$$

b) Since the generator speed is 180 r/min, $p = 40$. At $i_f = 1200$, $I_F = 3348$, and from Eq. (6.92)

$$T = \frac{3}{2\pi \times 60} \times \frac{40}{2}\left[-\frac{2.38 - 1.33}{2 \times 2.38 \times 1.33} \sin 2\beta_{0e} - 7967 \times 3348 \sin \beta_{0e}\right]$$

$$= -1.676 \sin 2\beta_{0e} - 4.245 \sin \beta_{0e} \quad MN\cdot m$$

To determine the maximum torque, let

$$\frac{dT}{d\beta_{0e}} = 0 = -3.352 \cos 2\beta_{0e} - 4.245 \cos \beta_{0e}$$

By trial and error, $\beta_{0e} = \pm 62.7°$. For a generator, the angle is $\beta_{0e} = +62.7°$. Inserting this value into the torque expression gives

$$T = 1.676 \sin (2 \times 62.7°) - 4.245 \sin 62.7°$$

$$= -5.138 \quad MN\cdot m$$

At this torque, the mechanical power is

$$P_{mech} = T \times 2\pi \times \frac{180}{60} = 96.85 \quad MW$$

*6.2 SOLID-STATE DRIVES USING SYNCHRONOUS MOTORS

Accurate speed control may be obtained using some type of synchronous motor excited by solid-state power converters similar to the system shown for an induction motor in Fig. 5.38. As for the induction motor, a constant-frequency, constant-potential source is assumed. Any type of synchronous motor—reluctance, permanent magnet, or wound-field—may be employed, each having advantages in particular applications. Wound-field synchronous motors are more expensive and larger than squirrel-cage induction motors of equal power, but the fact that they may be operated at a leading power factor makes them attractive. Modern permanent-magnet synchronous motors operate near unity power factor, and may have a large pullout torque for a given size. The great simplicty of construction of a reluctance motor is an advantage if its relatively low operating power factor can be accepted. When a large number of low-power drives are required to rotate in exact synchronism, reluctance motors are usually employed.

There are two modes of operation of synchronous-motor drives: (a) open-loop, where the stator potential difference and frequency are both directly controlled;

(b) closed-loop, where the stator potential is directly controlled but the frequency is governed by information obtained from a rotor position sensor.

*6.2.1 Open-Loop System

The relationship between the speed of rotation of a synchronous motor and the angular frequency of the stator excitation is

$$\omega_m = \frac{2}{p}\,\omega_s \quad \text{rad/s} \tag{6.93}$$

Since the frequency of operation of the inverter in Fig. 5.38 may be very accurately controlled by means of a crystal oscillator, a correspondingly accurate control of motor speed can be achieved.

To avoid saturation of the magnetic system, it is necessary to vary the inverter output potential difference at the same time as the frequency is varied. From Eq. (6.17), the air-gap torque of the machine is

$$T = -\frac{p}{2}\frac{3}{\omega_s}\,I_F V_a \sin \beta_{0e} \quad \text{N·m} \tag{6.94}$$

Thus in a wound-field motor if field current i_f, and consequently I_F, is held constant, then

$$T = -K \frac{V_a}{\omega_s} \sin \beta_{0e} \quad \text{N·m} \tag{6.95}$$

where K is a constant. When V_a and ω_s are the rated values for the motor, it will be operating at what may be called its base speed, an expression that has already been employed in relation to dc and induction motors. The ratio V_a/ω_s for the rated values may be maintained at lower speed, thus maintaining the magnitude of pull-out torque equal to that at rated speed. Pull out occurs when $\beta_{0e} = \pi/2$, and

$$T_{\max} = -K \frac{V_a}{\omega_s} \quad \text{N·m} \tag{6.96}$$

At quite low speed, the potential difference $R_s I_a$ in the winding resistance, which has been neglected in deriving the torque expression of Eq. (6.94), becomes significant in relation to the stator potential difference V_a. If the pull-out torque is to be maintained constant as speed is reduced, the stator flux must remain constant. This can be accomplished by keeping E_{ma}/ω_s constant, where E_{ma} is the per-phase emf induced in the stator winding. This necessitates increasing the ratio V_a / ω_s as zero speed is approached to compensate for the $R_s I_a$ drop.

Speed-torque characteristics for variable-frequency operation of a synchronous motor are shown in Fig. 6.40. The load driven by the motor may be decelerated by regenerative braking, provided that the inverter of Fig. 5.38 is supplied from a dual converter that is able to accept negative current. Operation in the third and fourth quadrants of Fig. 6.40 may be obtained by reversing the sequence of the thyristor

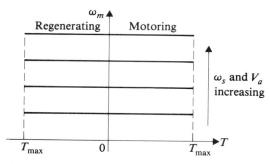

Fig. 6.40 Speed–torque characteristics for a synchronous motor.

gating signals applied to the inverter, thus reversing the phase sequence of the stator excitation.

For the machine to remain in synchronism, the relationship between the mechanical speed and the inverter frequency must remain that in Eq. (6.93). A change in speed requires acceleration or deceleration, which in turn is limited by the maximum torque—see Eq. (6.96)—and the polar moment of inertia of the motor and its load. There is therefore a maximum rate at which the speed of the open-loop synchronous motor drive can safely be changed. This limit is met by restricting the rate at which the inverter frequency is changed. Speed reversal is obtained by reversing the phase sequence of the inverter as its frequency passes at a controlled rate through zero. Because of the danger of losing synchronism, the open-loop drive is not suitable for loads in which the torque has large, sudden fluctuations.

Speeds in excess of the base speed may be obtained by increasing ω_s while maintaining V_a at the rated value. However, as may be seen from Eq. (6.96), the pull-out torque will decrease under these conditions.

A block diagram of an open-loop, synchronous-motor drive is shown in Fig. 6.41. For four-quandrant operation, the inverter output phase sequence must

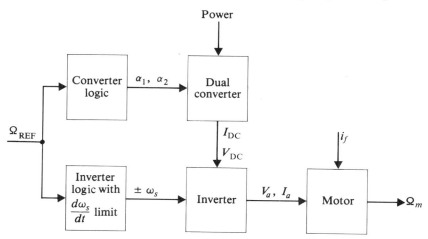

Fig. 6.41 Steady-state model of an open-loop synchronous motor drive.

reverse if the polarity of the speed command signal Ω_{REF} is reversed. For regeneration, the converter logic must apply gating signals that permit negative current I_{DC} at the dc terminals of the dual converter. Figure 6.42(a) shows the basic power circuit of a three-phase inverter that may be employed to supply the motor from the variable potential output of the dual converter. The thyristor symbol enclosed in a circle signifies a thyristor that may be turned on by a gating signal and commutated by circuit elements not shown. When any thyristor is turned on in conjunction with the antiparallel connected diode, it forms a short circuit, bringing one of the stator terminals to the potential of one of the inverter dc input terminals. The gating signals for the six thyristors along with the waveforms of the resulting line-to-line stator terminal potential differences, are shown in Fig. 6.42(b).

*6.2.2 Closed-Loop System

The chief shortcoming of the open-loop system is its liability to permit the motor to pull out of synchronism, so that the average motor torque becomes zero.

A dc motor, which has alternating current in the armature winding and direct current in the field winding, is not liable to a sudden loss of torque, since the mechanical commutator ensures that there is always a fixed relationship between the frequency of the armature currents and the motor speed, while the position of the brushes ensures that the angle δ between the rotor and stator mmf axes is always $\pi/2$ rad. This is the optimum position for developing torque as shown in Eq. (3.151), which states

$$T = -K \; \hat{\mathscr{F}}_{sg} \; \hat{\mathscr{F}}_{rg} \sin \delta \quad \text{N·m} \tag{6.97}$$

A similar relationship between stator frequency and shaft speed, as well as control of the angle δ between the two mmf axes, can be obtained in a synchronous machine excited from an inverter by making the gating signals functions of rotor position. The method of doing this is illustrated in Fig. 6.43.

The three proximity detectors in Fig. 6.43 are spaced $\pi/3$ rad from one another. The three signals resulting from the presence or absence of the shutter of the position sensor thus uniquely define six 60° intervals of the complete revolution of the two-pole rotor. (In a p-pole machine, the detectors would be spaced at $\pi/3$ electrical radians and the shutter would have a number of equally spaced large diameter sectors equal to the number of pairs of poles.) Using the position sensor it is possible to devise a logic system that, for any given rotor position, will cause those coils on the stator that exert a torque in a positive direction on the rotor to be energized. The position sensor, logic circuits, and thyristor circuits perform the functions of the brushes and commutator of the conventional dc machine. Indeed, this combination of inverter and synchronous machine is sometimes called a "commutatorless dc machine." If the sensor shutter is correctly positioned relative to the rotor poles, the rotor axis is maintained in space quadrature with the axis of the resultant field in the air gap. Thus the maximum (pull-out) torque of the machine is available at all speeds.

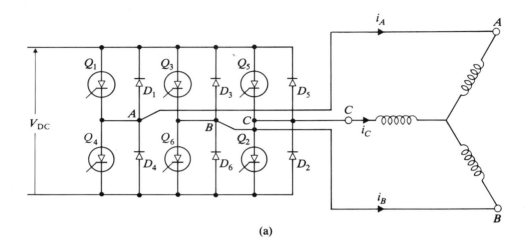

(a)

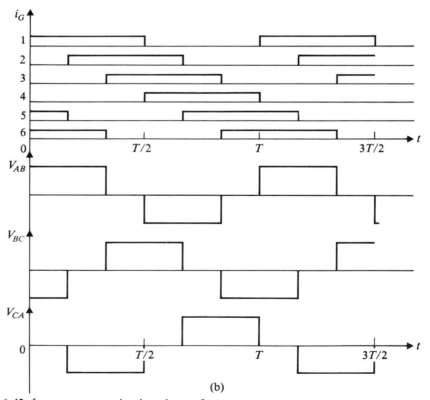

(b)

Fig. 6.42 Inverter power circuit and waveforms.

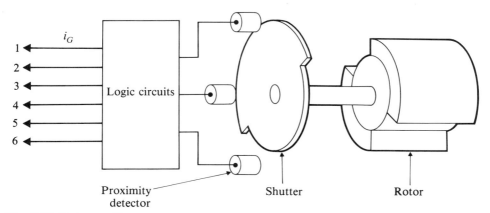

Fig. 6.43 Rotor position sensor.

The steady-state speed of the motor is controlled by adjusting the input potential difference of the inverter in much the same manner as the armature potential difference of a separately excited dc motor. The steady-state, speed–torque characteristic is similar to that of a separately excited dc motor. If the load torque is suddenly increased, the excess of load torque over the torque produced at the motor air gap causes the motor to decelerate. The rotor position sensor and inverter logic then produce a simultaneous reduction of the inverter frequency at the correct rate to maintain synchronism. This drive is therefore suitable for loads with suddenly varying torque. Four-quadrant operation is easily achieved, since the sequence of pulses from the rotor position sensor automatically changes as the speed passes through zero. The general behavior of this system may also be compared to that of the induction motor with slip-frequency control (see Section 5.3.2).

A block diagram of a closed-loop, synchronous-motor drive is illustrated in Fig. 6.44.

*6.3 LINEAR SYNCHRONOUS MOTORS

The operating principle of any form of rotating electric machine may also be applied in translational or linear machines. Thus the linear induction motor, which has many practical applications, has been discussed in Section 5.6. At present, linear dc machines have few, if any, practical applications, and have therefore been passed over without comment. The linear synchronous machine (LSM), while not at present widely used, exhibits such potential, particularly in the field of transportation, that a brief description of the mode of operation of two principal types is in order.

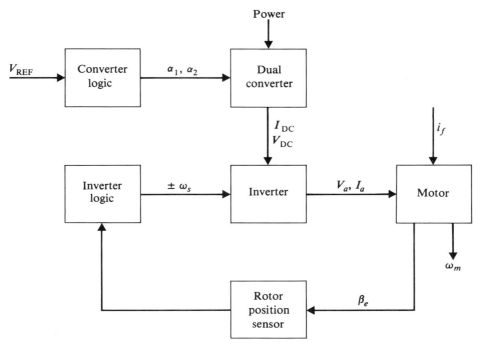

Fig. 6.44 Steady-state model of a closed-loop synchronous-motor drive.

*6.3.1 Iron-Cored LSM's

The process of cutting and unrolling visualized in the case of the cylindrical induction motor and illustrated in Fig. 5.62 may also be visualized for the synchronous machine. The cost considerations that, for most purposes, precluded the choice of the wound stator as the member to be extended the required distance, apply to the synchronous motor just as they did to the induction motor. An iron-cored LSM is therefore usually of the short-primary type, although long-primary LSM's have been built for use in experimental traction systems. The rotor of a normal synchronous machine cannot be reduced to a form as simple as the limiting case of the squirrel-cage induction motor shown in Fig. 5.62. A long array of discrete wound poles excited with dc would be prohibitively expensive. A practical solution is to be found, however, in the Lorenz inductor alternator, a homopolar machine in which both ac and dc windings are located on the stator.

A laboratory homopolar synchronous machine is shown in Fig. 6.45 and 6.46. The short primary, which would normally be the moving member, is fixed at the base of the machine. The extended secondary, which would normally be installed on the horizontal guideway in a practical LSM, is built in the form of a large wheel

Fig. 6.45 Experimental linear synchronous motor. *(University of Toronto)*

Fig. 6.46 Close view of primary windings. *(University of Toronto)*

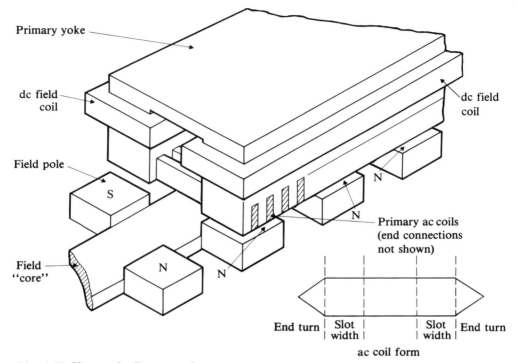

Fig. 6.47 Homopolar linear synchronous motor.

that may be driven by the fixed primary. The close view of the primary member in Fig. 6.46 shows the 3-phase ac winding in slots close to the air gap, while the dc winding, divided for convenience into two coils, is located beneath the ac winding. A diagram of this homopolar LSM as it would be in the truly linear form is shown in Fig. 6.47. The field coil mmf establishes flux down one side of the primary, across the secondary pole structure, and up the other side of the primary. The field poles modulate the flux, causing a rate of change of flux linkage in the primary coils as the primary moves relative to the secondary.

When applying an iron-cored LSM for traction, it is necessary to mount not only the motor primary but also the inverter on the vehicle. While the mass of the LSM primary may be greater than that of a linear induction motor capable of the same power output, both the efficiency and the power factor of the LSM are better than those of the LIM. Thus a smaller inverter rating may be used.

*6.3.2 Air-Cored LSM's

Even in the LSM of the form illustrated in Fig. 6.47, the amount of ferromagnetic material laid in the guideway would be great, as also would be the weight of the ferromagnetic core embodied in the primary mounted on the vehicle. Moreover, the use of such a machine for very high speeds would produce high core losses and consequent inefficient operation.

An alternative method of producing the field flux required in a LSM that does not demand the provision of a ferromagnetic system is made possible by the enormous mmf's that may be maintained virtually indefinitely by superconducting magnet coils. In this type of LSM, illustrated in Fig. 6.48, the superconducting field coils are mounted on the underside of the vehicle, while the 3-phase (and now appropriately named) stator winding is embedded in the top surface of a concrete guideway. The kind of stator winding required may be visualized as an extension of the developed 3-phase winding shown in Fig. 5.8(b). The superconducting magnet coils are so excited as to produce alternate N and S poles on the underside of the vehicle.

The guideway winding for a high-speed system would be supplied from regularly spaced track-side inverters. One of the advantages of this type of motor is that it does not require a power pick-up system on the vehicle, whose reliability would be questionable at high speed.

Although linear synchronous motors can be operated under open-loop conditions, sudden changes in load—due, for instance, to wind gusts or tunnel entry—may cause loss of synchronism. Pole-position sensing may be used to form a closed-loop system as described for rotating synchronous motors. Alternative methods of closed-loop control of load angle that do not require position sensing have also been developed.

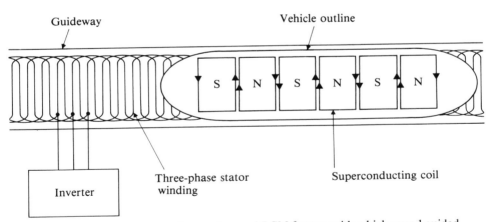

Fig. 6.48 Winding arrangement for an air-cored LSM for use with a high-speed guided vehicle.

*6.4 STEPPING MOTORS

Stepping motors have been developed in response to the demand for a device capable of producing a definite angular displacement in a driven shaft and holding its position against a torque applied to the driven shaft. For example, in machine-tool applications, such motors could, through gearing, position both the cutting tool and the workpiece very accurately, and could hold them in those positions during the subsequent cutting operation. Moreover, since the stepping motor is operated by current pulses, it may be controlled by command signals stored digitally. It is therefore suitable for computer-controlled operations.

*6.4.1 Variable-Reluctance Stepping Motors

The operation of this type of motor depends upon the basic principles already discussed in Section 3.1.3. One method of construction is illustrated in Fig. 6.49, which shows a machine having a rotor with eight poles and three separate 8-pole stators arranged along the rotor. The poles of stator phase "a" are energized with alternate polarity by a set of series-connected coils carrying current i_a. In this condition, the rotor poles tend to align with the poles of stator phase "a" and are shown so aligned in Fig. 6.49. If β is the angular displacement of the rotor measured, as usual, in a

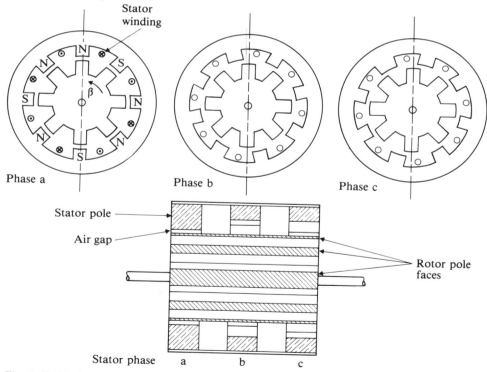

Fig. 6.49 Variable-reluctance stepping motor—phase "a" stator energized (stator sections viewed from the left).

counterclockwise direction from the position shown in the diagram, then the air-gap torque exerted upon the rotor with current i_a flowing is

$$T = \frac{1}{2} i_a^2 \frac{dL_a}{d\beta} \quad \text{N} \cdot \text{m} \tag{6.98}$$

where L_a is the inductance of the stator winding on phase "a." If a torque is applied to the rotor shaft, causing a positive or counterclockwise rotation β, the inductance L_a will be reduced and a negative or clockwise torque T will be developed by the motor.

As may be seen from Fig. 6.49 the phase "b" stator is identical with the phase "a" stator, except that its poles are displaced through 15° in a counterclockwise direction. If current i_a is now set to zero, and i_b is established, the motor will develop a torque rotating the rotor counterclockwise through 15°. Stator "c" has its poles displaced a further 15° counterclockwise with respect to stator "b." Interruption of current i_b and establishment of current i_c will therefore result in a further 15° counterclockwise rotation of the rotor. Finally, interruption of current i_c and reestablishment of current i_a result in the completion of a 45° rotation of the rotor, that is, one complete step. Further current pulses in the a-b-c sequence produce further counterclockwise steps. Reversal of the current pulse sequence to a-c-b produces reversed rotation.

The inductance of the phase "a" windings on the stator will have a maximum or direct-axis value when the rotor poles are aligned with the stator poles. It will have a minimum or quadrature-axis value when the rotor poles are midway between the stator poles. If it is assumed that the variation between these two limiting values of inductance is sinusoidal, then for the motor of Fig. 6.49, L_a may be expressed as a function of β by the relation

$$L_a = L_0 + L \cos 8\beta \quad \text{H} \tag{6.99}$$

Thus from Eq. (6.98), the torque due to current i_a is

$$T = -4i_a^2 L \sin 8\beta \quad \text{N} \cdot \text{m} \tag{6.100}$$

It will be noted that this torque is zero at $\beta = 0$; thus any load torque applied to the rotor shaft will introduce an alignment error.

The rotor and the mechanical system driven by the stepping motor possess rotational inertia. The potential-energy storage in the magnetic field of the motor in conjunction with the kinetic-energy storage in the inertia of the driven mechanism produce a second-order system that tends to oscillate as each step movement of the rotor is made. If the driven mechanism possessed no appreciable friction or output in the form of mechanical work, then the motor and mechanism would form an underdamped oscillatory system. Since this is undesirable, it is necessary to introduce some additional mechanical or electrical damping.

A stiffer and more powerful system will result if two of the stator phases are excited simultaneously in the sequence ab-bc-ca for positive rotation. By analogy with Eq. (6.99), the inductance of the phase "b" stator windings may be expressed by

$$L_b = L_0 + L \cos 8(\beta + 15°) \quad \text{H} \tag{6.101}$$

The torque due to the simultaneous excitation of phases "a" and "b" for movement between the phase "a" and "b" poles is then

$$T_{ab} = \frac{i_a^2}{2} \left[\frac{dL_a}{d\beta} + \frac{dL_b}{d\beta} \right] \quad \text{N·m} \tag{6.102}$$

since $i_a = i_b$. Substitution for L_a and L_b in Eq. (6.102) then yields

$$T_{ab} = - 4\sqrt{3} \, i_a^2 L \sin 8(\beta - 7.5°) \quad \text{N·m} \tag{6.103}$$

The zero value of this torque occurs at $\beta = 7.5°$, and the peak value is $\sqrt{3}$ times as great as that due to one pair of stator poles acting alone. For the same steady-state error in rotor angle, the torque exerted by the simultaneously excited system is $\sqrt{3}$ times as great as when only one phase of the motor is excited. With this increased stiffness, the frequency of the oscillation will also be increased. It may be noted from Eq. (6.103) that the maximum restoring torque occurs at a displacement of 12.5° from the zero-torque position for the 8-pole motor of Fig. 6.49. Malfunction due to any steady load below this maximum value is not possible.

As will be appreciated from the preceding discussion, the step angle of a motor is determined by the number of poles. Typical step angles are 15°, 5°, 2°, and 0.72°. The choice of step angle depends on the angular resolution required for the application. The speed at which a stepping motor can operate is limited by the degree of damping existing in the system. Speeds up to 200 steps per second are typically attainable. A steady and continuous speed of rotation (slewing) greater than this value can be achieved, but the motor is then unable to stop the system in a single step, since the kinetic energy of the system has become too great.

*6.4.2 Permanent-Magnet Stepping Motors

The torque developed by a stepping motor of a given size may be increased if a permanent-magnet rotor is employed. In addition, by reversal of stator current, a force of repulsion may also be exerted upon the rotor.

Figure 6.50 shows the stator arrangement for a 4-pole, 2-phase, permanent-magnet motor. The rotor is magnetized in the manner shown in Fig. 1.66(b) and is typically made of ferrite. Energizing the phase "a" winding only with the polarity shown in Fig. 6.50 will hold the rotor in the position illustrated. The additional energizing of phase "b" winding with the polarity shown in Fig. 6.50 will result in a movement of 22½° in a positive direction. De-energizing the phase "a" winding

Phase a

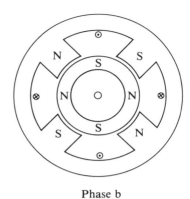

Phase b

Fig. 6.50 Permanent-magnet stepping motor.

will then result in a further 22½° movement. Reversal of the phase "a" winding current will now produce a further 22½° movement, and so on.

Double coils are usually employed in each phase to simplify the switching operation required to reverse stator polarities. A suitable circuit for the motor of Fig. 6.50 is shown in Fig. 6.51. The switches shown in that circuit would be electronic. A typical permanent-magnet motor is shown disassembled in Fig. 6.52. This is a 12-pole, 2-phase machine, as may be seen from the stator construction. Each stator is energized by a pair of circumferentially wound toroidal coils. Current in one of the coils energizes all twelve poles of one of the stators with the desired polarity. Each stator, therefore, has four leads. Axial flux is produced through each coil and passes into the annular ferromagnetic plates on each side of it. Each plate is equipped with six poles set at right angles to the plane of the plate so that the two sets of six poles are positioned alternately around the rotor.

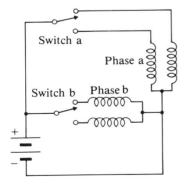

Fig. 6.51 Circuit for the motor of Fig. 6.50.

Fig. 6.52   Permanent-magnet stepping motor.

PROBLEMS

6.1 A source of 3-phase, 400-Hz potential is required in an industrial operation. The only utility supply available is 3 phase at 60 Hz. The frequency conversion is to be accomplished by use of a synchronous motor driving a synchronous generator. A variation of $\pm 3\%$ about the 400 Hz frequency is permissible. Determine a suitable number of poles for each of the synchronous machines and the speed at which the set will run.   *(Section 6.1)*

6.2 A 3-phase, 440-V, 10-kV·A, 60-Hz, 1200-r/min synchronous machine has the following parameters:

$$X_{ms} = 48 \quad \Omega \qquad\qquad R_s \text{ negligible}$$
$$X_{ls} = 2.8 \quad \Omega \qquad\qquad P_{\text{rot}} \text{ negligible}$$

A field current of 6 A is required to produce rated potential on open circuit. Magnetic saturation may be ignored.

   a) Draw an equivalent circuit of the form of Fig. 6.3(a), inserting all parameter values, including n'.

   b) Determine the synchronous reactance of the machine in ohms and in per-unit of the rated impedance of the machine.

   c) Draw an equivalent circuit like that in of Fig. 6.3(d), including n''.   *(Section 6.1.1)*

6.3 The synchronous machine of Problem 6.2 is to be used as a motor connected to a 440-V, 60-Hz supply.

a) If the mechanical output power of the motor is 10 kW, what field current is required to give unity power factor?

b) If the mechanical load in (a) is removed from the motor and the field current is unchanged, what are the input current and power factor?

c) What is the maximum torque that the motor can deliver at the field current found in part (a)?

d) What is the line current for the conditions of (c)?   *(Section 6.1.2)*

6.4 The synchronous machine of Problem 6.2 is used as a generator driven by an internal combustion engine. It is connected to a 440-V, 60-Hz, 3-phase supply.

a) Determine the field current necessary to supply 10 kW to the bus at unity power factor.

b) Using the field current of part (a), determine the maximum torque that can be exerted by the engine before the generator pulls out of synchronism.   *(Section 6.1.2)*

6.5 A 3-phase, 60-Hz, 2-pole, 60-MVA synchronous generator is connected to a 13.2 kV, line-to-line distribution system. This line potential is maintained constant. The generator has 1.2 per unit synchronous reactance based on the machine rating. Generator losses may be ignored. An initial field current of 1000 A is required to give unity power factor operation when the prime-mover shaft input is 50 MW.

a) Determine the lower value of field current required to make the generator deliver rated current at this shaft input power.

b) Determine the lowest value to which the field current can be reduced without loss of synchronism at the same shaft input power.

c) Determine the line current for the conditions established in (b).   *(Section 6.1.2)*

6.6 A 1000-hp, 2300-V, 60-Hz, 900 r/min, 3-phase synchronous motor has a synchronous reactance to 1.9 Ω/phase. On a no-load test a field current of 15 gave a no-load terminal potential difference of 2300 V. The load and field current are adjusted until the line current is 200 A. Stator winding resistance may be neglected.

Determine the field current and the angle between the stator and rotor mmf axes when the field current is adjusted to give a power factor of

a) unity,   b) 0.8 lagging,   c) 0.8 leading.   *(Section 6.1.3)*

6.7 With no load on the motor of Problem 6.6, the field current is adjusted so that the motor draws minimum current from the line. The excitation is then held constant, and the motor is loaded until the line current is 250 A. Stator winding resistance and all rotational losses may be neglected. Draw a phasor diagram showing this operating condition, and determine the angle through which the rotor is retarded when the load is applied.   *(Section 6.1.3)*

6.8 A 1-MW, 6600-V, 10-pole, 60-Hz, 3-phase synchronous motor has a synchronous reactance of 15 Ω/phase. The field current and load are such that the motor operates at a power factor 0.8 leading and draws a line current of 95 A. Stator winding resistance and all rotational losses may be neglected.

a) Draw a phasor diagram for this condition of operation.

b) Calculate the shaft torque of the motor.

c) Determine the reactive kVA the motor is drawing from the supply.   *(Section 6.1.3)*

6.9 A certain factory has the following loads: 1050 horsepower of induction motor at an average power factor 0.75, 75 kW of lighting and heating, and a fully loaded, 150-horsepower synchronous motor. Assume that the average induction motor efficiency is 85% and that the synchronous motor operates at a leading power factor of 0.8. Losses in the synchronous motor may be neglected.

Determine the overall power factor of the load on the supply system presented by the entire factory

a) When the synchronous motor is not running.

b) When the synchronous motor is running. *(Section 6.1.4)*

6.10 A 400-kW, 6600-V, 60-Hz, 8-pole, 3-phase synchronous motor has a synchronous reactance of 22.0 Ω/phase. Neglecting losses in the motor, determine the pull-out torque with the motor operating at full load and the field current adjusted to give

a) Unity power factor,

b) A power factor of 0.8 leading. *(Section 6.1.4)*

6.11 When a 50-kVA, 3-phase, 440-V, 60-Hz synchronous generator is driven at its rated speed, it is found that the open-circuit terminal potential difference is 440 V line-to-line with a field current of 7 A. When the stator terminals are short circuited, rated current is produced by a field current of 5.5 A.

a) Determine the synchronous reactance per phase.

b) Determine the per-unit synchronous reactance. The machine rating should be employed as a base. *(Section 6.1.5)*

6.12 The following test results were obtained from a 3-phase, 60-Hz, 6-pole, 15-kW, 220-V synchronous motor.

Open-Circuit Test	Short-Circuit Test	No-Load Test
Field current = 6.7 A Line-to-line stator terminal potential difference = 220 V	Field current = 6.7 A Average line current = 104 A	Three-phase input power = 0.8 kW

The motor is designed to operate at a full-load leading power factor of 0.8.

a) Draw a phasor diagram representing the full-load operating condition at rated power factor, and determine the required field current.

b) Neglecting stator copper losses, determine the reactive power the motor could absorb when supplying half of its rated mechanical power at rated current. In addition, determine the required field current. *(Section 6.1.5)*

*** 6.13** A 10-kVA, 110-V, 60-Hz, 3-phase ac generator is driven at its rated speed by a diesel engine. No-load and short-circuit tests on the generator gave the following results:

No-Load Test	Short-Circuit Test
Terminal potential difference = 110 V Field current = 3.45 A	Line current = 124 A Field current = 3.45 A

The generator is required to supply a mixed but balanced load consisting of a 5 hp induction motor with a power factor of 0.75 and 4.25 kW of lighting. The efficiency of the motor is 80%. The stator winding resistance of the generator may be neglected.

a) Assume, the no-load terminal potential difference of the generator is adjusted to 110 V. Determine the field current required when

   i) The lighting load and capacitors are switched on,
   ii) The combined load is switched on.

b) Draw a phasor diagram for each of the two operating conditions in (a). *(Section 6.1.6)*

*6.14 For the generator and load of Problem 6.13 an attempt is made to reduce the regulation by connecting three 750-μF capacitors in wye across the terminals of the generator.

a) Determine the terminal potential difference of the generator if the field current is held constant at 3.54 A and

   i) All load is switched on,
   ii) The lighting load only is switched on,
   iii) The capacitors are disconnected and the lighting load only is switched on.

b) Draw an equivalent circuit and a phasor diagram for each of the three conditions of operation specified in (a). *(Section 6.1.6)*

*6.15 The synchronous generator of Problem 6.11 is used to supply an independent load of 40 kW with 0.85 lagging power factor at a potential of 440 V.

a) Determine the field current required.
b) If the load is reduced to 20 kW at 0.75 lagging power factor, to what value will the field current have to be reduced to maintain rated load potential? *(Section 6.1.6)*

*6.16 To determine the equivalent circuit parameters for a 4400-V, 3-phase, 3000-kVA synchronous machine, an open-circuit test and a zero-power-factor load test were made. This last test was made with a capacitive load adjusted to take rated stator current at each value of potential. The test results were

Field Current (A)	Terminal Potential	
	Open Circuit (V)	Zero Power Factor (V)
0	50	4750
20	1450	5500
40	2880	6060
60	3960	6420
70	4400	6540
80	4740	6650
100	5300	
120	5680	
140	5920	

a) Determine the stator leakage reactance in Ω/phase and in per unit of rating.
b) Find the value of the current ratio n'.
c) Determine the magnetizing reactance in Ω/phase and in per unit when the air-gap emf E_{ma} is 1.0 per unit.

d) Assume this machine is operated as a generator supplying rated potential and rated current to a partially inductive load with a power factor of 0.75. Calculate the required field current.

e) Assume that the machine is operated as a motor with rated terminal potential and that the mechanical load and field current are adjusted so that the machine takes a rated 0.75 power factor leading current from the supply. What is the required field current.   *(Section 6.1.7)*

*6.17 Suppose that a balanced capacitive load of 6.0 Ω/phase is connected to the stator terminals of the synchronous generator described in Problem 6.16 and that the field current is zero. The machine is driven at rated speed.

a) To what value will the terminal potential rise?

b) Can the generator deliver power to a resistive load that is connected in parallel with the capacitor?   *(Section 6.1.7)*

*6.18 Suppose the machine described in Problem 6.16 is to be used as a synchronous reactor. It has no mechanical load and its mechanical losses may be ignored. It is connected to a 4400-V, 3-phase, 60-Hz supply.

a) What total value of reactive power will be taken by the machine when its field current is reduced to zero?

b) What field current is required when the machine is acting as a 3000-kVA capacitive load on the system?   *(Section 6.1.7)*

*6.19 A 3-phase synchronous generator has a stator leakage inductance of 0.8 mH and a magnetizing inductance of 6 mH. The stator resistance is negligible. A field current of 1.5 A is required in each of its four 500-turn field coils to produce a pole flux of 2.5 mWb, which in turn produces a rated terminal potential of 125 V per phase on no load at the rated frequency of 400 Hz. The rotor of the machine is then replaced by a permanent-magnet rotor of the form shown in Fig. 6.29 that has the same diameter and shape at the air gaps. Ferrite C material, as described in Fig. 1.65, is used, each block having a length of 0.02 m and a cross section of 0.008 m². The soft iron parts of the rotor have negligible reluctance.

a) Develop an equivalent circuit for the original machine of the form shown in Fig. 6.28(a).

b) Develop an equivalent magnetic circuit for each pole of the permanent-magnet rotor.

c) Develop an equivalent electric circuit for the permanent magnet machine.

d) Determine the short-circuit current from the equivalent circuit.

e) Establish whether the short-circuit current in part (d) would demagnetize the magnets.   *(Section 6.1.8)*

*6.20 A 4-pole, 60-Hz, 3-phase reluctance motor has a direct-axis reactance of 8.0 and a quadrature-axis reactance of 3.0 Ω/phase as seen from its stator terminals. Stator resistance and rotational losses may be ignored. The motor is connected to a 550-V supply.

a) Determine the maximum shaft torque that the motor can supply.

b) Since the power factor of reluctance machines is always low, it is advisable to operate the motor at its maximum power factor condition. Determine the power factor and the power output for this condition.   *(Section 6.1.10)*

*6.21 A salient-pole synchronous motor is rated at 10 MW, 13.8 kV, 60 Hz and 720 r/min. Its direct axis reactance is 24 Ω, and its quadrature axis reactance is 13 Ω. The stator resistance and the rotational losses may be ignored. A field current of 210 A is required to produce rated terminal potential on no load.

a) Determine the torque when the field current is 340 A and the angle β_{0e} is 25°.
b) Estimate the maximum steady-state torque that the motor can supply with a field current of 340 A before synchronism is lost.   *(Section 6.1.11)*

*6.22 A salient-pole synchronous machine has a direct-axis synchronous reactance of 1.0 per unit and a quadrature-axis synchronous reactance of 0.6 per unit. The machine is operated as a generator with a terminal potential of 1.0 per unit supplying a current of 1.0 per unit to a load having a lagging power factor of 0.8.

a) Determine the angle β_{0e} of the machine for this condition. (Hint: Consider resolving the current components in Fig. 6.39 about an axis at angle β_{0e}.)
b) Determine the field current required for this load in per unit of the field current required to produce 1.0 per unit terminal potential on no-load.   *(Section 6.1.11)*

*6.23 A salient-pole synchronous machine is used as a synchronous reactor. It has no mechanical load and negligible losses. It is connected to a supply having a potential difference of V_a per phase. Let i_{f0} be the value of field current that produces essentially zero stator current.

a) Suppose the field current is reduced to zero. Show that the complex power taken per phase is given by

$$\bar{S} = -j \frac{|V_a|^2}{X_d} \quad \text{VA}$$

b) To increase the inductive load the machine can take, consider reversing the direction of the field current. The rotor will remain in synchronism as long as a small shift in rotor angle β_{0e} away from $\beta_{0e} = 0$ will produce a torque tending to restore β_{0e} to zero. Show that the machine can remain in synchronism until the field current is reduced to

$$i_f = -\frac{X_d - X_q}{X_q} i_{f0} \quad \text{A}$$

c) Show that the complex power *per phase* into the machine for the field current of (b) is

$$\bar{S} = -j \frac{|V_a|^2}{X_q} \quad \text{VA}$$

(Section 6 1.11)

Appendix A /
Review of Fundamental Relationships

The object of this appendix is to remind the reader of the basic units and relationships employed in the text.

A.1 CHARGE AND CURRENT

Quantitity of electricity, or *charge,* is given the symbol q. The SI unit is the *coulomb,* and the unit designation is C. The rate of flow of charge in a conductor is the *current,* which is given the symbol i. The SI unit of current is the *ampere,* and the unit designation is A. Thus

$$i = \frac{dq}{dt} \quad \text{A} \tag{A.1}$$

where the symbol t signifies *time* expressed in *seconds,* unit designation s. Thus a constant current of one ampere exists in a conductor if charge flows through a cross section of the conductor at a constant rate of one coulomb per second.

Equation (A.1) also applies to instantaneous conditions when an irregular current is flowing. Thus for any current, constant or varying,

$$q = \int i dt \quad \text{C} \tag{A.2}$$

A.2 FORCE CAUSED BY MAGNETIC INDUCTION

A particle carrying electric charge and moving in a magnetic field is subjected to a *force,* symbol F. The SI unit of force is the newton, unit designation N. The *velocity* of the particle, symbol v, is expressed in *meters per second,* for which the SI unit designation is ms^{-1}, or m/s. The basic magnetic field quantity is the *magnetic induction,* which is given the symbol B. The SI unit of induction is the *tesla,* unit designation T.

The force on the moving particle is given by the relation

$$\vec{\mathbf{F}} = q(\vec{\boldsymbol{v}} \times \vec{\mathbf{B}}) \quad \text{N} \tag{A.3}$$

and the magnitude of the force may be expressed by

$$F = qvB \sin \theta \quad \text{N} \tag{A.4}$$

where θ is the angle between vectors $\vec{v}$ and $\vec{B}$. The vector force $\vec{F}$ is perpendicular to the plane in which $\vec{v}$ and $\vec{B}$ lie, and its direction is obtained by applying the right-hand-screw rule. If $\vec{v}$, the first vector of the cross product in Eq. (A.3), is rotated into alignment with $\vec{B}$, turning a right-hand screw as it moves, then the resulting motion of the screw gives the direction of $\vec{F}$.

For the simple situation illustrated in Fig. A.1, where a positive charge is moving at constant velocity at right angles to a uniform magnetic field, the three vectors are mutually perpendicular, and

$$F = qvB \quad \text{N} \tag{A.5}$$

since θ in Eq. (A.4) is 90°.

The moving charge of Fig. A.1 may form part of a current flowing in a conductor. Figure A.2 illustrates such a system, where the cross on one end of the section of conductor indicates that current is flowing from that end to the other end. For the case in which vectors are orthogonal, or mutually perpendicular, from Eq. (A.5)

$$F = qvB = q \frac{dl}{dt} B \quad \text{N} \tag{A.6}$$

where displacement l of the particle is expressed in meters. For an elementary length of conductor dl, measured in the direction of the current,

$$dF = \frac{dq}{dt} dl \, B \quad \text{N} \tag{A.7}$$

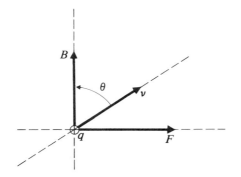

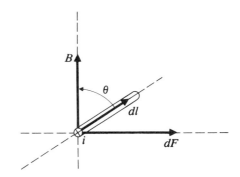

Fig. A.1 Force on a charged particle in a magnetic field.

Fig. A.2 Force on a conductor in a magnetic field.

Since, from Eq. (A.1), $dq/dt = i$, then for the general case, it follows that

$$\vec{dF} = i(\vec{dl} \times \vec{B}) \quad N \tag{A.8}$$

If B is uniform over a length of conductor l and perpendicular to it, then Eq. (A.8) becomes simply

$$F = ilB \quad N \tag{A.9}$$

A.3 MAGNETIC MOMENT

Oersted discovered that a current-carrying conductor has a magnetic field associated with it, and that the direction of the magnetic induction is tangential to a circle centered on the conductor and lying in a plane normal to its center line. This fact is illustrated in Fig. A.3, where the X on the cross section of the conductor indicates that the current is flowing into the plane of the diagram. The relationship between the direction of i and that of $\vec{B}$ is expressed by the right-hand rule: If a right-hand screw is turned to give a motion in the direction of the current, then the direction in which it is turned is that of $\vec{B}$.

If a closed loop of conductor lying in a plane carries a current, then a magnetic field is induced that may be illustrated by lines of induction. If the lines are close together, a high value of B is indicated; wide spacing indicates a low value of B. Figure A.4 shows a cross section of a circular loop of conductor carrying current i; broken lines indicate induction. The magnetic induction $\vec{B}$ at any point in space is tangential to a line of induction. In the plane of the flat loop, $\vec{B}$ is normal to the plane.

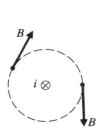

Fig. A.3 Magnetic induction due to current in a conductor.

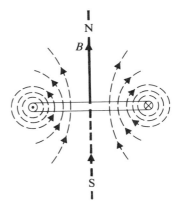

Fig. A.4 Magnetic induction due to a current-carrying loop.

The loop thus constitutes a simple electromagnet with a *north pole* (north-seeking pole) above the loop and a *south pole* beneath it. Although the name "pole" may appear to imply a point, it must be realized that *the north pole extends over the entire area* in which lines of induction emerge from the coil. Conversely, the south pole extends over the entire area where lines of induction enter the coil. In this feature of possessing an extended north (N) and south (S) pole, the loop may be compared to a very short bar magnet.

A property of the loop, called the *magnetic moment p_m*, is defined by the relationship

$$\vec{p_m} = i\,\vec{A} \quad A\cdot m^2 \tag{A.10}$$

where $\vec{A}$ is a vector normal to the plane of the loop in the direction of the magnetic field produced by the loop and of a magnitude equal to the area enclosed by the loop. Thus the unit of magnetic moment is the ampere·meter². If, instead of a single loop, an N-turn coil is considered, in which the turns are so closely packed that their total cross-section area is small compared to the area enclosed by the coil, then the magnetic moment of the coil is N times that of one of its turns.

If a current-carrying loop is placed in a uniform magnetic field of induction $\vec{B}$, as is illustrated in Fig. A.5, then the loop will be subjected to a *torque* tending to turn it so that vector $\vec{p_m}$ moves into alignment with vector $\vec{B}$. This torque is expressed by the relationship

$$\vec{T} = \vec{p_m} \times \vec{B} \quad N\cdot m \tag{A.11}$$

The direction of the torque vector is given by the right-hand-screw rule, already discussed in relation to Eq. (A.3). Thus the torque vector $\vec{T}$ cannot be shown in Fig. A.5, since it is directed into the plane of the diagram. It may be interpreted as a "twisting" about the axis of the vector as shown. In physical terms, this torque

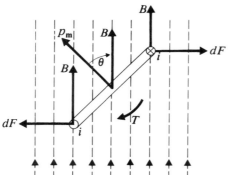

Fig. A.5 Torque on a current-carrying loop.

may be considered to arise out of the forces on the conductor due to the field. The elementary forces $\overrightarrow{dF}$ at the two diametrically opposite points on the loop illustrated in Fig. A.5 may be seen to exert a torque tending to align $\overrightarrow{p_m}$ with $\overrightarrow{B}$.

Example A.1   A closely packed 10-turn circular coil, 100-mm in diameter, is suspended in a uniform magnetic field of flux density $B = 0.2$ T. The coil is carrying a current of 0.25 A. Determine the maximum torque that will be exerted on this coil as it is rotated about a diameter that is perpendicular to the direction of magnetic induction.

Solution   The magnetic moment of the coil is

$$p_m = 0.25 \times 10 \times \frac{\pi \times 0.1^2}{4} = \frac{\pi}{160} \quad \text{A} \cdot \text{m}^2$$

The maximum torque will occur when θ in Fig. A.5 is 90°. Then from Eq. (A.11),

$$T = p_m B \sin \theta = \frac{\pi}{160} \times 0.2 = 3.927 \times 10^{-3} \quad \text{N} \cdot \text{m}$$

A.4 MAGNETIC FLUX AND FLUX DENSITY

The integration of a field vector quantity such as $\overrightarrow{B}$ over an area yields a quantity called "flux." *Magnetic flux* is given the symbol ϕ. The SI unit of flux is the *weber*, unit symbol Wb. It is the magnetic induction which would pass normally through an area of one square meter if $\overrightarrow{B}$ were of unit magnitude.

If this concept is applied to the magnetic induction through a closed path or a loop of conductor in a magnetic field as illustrated in Fig. A.6, then the magnetic flux through the loop is given by

$$\phi = \int_A \overrightarrow{B} \cdot \overrightarrow{dA} \quad \text{Wb} \tag{A.12}$$

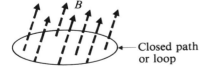

Closed path or loop

Fig. A.6 Magnetic induction through a closed path or loop of conductor.

The expression on the right-hand side of Eq. (A.12) is a surface integral of the scalar product of magnetic induction and area taken over any surface bounded by the path or loop. The meaning of this integral is illustrated in Fig. A.7. Vector $\overrightarrow{dA}$ is normal

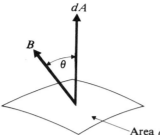

Area dA Fig. A.7 Illustration of a surface integral.

to the elementary area dA, and $\vec{B}$ represents the magnitude and direction of magnetic induction through the elementary area dA. Thus when B is perpendicular to the surface, from equation (A.12),

$$\phi = BA \quad \text{Wb} \tag{A.13}$$

from which

$$B = \frac{\phi}{A} \quad \text{T or Wb/m}^2 \tag{A.14}$$

Equation (A.14) shows that B expresses the magnetic flux per unit area; consequently the most commonly employed name for B is *magnetic flux density*. Equation (A.14) also shows that B may be expressed in webers per square meter, which is the same unit as the tesla.

A further property of magnetic flux density B is expressed by the surface integral taken over a *closed* surface in a magnetic field. For such a surface,

$$\int_A \vec{B} \cdot \vec{dA} = 0 \tag{A.15}$$

Equation (A.15) expresses an important physical fact, namely, that lines of induction follow closed paths. All magnetic flux that enters the closed surface must also leave it again.

A.5 MAGNETIC FIELD INTENSITY

Magnetic flux may be produced in one of three ways: (a) by electric currents, (b) by permanent magnets, (c) by a changing electric field. The last of these is of importance only when the electric fields are changing very rapidly. Such situations do not normally arise in electromagnetic energy conversion devices and this effect is therefore neglected here.

In discussing the production of magnetic fields by means of electric currents, it is convenient to consider that a physical property exists that is intermediate between current and flux density. This property is the *magnetic field intensity*, $\vec{H}$.

Thus the existence of a current i gives rise to magnetic field intensity H, which results in the production of flux density B. In a vacuum (and, to a very close approximation, in air) the relationship between B and H is a simple one, expressed by

$$\vec{B} = \mu_0 \vec{H} \quad \text{T} \tag{A.16}$$

where μ_0 is a physical constant called the *magnetic constant*. The value of μ_0 is $4\pi \times 10^{-7}$H/m. Vectors $\vec{B}$ and $\vec{H}$ are colinear in free space, and either may be employed to express the magnitude of a magnetic field in vacuum. Thus if, as shown in Fig. A.3, $\vec{B}$ is tangential to a circle centered on a current-carrying conductor, then $\vec{H}$ may be similarly illustrated, as shown in Fig. A.8.

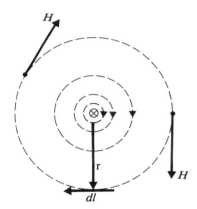

Fig. A.8 Magnetic field intensity around a current-carrying conductor.

Magnetic field intensity is related to the current producing it by *Ampère's circuital law*. This law may be expressed by the equation

$$\oint \vec{H} \cdot \vec{dl} = \int_A \vec{J} \cdot \vec{dA} \quad \text{A} \tag{A.17}$$

(An additional term would be needed to account for the effect, neglected here, of a changing electric field.) The significance of Eq. (A.17) is illustrated in Fig. A.9. The lines of magnetic induction shown there represent a magnetic field that is nonuniform in three dimensions. The closed path does not lie in one plane, nor yet is it disposed normally to the lines of magnetic induction. The magnetic field is produced by the current-carrying conductors, some of which are surrounded by the closed path. These conductors are not parallel, nor do they all carry current in the same direction, up or down through the path. The currents may be of unequal magnitude.

The expression on the left-hand side of Eq. (A.17) is a line integral around the closed path. At points p_1 and p_2 on the path are shown vectors $\mathbf{H}$ giving the direc-

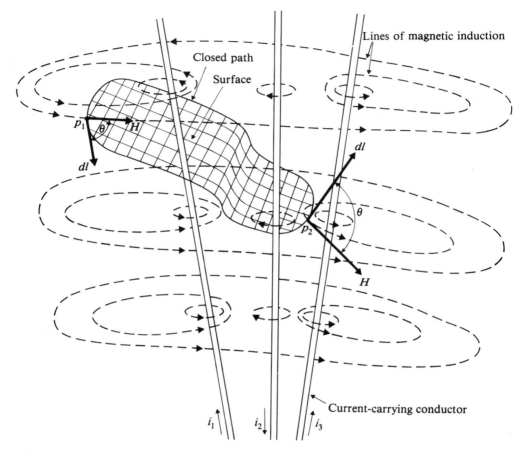

Fig. A.9 Illustration of Ampére's circuital law.

tion and magnitude of magnetic field intensity, and $\vec{\mathbf{dl}}$, representing an incremental distance in the direction of integration.

The expression on the right-hand side of Eq. (A.17) is a surface integral over any surface bounded by the closed path. This surface is penetrated by some of the conductors producing the magnetic field, but these penetrations are not in general normal to the surface. J is the *current density* expressed in amperes per square meter at any point in the conductors, so that the surface integral signifies several integrations carried out over separate conductor cross sections, as shown in Fig. A.10.

For the simple physical system of Fig. A.8, the closed path may be a circle of radius r centered on the single conductor, which is normal to the flat surface

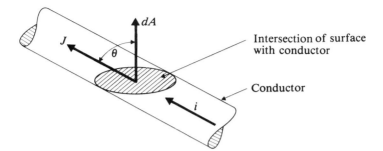

Fig. A.10 Illustration of the surface integral in Eq. A.17.

bounded by the path and which carries current i. Under these circumstances, Eq. (A.17) becomes

$$H(2\pi r) = i \quad \text{A} \tag{A.18}$$

or

$$H = \frac{i}{2\pi r} \quad \text{A/m} \tag{A.19}$$

From this expression it is seen that H is measured in units of amperes per meter.

It should be noted that the constant μ_0, which relates B and H in free space— see Eq. (A.6), has the dimensions of tesla·meter per ampere.

Example A.2 Two parallel wires shown in Fig. A.11 carry equal currents i in the directions indicated. Determine whether the force acting between the two wires is one of attraction or repulsion, and express the magnitude of that force in terms of the dimensions of the system.

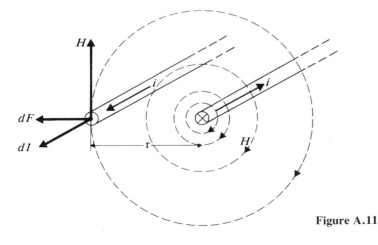

Figure A.11

Solution The force between the two wires may be considered as that exerted on the left-hand wire by the field produced by the current in the right-hand wire. Thus at the left-hand conductor, from Eq. (A.19), the magnetic field intensity H produced by the right-hand conductor is

$$H = \frac{i}{2\pi r} \quad \text{A/m}$$

and

$$B = \mu_0 H \quad \text{T}$$

From Eq. (A.8), the force on an element of length dl of the left-hand conductor carrying current i is

$$\overrightarrow{\mathbf{dF}} = i(\overrightarrow{\mathbf{dl}} \times \overrightarrow{\mathbf{B}}) \quad \text{N}$$

where dl is measured in the direction of current i in the left-hand conductor. From this vector relationship it is seen that the force acts to the left and is one of repulsion. Since the system is orthogonal, from Eq. A.9,

$$F = ilB \quad \text{N}$$

and by substitution, the force in newtons per meter of conductor is

$$\frac{F}{l} = \frac{\mu_0 i^2}{2\pi r} \quad \text{N/m}$$

This quantity is directly measurable and is one basis for defining unit current.

A.6 INDUCED ELECTROMOTIVE FORCE AND INDUCTANCE

Faraday showed that if a loop of conductor is situated in a magnetic field, and if the amount of magnetic flux passing through that loop is changed, then an *electromotive force (emf)* is induced in the loop. The symbol for emf is e, and the SI unit of emf is the *volt,* unit designation V. The magnitude and direction of emf may be determined by measuring the *electrical potential difference* appearing at the terminals of the slightly opened loop. The symbol for potential difference is v, and the SI unit is the volt. The change in the amount of magnetic flux passing through the loop may be brought about by moving the loop, changing the size of the loop, or changing the flux density of the field.

The flux within the loop may be the result of a current flowing in the loop itself, the situation illustrated in Fig. A.12, where a source of potential difference v results in current i that, in turn, produces a magnetic field indicated by the lines of induction. If B is integrated over the area of a surface bounded by the loop, then, from Eq. (A.12), the result obtained will be ϕ, the flux passing through, or *linking,* the loop.

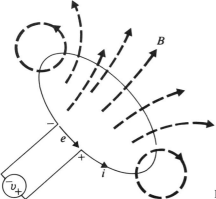

Fig. A.12 Illustration of Faraday's law.

If the current i in the loop is increased by adjusting the source, the flux linking the loop will increase, and consequently an emf e will be induced in the loop. By Lenz's law, e will have a direction opposing the increase of current. In the loop of Fig. A.12, e is positive when i is increasing. Faraday's law states

$$e = \frac{d\phi}{dt} \quad \text{V} \tag{A.20}$$

If the single loop of conductor shown in Fig. A.12 were replaced by a closely packed coil of N turns, then equal emf's would be induced in all turns of the coil, and as a consequence the coil emf would be

$$e = N\frac{d\phi}{dt} \quad \text{V} \tag{A.21}$$

Under normal circumstances, the coil conductor will possess some *electrical resistance*. The effect of this property of the conductor is that a constant current can be maintained in the coil only if a potential difference exists between its terminals. The symbol for resistance is R and the SI unit is the *ohm*, unit designation Ω. To produce constant current in the coil, the required potential difference at the terminals is given by

$$v = Ri \quad \text{V} \tag{A.22}$$

From Eq. (A.22) it is seen that

$$R = \frac{v}{i} \quad \Omega \tag{A.23}$$

so that the unit of resistance could also be expressed in volts per ampere.

Equation (A.22) also applies to instantaneous conditions where i is varying, but the potential difference applied to the coil terminals must then be such as to match the combined effects of the emf induced in the coil and of the coil resistance. Under these circumstances,

$$v = Ri + N\frac{d\phi}{dt} \quad \text{V} \tag{A.24}$$

PROBLEMS

A.1 A raindrop carrying an electric charge of 25×10^{-12} C falls at a velocity of 5 m/s between the poles of a permanent magnet. It passes through a 50-mm-long region in which a magnetic flux density of 0.9 T is directed horizontally.

  a) Find the force on the raindrop.
  b) Assume the raindrop has a diameter of 1 mm, and estimate the horizontal component of raindrop velocity after it passes through the magnetic field.   *(Section A.2)*

A.2 In an oscilloscope tube, electrons are accelerated through a potential difference of 2000 V from the emitting cathode to an anode. They then pass through a 15-mm-long region in which there is a magnetic flux density directed perpendicularly to the electron path. Finally the electrons travel through a distance of 150 mm in field-free vacuum to the fluorescent screen.

   Determine the magnetic flux density necessary to deflect the electron beam 30 mm away from the axis of the oscilloscope tube.   *(Section A.2)*

A.3 A loudspeaker has a permanent magnet that produces a flux density of 0.85 T directed radially inward across a cylindrical air gap. A coil with 20 turns, each of 25-mm diameter, is located in the air gap and mechanically connected to the loudspeaker's vibrating cone.

  a) Determine the magnitude of the force on the coil in Fig. A.13 when its current is 0.5 A.
  b) If the coil is moving to the left with a velocity of 8 m/s, what is the mechanical power being supplied to the speaker cone?   *(Section A.2)*

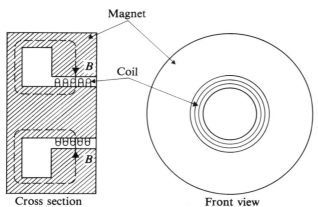

Cross section   Front view   **Figure. A.13**

A.4 The rectangular single-turn loop *abcd* of the conductor illustrated in Fig. A.14 carries a current $i = 15$ A and is located in a magnetic field of flux density $B = 0.2$ T perpendicular to the axis of the loop. A perpendicular to the plane of the loop is at angle $\theta = 60°$ to the direction of the magnetic field.

   a) Determine the magnitude and direction of the force on each of the four sides of the loop.

   b) Show that the net translational force on the loop is zero.

   c) Determine the torque on the loop from the forces of (a).

   d) Determine the magnetic moment of the loop.

   e) Show that the torque in (c) is equal to that obtained by Eq. (A.11).   (*Section A.3*)

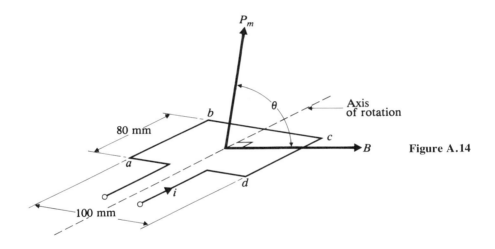

Figure A.14

A.5 A long solid cylindrical conductor 40 mm in diameter has a uniform current density of 4×10^6 A/m². The return path of the current is a considerable distance away.

   Determine the magnetic field intensity and magnetic flux density at the surface of the conductor.   (*Section A.5*)

A.6 A power distribution line has two parallel conductors, as illustrated in Fig. A.11. The spacing between the conductors is 1.2 m, and the span between supporting poles is 150 m. The rated sinusoidal load current of the line is 1100 A (rms).

   a) Determine the magnitude and direction of the average force acting on one conductor over one span length under rated load conditions.

   b) When the line is short-circuited, the current rises to nine times the rated value. Determine the force per span under such fault conditions.

   c) Assuming that the conductor has a diameter of 20 mm and a density of 8000 kg/m³, determine the ratio of the force in (b) to the gravitational force on the conductor.   (*Section A.5*)

A.7 The essential features of an instrument to measure magnetic flux density are shown in Fig. A.15. A small 200-turn coil enclosing an area of 10×10^{-6} m² is rotated at an angular velocity of 400 rad/s by an electric motor. The coil leads are brought along the rotating shaft to two slip rings. The brushes on the rings are connected to a voltmeter.

    a) Determine the calibration factor for the instrument in rms volts per tesla.

    b) Show how the instrument may be used to determine the direction of the magnetic field.   *(Section A.6)*

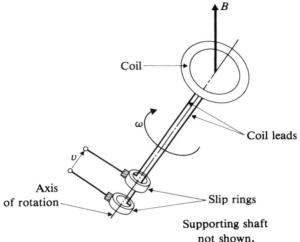

Figure A.15

Appendix B /
Measurement of Magnetic Flux and Flux Density

Magnetic flux may be measured by applying Faraday's law. In the toroidal system shown in Fig. B.1, two coils of N_1 and N_2 turns respectively have been wound on a torus. Any magnetic flux produced in coil 1 by current i in that coil links all of the N_2 turns of coil 2. If the current is changed, then the flux will be changed, and the emf induced in coil 2 will be

$$e_2 = \frac{d\lambda_2}{dt} = N_2 \frac{d\phi}{dt} \quad \text{V} \tag{B.1}$$

where flux linkage λ is defined in Section 1.1 of this textbook. If the current in coil 1 is increased from zero to i, then the flux will increase correspondingly from zero to ϕ. While this increase is taking place, an emf will be induced in coil 2. If coil 2 is open-circuited, this emf produces a potential difference at the coil terminals

$$v_2 = e_2 = N_2 \frac{d\phi}{dt} \quad \text{V} \tag{B.2}$$

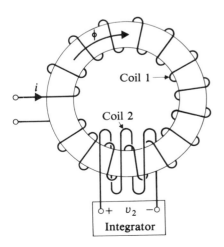

Fig. B.1 Measurement of flux.

The magnetic flux may therefore be determined from the terminal potential difference as

$$\phi = \frac{1}{N_2} \int_0^t v_2 \, dt \quad \text{Wb} \tag{B.3}$$

Any instrument that can measure the integral of a potential difference applied to its terminals may thus be used to measure flux. This is conveniently done electronically. Alternatively, a ballistic galvanometer, which measures the integral of the current flowing through it, may be employed. A fluxmeter is a special type of ballistic galvanometer designed and calibrated for flux measurement.

If the coils of Fig. B.1 are wound on a torus of ferromagnetic material, then the material's measured magnetization curve, similar to those shown in Fig. 1.7, may be obtained by determining a series of sets of values for B and H over a suitable range of coil current.

The method by which a point on the curve is obtained is as follows. The coil current is adjusted to a chosen value i_1. By means of a reversing switch, the current is then changed to $-i_1$, and the resulting change in flux $\Delta\phi$ in the torus is observed. The current is reversed repeatedly and the flux change observed for each reversal until the flux change reaches a steady value, which is recorded. Half of this flux change is taken to be the flux in the torus produced by i_1. From the number of turns and the dimensions of the torus, the values of H and B corresponding to the observed values of i_1 and $\Delta\phi/2$ may be calculated. They provide one point on the magnetization curve. Other points for different currents are similarly obtained. (The reason for this procedure will be understood after Section 1.4 on hysteresis has been read.)

Flux density may be measured by means of the Hall Effect. Figure B.2 illustrates a small strip of semiconductor in which the current consists of a movement

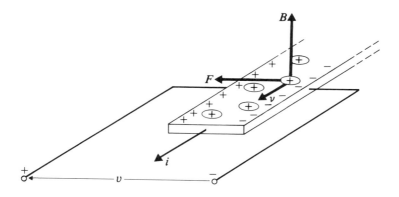

Fig. B.2 Measurement of flux density.

of positive charge carriers (holes). The drift velocity v of the charge carriers may be determined from the current density in the material and the number of charge carriers per unit volume. This last factor depends upon the nature of the material. If the current-carrying strip is placed in a magnetic field B, perpendicular to the direction of current, then from Eq. (A.3) of Appendix A the charges will be acted upon by a force in the direction shown in Fig. B.2. As all charge carriers are driven toward the left-hand side of the conductor, the concentration of positive charge at that side and the resulting concentration of negative charge at the right-hand side give rise to a potential difference v across the width of the conductor. For a specified current in a strip of known dimensions and composition, v provides a measure of B.

PROBLEMS

B.1 A toroidal coil of 200 turns is wound on a plastic ring of rectangular cross section. The internal diameter of the ring is 300 mm, and the external diameter is 400 mm. The depth of the ring is 100 mm. A search coil of twelve turns is wound on top of the toroidal coil, and a switch is arranged to reverse a current of 50 A flowing in the main coil.

a) If the current is reversed at a constant rate of 10 kA/s, what emf will be induced in the search coil?

b) If an integrator calibrated in volt-seconds is available, by what factor should its reading be multiplied to give the flux in the core prior to reversal?

B.2 A Hall-effect probe for the measurement of flux density, as shown in Fig. B.2, is 3 mm wide and 0.5 mm thick. The material is a semiconductor that has 5×10^{22} charge carriers per cubic meter, each carrying a quantum of positive charge.

a) If the current i is 0.2 A, what is the average velocity of the charge carriers?

b) If the direction of flux is perpendicular to the upper surface of the probe, what is the calibration factor (in volts per tesla) for the instrument?

Appendix C /
Complex Power

The method of representing alternating emf's, potential differences, and currents by phasors that may be written as complex quantities is usually fully discussed in circuits textbooks. The very convenient method of representing apparent power in a circuit as a complex quantity and expressing it in terms of phasors is sometimes overlooked.

Figure C.1 shows a phasor quantity

$$\overline{A} = A \underline{|\phi_A} = A \epsilon^{j\phi_A} \tag{C.1}$$

The conjugate of this quantity is

$$\overline{A}^* = A \underline{|-\phi_A} = A \epsilon^{-j\phi_A} \tag{C.2}$$

This also is shown in Fig. C.1.

The single-phase circuit shown in Fig. C.2 is excited by rms potential difference V and carries rms current I. Their phasors are

$$\overline{V} = V \underline{|\phi_V} = V \epsilon^{j\phi_V} \quad \text{V} \tag{C.3}$$

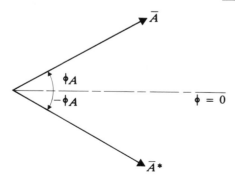

Fig. C.1 Phasor and conjugate.

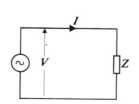

Fig. C.2 Single-phase circuit.

$$\bar{I} = I \lfloor \phi_I = I\epsilon^{j\phi_I} \quad \text{A} \tag{C.4}$$

These phasors are shown in Fig. C.3.

The average active power developed in the single-phase circuit of Fig. C.2 is

$$P = VI \cos \phi \quad \text{W} \tag{C.5}$$

where

$$\phi = \phi_I - \phi_V \quad \text{rad} \tag{C.6}$$

(Under the convention established by the International Electrotechnical Commission, ϕ is taken to be positive when the current leads the voltage.) Correspondingly, the reactive power developed is

$$Q = VI \sin \phi \quad \text{VAR} \tag{C.7}$$

The apparent power developed in the circuit may thus be expressed as

$$S = P + jQ = VI \cos \phi + jVI \sin \phi \quad \text{VA} \tag{C.8}$$

This equation is illustrated in the power triangle of Fig. C.4. Thus,

$$S = VI(\cos \phi + j \sin \phi) = VI\epsilon^{j\phi} = VI\epsilon^{j(\phi_I - \phi_V)}$$
$$= I\epsilon^{j\phi_I} \times V\epsilon^{-j\phi_V} = \bar{I} \cdot \bar{V}^* \quad \text{VA} \tag{C.9}$$

and the active power is

$$P = \mathscr{Re} (\bar{I} \cdot \bar{V}^*) \quad \text{W} \tag{C.10}$$

where $\mathscr{Re}$ signifies "real part of" and the asterisk denotes the conjugate.

The foregoing definition of complex power differs from that normally employed by power-systems engineers in the United States, which is

$$S = \bar{V} \cdot \bar{I}^* \quad \text{VA} \tag{C.11}$$

This controversy need cause no concern in the restricted context in which the concept is employed in this textbook.

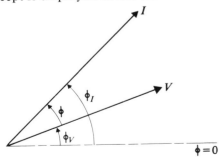

Fig. C.3 Phasor diagram for the circuit of Fig. C.2.

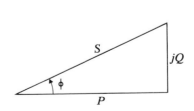

Fig. C.4 Power triangle for the circuit of Fig. C.2.

Appendix D /
SI Unit Equivalents

Property Measured	SI Unit	Equivalents
Length	1 meter (m)	3.281 feet (ft)
		39.36 inches (in.)
Angle	1 radian (rad)	57.30 degrees
Mass	1 kilogram (kg)	0.0685 slugs
		2.205 pounds (lb)
		35.27 ounces (oz)
Force	1 newton (N)	0.2248 pounds (lbf)
		7.233 poundals
		10^5 dynes
		102 grams
Torque	1 newton-meter (N·m)	0.738 pound-feet (lbf·ft)
		141.7 oz-in.
		10^7 dyne-centimeter
		1.02×10^4 gram-centimeter
Moment of inertia	1 kilogram-meter² (kg·m²)	0.738 slug-feet²
		23.7 pound-feet² (lb·ft²)
		5.46×10^4 ounce-inches²
		10^7 gram-centimeter² (g·cm²)
Energy	1 joule (J)	1 watt-second
		0.7376 foot-pounds (ft·lb)
		2.778×10^{-7} kilowatt-hours (kWh)
		0.2388 calorie (cal)
		9.48×10^{-4} British Thermal Units (BTU)
		10^7 ergs

Property Measured	SI Unit	Equivalents
Power	1 watt (W)	0.7376 foot-pounds/second 1.341×10^{-3} horsepower (hp)
Resistivity	1 ohm-meter ($\Omega \cdot$ m)	6.015×10^8 ohm-circular mil/foot 10^8 micro-ohm/centimeter
Magnetic flux	1 weber (Wb)	10^8 maxwells or lines 10^5 kilolines
Magnetic flux density	1 tesla (T)	10^4 gauss 64.52 kilolines/in.^{2}
Magnetomotive force	1 ampere-(turn) (A)	1.257 gilberts
Magnetic field intensity	1 ampere/meter (A/m)	2.54×10^{-2} ampere/in. 1.257×10^{-2} oersted

Appendix E /
Physical Constants

Quantity	Symbol	Value	Unit
Electric constant	ϵ_0	8.854×10^{-12}	coulomb^2/newton-meter^2
Magnetic constant	μ_0	$4\pi \times 10^{-7}$	newton/ampere^2
Gravitation acceleration constant	g_0	9.807	meter/second^2
Magnitude of proton or electron charge	Q_e	1.603×10^{-19}	coulomb
Electron mass	m_e	9.1×10^{-31}	kilogram
Proton mass	m_p	1.67×10^{-27}	kilogram
Bohr magneton	p_{Bohr}	9.27×10^{-24}	ampere-meter^2

Appendix F /
Resistivity and Temperature Coefficient
of Resistivity of Some Conductive Materials

Material	ρ_{20} (ohm-meter)	α_{20} $(^\circ C)^{-1}$
Copper (annealed)	1.72×10^{-8}	3.93×10^{-3}
Copper (hard drawn)	1.78×10^{-8}	3.82×10^{-3}
Aluminum	2.7×10^{-8}	3.9×10^{-3}
Sodium	4.65×10^{-8}	5.4×10^{-3}
Nickel	7.8×10^{-8}	5.4×10^{-3}
Lead	2.2×10^{-7}	4.0×10^{-3}
Tungsten	5.5×10^{-8}	4.5×10^{-3}
Iron	9.8×10^{-8}	6.5×10^{-3}
Mercury	9.58×10^{-7}	8.9×10^{-4}

Appendix G /
Wire Table

Gauge No. (B & S)	Diameter Millimeters	Inches
0	8.252	0.3249
1	7.348	0.2893
2	6.543	0.2576
3	5.827	0.2294
4	5.189	0.2043
5	4.620	0.1819
6	4.115	0.1620
7	3.665	0.1443
8	3.264	0.1285
9	2.906	0.1144
10	2.588	0.1019
11	2.304	0.0907
12	2.052	0.0808
13	1.829	0.0720
14	1.628	0.0641
15	1.450	0.0571
16	1.290	0.0508
17	1.151	0.0453
18	1.024	0.0403
19	0.912	0.0359
20	0.813	0.0320

Gauge No. (B & S)	Diameter Millimeters	Inches
21	0.724	0.0285
22	0.643	0.0253
23	0.574	0.0226
24	0.511	0.0201
25	0.455	0.0179
26	0.404	0.0159
27	0.361	0.0142
28	0.320	0.0126
29	0.287	0.0113
30	0.255	0.01003
31	0.227	0.00893
32	0.202	0.00795
33	0.180	0.00708
34	0.160	0.00630
35	0.142	0.00559
36	0.127	0.00500
37	0.114	0.00449
38	0.102	0.00402
39	0.089	0.00350
40	0.079	0.00311

Appendix H /
The Improved Induction Motor Equivalent Circuit

The improved equivalent circuit may be obtained by means of a short series of transformations of the circuit of Fig. 5.19 followed by a minor approximation. This procedure results in a simple circuit that yields results differing usually by an error of less than 1% from those obtained from the circuit of Fig. 5.19.

In Fig. 5.19, E_a is the emf induced in phase "a" of the stator by all flux linking that winding. It is possible to replace that part of the equivalent circuit lying to the right of points x and y by an exact equivalent circuit consisting of two branches in parallel. The part of the circuit to be replaced is shown in Fig. H.1(a), where as before,

$$X_{ls} = \omega_s L_{ls} \quad \Omega \tag{H.1}$$

$$X_{ms} = \omega_s L_{ms} \quad \Omega \tag{H.2}$$

$$X'_{lr} = \omega_s L'_{lr} \quad \Omega \tag{H.3}$$

If E_a is considered to be a source supplying the circuit in Fig. H.1(a), then by applying Thévenin's theorem at points m and n, the circuit of Fig. H.1(a) can be transformed into that shown in Fig. H.1(b), where

$$E_T = \frac{X_{ms}}{X_{ms} + X_{ls}} E_a \quad \text{V} \tag{H.4}$$

and

$$X_T = \frac{X_{ms} X_{ls}}{X_{ms} + X_{ls}} \quad \Omega \tag{H.5}$$

Now let

$$\frac{X_{ms}}{X_{ms} + X_{ls}} = k \tag{H.6}$$

Then from Fig. H.1(b),

$$I'_a = \frac{E_T}{(R'_r/s) + j(X_T + X'_{lr})} = \frac{kE_a}{(R'_r/s) + j(kX_{ls} + X'_{lr})} \tag{H.7}$$

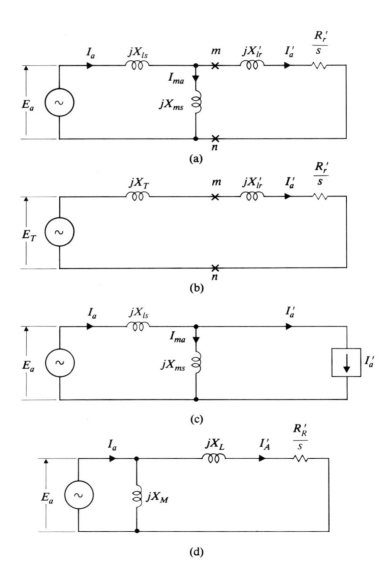

Fig. H.1 Transformation of the equivalent circuit of Fig. 5.19.

In Fig. H.1(a), the rotor-branch impedance may now be replaced by a current source delivering I'_a, as shown in Fig. H.1(c). In this circuit, stator current I_a may be expressed as the sum of two components, one due to the emf source E_a, the other due to the current source I'_a. Thus, when the current source is set to zero (open circuit), the first component is

$$\bar{I}_{a1} = \frac{\bar{E}_a}{j(X_{ms} + X_{ls})} = \frac{k\bar{E}_a}{jX_{ms}} \quad \text{A} \tag{H.8}$$

When the emf source is set to zero (short circuit), the second component is

$$\bar{I}_{a2} = \frac{X_{ms}}{X_{ms} + X_{ls}} \bar{I}'_a = k\bar{I}'_a \quad \text{A} \tag{H.9}$$

Thus by superposition,

$$\bar{I}_a = \bar{I}_{a1} + \bar{I}_{a2} = \frac{k\bar{E}_a}{jX_{ms}} + k\bar{I}'_a \quad \text{A} \tag{H.10}$$

By substitution from Eq. (H.6) and rearrangement,

$$\bar{I}_a = \frac{\bar{E}_a}{jX_{ms}/k} + \frac{\bar{E}_a}{R'_r/k^2 s + j(X_{ls}/k + X'_{ls}/k^2)} = \frac{\bar{E}_a}{jX_M} + \frac{\bar{E}_a}{R'_R/s + jX_L} \quad \text{A} \tag{H.11}$$

where

$$X_M = \frac{X_{ms}}{k} \quad \Omega \tag{H.12}$$

$$R'_R = \frac{R'_r}{k^2} \quad \Omega \tag{H.13}$$

$$X_L = \frac{X_{ls}}{k} + \frac{X'_{lr}}{k^2} \quad \Omega \tag{H.14}$$

Equation (H.11) describes the current resulting from a source E_a applied to the circuit of Fig. H.1(d).

If the circuit in Fig. H.1(d) is now used to replace that part of the circuit of Fig. 5.19 to the right of points x and y, the result is shown in Fig. H.2. This is an *exact*

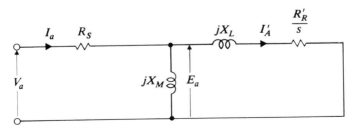

Fig. H.2 Exact equivalent of the circuit of Fig. 5.19.

equivalent of the circuit in Fig. 5.19. In particular, since $\overline{E}_a$ and $\overline{I}_a$ are unchanged, the energy dissipated to the right of points x and y is unchanged, so that

$$R'_r(I'_a)^2 = R'_R(I'_A)^2 \quad \text{W} \tag{H.15}$$

and Fig. H.2 may be employed to calculate the power developed in the machine. From Eqs. (5.57) and (H.15), therefore, the power transferred to the rotor across the air gap is

$$P_{ma} = R'_R(I'_A)^2 + \frac{1-s}{s} R'_R(I'_A)^2 \quad \text{W} \tag{H.16}$$

where the first term on the right-hand side of Eq. (H.16) represents the per-phase resistive loss in the rotor winding, and the second term represents the per-phase mechanical power developed in the machine. That is,

$$P_{amech} = \frac{1-s}{s} R'_R(I'_A)^2 \quad \text{W} \tag{H.17}$$

It is now convenient to make a small approximation by modifying the equivalent circuit of Fig. H.2 to the form shown in Fig. 5.20. If Eqs. (H.16) and (H.17) are applied to the circuit of Fig. 5.20, errors less than 1% are incurred, as compared with the results that would be obtained from Fig. 5.19.

This justifies the use of Eqs. (5.61) to (5.66).

Answers

Gaps in the numbering sequence occur when answers are not numeric.

CHAPTER 1 / MAGNETIC SYSTEMS

1.1 a) 0.63 T, b) 1.44 T. **1.2** b) 60×10^3 A/m. **1.3** a) 1620 turns; b) 2.37 mm, 2.29 Ω; c) 6.54 ms; e) 1.74×10^6 A/Wb, 0.185 mWb. **1.4** a) 3060 A/m, b) 3.84 μWb, c) 3.84×10^{-3} T, d) 2.30 mH, e) 1.47 mWb, f) 382, g) 0.88 H. **1.5** a) 1.54 A, b) 2680, 1.38 H. **1.6** a) 542 J/m³, b) 2.98 J. **1.7** a) 0.22 J/cycle, b) 4.40 W, c) 177 turns. **1.8** a) 87 V, b) 67.5 mA, c) 87 V. **1.9** 8 V. **1.10** a) 379 W, b) 253 W. **1.11** a) 21.3 ms, b) 0.094 A, c) 5.33 V, d) 8 mA. **1.12** a) 500 V, c) $0.095 \sin 600\pi t - 0.0713 \sin 1800\pi t + 0.0596 \sin 3000\pi t$ A. **1.13** $(0.060/\pi)(\sin 360\pi t + (1/3)\sin 1080\pi t + (1/5)\sin 1800\pi t)$ A. **1.14** a) 595×10^3 A/Wb, 429×10^3 A/Wb, b) 77.9 mA. **1.15** a) 21.1 mm, 13.6 mm; b) 0.703 A; c) 0.322×10^6 A/Wb, 99.8×10^3 A/Wb. **1.16** a) 212×10^3, 292×10^3, 159×10^3 A/Wb; b) 0.159 Wb, 0.11 Wb. **1.17** a) 318×10^3 A/Wb, b) 3.14 H, c) 0.786 T. **1.18** a) 174 A, b) 0.95 mWb. **1.19** a) 38.8×10^3, 265 mA/Wb; b) 0.608 A; c) 485×10^{-3}, 33.1 mJ; d) 0.205 H; e) 1.09 A. **1.20** a) 531×10^3, 398×10^3 A/Wb; b) 0.754, 1.01 mWb; c) 10 Ω, 0.302 H, 0.402 H. **1.21** a) 194 turns, b) 0.482 A. **1.22** 150 mm. **1.23** One possible set of answers is: N = 2190; 2 air gaps 0.825 mm each; E = 37 mm. **1.24** Assuming that losses are minimized: PF = 0.008; W = 133 J. **1.25** 20 mm. **1.26** a) 56×10^{-3} m³, 2×10^3 mm² × 44.4 mm; b) 0.140 m³, 5×10^3 mm² × 44.4 mm. **1.27** 1.23 T. **1.28** 77 mm³, 2×10^3 mm² × 15.9 mm. **1.29** b) 3.05 mm, c) 40×10^6 A/Wb, e) 2.05 mm. **1.30** a) 40 A, b) 4.2 mJ, c) 84 V. **1.31** a) $ab = 300$ mm²; $cd = 300$ mm². b) 0.288 kg.

CHAPTER 2 / TRANSFORMERS

2.1 b) 3 turns. **2.2** a) 550 V, 18.2 A, 45.5 A; b) 4.84 Ω. **2.3** $2.75\underline{|-25.8°}$ Ω. **2.4** a) 6.91, b) 177 V, 5.7A; 25.6 V, 39.4 A. **2.5** 11.2. **2.6** b) 3.6 H, 540 Ω, 4.17 μF. **2.7** a) $R'_2 = 0.432$ Ω; $L'_{l2} = 3.17$ mH. b) $R''_1 = 34.7$ Ω; $L''_{l1} = 0.243$ H; $L''_m = 375$ H.

2.8 L_{11} = 184 H; L_{22} = 0.46 H; L_{12} = 9.2 H. **2.9** L_{l1} = 17 mH; L_{l2} = 0; L'_m = 0.284 H.
2.10 10,700 V, 0.746. **2.11** a) 200 V, 5.49 A; b) 177 W; c) 5.41 A; d) 1080 VAR.
2.12 a) 1990 W, b) 2000 W. **2.13** a) 0.75 pu, 1.25 pu, b) 16 MVA, c) Yes.
2.14 a) R''_1 = 0.103 Ω; R_2 = 0.113 Ω; X''_{l1} = X_{l2} = 0.284 Ω; X''_m = 344 Ω; R''_c = 484 Ω.
b) 94.6%. **2.15** a) R_1 = 2.51 Ω; R'_2 = 3.11 Ω; X_{l1} = X'_{l2} = 10.9 Ω; X'_m = 16,600 Ω;
R'_c = 30,100 Ω. b) 96.0%; c) 59.0%. **2.16** a) B, Loss$_A$ = 1600 kWh/day;
Loss$_B$ = 1440 kWh/day. b) \$1460. **2.17** a) R''_1 = 0.243 Ω; R_2 = 0.288 Ω;
X''_{l1} = X_{l2} = 1.25 Ω; X''_m = 1440 Ω; R''_c = 4940 Ω. b) 3.85%; c) 4.81. **2.18** a) 2300 V,
87.0 A, 26.5 Ω, 200 kW; 230 V, 870 A, 0.265 Ω, 200 kW. b) 0.0091 pu, 0.605 pu.
c) 0.0079 pu. d) 0.0214 pu. e) 0.0170 pu. d) 0.0214 pu. e) 0.0170 pu.
2.19 All values in per unit (pu) of respective base ratings:

	1.5 kVA	20 kVA	100 kVA
I_e	0.029	0.017	0.035
R_1	0.013	0.010	0.005
X_{l1} = X_{l2}	0.035	0.045	0.026
R_2	0.014	0.013	0.006
R_c	60.0	68.6	102.0
X_m	41.4	124.0	29.8

2.20 a) 97.6%, b) 3.89%. **2.21** b) 4300 A, c) 5470 A. **2.22** b) 22.5 Ω, 87.2 Ω.
2.23 a) 90,400, b) 1.04 kVA, c) 26.2 W, d) 30.0 W. **2.24** k^2, k^2, k^4, k^3, k, k^3, k^{-1}, k^3,
k, k^{-1}, k, k. **2.25** One possible set of answers is: N_1 = 3190; N_2 = 2 $\times$ 160;
E = 30.7 mm; F = D = 61.4 mm; G = 123 mm. **2.26** 258 W, 0.0328 m^2, 7.85 kW/m^2.
2.27 a) 13.3 $\times$ 10^3, 72.9 $\times$ 10^3, 99.5 $\times$ 10^3 A/Wb; b) 45.2 mH; c) 2.35 V.
2.28 a) 86 mH, 0.236 H, 0.123 H; b) 830 Ω; c) 64.4 Ω. **2.29** a) L''_g = 0.34 mH;
b) 63.2 V, 492 A; c) 15.6 kW. **2.30** R_1 = 9.6 Ω; L_1 = 21.7 mH; R'_2 = 5.76 Ω;
L'_2 = 72.4 mH. **2.31** b) 119 kW. **2.32** b) 212 μF. **2.33** b) 95.5, 91,700 rad/s.
2.35

Y–Y:	220$\sqrt{3}$ V	110$\sqrt{3}$ V	9.09 A	18.2 A	2	0.5
Δ–Y:	220 V	110$\sqrt{3}$ V	9.09$\sqrt{3}$ A	18.2 A	2/$\sqrt{3}$	$\sqrt{3}/2$
Y–Δ:	220$\sqrt{3}$ V	110 V	9.09 A	18.2$\sqrt{3}$ A	2$\sqrt{3}$	0.5$\sqrt{3}$
Δ–Δ:	220 V	110 V	9.09$\sqrt{3}$ A	18.2$\sqrt{3}$ A	2	0.5

2.36 2300 V, 115 V; 2300 V, 115$\sqrt{3}$ V; 3.48 A, 69.6 A; 6.02 A, 69.6 A.
2.37 a) 15.8 kV, 345/$\sqrt{3}$ kV; 2110 A, 168 A; b) 17.3 kV.
2.38 2490 V. **2.39** 254 V. **2.40** 17.4 MW. **2.41** b) 1.06$\lfloor$19.1°.
2.44 210:116, 242:116. **2.45** a) 125 A, 83.3 A; b) 200 kVA, 200 kVA. **2.46** 98 turns,
0.108 m. **2.47** a) 160 turns, b) 32.1 Ω, c) 0.187%, d) 0.124°. **2.48** a) 320 V,
b) 0.5 A, c) 1.59 mA. **2.49** a) 194 turns, b) 3070 turns, c) 4920 V.

CHAPTER 3 / BASIC PRINCIPLES OF ELECTRIC MACHINES

3.1 a) 0.329 mH, b) 0.0533 J. **3.2** a) 40π $\times$ 10^{-9} z H, b) 200π N, c) 62.8 N/m^2.
3.3 0.48 J. **3.4** 0.18 N. **3.5** -6.63 N. **3.6** c) 853 N, e) 64.1 N. **3.7** 237 N.

3.8 408 kN.   **3.9** a) 503 N,   b) 757 N,   c) 537 N,   d) 736 N.   **3.10** a) 41.9 mm,
b) 5740 N,   c) 579 kg,   d) 0.2 A,   e) 11.8 m/s².   **3.11** 11.6 N.   **3.12** b) $y \le 3.53$ mm;
c) 8.90 J;   d) 191 J.   **3.13** b) 1.43 rad,   c) 85 mV.   **3.14** a) 4.78 A,   b) 0.919 N·m,
c) 1.44 J.   **3.15** b) -35.3 N·m; $0 < \alpha < 60°$.   **3.16** b) $i_1 = i_2 = 0.597$ A; 3.75 N·m/A.
3.17 b) 0.60 N·m.   **3.18** a) 27.5 mWb,   b) ± 377 rad/s,   c) 14.4 N·m,   d) 5410 W.
3.19 a) 157 rad/s, ± 314 rad/s;   b) 0.5 N·m, 0.375 N·m;   c) 78.5 W, 118 W.
3.20 ± 189 rad/s, 0.829 N·m.   **3.21** a) 41.1 cos θ μH,   b) -1.65×10^{-3} N·m, 90°,
c) 0.   **3.22** a) -0.75×10^{-3} sin θ N·m,   b) 0.358 $\times 10^{-3}$ N·m/rad,   c) 0.269 A.
3.23 76, 65, 44, 15 turns.   **3.24** a) -0.474 N·m, 0 rad/s, 0.948 N·m;   b) -0.474 N·m,
0 or 2ω rad/s, 0.474 N·m;   c) 0 N·m, 70π or 170π rad/s, 0.474 N·m;   d) 0 N·m, 50π rad/s,
0.670 N·m.   **3.25** a) 100 V,   b) 0 or 2ω rad/s,   c) 1.25 N·m.   **3.26** a) 50 Hz,
b) 1231 V   **3.27** d) 0.795 A.   **3.28** b) $e_r = \omega_s E$ cos $(2\omega_s t + \alpha_s + \beta_0)$, where
$E = \mu_o N_s N_r l r i_{sm}/g$.   **3.30** a) $\mathscr{F}_\theta = N_{se} i$ sin θ A;   b) $B_\theta = 0.754$ sin ωt sin θ T;
c) $\lambda_{aa} = (\mu_o N_{se}^2 i a l r \pi / 4g')$ Wb;   d) $L_{aa} = (\mu_o N_{se}^2 l r \pi / 4g')$ H.   e) $L_{ab} = L_{aa}$ cos $(2\pi/3)$ H,
$L_{ac} = L_{aa}$ cos $(4\pi/3)$ H.   f) $B_\theta = 1.31$ cos $(\omega_s t - \theta)$ T.   **3.31** a) (i) $7200/p$ r/min;
(ii) $6000/p$ r/min;   b) 600 r/min.   **3.32** a) 35 Hz, 85 Hz;   b) 60 Hz.   **3.33** a) 90 Hz,
30 Hz;   b) 3:1.   **3.34** a) Yes;   b) 0.333 Hz, 3.33 Hz, 1:10;   c) 0.056:1.

CHAPTER 4 / DIRECT-CURRENT MACHINES

4.1 7.85 mWb/A.   **4.2** b) 98.4 H,   c) 8.20 mWb,   d) 2.73 H,   e) 8.20 mWb.
4.3 b) 2.80 mm,   c) 2.75 mm.   **4.4** a) 221 $\times 10^3$ A/Wb;   b) 5.82 mm, 257 $\times 10^3$ A/Wb;
c) 11.7 mWb,   d) 21.7 mWb.   **4.5** a) 70.7,   b) 234 V,   c) 1490 N·m,   d) 141, 487 V,
2980 N·m.   **4.5** 114 mWb.   **4.7** 789 r/min.   **4.8** a) 86.6 $\times 10^3$ A/Wb,
b) 14.8 mWb/A,   c) 0.546,   d) 67.1 mWb,   f) 2600 N·m,   g) 396 V, 326 kW.
4.9 d) 0.795 A.   **4.10** a) 500 V,   b) 955 N·m,   c) 50 kW.   **4.11** a) 288 V,
6.25 kW;   b) 115 V, 5 kW;   c) 460 V, 10 kW, 0.5 A.   **4.12** 564 hp.   **4.13** a) 60 Ω,
34.7 H, 0.720, 13.5 g·m²;   b) $V_f = 60 i_f + 34.7$ $(d i_f / dt)$ V; $V_t = 0.720 i_f \omega_m - 0.6 i_a$ V;
c) $T_{shaft} = 0.720$ $i_f i_a + (13.5/10^3)$ $(d\omega_m / dt)$ N·m.   **4.14** 61.8 N·m.   **4.15** 207 W.
4.16 $1.96 \le i_f \le 3.32$ A.   **4.17** a) 212 V, 171 V;   b) 193 V;   c) 220 V, 125 V;
d) 187 V;   e) 1030 A, 0.0777 Ω;   f) 165 A.   **4.18** b) 3.33 Ω, 120 V, 2.33 A.   **4.19** 90 V,
38.1 A, 2.36 A; 25.7 V, 39.8 A, 0.72 A; 2.36 Ω.   **4.20** b) 4.62 Ω,   c) $2 < I_L < 19.5$ A,
d) 51 A.   **4.21** a) 117 A,   b) 168 N·m.   **4.22** a) 8.48:1,   b) 2.55,   c) 167 A,   d) 1.59.
4.24 1900 r/min.   **4.25** a) 274 N·m,   b) 1420 r/min.   **4.26** 0.23 Ω, 0.03 Ω.
4.27 221 N·m.   **4.28** 12.9 Ω.   **4.29** 119 rad/s, 33.6 A.   **4.30** 2580 r/min.
4.31 1260 r/min.   **4.32** a) 12.6 N·m,   b) 295 W,   c) 0.919.   **4.33** a) 1.20 ms,   b) 1.93 ms.
4.34 a) 1.72;   b) 23.1°;   c) (i) $\alpha_p = 52.1°$, (ii) $\alpha_n = 136°$.   **4.35** 0.801 pu.
4.36 0.797 pu.   **4.37** 5.86%.   **4.38** 2670 r/min.   **4.39** 76 mV/rad/s.   **4.40** a) 3.45 mm,
b) 157 mWb, c) 52.6 V,   d) 595 W.

CHAPTER 5 / INDUCTION MACHINES

5.1 10.9 A   **5.2** 143 turns.   **5.4** 97.0$\lfloor -30°$ V, $j0.125$ Ω.   **5.5** 2.36 $\lfloor 66.6°$ pu.
5.6 33.4$\lfloor -58.3°$ V.   **5.7** 0.550.   **5.8** a) 97.6 $\times 10^{-3}/(1 - 0.952$ cos $\beta_{0e})$ Ω,   b) $\pm 31.1°$.
5.9 b) $j0.195$ Ω,   d) 157°.   **5.10** a) $800 \le n \le -1800$ r/min;   b) $73.3 \le V_t \le 550$ V.
5.11 21.6 A.   **5.12** a) 6.97 kW, 1.03 kW;   b) 5.55 kW, 3.11 kW;   c) 2.41 kW, -1.75 kW.
5.13 a) 3.27 $\lfloor 60.4°$ Ω,   b) 7320 W,   c) 5180 W,   d) 2140 W,   e) 58.3 N·m.

5.14 a) 2.83 pu, 1.69 pu; b) 3900 W, 100 W, 195 W.   **5.15** 2.47 N·m, 0.272.
5.16 −12.0 N·m, 4520 r/min.   **5.17** a) 86.1 N·m, b) 326 N·m, c) 1040 r/min,
d) 4850 N·m per unit of slip, e) 1140 r/min, f) 18.5 rad/s, g) 0.951 pu.
5.18 c) 2.55 N·m, 7.36 A, 0.702; 2.20 N·m, 6.52 A, 0.685.   **5.19** a) 748 W; b) 0.872;
c) 8.75 N·m, 2660 r/min; d) 4.54 N·m.   **5.21** a) 76.2 N·m, b) 1260 r/min, c) 41.9 N·m,
d) T_{rated}(50 Hz)/T_{rated}(60 Hz) = 0.873.   **5.22** b) 619 r/min, c) 111 μF.   **5.23** a) 38.6 A,
b) 0.355, c) 0.914, d) 1530 W.   **5.24** 0.123 Ω.   **5.25** a) 1.41 Ω, b) 247 r/min,
c) 20.6%.   **5.26** a) 1120 r/min, 25.7 kW, b) 1070 r/min, 22.2 kW.   **5.27** a) 150 r/min,
100×10^3 N·m, 90 r/min; c) 915 kW, 116 kW.   **5.28** b) 15.6 Hz, c) 57.3 V.
5.29 a) 0.305 Ω; b) 3.51 Ω, 4.70 Ω; c) 83.7 Ω; d) 480 W; f) 4.86 A, 11.8 N·m, 77.0%.
5.30 a) R_s = 0.251 Ω; R'_R = 0.475 Ω; X_L = 0.807 Ω; X_M = 18.4 Ω. b) 1620 r/min.
5.31 R_s = 0.3 Ω; X_{ls} = 2.07 Ω; X_{ms} = 74.0 Ω; R_r = 0.824 Ω; N_{se}/N_{re} = 1.99.
5.32 a) 254 V, 21.0 Ω, 126 rad/s, 1770 VA, 5320 VA.   b) V_a = 1.0; R_s = 0.007;
X_L = 0.205; X_M = 2.29; R'_R = 0.019 pu.   **5.33** 0.897, 0.970 pu.   **5.34** 5.61 A, 0.555,
433 W, 0.632 pu.

CHAPTER 6 / SYNCHRONOUS MACHINES

6.1 p_{60} = 4; p_{400} = 26; 1800 r/min.   **6.2** b) 50.8 Ω, 2.62 pu.   **6.3** a) 16.9 A;
b) 9.04$\underline{/90°}$ A, 0; c) 85.1 N·m; d) 14.9$\underline{/-19.6°}$ A.   **6.4** a) 16.8 A, b) 85.1 N·m.
6.5 a) 747 A, b) 707 A, c) 3090 A.   **6.6** a) 15.6 A, −106°; b) 12.9 A, −68.5°;
c) 17.9 A, −138°.   **6.7** 20.6°.   **6.8** b) 11.5 × 10³ N·m, c) 652 kVAR.   **6.9** a) 0.775
lagging, b) 0.837 lagging.   **6.10** a) 21.4 × 10³ N·m, b) 24.6 × 10³ N·m.   **6.11** a) 3.04 Ω,
b) 0.786 pu.   **6.12** a) 8.97 A, b) 16.8 kVAR, 9.65 A.   **6.13** a) 3.51 A, b) 4.26 A.
6.14 a) 103 V, b) 129 V, c) 111 V.   **6.15** a) 10.6 A, b) 9.10 A.   **6.16** a) 1.16 Ω,
0.182 pu; b) 6.56; c) 5.53 Ω, 0.857 pu; d) 138 A; e) 138 A.   **6.17** a) 3580 V.
6.18 a) 2.54 MVAR, b) 154 A.   **6.19** d) 34 A.   **6.20** a) 167 N·m; b) 0.455, 28 kW.
6.21 a) 106 × 10³ N·m, b) 170 × 10³ N·m.   **6.22** a) 19.5°, b) 1.78 pu.

APPENDIX A / A REVIEW OF FUNDAMENTAL RELATIONSHIPS

A.1 a) 0.1125 × 10⁻⁹ N, b) 2.15 × 10⁻⁶ m/s.   **A.2** 2.01 × 10⁻³ T.   **A.3** a) 0.67 N,
b) 5.34 W.   **A.4** a) $|f_{ab}|$ = $|f_{cd}|$ = 0.24 N, $|f_{bc}|$ = $|f_{ad}|$ = 0.15 N; c) 0.0208 N·m;
d) 0.12 A·m².   **A.5** 40 × 10³ A/m, 0.0503 T.   **A.6** a) 30.25 N, b) 2450 N, c) 0.66.
A.7 a) 0.566 V/T.

APPENDIX B / MEASUREMENT OF MAGNETIC FLUX AND FLUX DENSITY

B.1 a) 0.69 V, b) 0.0416   **B.2** a) 16.6 m/s, b) 0.05 V/T

Index

FREQUENTLY-EMPLOYED SUBSCRIPTS AND SUPERSCRIPTS*

Subscripts (Used singly or in combinaton)

a	armature
av	average
c	core
C	capacitive
e, e	exciting, electrical, effective
eq	equivalent
ex, EX	external
f	field
g	air-gap
hp, HP	high potential
l	leakage
L	load, inductive
lp, LP	low potential
m	magnetizing
max/min	maximum/minimum
mech	mechanical
oc	open circuit
r	rotor
s	stator
sc	short circuit
t	terminal

---

*See front endpapers for Glossary of Symbols.